Springer Monographs in Mathematics

More information about this series at http://www.springer.com/series/3733

Marek Jarnicki • Peter Pflug

Continuous Nowhere Differentiable Functions

The Monsters of Analysis

Springer

Marek Jarnicki
Institute of Mathematics
Jagiellonian University
Kraków, Poland

Peter Pflug
Insitute for Mathematics
Carl von Ossietzky University Oldenburg
Oldenburg, Germany

ISSN 1439-7382 ISSN 2196-9922 (electronic)
Springer Monographs in Mathematics
ISBN 978-3-319-12669-2 ISBN 978-3-319-12670-8 (eBook)
DOI 10.1007/978-3-319-12670-8

Library of Congress Control Number: 2015955827

Springer Cham Heidelberg New York Dordrecht London

Printed on acid-free paper

Springer International Publishing AG Switzerland is part of Springer Science+Business Media (www.springer.com)

Preface

As its title indicates, this book aims to be a comprehensive, self-contained compendium of results on continuous nowhere differentiable functions, collecting many results hitherto accessible only in the scattered literature.

Motivation for Writing This Book

Why did the authors, both specialists in several complex variables, decide to write a book on continuous nowhere differentiable functions? Let us try to answer this question:

(a) Whenever we would give a lecture on real analysis, we felt unsatisfied, since there was almost no time to discuss continuous nowhere differentiable functions in detail. Therefore, we could only mention the existence of such functions in most of our lectures. Moreover, whenever we wanted to deal with such functions in a proseminar, it was difficult to find a source book. Some information could be found in a master's thesis by J. Thim (see [Thi03]), which presented a more detailed description of these functions. Later, during the writing of this book, we found another survey article by A.N. Singh (see [Sin35]). With few sources available, we thought that a modern and complete description of how these functions appeared would be of great use, both for students and for colleagues creating their lectures and preparing proseminars.
(b) Looking back to the middle of the nineteenth century, we see that that was an important time in the history of mathematics, when many arguments turned from being based more or less on heuristics into being grounded in precise definitions and proofs. We are still experiencing the consequences of this birth of mathematical precision. It is interesting to see how the methods used to discuss continuous nowhere differentiable functions has changed over time and to observe that there are still problems that have not been solved.

We hope that the reader will accept our motivation and that our book can be used for learning some very nice mathematics or for preparing proseminars or lectures on analysis.

Remarks for the Reader

To make a big part of the material accessible even to high-school graduates, we ordered the content into four main parts:

- Part I: Classical results.
  In this part are collected all results from the middle of the nineteenth century up to about 1950. The proofs are based on complicated arguments, but to understand them requires only some basic facts from analysis.
- Part II: Topological methods.
  This part is based on standard techniques from functional analysis that are certainly taught in any beginning course.
- Part III: Modern approach.
  This part requires some more highly developed ideas from analysis, such as measure theory and Fourier transforms.
- Part IV: The Riemann function.
  This part is in some sense unusual. On the one hand, it does not directly follow the theme of the book, since the Riemann function discussed here does not belong to the class of nowhere differentiable functions. On the other hand, it is more difficult and requires knowledge from several different fields of mathematics. To help the reader, we have placed such information in an appendix.

Nevertheless, we are convinced that at least 10 % of the book may be understood by high-school graduates, 40 % by students of mathematics who have completed a first analysis course, and the remainder by master's-level students.

We did not include any exercises, as they can be found in many textbooks. But the reader will find the word EXERCISE at different places in the text. It is at such points that the reader is asked to stop reading and to extend our arguments into greater detail.

Moreover, whenever some function is discussed in the book, the reader is asked to continue its study. For example, if f is claimed to be nowhere differentiable on the interval $[0, 1]$ and nothing, even later in the text, is said about infinite derivatives, then the reader should try to discuss this question on his own. In any case, any additional information in such directions that we have found in the literature has been added to the text.

Each chapter begins with a brief summary of its content. Moreover, the reader will find open problems in some chapters. They are indicated by the sign $\boxed{?}\dots\boxed{?}$. All these problems are collected at the end of the book, see List of Problems section in Appendix C. The reader is asked to work on these questions, although they do not seem to be simple to solve. For notation that may appear in the text without explanation, the reader is asked to consult Sect. B.1.

We wish to thank all our colleagues who told us about gaps in this book during its writing. In particular, we thank Dr. P. Zapałowski for all the corrections he made. It would not have been possible to reach the current level of presentation without his precise and detailed observations. Nevertheless, according to our experiences with our former books, we are sure that many errors have remained, and we are responsible for not detecting them.

We will be pleased if readers inform us about any observations they may have while studying the text. Please use the following e-mail addresses:

- `Marek.Jarnicki@im.uj.edu.pl`
- `Peter.Pflug@uni-oldenburg.de`

Finally, it is our pleasure to thank the following institutions for their support during the writing of this book:
Jagiellonian University in Kraków,
Carl von Ossietzky Universität Oldenburg,
Polish National Science Center (NCN)—grant UMO-2011/03/B/ST1/04758.

Kraków, Poland
Oldenburg, Germany

Marek Jarnicki
Peter Pflug

Contents

Chapter 1
Introduction: A Historical Journey

Isaac Newton (1643–1727) first developed calculus having been inspired by the physical world: the orbit of a planet, the swing of a pendulum, perhaps even, as legend has it, the motion of falling fruit. His thinking led to a geometric intuition about mathematical structures. They should make sense in the same way that a physical object would. As a result, many mathematicians concentrated on "continuous" functions. Conceptually, these are the functions that can be drawn without taking pen away from paper. There will be no gaps or sudden jumps.

A first "definition" of a function was given by Leonhard Euler (1707–1783) in [Eul48], page 4: "A function of a real variable is an analytic expression that is built from the variable, numbers, and constants."[1] Functions in that sense are automatically everywhere continuous (in the modern sense) up to possibly a discrete set of discontinuities.

Nevertheless, the notion of a function remained a vague one for a long time. It seems that in 1873, Lejeune Dirichlet (1805–1859) became the first to give a precise definition (see [DS00], §1): "Fix two values a and b. Then x may be thought as a quantity that may take all values between a and b. Assume that to every x a value $y = f(x)$ is associated such that if x runs continuously through the interval from a to b, then $y = f(x)$ changes also in a continuous way. Then y is called a *continuous function of x on the interval.* It is not necessary that y be built according to one law for each x; even more, there is no need to think of this relation in the form of a mathematical operation."[2]

Even more, Dirichlet pointed out that his definition does not require a common rule regarding how such a function should be built. It is allowed that the function may be constructed from different pieces or even more, it may be given without a common rule for its pieces.[3]

Note that Dirichlet defines a "continuous function," but it is clear how the term function has to be understood out of his definition. It is important and new that a function is no

---

[1] "Functio quantitatis variabilis est expressio analytica quomodocunque composita ex illa quantita variabili et numeris seu quantitatibus constantibus."

[2] "Man denke sich unter a und b zwei feste Werthe und unter x eine veränderliche Grösse, welche nach und nach alle zwischen a und b liegenden Werthe annehmen soll. Entspricht nun jedem x ein einziges endliches y und zwar so, dass, während x das Intervall von a bis b stetig durchläuft, $y = f(x)$ sich ebenfalls allmählich verändert, so heisst y eine stetige oder continuirliche Function von x für dieses Intervall. Es ist dabei gar nicht nöthig, dass y in diesem ganzen Intervalle nach demselben Gesetze von x abhängig sei, ja man braucht nicht einmal an eine durch mathematische Operationen ausdrückbare Abhängigkeit zu denken."

[3] "Diese Definition schreibt den einzelnen Theilen der Curve kein gemeinsames Gesetz vor; man kann sich dieselbe aus den verschiedenartigsten Theilen zusammengesetzt oder ganz gesetzlos gezeichnet denken." See [DS00], § 153.

M. Jarnicki, P. Pflug, *Continuous Nowhere Differentiable Functions*, Springer Monographs in Mathematics, DOI 10.1007/978-3-319-12670-8_1

longer something that is given by a closed analytic expression. It is the above definition that is familiar to today's mathematicians: to any point x of a certain set X one and only one value $f(x)$ is given, and the whole association is called the function f.

Nevertheless, the experiences at that time made people believe that for every continuous curve, it was possible to find the slope at all but a finite number of points. This seemed to match intuition: a line might have a few jagged bits, but there would always be a few sections that were "smooth." The French physicist and mathematician André-Marie Ampère (1775–1836) even published a proof of this claim (see [Amp06]). His argument was built on the "intuitively evident" fact that a continuous curve must have sections that increase, decrease, or remain flat. This meant that it must be possible to calculate the slope in those regions. Ampère did not think about what happened when the sections became infinitely small, but he claimed that he did not need to. His approach was general enough to avoid having to consider things that were "infiniment petits." Most mathematicians were happy with his reasoning. By the middle of the nineteenth century, almost every calculus textbook quoted Ampère's proof.

But during the 1860s, rumors began circulating about a strange function that contradicted Ampère's theorem. In Germany, the great Bernhard Riemann (1826–1866) told his students that he knew of a continuous function that had no smooth sections, and for which it was impossible to calculate the derivative of the function at any point. Riemann did not publish a proof, and neither did Charles Cellérier (1818–1889), at the University of Geneva, who—despite writing that he had discovered something "very important and I think new"—stuffed the work into a folder that would become public only after his death decades later (see [Cel90]). Over the years, it was found that the function Riemann proposed does not fulfill the property of being nowhere differentiable. Although his function is, in fact, somewhere differentiable, we decided to put an extensive discussion of this function into our book, showing the current state of knowledge (see Chap. 13).

Such a monster of a function was finally publicly accessible in 1872, when Karl Weierstrass (1815–1897) announced in a lecture in front of the Königliche Akademie der Wissenschaften, Berlin, that he had found a function that was continuous everywhere and yet not smooth at any point. He had constructed it by adding together an infinitely long sequence of cosine functions. To be more precise, it is given by the following formula:

$$f(x) = \sum_{n=1}^{\infty} a^n \cos(b^n \pi x), \quad x \in \mathbb{R},$$

where $a \in (0,1)$, b is an odd integer, and $ab > 1 + \frac{3}{2}\pi$.

As a function, it was ugly and awkward. It was not even clear what it would look like when plotted on a graph. But that did not matter to Weierstrass. His proof consisted of equations rather than shapes, and that is what made his announcement so powerful. Not only has he created a *monster*, he has built it from concrete logic. He had taken his new, rigorous definition of a derivative and shown that it was impossible to calculate one anywhere for this new function.

The lecture by Weierstrass was not immediately published, but it seems that his example reached many mathematicians at that time. Thus Paul du Bois-Reymond (1831–1889) wrote to Weierstrass asking for details. After Weierstrass had sent him his notes, Bois-Reymond published the example (see [BR74]). Bois-Reymond added the following comment, showing the influence that this example had had on him: "There is not only no implication between continuity and differentiability at one point, but it is an exciting result that there exists a continuous function in an interval having no differential quotient at any point of it."[4] This is the first example of a continuous nowhere differentiable function published in a mathematical journal.

After the Weierstrass lecture and before its publication by Bois-Reymond, Gaston Darboux (1842–1917) also observed another similar monster. He showed that the function

$$f(x) = \sum_{n=1}^{\infty} \frac{\sin((n+1)!x)}{n!}, \qquad x \in \mathbb{R},$$

is continuous but nowhere differentiable (see [Dar75, Dar79]). His proof in the first cited paper is very sketchy, while the second paper contains more details of the proof. It is interesting to observe that in his preface to the first paper, he mentioned names like Riemann, Hankel, Schwarz, and Klein, but omitted to cite Weierstrass. This was also the case in the second paper, even though Weierstrass had protested in a letter to Bois-Reymond, claiming that the first examples were due to him (see [Wei23], page 211).

Also Ulisse Dini (1845–1918) published in 1877 a paper (see [Din77]) in which he presented another example, namely

$$F(x) = \sum_{n=1}^{\infty} \frac{a^n}{1 \cdot 3 \cdot 5 \cdots (2n-1)} \cos(1 \cdot 3 \cdot 5 \cdots (2n-1)x), \qquad x \in \mathbb{R},$$

which is continuous but nowhere differentiable if $a > 1 + \frac{3}{2}\pi$. He referred to the example of Weierstrass, but his aim was to find other such strange functions.

This result[5] threw the mathematics community into a state of shock. The French mathematician Émile Picard (1856–1941) pointed out that if Newton had known about such functions, he would have never created calculus. Rather than harnessing ideas about the physics of nature, he would have been stuck trying to clamber over rigid mathematical obstacles. The monster also began to trample over previous research. Results that had been "proven" began to buckle. Ampère had used the vague definitions favored by Cauchy to prove his smoothness theorem. Now his arguments began to collapse. The vague notions of the past were hopeless against the monster. Worse, it was no longer clear what constituted a mathematical proof. The intuitive geometry-based arguments of the previous two centuries seemed to be of little use. If mathematics tried to wave the monster away, it would stand firm. With one bizarre equation, Weierstrass had demonstrated that physical intuition was not a reliable foundation on which to build mathematical theories. So this new mathematics (arithmetic analysis) led to a breaking away from trusting one's intuition, geometric or otherwise.

Established mathematicians tried to brush the result aside, arguing that it was awkward and unnecessary. They feared that pedants and troublemakers were hijacking their beloved subject. At the Sorbonne, Charles Hermite (1822–1901) wrote to Stieltjes (see [BB05], page 318): "I turn with terror and horror from this lamentable scourge of functions with no derivatives."[6] Henri Poincaré (1854–1912)—who was the first to call such functions monsters—

[4] "Mit der Existenz eines Differentialquotienten hat die Bedingung der Stetigkeit nicht allein für einen einzelnen Punkt nichts zu schaffen, sondern es ist eines der ergreifendsten Ergebnisse der neueren Mathematik, dass eine Funktion in *allen* Punkten eines Intervalles stetig sein kann, ohne für einen Punkt dieses Intervalles einen bestimmten Differentialquotienten zu ergeben."

[5] The present paragraph and others as well are taken from the lovely article [Kuc14], sometimes word for word (see also [Vol1987, Vol1989]).

[6] "Je me détourne avec effroi et horreur de cette plaie lamentable des fonctions continues qui n'ont point de dérivées."

denounced Weierstrass's work as "an outrage against common sense." He claimed that the functions were an arrogant distraction, and of little use to the subject. "They are invented on purpose to show that our ancestors' reasoning was at fault," he said, "and we shall never get anything more out of them." See [Poi99], page 159.

Many of the old guard wanted to leave Weierstrass's monster in the wilderness of mathematics. It did not help that nobody could visualize the shape of this strange function they were dealing with—only with the advent of computers did it become possible to plot it. Its hidden form made it hard for the mathematics community to grasp how such a function could exist. Weierstrass's style of proof was also unfamiliar to many mathematicians. His argument involved dozens of logical steps and ran to several pages. The trail of ideas was subtle and technically demanding, with no real-life analogies to guide the way. The general instinct was to avoid it.

But with the dawn of the twentieth century, situation changed. Even physicists began to discuss strange curves like the Ludwig Boltzmann (1844–1906) nonrectifiable H-curve, which was used to describe the movement of particles in statistical mechanics. In fact, much later, Norbert Wiener (1894–1964) was able to prove that the trajectory of a particle, in view of Brownian motion, is not rectifiable. The twentieth century has forced upon us the inadequacy of so-called ordinary curves to represent the facts of nature. Let us quote the French physicist Jean Baptiste Perrin (1870–1942), who helped to prove that atoms and molecules exist, an achievement that earned him the 1926 Nobel Prize in physics. In his 1913 book *Les atomes*, about the motion of atoms (see the English translation [Per16]), he writes in the introduction: "I wish to offer a few remarks designed to give objective justification for certain logical exigencies of the mathematicians. It is well known that before giving accurate definitions we show beginners that they already possess the idea of continuity. We draw a well-defined curve and say to them, holding a ruler against the curve, 'You see that there is a tangent at every point.' Or again, in order to impart the more abstract notion of the true velocity of a moving object at a point in its trajectory, we say, 'You see, of course, that the mean velocity between two neighbouring points on this trajectory does not vary appreciably as these points approach infinitely near to each other.' And many minds, perceiving that for certain familiar motions this appears true enough, do not see that there are considerable difficulties in this view. To mathematicians, however, the lack of rigour in these so-called geometrical considerations is quite apparent, and they are well aware of this childishness of trying to show, by drawing curves, for instance, that every continuous function has a derivative. Though derived functions are the simplest and the easiest to deal with, they are nevertheless exceptional; to use geometrical language, curves that have no tangents are the rule, and regular curves, such as the circle, are interesting though quite special cases. At first side the consideration of such cases seems merely an intellectual exercise, certainly ingenious but artificial and sterile in application, the desire for absolute accuracy carried to a ridiculous pitch. And often those who hear of curves without tangents, or underived functions, think at first that Nature presents no such complications, nor even offers any suggestion of them. The contrary, however is true, and the logic of mathematicians has kept them nearer to reality than the practical representations employed by physicists."

Or consider Grace Chisholm Young's (1868–1944) apologia (see [You16a], §18) of continuous nowhere differentiable functions, in which she says, "We of the twentieth century are bound to recognise it in its full importance. These curves (i.e. such without tangents) afford us a means of rendering more veracious the representation of the physical universe by the realm of Mathematics." So the last resistance to this kind of new function gradually disappeared.

In addition to Cellérier, another mathematician, Bernard Bolzano (1781–1841), found a function continuous but not differentiable at many points. This function is contained in Bolzano's book *Functionenlehre*, written around 1834, but published only in 1930.

The function itself remained unpublished until 1921, when it was discovered by the young Czech mathematician M. Jasek, who was asked by the Bohemian Academy of Sciences to go through Bolzano's manuscripts. Bolzano's function is the limit of a sequence of effectively given piecewise linear functions. Bolzano himself comments thus on his function: "The function F_x considered in I, §75, changes its increasing and decreasing behavior so many times that for no value of x does there exist a small enough w so that it is possible to believe that F_x is continuously increasing or continuously decreasing between x and $x \pm w$. This function gives us a proof that even a continuous function can have no derivative for so many values of the variable that between each two such points there is a third one for which there is also no derivative to be found."[7]

A precise proof that his function is continuous and even nowhere differentiable was given by Karel Rychlik (1855–1968) in 1922 (see his comment in [Ryc23]) and by Vojtěch Jarník (1897–1970) (see [Jar22]). Because of its late publication, this kind of function did not have as great an influence on the early discussions about continuous but nowhere differentiable functions as did the example of Weierstrass.

A number of papers dealing with new examples of continuous nowhere differentiable functions appeared. In fact, in the bibliography of Emde-Boas (see [Boa69]) there are eight articles listed before 1900 and 33 papers during the period 1901–1931; see also the bibliography in [Sin35] and the one for this book. Even more, the Weierstrass example began to appear in several textbooks, for example in U. Dini: Grundlagen für eine Theorie der Funktionen einer veränderlichen reellen Grösse (see [Din92]), F. Klein: Anwendungen der Differential- und Integralrechnung auf Geometrie. Eine Revision der Prinzipien (see [Kle02]), M. Pasch: Veränderliche und Funktion (see [Pas14]), E.W. Hobson: The theory of functions of a real variable and the theory of Fourier series (see [Hob26]). For example, let us quote U. Dini from his book, §145: "The theorems proved in the last paragraphs should be able to reject, at least from the better books, the belief up to now that a continuous function has to have a derivative."[8] Finally, modern mathematics, such as the theory of fractals, has sufficiently proved the importance of the existence of these monster functions.

In developing the discussion of these monster functions, there are first examples that, under certain restrictions on their parameters, can be handled by simple means. The discussion of these particular functions is exactly the content of Part I. Later on, mathematicians became interested in understanding the role of the parameters that lead to a function being nowhere differentiable. More difficult reasoning became necessary to study such functions. Moreover, one-sided derivatives and also infinite derivatives became of interest. Results of this kind will be discussed in Part III.

But apart from all these examples, more is true, namely that most of the continuous functions are monster functions. This kind of investigation has its basis in the theorem of Baire. It was Stefan Banach (1892–1945) who proved that the complement of the set of continuous nowhere differentiable functions is of first category, i.e., is a rather small set. As it turned out, most continuous functions behave in a strange way and are thus themselves monsters of various types. This is the content of Part II. Note that this abstract approach

---

[7] "Die in I, §75, betrachtete Function F_x, bey welcher das Steigen und Fallen so vielmals abgewechselt, dass es zu keinem Werthe von x ein w klein genug gibt, um behaupten zu können, dass F_x innerhalb x und $x \pm w$ fortwährend wachse oder fortwährend abnehme, gibt uns einen Beweis, dass eine Function sogar stetig seyn könne und doch keine abgeleitete hat für so viele Werthe ihrer Veränderlichen, dass zwischen je zwey derselben sich noch ein dritter, für welchen sie abermahls keine abgeleitete hat nachweisen."

[8] "Den in den letzten Paragraphen bewiesenen Sätzen dürfte, wie uns scheint, die Aufgabe zufallen, künftig aus den bessern Lehrbüchern den bis in die neueste Zeit als Grundlage der Differentialrechnung figurirenden Leitsatz zu verdrängen, nach welchem die Existenz der Derivierten jeder endlichen und stetigen Function wenigstens im Allgemeinen ausser Zweifel sein sollte."

does not give any effective example of such a function. Thus it makes the study of concrete examples not superfluous at all. The notion of being of first category has certain refinements such as porosity. Looking at even stranger monsters such as continuous functions having nowhere finite or infinite one-sided derivatives ended with a negative result: those functions are rare among the continuous ones. Such functions, as was shown by Stanisław Saks (1897–1942) in 1932, are of first category among all continuous functions. So there was no immediate deduction that such functions exist. Earlier, in 1924, Abram Samoilovitch Besicovitch (1891–1970) had already constructed such an example using very difficult geometric reasoning. In Chap. 11, we will present, in addition to concrete examples, a categorial argument showing, in fact, that there are many of those monsters.

Later, at the end of the twentieth century and into the current one, there appeared authors who have constructed Weierstrass-type monsters with additional pathologies. It has been a generalized trend in mathematics toward the search for large algebraic structures of pathological objects such as the continuous nowhere differentiable functions. The lineability of this type of functions has been thoroughly studied in recent years. Recall that a subset M of a topological vector space X is called *lineable* (resp. *spaceable*) in X if there exists an infinite-dimensional linear space (resp. an infinite-dimensional closed linear space) $Y \subset M \setminus \{0\}$. These notions of lineable and spaceable were originally coined by V.I. Gurariy (1935–2005). The very first result in this direction was also due to him (see [Gur67, Gur91]). He showed that the set of continuous nowhere differentiable functions on $[0, 1]$ is lineable. Further, V.P. Fonf, V.I. Gurariy, and M.I. Kadets (see [FGK99]) proved that the set of nowhere differentiable functions on $[0, 1]$ is spaceable. To give the reader a feeling for such results, we discuss some of them in Chap. 12.

We close this discussion by emphasizing that we have given only our own historical journey. We do not claim that it is a complete survey.

Part I
Classical Results

Chapter 2
Preliminaries

Summary. This chapter contains definitions and auxiliary results related to various notions of nowhere differentiability. In particular, in § 2.3, we present a proof of the famous Denjoy–Young–Saks theorem, which may permit the reader to understand better the sense of nowhere differentiability.

2.1 Derivatives

Let $I \subset \mathbb{R}$ be an arbitrary interval containing at least two distinct points.

Definition 2.1.1. For a function $\varphi : I \longrightarrow \mathbb{C}$, set

$$\boldsymbol{\Delta}\varphi(t,u) := \frac{\varphi(u) - \varphi(t)}{u - t}, \quad t, u \in I,\ t \neq u.$$

Recall that φ has a *(finite) derivative* $\varphi'(t)$ at a point $t \in I$ if the limit

$$\varphi'(t) := \lim_{I \ni u \to t} \boldsymbol{\Delta}\varphi(t,u)$$

exists and is finite. In the case $\varphi : I \longrightarrow \mathbb{R}$, we may also consider an *infinite derivative* $\varphi'(t)$ if the limit

$$\varphi'(t) := \lim_{I \ni u \to t} \boldsymbol{\Delta}\varphi(t,u)$$

exists but is infinite, i.e., $\varphi'(t) \in \{-\infty, +\infty\}$.

Remark 2.1.2. If $\varphi : I \longrightarrow \mathbb{C}$, then

$$\boldsymbol{\Delta}\varphi(u_1,u_2) = \frac{u_2 - t}{u_2 - u_1}\boldsymbol{\Delta}\varphi(t,u_2) + \frac{t - u_1}{u_2 - u_1}\boldsymbol{\Delta}\varphi(t,u_1),$$
$$t, u_1, u_2 \in I,\ u_1 < t < u_2.$$

M. Jarnicki, P. Pflug, *Continuous Nowhere Differentiable Functions*, Springer Monographs in Mathematics, DOI 10.1007/978-3-319-12670-8_2

Consequently:

(a) If a finite derivative $\varphi'(t)$ exists at an interior point $t \in \operatorname{int} I$, then

$$\varphi'(t) = \lim_{\substack{u_1,u_2\to t\\ u_1<t<u_2}} \boldsymbol{\Delta}\varphi(u_1,u_2);$$

note that this fact was already known to T.J. Stieltjes (cf. [Sti14]).

(b) If $\varphi : I \longrightarrow \mathbb{R}$, then

$$\min\{\boldsymbol{\Delta}\varphi(t,u_2), \boldsymbol{\Delta}\varphi(t,u_1)\} \le \boldsymbol{\Delta}\varphi(u_1,u_2) \le \max\{\boldsymbol{\Delta}\varphi(t,u_2), \boldsymbol{\Delta}\varphi(t,u_1)\},$$
$$t, u_1, u_2 \in I,\ u_1 < t < u_2.$$

In particular, if an infinite derivative $\varphi'(t)$ exists at an interior point $t \in \operatorname{int} I$, then

$$\varphi'(t) = \lim_{\substack{u_1,u_2\to t\\ u_1<t<u_2}} \boldsymbol{\Delta}\varphi(u_1,u_2).$$

Definition 2.1.3. Let $\varphi : I \longrightarrow \mathbb{C}$, $t \in I$. We say that φ has a *finite right-* (resp. *left-*) *sided derivative* $\varphi'_+(t)$ (resp. $\varphi'_-(t)$) at t if the limit

$$\varphi'_+(t) := \lim_{\substack{I\ni u\to t\\ u>t}} \boldsymbol{\Delta}\varphi(t,u) = \lim_{I\ni u\to t+} \boldsymbol{\Delta}\varphi(t,u)$$

$$\Big(\text{resp. } \varphi'_-(t) := \lim_{\substack{I\ni u\to t\\ u<t}} \boldsymbol{\Delta}\varphi(t,u) = \lim_{I\ni u\to t-} \boldsymbol{\Delta}\varphi(t,u)\Big)$$

exists and is finite. In the case $\varphi : I \longrightarrow \mathbb{R}$, we allow *infinite one-sided derivatives* $\varphi'_\pm(t) \in \{-\infty, +\infty\}$. Notice that:

- if $t \in I$ is the right endpoint of the interval, then $\varphi'_+(t)$ is not defined and $\varphi'_-(t) = \varphi'(t)$;
- if $t \in I$ is the left endpoint of the interval, then $\varphi'_-(t)$ is not defined and $\varphi'_+(t) = \varphi'(t)$.

One-sided derivatives are also called *unilateral derivatives.*

Remark 2.1.4. Let $\varphi : I \longrightarrow \mathbb{C}$.

(a) If a finite $\varphi'_+(t)$ exists, then for every $C > 0$, we have

$$\varphi'_+(t) = \lim_{\substack{I\ni u',u''\to t,\ t<u'<u''\\ |\frac{u''-t}{u''-u'}|\le C}} \boldsymbol{\Delta}\varphi(u',u'').$$

Indeed, we have $\varphi(u) = \varphi(t) + \varphi'_+(t)(u-t) + \alpha(u)(u-t)$, $t < u \in I$, where $\lim_{u\to t+} \alpha(u) = 0$. Hence

$$\begin{aligned}\boldsymbol{\Delta}\varphi(u',u'') &= \frac{\varphi(t) + \varphi'_+(t)(u''-t) + \alpha(u'')(u''-t)}{u''-u'}\\ &\quad - \frac{\varphi(t) + \varphi'_+(t)(u'-t) + \alpha(u')(u'-t)}{u''-u'}\\ &= \varphi'_+(t) + \frac{u''-t}{u''-u'}\alpha(u'') - \frac{u'-t}{u''-u'}\alpha(u') \underset{\substack{I\ni u',u''\to t\\ t<u'<u''}}{\longrightarrow} \varphi'_+(t),\end{aligned}$$

provided $\frac{u''-t}{u''-u'}$ is bounded.

(b) An analogous result may be easily obtained for finite left derivatives.

(c) Notice that (a) is not true for infinite unilateral derivatives.
For example, let $n_1 = 2$, $n_{k+1} = n_k^2$, $k \in \mathbb{N}$. Define $\varphi : \left[0, \frac{1}{4}\right] \longrightarrow \mathbb{R}$, $\varphi(0) := 0$,

$$\varphi(u) := \frac{1}{n_k}, \ u \in \left[\frac{1}{n_k^3}, \frac{1}{n_k^2}\right], \quad \varphi(u) := n_{k+1}u, \ u \in \left[\frac{1}{n_{k+1}^2}, \frac{1}{n_k^3}\right], \quad k \in \mathbb{N}.$$

Observe that φ is continuous and $\varphi'_+(0) = +\infty$. In fact, for $u \in \left[\frac{1}{n_k^3}, \frac{1}{n_k^2}\right]$, we have $\boldsymbol{\Delta}\varphi(0, u) = \frac{1}{n_k u} \geq n_k$. For $u \in \left[\frac{1}{n_{k+1}^2}, \frac{1}{n_k^3}\right]$, we have $\boldsymbol{\Delta}\varphi(0, u) = n_{k+1}$.
Take $u'_k := \frac{1}{n_k^3}$, $u''_k := \frac{1}{n_k^2}$. Then $\boldsymbol{\Delta}\varphi(u'_k, u''_k) = 0$ and $\frac{u''_k - 0}{u''_k - u'_k} \leq 2$.

(d) A finite derivative $\varphi'(t)$ exists at an interior point $t \in \operatorname{int} I$ iff

$$\forall_{\varepsilon>0} \ \exists_{\delta>0} \ \forall_{\substack{t-\delta \leq a_i \leq t \leq b_i \leq t+\delta \\ a_i, b_i \in I, \ a_i < b_i, \ i=1,2}} : |\boldsymbol{\Delta}\varphi(a_1, b_1) - \boldsymbol{\Delta}\varphi(a_2, b_2)| < \varepsilon.$$

Indeed, if the above condition is satisfied, then taking $a_1 = a_2 = t$ (resp. $b_1 = b_2 = t$), we conclude that a finite one-sided derivative $\varphi'_+(t)$ (resp. $\varphi'_-(t)$) exists. Taking $a_1 = b_2 = t$, we get $\varphi'_+(t) = \varphi'_-(t)$. Conversely, if $\varphi'(t) \in \mathbb{R}$ exists, then we use Remark 2.1.2(a).

We will use also the following more general derivatives, introduced, e.g., by U. Dini in [Din92].

Definition 2.1.5. Let $\varphi : I \longrightarrow \mathbb{R}$, $t \in I$. The *lower* (resp. *upper*) *right Dini derivative* $D_+\varphi(t)$ (resp. $D^+\varphi(t)$) of φ at t is defined as

$$D_+\varphi(t) := \liminf_{I \ni u \to t+} \boldsymbol{\Delta}\varphi(t, u) \in \overline{\mathbb{R}}$$
$$\left(\text{resp. } D^+\varphi(t) := \limsup_{I \ni u \to t+} \boldsymbol{\Delta}\varphi(t, u) \in \overline{\mathbb{R}}\right).$$

Analogously, the *lower* (resp. *upper*) *left Dini derivative* $D_-\varphi(t)$ (resp. $D^-\varphi(t)$) of φ at t is defined as

$$D_-\varphi(t) := \liminf_{I \ni u \to t-} \boldsymbol{\Delta}\varphi(t, u) \in \overline{\mathbb{R}}$$
$$\left(\text{resp. } D^-\varphi(t) := \limsup_{I \ni u \to t-} \boldsymbol{\Delta}\varphi(t, u) \in \overline{\mathbb{R}}\right).$$

Similarly to the above, $D^+\varphi(t)$ and $D_+\varphi(t)$ (resp. $D^-\varphi(t)$ and $D_-\varphi(t)$) are not defined if $t \in I$ is the right (resp. left) endpoint of the interval.

Remark 2.1.6. (a) $\varphi'_+(t)$ exists iff $D^+\varphi(t) = D_+\varphi(t)$; $\varphi'_-(t)$ exists iff $D^-\varphi(t) = D_-\varphi(t)$.
(b) $D^-\varphi = -D_-(-\varphi)$, $D_+\varphi = -D^+(-\varphi)$.
(c) $D^-\overset{\vee}{\varphi}(t) = -D_+\varphi(-t)$, $D_-\overset{\vee}{\varphi}(t) = -D^+\varphi(-t)$, where $\overset{\vee}{\varphi}(t) := \varphi(-t)$ (provided that $-I = I$).

Remark 2.1.7. If $\varphi : I \longrightarrow \mathbb{R}$ is continuous, then the functions $D^+\varphi$, $D_+\varphi$, $D^-\varphi$, $D_-\varphi$ are Borel measurable.

We will prove that $D^+\varphi$ is Borel measurable (the remaining cases are left to the reader as an EXERCISE). We may assume that the right endpoint of I does not belong to I. It suffices to show that for every $C \in \mathbb{R}$, the set $A_C := \{t \in I : D^+\varphi(t) < C\}$ is Borel measurable. Fix a $C \in \mathbb{R}$. Let $N \in \mathbb{N}$ be such that $I_n := \{t \in I : t + \frac{1}{n} \in I\} \neq \varnothing$ for $n \geq N$. Now we need only observe that in view of the continuity of φ, we have

$$A_C = \bigcup_{n\in\mathbb{N}_N,\, k\in\mathbb{N}} \bigcap_{h\in\mathbb{Q}\cap(0,\frac{1}{n})} \Big\{t \in I_n : \frac{\varphi(t+h)-\varphi(t)}{h} \leq C - \frac{1}{k}\Big\}.$$

Notice that the result remains true for arbitrary Borel-measurable functions $\varphi : I \longrightarrow \mathbb{R}$ (cf. [Ban22]).

2.2 Families of Continuous Nowhere Differentiable Functions

Recall that our principal aim is to discuss *continuous* nowhere differentiable functions. To simplify notation related to nowhere differentiability, we define the following classes of continuous nowhere differentiable functions.

- $\mathfrak{ND}(I)$:= the set of all $\varphi \in \mathcal{C}(I,\mathbb{C})$ that are nowhere differentiable in the finite sense;
- $\mathfrak{ND}^\infty(I)$:= the set of all $\varphi \in \mathcal{C}(I)$ that are nowhere differentiable in the finite or infinite sense;
- $\mathfrak{ND}_\pm(I)$:= the set of all $\varphi \in \mathcal{C}(I,\mathbb{C})$ such that for every $t \in I$, there is neither a finite right nor a finite left derivative at t;
- $\mathfrak{ND}_\pm^\infty(I) = \mathfrak{B}(I)$:= the set of all *Besicovitch functions*, i.e., the set of all $\varphi \in \mathcal{C}(I)$ such that for every $t \in I$, there is neither a finite or infinite right nor a finite or infinite left derivative at t (cf. § 7.5);
- $\mathfrak{M}(I)$:= the set of all *Morse functions*, i.e., the set of all $\varphi \in \mathcal{C}(I)$ such that

$$\max\{|D^+\varphi(t)|, |D_+\varphi(t)|\} = \max\{|D^-\varphi(t)|, |D_-\varphi(t)|\} = +\infty, \quad t \in I;$$

  we skip the left (resp. right) max{...} if t is the right (resp. left) endpoint of the interval;
- $\mathfrak{BM}(I) = \mathfrak{B}(I) \cap \mathfrak{M}(I)$:= the set of all *Besicovitch–Morse functions* (cf. § 11.1).

Notice that

$$\mathfrak{BM}(I) \subset \mathfrak{M}(I) \subset \mathfrak{ND}_\pm(I) \subset \mathfrak{ND}(I),$$
$$\mathfrak{BM}(I) \subset \mathfrak{B}(I) = \mathfrak{ND}_\pm^\infty(I) \subset \mathfrak{ND}^\infty(I).$$

Remark 2.2.1. Observe that if I is an open interval, then there exists a real-analytic increasing diffeomorphism $\sigma : \mathbb{R} \longrightarrow I$. In particular, if a continuous function $\varphi : I \longrightarrow \mathbb{C}$ belongs to one of the above classes of nowhere differentiable functions on I, then the function $\varphi \circ \sigma$ belongs to the corresponding class on $\mathbb{R}$.

The above remark permits us to transport many results from I to $\mathbb{R}$ and vice versa.

2.3 The Denjoy–Young–Saks Theorem

The following result may give some feelings for the general behavior of functions with respect to their differentiability. *On a first reading, the reader may skip the proof.*

Theorem 2.3.1 (Denjoy–Young–Saks). *Let $I \subset \mathbb{R}$ be an arbitrary nontrivial interval. Let $f : I \longrightarrow \mathbb{R}$. Then there exists a set $E \subset I$ of Lebesgue measure zero such that for every $x \in I \setminus E$, either*

- *a finite $f'(x)$ exists, or*
- *$D^+f(x) = D_-f(x) \in \mathbb{R}$ and $D_+f(x) = -\infty$, $D^-f(x) = +\infty$, or*

- $D^{-}f(x) = D_{+}f(x) \in \mathbb{R}$ *and* $D^{+}f(x) = +\infty$, $D_{-}f(x) = -\infty$, *or*
- $D^{-}f(x) = D^{+}f(x) = +\infty$ *and* $D_{-}f(x) = D_{+}f(x) = -\infty$.

Remark 2.3.2. Symbolically, for $x \in I \setminus E$ we have the following four possibilities:

$$\begin{array}{r|l} & \\ * & * \\ & \end{array} \qquad \begin{array}{r|l} +\infty & \\ * & * \\ & -\infty \end{array} \qquad \begin{array}{r|l} & +\infty \\ * & * \\ -\infty & \end{array} \qquad \begin{array}{r|l} +\infty & +\infty \\ & \\ -\infty & -\infty \end{array}$$

If f is continuous, the result was first proved by A. Denjoy in [Den15]. The case in which f is measurable was solved by G.C. Young in [You16b]. Finally, the general case was proved by S. Saks in [Sak24]. Our elementary proof is due to E.H. Hanson [Han34].

Corollary 2.3.3. *Let* $f : I \longrightarrow \mathbb{R}$, $f \in \mathfrak{ND}(I)$. *Then at almost all points of* I, *the function* f *has no one-sided (finite or infinite) derivatives.*

The following two classical results from measure theory will be important for the proof.

Theorem 2.3.4 (Vitali Covering Theorem; Cf. [KK96], Theorem 0.3.2)**.** *Let* $S \subset \mathbb{R}$ *be bounded and let* $\mathcal{F}$ *be a family of bounded closed intervals, none consisting of a single point, such that for every* $x \in S$ *and* $\varepsilon > 0$, *there exists a* $P \in \mathcal{F}$ *such that* $x \in P$ *and* $\operatorname{diam}(P) \leq \varepsilon$. *Then there exists an at most countable subfamily* $\mathcal{F}^0 \subset \mathcal{F}$, *consisting of pairwise disjoint intervals, such that*

$$\mathcal{L}\Big(S \setminus \bigcup_{P \in \mathcal{F}^0} P\Big) = 0,$$

where $\mathcal{L}$ *denotes the Lebesgue measure on* $\mathbb{R}$.

Theorem 2.3.5 (Lebesgue Density Theorem; Cf. [KK96], Theorem 2.2.1)**.** *Let* $A \subset \mathbb{R}$. *Then for almost all* $x \in A$ *and for every sequence* $(P_s)_{s=1}^{\infty}$ *of bounded intervals with* $x \in P_s$ *and* $0 < \operatorname{diam}(P_s) \longrightarrow 0$, *we have*

$$\lim_{s \to +\infty} \frac{\mathcal{L}^*(A \cap P_s)}{\mathcal{L}(P_s)} = 1,$$

where $\mathcal{L}^*$ *stands for the outer Lebesgue measure on* $\mathbb{R}$.

Proof of Theorem 2.3.1. Using Remark 2.2.1, we may assume that $I = \mathbb{R}$.

Step 1°. It suffices to prove that there exists a zero-measure set $E_0 = E_0(f)$ such that for every $x \in \mathbb{R} \setminus E_0$, either

- $D^{+}f(x) = D_{-}f(x) \in \mathbb{R}$, or
- $D^{+}f(x) = +\infty$ and $D_{-}f(x) = -\infty$.

Indeed, then we put $E := E_0(f) \cup E_0(-f)$.

Step 2°. The main idea of the proof is to show that:

(a) the set $E_1 := \{x \in \mathbb{R} : D^{+}f(x) = +\infty,\ D_{-}f(x) \neq -\infty\}$ is of measure zero,
(b) the set $E_2 := \{x \in \mathbb{R} : D_{-}f(x) = -\infty,\ D^{+}f(x) \neq +\infty\}$ is of measure zero,
(c) the set $E_3 := \{x \in \mathbb{R} : D^{+}f(x) < D_{-}f(x) \text{ or } D^{-}f(x) < D_{+}f(x)\}$ is at most countable,
(d) the set $E_4 := \{x \in \mathbb{R} : D^{+}f(x), D_{-}f(x) \in \mathbb{R},\ D^{+}f(x) \neq D_{-}f(x)\}$ is of measure zero.

Observe that (b) follows from (a) applied to the function $-f$.

Suppose for a moment that the above properties are already proven. Put $E_0 := E_1 \cup E_2 \cup E_3 \cup E_4$ and fix an $x \in \mathbb{R} \setminus E_0$. By (d), we need to check only that if $D^{+}f(x)$ or $D_{-}f(x)$ is infinite, then $D^{+}f(x) = +\infty$ and $D_{-}f(x) = -\infty$. The configurations from (a) and (b) are excluded. Thus, their remains the case $D^{+}f(x) = -\infty$ (resp. $D_{-}f(x) = +\infty$), but then, in view of (c), $D_{-}f(x) = -\infty$ (resp. $D^{+}f(x) = +\infty$), which contradicts (b) (resp. (a)).

Step 3°. Proof of (a).

We have

$$E_1 = \bigcup_{r\in\mathbb{Q},\, n\in\mathbb{N}} A_{r,n},$$

where

$$A_{r,n} := \{x \in \mathbb{R} : D^+ f(x) = +\infty,\ \forall_{x'\in(x-\frac{1}{n},x)} :\ \boldsymbol{\Delta} f(x,x') > r\}.$$

We need to prove only that each set $A_{r,n}$ is of measure zero. Fix r, $n \in \mathbb{N}$, and $b \in A_{r,n}$. Let $a \in \mathbb{R}$ be such that $0 < b - a < \frac{1}{n}$. Put $S := A_{r,n} \cap (a,b)$. Take an arbitrary $t \in \mathbb{R}$ and let

$$\mathcal{F}_t := \{[p,q] : q > p,\ [p,q] \subset (a,b),\ p \in S,\ \boldsymbol{\Delta} f(p,q) > t\}.$$

It is clear that $(S, \mathcal{F}_t)$ satisfies the assumptions of the Vitali covering theorem. Thus there exists an at most countable subfamily $\mathcal{F}_t^0 \subset \mathcal{F}_t$, consisting of pairwise disjoint intervals, such that $\mathcal{L}(S \setminus \bigcup_{P\in\mathcal{F}_t^0} P) = 0$. Take $P_1, \dots, P_N \in \mathcal{F}_t^0$, $P_i = [p_i, q_i]$. Then $(a,b) \setminus \bigcup_{i=1}^N P_i = \bigcup_{j=1}^M (\alpha_j, \beta_j)$, where the intervals $(\alpha_1, \beta_1), \dots, (\alpha_M, \beta_M)$ are pairwise disjoint and $\beta_j \in A_{r,n}$, $j = 1, \dots, M$. In particular, $\boldsymbol{\Delta} f(\alpha_j, \beta_j) > r$. Consequently,

$$\begin{aligned} f(b) - f(a) &= \sum_{j=1}^M (f(\beta_j) - f(\alpha_j)) + \sum_{i=1}^N (f(q_i) - f(p_i)) \\ &> r \sum_{j=1}^M (\beta_j - \alpha_j) + t \sum_{i=1}^N (q_i - p_i) = (t - r) \sum_{i=1}^N \mathcal{L}(P_i) + r(b-a). \end{aligned}$$

Thus

$$f(b) - f(a) \geq (t-r) \sum_{P\in\mathcal{F}_t^0} \mathcal{L}(P) + r(b-a).$$

Observe that

$$\sum_{P\in\mathcal{F}_t^0} \mathcal{L}(P) = \mathcal{L}(\bigcup_{P\in\mathcal{F}_t^0} P) \geq \mathcal{L}^*(S).$$

Consequently, for $t > r$, we get

$$f(b) - f(a) \geq (t-r)\mathcal{L}^*(S) + r(b-a).$$

Letting $t \longrightarrow +\infty$, we conclude that $\mathcal{L}^*(S) = \mathcal{L}(A_{r,n} \cap (a,b)) = 0$. Hence, $\mathcal{L}(A_{r,n}) = 0$.

Step 4°. Proof of (c).

It suffices to prove that the set $A := \{x \in \mathbb{R} : D^+ f(x) < D_- f(x)\}$ is of measure zero (and then apply this result to $-f$). Observe that

$$A = \bigcup_{r\in\mathbb{Q},\, n\in\mathbb{N}} A_{r,n},$$

where

$$A_{r,n} := \{x \in \mathbb{R} : \forall_{x'\in(x-\frac{1}{n},x),\ x''\in(x,x+\frac{1}{n})} :\ \boldsymbol{\Delta} f(x,x') < r < \boldsymbol{\Delta} f(x,x'')\}.$$

It is clear that if $x, y \in A_{r,n}$, then $|x - y| \geq \frac{1}{n}$. Consequently, $A_{r,n}$ is at most countable.

Step 5°. Proof of (d).

We have

$$E_4 \setminus E_3 = \bigcup_{\substack{r_1,r_2,r_3,r_4 \in \mathbb{Q} \\ r_1 > r_2 > r_3 > r_4,\ n \in \mathbb{N}}} A_{r_1,r_2,r_3,r_4,n},$$

where

$$A_{r_1,r_2,r_3,r_4,n} := \{x \in \mathbb{R} : r_4 < D_- f(x) < r_3 < r_2 < D^+ f(x) < r_1, \\ \forall_{x' \in (x-\frac{1}{n},x)} : \boldsymbol{\Delta} f(x,x') > r_4,\ \forall_{x'' \in (x,x+\frac{1}{n})} : \boldsymbol{\Delta} f(x,x'') < r_1\}.$$

Fix $r_1 > r_2 > r_3 > r_4$, $n \in \mathbb{N}$, and $a, b \in A_{r_1,r_2,r_3,r_4,n}$ such that $0 < b - a < \frac{1}{n}$. Put $S := A_{r_1,r_2,r_3,r_4,n} \cap (a,b)$. In view of the proof of Step 3° with $(r,t) = (r_4, r_2)$, we get

$$f(b) - f(a) \geq (r_2 - r_4)\mathcal{L}^*(S) + r_4(b-a).$$

Let

$$\mathcal{F} := \{[p,q] : q > p,\ [p,q] \subset (a,b),\ q \in S,\ \boldsymbol{\Delta} f(p,q) < r_3\}.$$

It is clear that $(S, \mathcal{F})$ satisfies the assumptions of the Vitali covering theorem. Thus there exists an at most countable subfamily $\mathcal{F}^0 \subset \mathcal{F}$, consisting of pairwise disjoint intervals, such that $\mathcal{L}^*\Big(S \setminus \bigcup_{P \in \mathcal{F}^0} P\Big) = 0$.

Take $P_1, \dots, P_N \in \mathcal{F}_t^0$, $P_i = [p_i, q_i]$. Then $(a,b) \setminus \bigcup_{i=1}^N P_i = \bigcup_{j=1}^M (\alpha_j, \beta_j)$, where the intervals $(\alpha_1, \beta_1), \dots, (\alpha_M, \beta_M)$ are pairwise disjoint and $\alpha_j \in A_{r_1,r_2,r_3,r_4,n}$, $j = 1, \dots, M$. In particular, $\boldsymbol{\Delta} f(\alpha_j, \beta_j) < r_1$. Consequently,

$$f(b) - f(a) \leq (r_3 - r_1) \sum_{P \in \mathcal{F}_t^0} \mathcal{L}(P) + r_1(b-a) \leq (r_3 - r_1)\mathcal{L}^*(S) + r_1(b-a).$$

Hence

$$\frac{\mathcal{L}^*(S)}{b-a} = \frac{\mathcal{L}^*(A_{r_1,r_2,r_3,r_4,n} \cap [a,b])}{\mathcal{L}([a,b])} \leq \frac{r_1 - r_4}{r_1 - r_4 + r_2 - r_3} < 1. \tag{2.3.1}$$

Suppose that $\mathcal{L}^*(A_{r_1,r_2,r_3,r_4,n}) > 0$. Then by the Lebesgue density theorem, there exists a point $b \in A_{r_1,r_2,r_3,r_4,n}$ such that

$$\lim_{a \to b-} \frac{\mathcal{L}^*(A_{r_1,r_2,r_3,r_4,n} \cap [a,b])}{\mathcal{L}([a,b])} = 1. \tag{2.3.2}$$

In particular, in view of (2.3.1), there are no sequences $(a_s)_{s=1}^\infty \subset A_{r_1,r_2,r_3,r_4,n}$ such that $0 < b - a_s < \frac{1}{n}$ and $a_s \longrightarrow b$. Thus $A_{r_1,r_2,r_3,r_4,n} \cap (b, b - \frac{1}{s}) = \varnothing$ for $s \gg 1$, which contradicts (2.3.2). □

2.4 Series of Continuous Functions

Many of the functions discussed in this book will be of the form

$$\varphi(t) := \sum_{n=0}^{\infty} \varphi_n(t), \quad t \in I,$$

where $\varphi_n : I \longrightarrow \mathbb{C}$ is continuous, $n \in \mathbb{N}_0$, and the series is *normally convergent*, i.e.,

$$A := \sum_{n=0}^{\infty} (\sup_{t \in I} |\varphi_n(t)|) < +\infty.$$

In particular, such a series is *uniformly convergent*, and therefore, the function φ is continuous. Obviously, φ is bounded and $|\varphi(x)| \le A$, $x \in I$.

Remark 2.4.1. It is well known that if, moreover, each function $\varphi_n : I \longrightarrow \mathbb{C}$ is differentiable and the series $\sum_{n=0}^{\infty} \varphi_n'$ is uniformly convergent (e.g., normally convergent) in I, then φ is differentiable and $\varphi'(t) = \sum_{n=0}^{\infty} \varphi_n'(t)$, $t \in I$.

2.5 Hölder Continuity

Definition 2.5.1. Let $\alpha \in (0, 1]$. We say that a continuous function $\varphi : I \longrightarrow \mathbb{C}$ is:

- *α-Hölder continuous at a point $t \in I$* ($\varphi \in \mathcal{H}^\alpha(I; t)$) if

$$\exists_{c,\,\delta>0}\ \forall_{h \in (-\delta,\delta) \cap (I-t)} : \ |\varphi(t+h) - \varphi(t)| \le c|h|^\alpha;$$

- *Lipschitz at a point $t \in I$* if $\varphi \in \mathcal{H}^1(I; t)$;
- *α-Hölder continuous* ($\varphi \in \mathcal{H}^\alpha(I)$) if

$$\exists_{C>0}\ \forall_{t,u \in I} : \ |\varphi(u) - \varphi(t)| \le C|u - t|^\alpha;$$

- *Lipschitz continuous* if φ is 1-Hölder continuous;
- *M-Lipschitz at a point $t \in I$* (where $M > 0$) if

$$\forall_{u \in I} : \ |\varphi(u) - \varphi(t)| \le M|u - t|.$$

Remark 2.5.2. (a) Observe that if $\varphi : I \longrightarrow \mathbb{C}$ is a bounded continuous function, then φ is α-Hölder continuous at t iff

$$\exists_{c>0}\ \forall_{u \in I} : \ |\varphi(u) - \varphi(t)| \le c|u - t|^\alpha \text{ (EXERCISE)};$$

in particular, φ is 1-Hölder continuous at t iff φ is M-Lipschitz at t for some $M > 0$.

(b) If a finite derivative $\varphi'(t)$ exists, then φ is Lipschitz at t.

(c) It is known (cf. [KK96], Theorems 1.2.8, 6.1.5, 6.1.15) that if $\varphi : I \longrightarrow \mathbb{C}$ is Lipschitz continuous, then there exists a zero-measure set $S \subset I$ such that $\varphi'(t)$ exists for all $t \in I \setminus S$.

(d) Assume that I is a bounded closed interval and let T_M denote the set of all $\varphi \in \mathcal{C}(I, \mathbb{C})$ such that for every $t \in I$, the function φ is not M-Lipschitz at t. Consider $\mathcal{C}(I, \mathbb{C})$ as a metric space endowed with the distance $d(\varphi, \psi) := \max_I |\varphi - \psi|$. Then T_M is open in $\mathcal{C}(I, \mathbb{C})$[1] (EXERCISE). Consequently, the set $T := \bigcap_{M \in \mathbb{Q}_{>0}} T_M$ of all functions that are nowhere Lipschitz on I is a Borel set. Observe that $T \subset \mathcal{ND}(I)$.

---

[1] Recall that a pair (X, d) is a *metric space* if $d : X \times X \longrightarrow \mathbb{R}_+$, $(d(x, y) = 0 \iff x = y)$, $d(x, y) = d(y, x)$, and $d(x, y) \le d(x, z) + d(z, y)$. A set $A \subset X$ is called *open* if for each $a \in A$, there exists an $r > 0$ such that $\{x \in X : d(x, a) < r\} \subset A$.

Definition 2.5.3. For $\alpha > 0$, we say that a continuous function $\varphi : I \longrightarrow \mathbb{C}$ is:

- *nowhere α-Hölder continuous* ($\varphi \in \mathcal{NH}^\alpha(I)$) if $\forall_{t\in I} : \varphi \notin \mathcal{H}^\alpha(I;t)$;
- *α-anti-Hölder continuous* if

$$\exists_{\varepsilon>0}\ \forall_{t\in I,\ \delta\in(0,1)}\ \exists_{\substack{h_\pm\in(0,\delta)\\ t\pm h_\pm\in I}} : |\varphi(t \pm h_\pm) - \varphi(t)| > \varepsilon\delta^\alpha;$$

we skip h_+ (resp. h_-) if t is the right (resp. left) endpoint of the interval;

- *weakly α-anti-Hölder continuous* if

$$\exists_{\varepsilon>0}\ \forall_{t\in I,\ \delta\in(0,1)}\ \exists_{h\in(-\delta,\delta)\cap(I-t)} : |\varphi(t+h) - \varphi(t)| > \varepsilon\delta^\alpha.$$

Remark 2.5.4. Let $\alpha \in (0, 1)$.

(a) If φ is α-anti-Hölder continuous, then $\varphi \in \mathcal{M}(I) \subset \mathcal{ND}_\pm(I)$.
(b) If φ is weakly α-anti-Hölder continuous, then φ is nowhere 1-Hölder continuous, and hence $\varphi \in \mathcal{ND}(I)$.

Chapter 3
Weierstrass-Type Functions I

Summary. The aim of this chapter is to present various classical methods of testing the nowhere differentiability of the Weierstrass-type function $x \longmapsto \sum_{n=0}^{\infty} a^n \cos^p(2\pi b^n x + \theta_n)$. More developed results will be discussed in Chap. 8.

3.1 Introduction

We will discuss the nowhere differentiability of the following *Weierstrass-type function*

$$\boldsymbol{W}_{p,a,b,\boldsymbol{\theta}}(x) := \sum_{n=0}^{\infty} a^n \cos^p(2\pi b^n x + \theta_n), \quad x \in \mathbb{R}, \tag{3.1.1}$$

where

$$p \in \mathbb{N}, \quad 0 < a < 1, \quad ab \geq 1, \quad \boldsymbol{\theta} := (\theta_n)_{n=0}^{\infty} \subset \mathbb{R}. \tag{3.1.2}$$

Throughout the chapter, we always assume that $p, a, b, \boldsymbol{\theta}$ *satisfy* (3.1.2) (cf. Figs. 3.1, 3.2, and 3.3).

Notice that the function $\boldsymbol{W}_{1,a,b,0}$ with $p = 1$, $b \in 2\mathbb{N} + 1$, and $ab > 1 + \frac{3}{2}\pi$, coincides with the original nowhere differentiable Weierstrass function presented by him to the Königliche Akademie der Wissenschaften on 18 July 1872; cf. [Wei86].

We will be mainly interested in a characterization of the parameters $p, a, b, \boldsymbol{\theta}$ for which the function $\boldsymbol{W}_{p,a,b,\boldsymbol{\theta}}$ belongs to one of the following three classes of nowhere differentiable functions: $\mathcal{ND}^{\infty}(\mathbb{R})$, $\mathcal{ND}_{\pm}(\mathbb{R})$, and $\mathcal{M}(\mathbb{R}) \cap \mathcal{ND}^{\infty}(\mathbb{R})$. Recall that $\mathcal{M}(\mathbb{R}) \subset \mathcal{ND}_{\pm}(\mathbb{R})$. We would like to point out that in general, most of the cases are not completely understood (even for $p = 1$ and $\boldsymbol{\theta} = 0$).

To simplify notation, we will use the following conventions:

- If $\theta_n = \theta$ for all $n \in \mathbb{N}_0$, then we simply write $\boldsymbol{\theta} = \theta$.
- If the parameters p, a, b are fixed, then $\boldsymbol{W}_{\boldsymbol{\theta}} := \boldsymbol{W}_{p,a,b,\boldsymbol{\theta}}$.

A special role is played by the cases in which $p = 1$ or/and ($\boldsymbol{\theta} = 0$ or $\boldsymbol{\theta} = -\frac{\pi}{2}$). In particular,

$$\boldsymbol{C}_{a,b}(x) := \boldsymbol{W}_{1,a,b,0}(x) = \sum_{n=0}^{\infty} a^n \cos(2\pi b^n x),$$

M. Jarnicki, P. Pflug, *Continuous Nowhere Differentiable Functions*, Springer Monographs in Mathematics, DOI 10.1007/978-3-319-12670-8_3

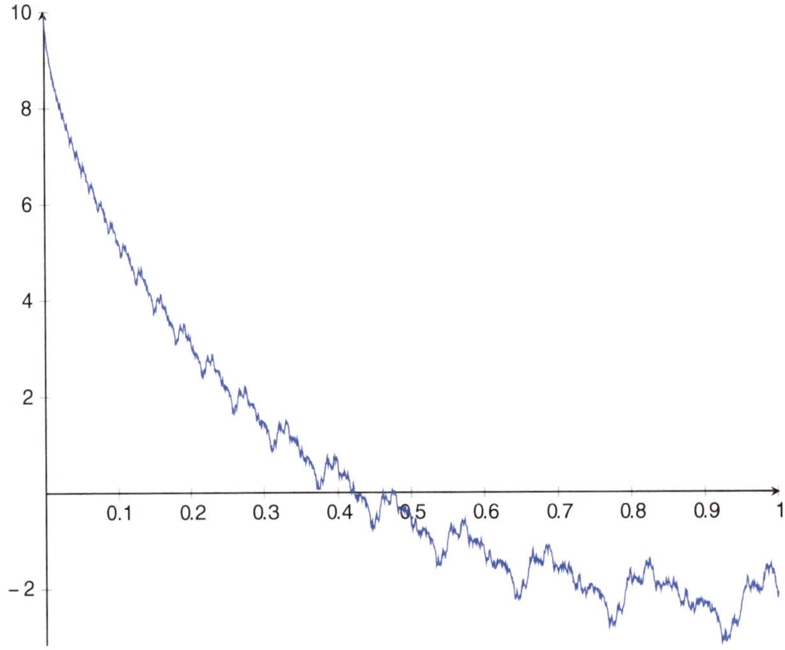

Fig. 3.1 Weierstrass-type function $\mathbb{I} \ni x \longmapsto \boldsymbol{W}_{1,0.9,1.2,0}(x)$

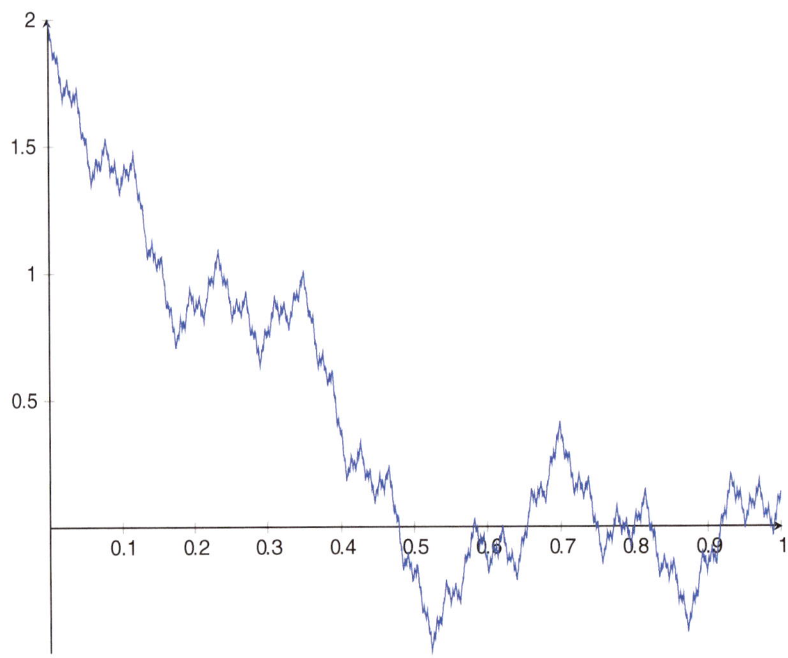

Fig. 3.2 Weierstrass-type function $\mathbb{I} \ni x \longmapsto \boldsymbol{W}_{1,0.5,3,0}(x)$

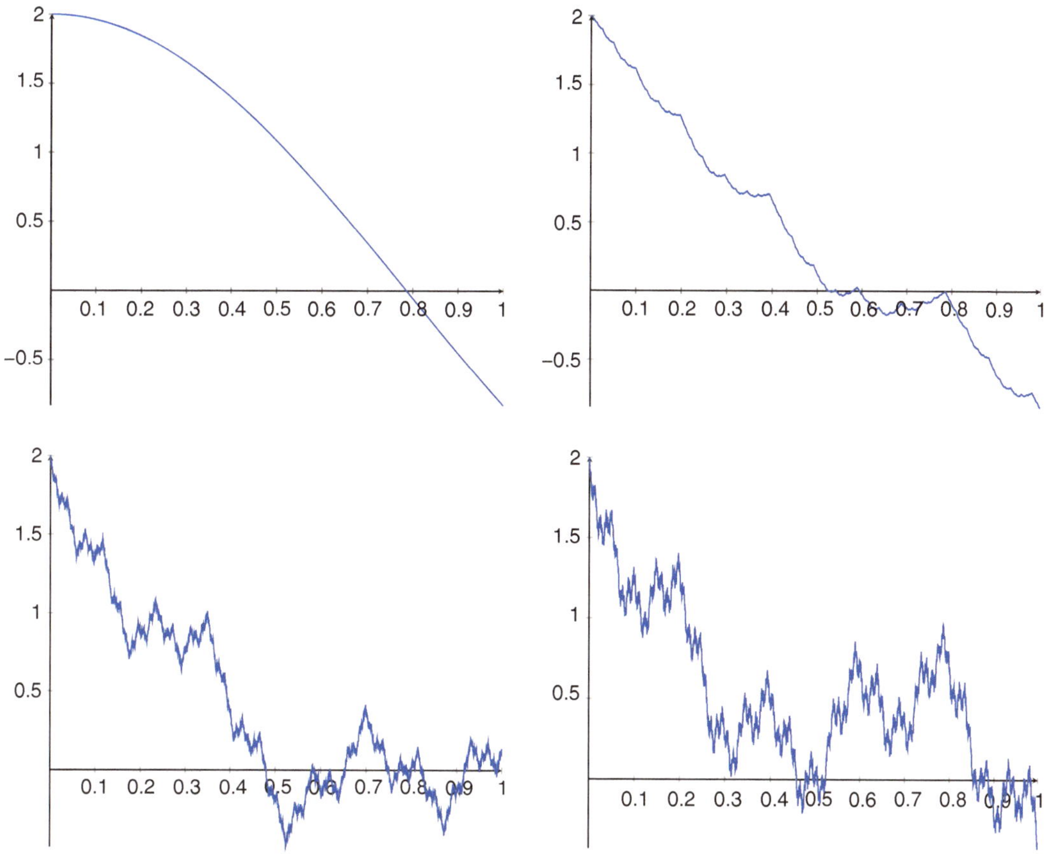

Fig. 3.3 Weierstrass-type functions $\boldsymbol{W}_{1,0.5,1,0}$, $\boldsymbol{W}_{1,0.5,2,0}$, $\boldsymbol{W}_{1,0.5,3,0}$, $\boldsymbol{W}_{1,0.5,4,0}$

$$\boldsymbol{S}_{a,b}(x) := \boldsymbol{W}_{1,a,b,-\frac{\pi}{2}}(x) = \sum_{n=0}^{\infty} a^n \sin(2\pi b^n x), \quad x \in \mathbb{R},$$

are the *classical Weierstrass functions* (cf. [BR74, Wei86]).

Remark 3.1.1. To give the reader an idea of the content of the chapter, we give below a list of results that will be presented. The list is organized in chronological order. We do not pretend that the list is complete. Most of the results will be presented in a somewhat more general form than in the original papers. Nowadays, most of these results have only historical significance. They will be essentially generalized and strengthened in Chap. 8. Nevertheless, they might give some insight into how over 120 years (1872–1992), the methods of studying nowhere differentiability have evolved.

(1) **1872:** If $b, p \in 2\mathbb{N}_0 + 1$ and $ab > 1 + \frac{3}{2}p\pi$, then $\boldsymbol{W}_{p,a,b,0} \in \mathfrak{M}(\mathbb{R}) \cap \mathfrak{ND}^\infty(\mathbb{R}) \subset \mathfrak{ND}_\pm(\mathbb{R}) \cap \mathfrak{ND}^\infty(\mathbb{R})$ (Theorem 3.5.1).
(2) **1890:** If $b \in 2\mathbb{N}$ and $b \geq 14$, then $\boldsymbol{W}_{1,1/b,b,\boldsymbol{\theta}} \in \mathfrak{ND}_\pm(\mathbb{R})$ (Theorem 3.6.1).
(3) **1892:** If ($a < a_1(p)$ and $b > \Psi_1(a)$) or ($a < a_2(p)$ and $b > \Psi_2(a)$) (the functions a_i, Ψ_i, $i = 1, 2$, are given by effective formulas), then $\boldsymbol{W}_{p,a,b,\boldsymbol{\theta}} \in \mathfrak{ND}_\pm(\mathbb{R})$. In particular, if ($a < \frac{1}{3}$ and $ab > 1 + \frac{3}{2}\pi\frac{1-a}{1-3a}$) or ($a < \frac{2}{9}$ and $ab^2 > 1 + \frac{21}{4}\pi^2\frac{1-a}{2-9a}$), then $\boldsymbol{W}_{1,a,b,\boldsymbol{\theta}} \in \mathfrak{ND}_\pm(\mathbb{R})$ (Theorem 3.7.1).

(4) **1908:** If $b \in 2\mathbb{N}+1$, $\boldsymbol{\theta} = \theta$ with $\theta \in (-\frac{\pi}{2}, \frac{\pi}{2})$, and $ab > 1 + \frac{3}{2}\frac{\pi}{\cos\theta}(1-a)$, then $\boldsymbol{W}_{1,a,b,\theta} \in \mathcal{ND}^\infty(\mathbb{R})$ (Theorem 3.8.1).
(5) **1916:** If $ab \geq 1$, then $\boldsymbol{C}_{a,b}, \boldsymbol{S}_{a,b} \in \mathcal{ND}(\mathbb{R})$ (Theorems 8.2.1, 8.2.12). This crucial result, due to G.H. Hardy [Har16], will be presented in Chap. 8. It will also follow from a more general theorem, Theorem 8.6.7.
(6) **1949:** We will present the following two groups of results obtained by F.A. Behrend in [Beh49] (which are typical of the "post Hardy" period).
- Extensions of the classical Weierstrass result:
  - If $b \in 2\mathbb{N} \setminus (3\mathbb{N})$ and $ab > 1 + \frac{16\pi}{9}(1-a)$, then $\boldsymbol{C}_{a,b} \in \mathcal{ND}^\infty(\mathbb{R})$ (Theorem 3.9.5).
  - If $b > 3$ and $ab > 1 + \frac{(3+2\varepsilon)\pi}{2\cos(\pi\varepsilon)}(1-a)$, where $\varepsilon := \frac{1}{b-1}$, then $\boldsymbol{C}_{a,b} \in \mathcal{ND}^\infty(\mathbb{R})$ (Theorem 3.9.9).
- Elementary proofs of some special cases of Hardy's results:
  - If $b \in \mathbb{N}_2$ and $ab \geq 1$, then $\boldsymbol{C}_{a,b} \in \mathcal{ND}(\mathbb{R})$ (Theorem 3.9.14).
  - If $b > 3$, $ab \geq 1$, $ab^2 > 1 + \frac{(3+2\varepsilon)(1+2\varepsilon)}{8\cos(\pi\varepsilon)}\pi^2(1-a)$, where $\varepsilon := \frac{1}{b-1}$, then $\boldsymbol{C}_{a,b} \in \mathcal{ND}(\mathbb{R})$ (Theorem 3.9.15).

(7) **1969:** $\boldsymbol{S}_{1/2,2} \in \mathcal{ND}(\mathbb{R})$ (an elementary proof; Theorem 3.10.1).
(8) **1992:** If $b \in 2\mathbb{N}+1$ and $ab > 1$, then $\boldsymbol{W}_{1,a,p,0} \in \mathcal{ND}(\mathbb{R})$ (Theorem 3.11.1).

At the beginning of § 8.1, the reader will find a list of the best results obtained so far (up to 2015).

3.2 General Properties of $\boldsymbol{W}_{p,a,b,\theta}$

We begin with a remark collecting elementary properties of $\boldsymbol{W}_{p,a,b,\boldsymbol{\theta}}$.

Remark 3.2.1. (a) Each term of the series (3.1.1),

$$\mathbb{R} \ni x \overset{w_n}{\longmapsto} a^n \cos^p(2\pi b^n x + \theta_n), \quad n \in \mathbb{N}_0,$$

is a real-analytic function.
(b) $\boldsymbol{W}_{p,a,b,\boldsymbol{\theta}} \in \mathcal{C}(\mathbb{R})$ and $|\boldsymbol{W}_{p,a,b,\boldsymbol{\theta}}(x)| \leq A := \frac{1}{1-a}$, $x \in \mathbb{R}$ (cf. § 2.4).
(c) The function $\boldsymbol{W}_{p,a,b,\boldsymbol{\theta}}$ may be formally defined for all $b > 0$. However, the case $ab < 1$ is from our point of view irrelevant, because if $ab < 1$, then $\boldsymbol{W}_{p,a,b,\boldsymbol{\theta}} \in \mathcal{C}^1(\mathbb{R})$ (cf. Remark 2.4.1; see Fig. 3.3).
(d) $\boldsymbol{W}_{\boldsymbol{\theta}}(x + x_0) = \boldsymbol{W}_{(2\pi b^n x_0 + \theta_n)_{n=0}^\infty}(x)$, $\boldsymbol{W}_{\boldsymbol{\theta}}(-x) = \boldsymbol{W}_{-\boldsymbol{\theta}}(x)$, $x, x_0 \in \mathbb{R}$.
(e) For every p, a, b, and $\beta \in (0, 1]$, the following conditions are equivalent:

(†)$\boldsymbol{W}_{\boldsymbol{\theta}}$ is *β-Hölder continuous uniformly with respect to $\boldsymbol{\theta}$*, i.e.

$$\exists_{c>0}\ \forall_{\boldsymbol{\theta}} : |\boldsymbol{W}_{\boldsymbol{\theta}}(x+h) - \boldsymbol{W}_{\boldsymbol{\theta}}(x)| \leq c|h|^\beta, \quad x, h \in \mathbb{R};$$

(‡)$\boldsymbol{W}_{\boldsymbol{\theta}}$ is *right-sided β-Hölder continuous at* 0 *uniformly with respect to $\boldsymbol{\theta}$*, i.e.,

$$\exists_{c,\,\delta_0>0}\ \forall_{\boldsymbol{\theta}} : |\boldsymbol{W}_{\boldsymbol{\theta}}(h) - \boldsymbol{W}_{\boldsymbol{\theta}}(0)| \leq ch^\beta, \quad h \in (0, \delta_0).$$

Indeed, if (‡) is satisfied, then, using (d), for $h \in (0, \delta_0)$ we get

$$|\boldsymbol{W}_{\boldsymbol{\theta}}(x \pm h) - \boldsymbol{W}_{\boldsymbol{\theta}}(x)| = |\boldsymbol{W}_{(2\pi b^n x+\theta_n)_{n=0}^\infty}(\pm h) - \boldsymbol{W}_{(2\pi b^n x+\theta_n)_{n=0}^\infty}(0)|$$
$$= |\boldsymbol{W}_{(\pm(2\pi b^n x+\theta_n))_{n=0}^\infty}(h) - \boldsymbol{W}_{(\pm(2\pi b^n x+\theta_n))_{n=0}^\infty}(0)| \le ch^\beta.$$

If $|h| \ge \delta_0$, then $|\boldsymbol{W}_{\boldsymbol{\theta}}(x \pm h) - \boldsymbol{W}_{\boldsymbol{\theta}}(x)| \le 2A \le \frac{2A}{\delta_0^\beta}|h|^\beta$.

(f) Let

$$W_m(x) := \sum_{n=0}^{m-1} a^n \cos^p(2\pi b^n x + \theta_n), \quad m \in \mathbb{N},\ x \in \mathbb{R}.$$

Assume that $ab > 1$. Then

$$\sup_{x_0 \in \mathbb{R},\ h \in \mathbb{R}_*} |\boldsymbol{\Delta} W_m(x_0, x_0 + h)| < 2p\pi \frac{(ab)^m}{ab-1}.$$

Indeed, by the mean value theorem, we get

$$|\boldsymbol{\Delta} W_m(x_0, x_0+h)|$$
$$= \Big|\sum_{n=0}^{m-1} a^n 2p\pi b^n \cos^{p-1}(2\pi b^n\xi + \theta_n)\sin(2\pi b^n \xi + \theta_n)\Big|$$
$$\le 2p\pi \sum_{n=0}^{m-1} (ab)^n = 2p\pi \frac{(ab)^m - 1}{ab-1} < 2p\pi\frac{(ab)^m}{ab-1}.$$

(g) If $ab > 1$ and $\alpha := -\frac{\log a}{\log b}$, then $\boldsymbol{W}_{\boldsymbol{\theta}}$ is α-Hölder continuous uniformly with respect to $\boldsymbol{\theta}$. Indeed, fix an $h \in (0,1)$ and let $N = N(h) \in \mathbb{N}_0$ be such that $b^N h \le 1 < b^{N+1}h$. Then (using (f)) we get

$$|\boldsymbol{W}_{\boldsymbol{\theta}}(h) - \boldsymbol{W}_{\boldsymbol{\theta}}(0)| \le |W_N(h) - W_N(0)| + 2\sum_{n=N}^\infty a^n$$
$$< 2p\pi\frac{(ab)^N}{ab-1}h + \frac{2a^N}{1-a} \le 2\Big(\frac{p\pi}{ab-1} + \frac{1}{1-a}\Big)a^N \le cph^\alpha,$$

where c depends only on a and b. Now by (e), we get the result.

(h) For every p, a, b, and $\beta \in (0,1]$, the following conditions are equivalent:

(†)$\boldsymbol{W}_{\boldsymbol{\theta}}$ is *β-anti-Hölder continuous uniformly with respect to $x \in \mathbb{R}$ and $\boldsymbol{\theta}$*, i.e., $\exists_{\varepsilon>0}\ \forall_{\boldsymbol{\theta},\ x\in\mathbb{R},\ \delta\in(0,1)}\ \exists_{h_\pm \in (0,\delta)} : |\boldsymbol{W}_{\boldsymbol{\theta}}(x \pm h_\pm) - \boldsymbol{W}_{\boldsymbol{\theta}}(x)| > \varepsilon\delta^\beta$ (cf. Definition 2.5.3);

(‡)$\exists_{\varepsilon,\ \delta_0>0}\ \forall_{\boldsymbol{\theta},\ \delta\in(0,\delta_0)}\ \exists_{h_+\in(0,\delta)} : |\boldsymbol{W}_{\boldsymbol{\theta}}(h_+) - \boldsymbol{W}_{\boldsymbol{\theta}}(0)| > \varepsilon\delta^\beta$.

Indeed, suppose that (‡) is satisfied. If $\delta_0 < 1$, then fix $0 < \delta' < \delta_0$ and let $h_+ \in (0, \delta')$ be associated to $(\boldsymbol{\theta}, \delta')$ via (‡). Then for $\delta \in [\delta_0, 1)$, we have $|\boldsymbol{W}_{\boldsymbol{\theta}}(h_+) - \boldsymbol{W}_{\boldsymbol{\theta}}(0)| > \varepsilon\delta'^\beta \ge (\varepsilon\delta'^\beta)\delta^\beta$. Hence we may assume that $\delta_0 \ge 1$.

Take $\boldsymbol{\theta}$, $x \in \mathbb{R}$, and $\delta \in (0,1)$. Let $h_+ \in (0,\delta)$ be associated to $((\pm(2\pi b^n x + \theta_n))_{n=0}^\infty, \delta)$ via (‡). Then

$$|\boldsymbol{W}_{\boldsymbol{\theta}}(x \pm h_+) - \boldsymbol{W}_{\boldsymbol{\theta}}(x)|$$
$$= |\boldsymbol{W}_{(\pm(2\pi b^n x+\theta_n))_{n=0}^\infty}(h_+) - \boldsymbol{W}_{(\pm(2\pi b^n x+\theta_n))_{n=0}^\infty}(0)| > \varepsilon\delta^\beta.$$

(i) For every p, a, b, and $\beta \in (0,1]$, the following conditions are equivalent (EXERCISE):

(†)$\boldsymbol{W_\theta}$ is *weakly β-anti-Hölder continuous uniformly with respect to $x \in \mathbb{R}$ and $\boldsymbol{\theta}$*, i.e., $\exists_{\varepsilon>0}\,\forall_{\boldsymbol{\theta},\,x\in\mathbb{R},\,\delta\in(0,1)}\,\exists_{h\in(-\delta,\delta)} : |\boldsymbol{W_\theta}(x+h) - \boldsymbol{W_\theta}(x)| > \varepsilon\delta^\beta$ (cf. Definition 2.5.3);

(‡)$\exists_{\varepsilon,\,\delta_0>0}\,\forall_{\boldsymbol{\theta},\,\delta\in(0,\delta_0)}\,\exists_{h\in(-\delta,\delta)} : |\boldsymbol{W_\theta}(h) - \boldsymbol{W_\theta}(0)| > \varepsilon\delta^\beta$.

(j) The following conditions are equivalent (EXERCISE):

(†)$\boldsymbol{W_\theta} \in \mathcal{ND}(\mathbb{R})$ (resp. $\boldsymbol{W_\theta} \in \mathcal{ND}^\infty(\mathbb{R})$) for every $\boldsymbol{\theta}$;

(‡)for every $\boldsymbol{\theta}$, a finite (resp. finite or infinite) derivative $\boldsymbol{W}'_{\boldsymbol{\theta}}(0)$ does not exist.

(k) The following conditions are equivalent (EXERCISE):

(†)$\boldsymbol{W_\theta} \in \mathcal{ND}_\pm(\mathbb{R})$ (resp. $\boldsymbol{W_\theta} \in \mathcal{ND}^\infty_\pm(\mathbb{R})$) for every $\boldsymbol{\theta}$;

(‡)for every $\boldsymbol{\theta}$, a finite (resp. finite or infinite) right-sided derivative $(\boldsymbol{W_\theta})'_+(0)$ does not exist.

3.3 Differentiability of $\boldsymbol{W_{p,a,b,\theta}}$ (in the Infinite Sense)

It seems that G.H. Hardy was the first to notice that in general, $\boldsymbol{W_{p,a,b,\theta}} \notin \mathcal{ND}^\infty(\mathbb{R})$.

Theorem 3.3.1 (Cf. [Har16]). *If $ab \geq 1$ and $a(b+1) < 2$, then $\boldsymbol{S}'_{a,b}(0) = +\infty$.*

Notice that $a(b+1) < 2$, provided that $ab = 1$.

Proof. Put $f := \boldsymbol{S}_{a,b}$. It suffices to show that $f'_+(0) = +\infty$. Take an $h \in \mathbb{R}$, $0 < h \leq \frac{1}{4}$. Let $N = N(h) \in \mathbb{N}$ be such that $b^{N-1}h \leq \frac{1}{4} < b^N h$. Then

$$\frac{f(h)}{h} = \frac{1}{h}\sum_{n=0}^{N-1} a^n \sin(2\pi b^n h) + \frac{1}{h}\sum_{n=N}^{\infty} a^n \sin(2\pi b^n h) =: f_1(h) + f_2(h).$$

We have

$$f_1(h) \geq 4\sum_{n=0}^{N-1}(ab)^n = \begin{cases} 4N, & \text{if } ab = 1 \\ 4\frac{(ab)^N - 1}{ab-1}, & \text{if } ab > 1 \end{cases}, \qquad |f_2(h)| \leq \frac{1}{h}\frac{a^N}{1-a}.$$

First observe that in the case $ab = 1$, we have

$$\frac{1}{h}\frac{a^N}{1-a} = \frac{1}{hb^N}\frac{1}{1-a} < \frac{4}{1-a},$$

and therefore,

$$f_1(h) + f_2(h) \geq 4N(h) - \frac{4}{1-a} \underset{h\to 0+}{\longrightarrow} +\infty.$$

Now assume that $ab > 1$. Then $\frac{1}{ab-1} - \frac{1}{1-a} > 0$, and consequently,

$$\begin{aligned} f_1(h) + f_2(h) &\geq 4\frac{(ab)^N - 1}{ab-1} - \frac{1}{h}\frac{a^N}{1-a} > 4\frac{(ab)^N - 1}{ab-1} - 4b^N\frac{a^N}{1-a} \\ &= 4(ab)^{N(h)}\Big(\frac{1-(ab)^{-N(h)}}{ab-1} - \frac{1}{1-a}\Big) \underset{h\to 0+}{\longrightarrow} +\infty. \end{aligned}$$

□

Theorem 3.3.2 (Cf. [Hat88b]). *If* $ab > 1$ *and*

$$1 + \frac{b}{ab-1}\sin\frac{\pi}{2b} > \frac{ab}{1-a}, \tag{3.3.1}$$

then $\boldsymbol{S}'_{a,b}(0) = +\infty$.

Remark 3.3.3. (a) If $ab > 1$ and $a(b+1) \le 2$, then (3.3.1) is satisfied. In particular, in the case $ab > 1$, Theorem 3.3.2 generalizes the Hardy's original criterion ($a(b+1) < 2$) from Theorem 3.3.1 (see also [Beh49] (the footnote on page 467)).

(b) The function $(\frac{1}{b}, 1) \ni a \longmapsto 1 + \frac{b}{ab-1}\sin\frac{\pi}{2b} - \frac{ab}{1-a}$ is strictly decreasing. In particular, there exists exactly one $a = \varphi(b) \in (\frac{1}{b}, 1)$ such that $1 + \frac{b}{ab-1}\sin\frac{\pi}{2b} > \frac{ab}{1-a} \iff \frac{1}{b} < a < \varphi(b)$. Note that $\varphi(b) > \frac{2}{b+1}$.

Proof of Theorem 3.3.2. Put $f := \boldsymbol{S}_{a,b}$. It suffices to show that $f'_+(0) = +\infty$. Take an $h \in \mathbb{R}$, $0 < h \le \frac{1}{4}$. Let $N = N(h) \in \mathbb{N}$ be such that $4b^N h \le 1 < 4b^{N+1}h$. Then

$$\begin{aligned}\boldsymbol{\Delta} f(0,h) &= \Big(\frac{1}{h}\sum_{n=0}^{N-1} a^n \sin(2\pi b^n h)\Big) + \frac{a^N}{h}\sin(2\pi b^N h) \\ &\quad + \Big(\frac{1}{h}\sum_{n=N+1}^{\infty} a^n \sin(2\pi b^n h)\Big) =: f_1(h) + f_2(h) + f_3(h).\end{aligned}$$

We have

$$|f_3(h)| \le \frac{1}{h}\frac{a^{N+1}}{1-a} < \frac{4(ab)^{N+1}}{1-a}, \qquad f_2(h) \ge \frac{1}{h}a^N 4b^N h = 4(ab)^N.$$

For $1 \le n \le N-1$, we have $2\pi b^n h \le \frac{\pi}{2b}$. Hence

$$f_1(h) \ge \frac{1}{h}\sum_{n=0}^{N-1} a^n\Big(\sin\frac{\pi}{2b}\Big)4b^{n+1}h = \frac{4b((ab)^N - 1)}{ab-1}\sin\frac{\pi}{2b}.$$

Finally,

$$\boldsymbol{\Delta} f(0,h) \ge 4\Big(1 + \frac{b(1-(ab)^{-N(h)})}{ab-1}\sin\frac{\pi}{2b} - \frac{ab}{1-a}\Big)(ab)^{N(h)} \underset{h\to 0}{\longrightarrow} +\infty,$$

provided that (3.3.1) is satisfied. □

Proposition 3.3.4. *Assume that* a, b *are such that* $\boldsymbol{S}'_{a,b}(0) = +\infty$. *Then:*

(a) *(cf. [Har16]) if* $b \in 4\mathbb{N}+1$, *then* $\boldsymbol{C}'_{a,b}(\frac{1}{4}) = +\infty$;

(b) *(cf. [Hob26], p. 407) if* $b \in \mathbb{N}$, *then* $\boldsymbol{S}'_{a,b}(\frac{k}{2^m}) = +\infty$ *for all* $k \in \mathbb{Z}$ *and* $m \in \mathbb{N}$.

Remark 3.3.5. We will prove that $\boldsymbol{S}_{a,b} \in \mathfrak{ND}(\mathbb{R})$ (Theorem 8.6.7). Moreover, in view of Theorem 2.3.1, the set

$$A := \{x \in \mathbb{R} : \text{an infinite derivative } \boldsymbol{S}'_{a,b}(x) \text{ exists}\}$$

is of measure zero. On the other hand, (b) states that if $b \in \mathbb{N}$ and $\boldsymbol{S}'_{a,b}(0) = +\infty$, then the set A is dense in $\mathbb{R}$.

Proof of Proposition 3.3.4.

$$\text{(a)}\quad \boldsymbol{C}_{a,b}(x-\tfrac{1}{4}) = \sum_{n=0}^{\infty} a^n \cos(2\pi b^n(x-\tfrac{1}{4})) = \sum_{n=0}^{\infty} a^n \sin(2\pi b^n x) = \boldsymbol{S}_{a,b}(x).$$

(b) Fix an $x_0 := \frac{k}{2^m}$ and let $g(x) := \boldsymbol{S}_{a,b}(x+x_0)$. We have to prove that $g'(0) = +\infty$. Let

$$\varphi(x) := \sum_{n=0}^{m-1} a^n \sin(2\pi b^n(x+x_0)), \quad x \in \mathbb{R}.$$

Then

$$\begin{aligned} g(x) &= \sum_{n=0}^{m-1} a^n \sin(2\pi b^n(x+x_0)) + \sum_{n=m}^{\infty} a^n \sin(2\pi b^n x + 2\pi b^{n-m} k) \\ &= \varphi(x) + \sum_{n=m}^{\infty} a^n \sin(2\pi b^n x) = \varphi(x) + a^m \sum_{n=0}^{\infty} a^n \sin(2\pi b^n b^m x) \\ &= \varphi(x) + a^m \boldsymbol{S}_{a,b}(b^m x), \quad x \in \mathbb{R}. \end{aligned}$$

Since φ is a differentiable function and $\boldsymbol{S}'_{a,b}(0) = +\infty$, the proof is complete. □

3.4 An Open Problem

In view of the results presented in the previous section, one may formulate the following natural problem. Given $b > 1$, estimate the numbers

$$\begin{aligned} \boldsymbol{\alpha}_C(b) &:= \inf\{a \in [1/b, 1) : \boldsymbol{C}_{a,b} \in \mathcal{ND}^\infty(\mathbb{R})\}, \\ \boldsymbol{\alpha}_S(b) &:= \inf\{a \in [1/b, 1) : \boldsymbol{S}_{a,b} \in \mathcal{ND}^\infty(\mathbb{R})\}. \end{aligned}$$

Remark 3.4.1. (a) We have seen that $\boldsymbol{\alpha}_S(b) \geq \frac{2}{b+1}$ (Theorem 3.3.1) and $\boldsymbol{\alpha}_C(b) \geq \frac{2}{b+1}$ for $b \in 4\mathbb{N}+1$ (Proposition 3.3.4(a)).

(b) A better lower estimate was given in Theorem 3.3.2: $\boldsymbol{\alpha}_S(b) \geq \varphi(b) > \frac{2}{b+1}$, where $a = \varphi(b) \in (\frac{1}{b}, 1)$ is a uniquely determined root of the equation $1 + \frac{b}{ab-1}\sin\frac{\pi}{2b} - \frac{ab}{1-a} = 0$.

(c) Theorem 3.8.1 (cf. Remark 3.1.1(4)) will show that $\boldsymbol{\alpha}_C(b) \leq \frac{1+\frac{3}{2}\pi}{b+\frac{3}{2}\pi}$, provided $b \in 2\mathbb{N}+1$.

(d) Theorem 3.9.9 (cf. Remark 3.1.1(6)) will give $\boldsymbol{\alpha}_C(b) \leq \frac{1+\varkappa(b)}{b+\varkappa(b)}$ for $b > 3$, where $\varkappa(b) := \frac{(3+2\varepsilon)\pi}{2\cos(\pi\varepsilon)}$, $\varepsilon := \frac{1}{b-1}$. Note that if $b \in 2\mathbb{N}_2+1$, then (d) is not better than (c).

(e) The best known (as of 2015) estimate will be proved in Theorem 8.7.6: $\boldsymbol{\alpha}_S(b), \boldsymbol{\alpha}_C(b) \leq \frac{H}{b}$, where $H := 1 + \frac{1}{\cos\psi^*}$, $\psi^* \in (0, \frac{\pi}{2})$ is such that $\tan\psi^* = \pi + \psi^*$; note that $\psi^* \approx 1.3518$, $H \approx 5.6034$.

(f) Observe that (Exercise)

$$\frac{1+\varkappa(b)}{b+\varkappa(b)} \geq \frac{H}{b} \text{ for } b \geq b_0 \approx 212.9669.$$

Thus for $b \geq 213$, the estimate (e) is better than (d).

(g) Moreover (EXERCISE),

$$\frac{1+\frac{3}{2}\pi}{b+\frac{3}{2}\pi} \geq \frac{H}{b} \text{ for } b \geq \frac{\frac{3}{2}\pi H}{\frac{3}{2}\pi + 1 - H} \approx 242.1373.$$

In particular, for $b \in 2\mathbb{N}+1$, $b \geq 243$, the estimate (e) is better than (c).

[?] Exact values of $\boldsymbol{\alpha}_C(b)$ and $\boldsymbol{\alpha}_S(b)$ are not known [?]

3.5 Weierstrass's Method

The following sections will present different attempts to get the nowhere differentiability of $\boldsymbol{W}_{p,a,b,\boldsymbol{\theta}}$ for certain configurations of the parameters $p, a, b, \boldsymbol{\theta}$. We will see that the case $ab = 1$ is the most difficult one. We point out that in general, these results are not optimal. Nevertheless, they perfectly illustrate various ways of attacking the problem.

Theorem 3.5.1 (Cf. [BR74, Her79, Wei86, Muk34]; see also [Mal09]). *Assume that $b, p \in 2\mathbb{N}_0 + 1$ and $ab > 1 + \frac{3}{2}p\pi$. Put $f := \boldsymbol{W}_{p,a,b,0}$. Then for every $x \in \mathbb{R}$,*

$$\textit{either } (D^+f(x) = +\infty \textit{ and } D_-f(x) = -\infty),$$
$$\textit{or } (D^-f(x) = +\infty \textit{ and } D_+f(x) = -\infty).$$

In particular, $f \in \mathfrak{M}(\mathbb{R}) \cap \mathfrak{ND}^\infty(\mathbb{R}) \subset \mathfrak{ND}_\pm(\mathbb{R}) \cap \mathfrak{ND}^\infty(\mathbb{R})$.

Remark 3.5.2. (a) According to P. Bois-Reymond (cf. [BR74], p. 31), Weierstrass himself conjectured that $\boldsymbol{C}_{a,b} \in \mathfrak{ND}(\mathbb{R})$ for all $0 < a < 1$ and $ab \geq 1$.
(b) Note that for $p = 1$, $b \in 2\mathbb{N}+1$, the inequality $ab > 1 + \frac{3}{2}\pi$ implies that $b \geq 7$.
(c) The case $p > 1$ was first considered by K. Hertz in [Her79]. His proof is a direct modification of the original Weierstrass proof for $p = 1$.

Proof of Theorem 3.5.1. Fix $x \in \mathbb{R}$ and $m \in \mathbb{N}$. Let $\alpha_m \in \mathbb{Z}$ be such that

$$h_m := 2b^m x - \alpha_m \in (-\tfrac{1}{2}, \tfrac{1}{2}].$$

Put $x_m^\pm := \frac{1}{2}(\alpha_m \pm 1)b^{-m}$ and observe that $x_m^\pm - x = \frac{1}{2}(\pm 1 - h_m)b^{-m}$. In particular, $x_m^- \longrightarrow x-$ and $x_m^+ \longrightarrow x+$. Then

$$\begin{aligned}\boldsymbol{\Delta} f(x, x_m^\pm) &= \sum_{n=0}^{m-1} a^n \frac{\cos^p(2\pi b^n x_m^\pm) - \cos^p(2\pi b^n x)}{x_m^\pm - x} \\ &+ \sum_{n=m}^{\infty} a^n \frac{\cos^p(2\pi b^n x_m^\pm) - \cos^p(2\pi b^n x)}{x_m^\pm - x} =: Q'_{m,\pm} + Q''_{m,\pm}.\end{aligned}$$

By Remark 3.2.1(f), we obtain $|Q'_{m,\pm}| < 2p\pi\frac{(ab)^m}{ab-1}$. For $n \geq m$, we have

$$\cos^p(2\pi b^n x_m^\pm) = \cos^p(\pi b^{n-m}(\alpha_m \pm 1)) = -(-1)^{\alpha_m},$$
$$\cos^p(2\pi b^n x) = \cos^p(\pi b^{n-m}(h_m + \alpha_m)) = (-1)^{\alpha_m}\cos^p(\pi b^{n-m} h_m).$$

Hence

$$Q''_{m,\pm} = 2\sum_{n=m}^{\infty} a^n \frac{-(-1)^{\alpha_m}(1+\cos^p(\pi b^{n-m}h_m))}{(\pm 1 - h_m)b^{-m}}$$
$$= \mp(-1)^{\alpha_m}(ab)^m 2\sum_{n=0}^{\infty} a^n \frac{1+\cos^p(\pi b^n h_m)}{1\mp h_m} = \mp(-1)^{\alpha_m}(ab)^m 2T_{m,\pm},$$

where

$$T_{m,\pm} \geq a^0 \frac{1+\cos^p(\pi h_m)}{1\mp h_m} \geq \frac{2}{3}.$$

Thus

$$\boldsymbol{\Delta} f(x, x_m^{\pm}) = \mp(-1)^{\alpha_m} 2(ab)^m \Big(\frac{p\pi}{ab-1}V_{m,\pm} + \frac{2}{3}U_{m,\pm}\Big), \tag{3.5.1}$$

where $U_{m,\pm} \geq 1$, $|V_{m,\pm}| \leq 1$. The condition $ab > 1 + \frac{3}{2}\pi$ implies that

$$\operatorname{sgn} \boldsymbol{\Delta} f(x, x_m^+) = -\operatorname{sgn} \boldsymbol{\Delta} f(x, x_m^-), \quad |\boldsymbol{\Delta} f(x, x_m^{\pm})| \underset{m\to+\infty}{\longrightarrow} +\infty.$$

Hence, either ($D^+f(x) = +\infty$ and $D_-f(x) = -\infty$) or ($D^-f(x) = +\infty$ and $D_+f(x) = -\infty$). □

The above idea of the proof may be used to obtain other results concerning nowhere differentiability of the function $\boldsymbol{W}_{1,a,b,0}$ (see below).

Theorem 3.5.3. *If $b \in 2\mathbb{N}+1$ and $ab > 1 + \frac{\pi}{2}$, then $f := \boldsymbol{W}_{1,a,b,0} \in \mathcal{ND}(\mathbb{R})$.*

Proof. (Some ideas are taken from [Wie81].) We keep the notation from the proof of Theorem 3.5.1. Fix an $x \in \mathbb{R}$. For $m \in \mathbb{N}$, define $x_m := \frac{\alpha_m + q_m}{2b^m}$, where $q_m := \begin{cases} -1, & \text{if } h_m < 0 \\ 1, & \text{if } h_m \geq 0 \end{cases}$. Note that $x_m \longrightarrow x$. Then we get

$$\boldsymbol{\Delta} f(x, x_m) = (-1)^{\alpha_m+1} q_m 2\pi(ab)^m \Big(\frac{V_m}{ab-1} + U_m\Big),$$

where $|V_m| \leq 1$ and

$$U_m := \sum_{n=0}^{\infty} (ab)^n \frac{1+\cos(\pi b^n|h_m|)}{\pi b^n(1-|h_m|)} \geq \frac{1+\cos(\pi|h_m|)}{\pi(1-|h_m|)} \geq \frac{2}{\pi}$$

(because $\cos t \geq 1 - \frac{2}{\pi}t$ for $t \in [0, \frac{\pi}{2}]$). Consequently, if $ab > 1 + \frac{\pi}{2}$, then $|\boldsymbol{\Delta} f(x, x_m)| \longrightarrow +\infty$. □

In order to continue, we need the following auxiliary function $\varphi : \mathbb{R}_{>0} \longrightarrow \mathbb{R}_+$:

$$\varphi(x) := \frac{1-\cos x}{x}, \quad x > 0.$$

Observe that:

- φ is increasing on $(0, \frac{\pi}{2}]$,
- $\lim_{x\to 0+} \varphi(x) = 0$.

Lemma 3.5.4.

$$\text{(a)}\quad \varphi(x)+\Big(1+\frac{1}{\varphi(\frac{3\pi}{2b})}\Big)\varphi(bx)\geq\varphi\Big(\frac{3\pi}{2b}\Big),\quad b\geq 3,\ x\in\Big[\frac{\pi}{2b},\frac{3\pi}{2b}\Big],$$

$$\text{(b)}\quad \Big\{b\in 2\mathbb{N}_3+1:\varphi\Big(\frac{3\pi}{2b}\Big)>\frac{2}{3\pi}\Big\}=\{7,9\}.$$

Proof. EXERCISE—use a computer. □

Theorem 3.5.5 (Cf. [You16a]). *Assume that*

$$b\in 2\mathbb{N}+1 \text{ and } ab>\max\Big\{1+\frac{3}{2}\pi,1+\frac{1}{\varphi(\frac{3\pi}{2b})}\Big\}$$
$$=\begin{cases}1+\frac{3}{2}\pi, & \text{if } b\in\{7,9\}\\ 1+\frac{1}{\varphi(\frac{3\pi}{2b})}, & \text{if } b\geq 11\end{cases}. \tag{3.5.2}$$

Let $A:=\{\frac{k}{2b^\ell}:k\in\mathbb{Z},\ \ell\in\mathbb{N}\}$. *Put* $f:=\boldsymbol{W}_{1,a,b,0}$.

(a) *If* $x\notin A$, *then*

$$\text{either } (D^+f(x)=+\infty,\ D_+f(x)=-\infty,$$
$$\text{and } \max\{|D^-f(x)|,|D_-f(x)|\}=+\infty)$$
$$\text{or } (D^-f(x)=+\infty,\ D_-f(x)=-\infty,$$
$$\text{and } \max\{|D^+f(x)|,|D_+f(x)|\}=+\infty).$$

(b) *If* $x\in A$, *then either* $f'_\pm(x)=\pm\infty$ *or* $f'_\pm(x)=\mp\infty$.

Remark 3.5.6. (i) Notice that the points $x\in\mathbb{R}$ with property (b) are called *cusps*.

(ii) Observe that in view of the Denjoy–Young–Saks theorem, Theorem 2.3.1, almost every $x\notin A$ is a *knot point* for f i.e., $D^+f(x)=D^-f(x)=+\infty$ and $D_+f(x)=D_-f(x)=-\infty$.

Proof of Theorem 3.5.5. (Some ideas are taken from [Wie81].) We keep the notation from the proof of Theorem 3.5.1.

(a) In view of Theorem 3.5.1 (with $p=1$), we need to prove only that

$$\text{either } (D^+f(x)=+\infty,\ D_+f(x)=-\infty),$$
$$\text{or } (D^-f(x)=+\infty,\ D_-f(x)=-\infty).$$

Assume that $(D^+f(x)=+\infty$ and $D_-f(x)=-\infty)$. The case $(D^-f(x)=+\infty$ and $D_+f(x)=-\infty)$ is left to the reader as an EXERCISE.

If the set $\{m\in\mathbb{N}:\alpha_m \text{ is even}\}$ is infinite, then equality (3.5.1) implies that $D_+f(x)=-\infty$, which finishes the proof. Thus, we may assume that α_m is odd for $m\geq m_0$.

Now observe that the set $M:=\{m\in\mathbb{N}_{m_0}:|h_m|>\frac{1}{2b}\}$ is infinite. Indeed, suppose that $|h_m|\leq\frac{1}{2b}$ for some $m\geq m_0$. Let $r\in\mathbb{N}_2$ be the minimal number such that $b^r|h_m|>\frac{1}{2b}$. We have $2b^{m+r}x_0-b^r\alpha_m=b^rh_m=b(b^{r-1}h_m)\leq b\frac{1}{2b}=\frac{1}{2}$. Thus $m+r\in M$, and therefore M is infinite.

Let $x_m := \frac{\alpha_m}{2b^m}$, $m \in M$. Then, using the same method as in the proof of Theorem 3.5.1, we get

$$\boldsymbol{\Delta} f(x, x_m) = \operatorname{sgn}(h_m) 2\pi (ab)^m \Big(\frac{V_m}{ab-1} + U_m\Big), \tag{3.5.3}$$

where $|V_m| \le 1$ and

$$U_m := \sum_{n=0}^{\infty} (ab)^n \frac{1-\cos(\pi b^n |h_m|)}{\pi b^n |h_m|} = \sum_{n=0}^{\infty} (ab)^n \varphi(\pi b^n |h_m|)$$
$$\ge \varphi(\pi |h_m|) + ab\varphi(\pi b |h_m|) =: T_m.$$

The main problem is to show that

$$T_m \ge \varphi\Big(\frac{3\pi}{2b}\Big), \quad m \in M. \tag{3.5.4}$$

Indeed, if (3.5.4) is satisfied, then (3.5.3) implies that either $D^- f(x) = +\infty$ or $D_+ f(x) = -\infty$, which finishes the proof.

We move to the proof of (3.5.4). Recall that $\frac{1}{2b} < |h_m| \le \frac{1}{2}$ for $m \in M$. If $\frac{3}{2b} \le |h_m| \le \frac{1}{2}$, then we have $T_m \ge \varphi(\pi |h_m|) \ge \varphi(\frac{3\pi}{2b})$, and we are done. Thus, we may assume that $\frac{1}{2b} < |h_m| < \frac{3}{2b}$, and then we can use Lemma 3.5.4(a).

(b) Fix an $x = \frac{k}{2b^\ell}$. Take an h such that $\frac{1}{2b} < 2b^m |h| \le \frac{1}{2}$ for some $m \in \mathbb{N}_\ell$ and write $x = \frac{kb^{m-\ell}}{2b^m}$. Then

$$\boldsymbol{\Delta} f(x, x+h) = (-1)^{k+1} \operatorname{sgn}(h) 2\pi (ab)^m \Big(\frac{V_m}{ab-1} + U_m\Big),$$

where $|V_m| \le 1$ and

$$U_m := \sum_{n=0}^{\infty} (ab)^n \frac{1-\cos(2\pi b^{n+m}|h|)}{2\pi b^{n+m}|h|}.$$

In view of the proof of (a), we have $U_m \ge \varphi(\frac{3\pi}{2b})$. Consequently,

- if k is odd, then $f'_+(x) = +\infty$ and $f'_-(x) = -\infty$;
- if k is even, then $f'_+(x) = -\infty$ and $f'_-(x) = +\infty$.

□

Theorem 3.5.1 (with $p = 1$) allowed Weierstrass to answer his question on the existence of a holomorphic function on $\mathbb{D}$, continuous on $\overline{\mathbb{D}}$, and holomorphically uncontinuable across $\partial\mathbb{D}$ (cf. [Wei86], p. 90).[1]

Proposition 3.5.7 (Cf. [Wei86], p. 90; see also Remark 8.5.3(g)). *Assume that $b \in 2\mathbb{N}+1$, $ab > 1 + \frac{3}{2}\pi$, and define*

[1] "Ich habe in meinen Vorlesungen über die Elemente der Functionenlehre von Anfang an zwei mit den gewöhnlichen Ansichten nicht übereinstimmende Sätze hervorgehoben, nämlich: (...) (2) dass eine Function eines complexen Arguments, welche für einen beschränkten Bereich des letzteren definirt ist, sich nicht immer über die Grenzen dieses Bereichs hinaus fortsetzen lasse; mit andern Worten, dass monogene Functionen einer Veränderlichen existiren, welche die Eigenthümlichkeit besitzen, dass in der Ebene der Veränderlichen diejenigen Stellen, für welche die Function nicht definirbar ist, nicht bloss einzelne Punkte sind, sondern auch Linien und Flächen bilden."

$$F(z) := \sum_{n=0}^{\infty} a^n z^{b^n}, \quad z \in \overline{\mathbb{D}}.$$

Then $F \in \mathcal{O}(\mathbb{D}) \cap \mathcal{C}(\overline{\mathbb{D}}, \mathbb{C})$, and $\mathbb{D}$ is the domain of existence of F.

Proof. Since $|a^n z^{b^n}| \le a^n$, $z \in \overline{\mathbb{D}}$, $n \in \mathbb{N}_0$, the function F is well defined, $F \in \mathcal{O}(\mathbb{D})$, and $F \in \mathcal{C}(\overline{\mathbb{D}}, \mathbb{C})$. Suppose that there exist $z_0 = e^{2\pi i x_0} \in \mathbb{T}$, $r > 0$, and $\widetilde{F} \in \mathcal{O}(\mathbb{D}(z_0, r))$ such that $\widetilde{F} = F$ in $\mathbb{D} \cap \mathbb{D}(z_0, r)$. Then $\boldsymbol{W}_{1,a,b,0}(x) = \operatorname{Re} F(e^{2\pi i x}) = \operatorname{Re} \widetilde{F}(e^{2\pi i x})$ for x in a neighborhood of x_0. Consequently, $\boldsymbol{W}_{1,a,b,0}$ is a real-analytic function near x_0, which contradicts Theorem 3.5.1. □

The above proposition may be easily generalized as follows.

Proposition 3.5.8. *Assume that $b \in \mathbb{N}_2$ and $\boldsymbol{W}_{1,a,b,\boldsymbol{\theta}} \in \boldsymbol{\mathcal{ND}}(\mathbb{R})$. Define*

$$F(z) := \sum_{n=0}^{\infty} a^n e^{i\theta_n} z^{b^n}, \quad z \in \overline{\mathbb{D}}.$$

Then $F \in \mathcal{O}(\mathbb{D}) \cap \mathcal{C}(\overline{\mathbb{D}}, \mathbb{C})$, and $\mathbb{D}$ is the domain of existence of F.

3.5.1 Lerch's Results

The class of the Weierstrass-type functions $\boldsymbol{W}_{p,a,b,\boldsymbol{\theta}}$ may be extended in the following natural way. We define

$$\boldsymbol{W}_{p,\boldsymbol{a},\boldsymbol{b},\boldsymbol{\theta}}(x) := \sum_{n=0}^{\infty} a_n \cos^p(2\pi b_n x + \theta_n), \quad x \in \mathbb{R},$$

where $p \in \mathbb{N}$, $\boldsymbol{a} := (a_n)_{n=0}^{\infty} \subset \mathbb{C}_*$ with $\sum_{n=0}^{\infty} |a_n| < +\infty$, $\boldsymbol{b} := (b_n)_{n=0}^{\infty} \subset \mathbb{R}_{>0}$, and $\boldsymbol{\theta} := (\theta_n)_{n=0}^{\infty} \subset \mathbb{R}$. Functions of the above type will be studied in § 8.5. Now we mention only two results due to M. Lerch in [Ler88] that extend Theorem 3.5.1 (with $p = 1$ and $\boldsymbol{\theta} = 0$).

Theorem 3.5.9 (Cf. [Ler88]). *Let $f := \boldsymbol{W}_{1,\boldsymbol{a},\boldsymbol{b},0}$, where:*
- *$\boldsymbol{a} = (a_n)_{n=0}^{\infty} \subset \mathbb{R}_*$ is such that $\sum_{n=0}^{\infty} |a_n| < +\infty$,*
- *$\boldsymbol{b} = (b_n)_{n=0}^{\infty} \subset \mathbb{N}$ is such that there exists a sequence $(q_m)_{m=0}^{\infty} \subset \mathbb{N}$ for which $\frac{b_n}{q_m} =: p_{n,m} \in \mathbb{N}$ for $n > m$.*

Let $x_0 := \frac{\ell}{2q_m}$ for some $\ell \in \mathbb{Z}$ and $m \in \mathbb{N}_0$. Assume that there is a $\mu > m$ such that $\operatorname{sgn}(-1)^{\ell p_{n,m}} = \varepsilon \in \{-1, 1\}$ for $n \ge \mu$. Then $f'(x_0)$ exists iff $f'(0)$ exists. Moreover, if $f'(x_0)$ exists, then

$$f'(x_0) = -2\pi \sum_{n=0}^{\infty} a_n b_n \sin(2\pi b_n x_0)$$

(i.e., we can formally differentiate under the summation sign).

Remark 3.5.10. (a) Since f is an even function, if $f'(0)$ exists, then $f'(0) = 0$.

(b) Consider, for example, $b_n := n!$, $q_m := (m+1)!$. Then $p_{n,m} := (m+2)\cdots n$ ($p_{m+1,m} = 1$) and $(-1)^{\ell p_{n,m}} = 1$ for $n \ge m+3$.

Proof of Theorem 3.5.9. We have

$$f(x_0+h)=\sum_{n=0}^{m} a_n\cos(2\pi b_n(x_0+h))+\sum_{n=m+1}^{\infty} a_n(-1)^{\ell p_{n,m}}\cos(2\pi b_n h).$$

Hence,

$$\begin{aligned}&\boldsymbol{\Delta} f(x_0,x_0+h)-\sum_{n=0}^{m} a_n\frac{\cos(2\pi b_n(x_0+h))-\cos(2\pi b_n x_0)}{h}\\&=\sum_{n=m+1}^{\infty} a_n(-1)^{\ell p_{n,m}}\frac{\cos(2\pi b_n h)-1}{h}=\varepsilon\boldsymbol{\Delta} f(0,h)\\&-\varepsilon\sum_{n=0}^{m} a_n\frac{\cos(2\pi b_n h)-1}{h}+\sum_{n=m+1}^{\mu-1} a_n((-1)^{\ell p_{n,m}}-\varepsilon)\frac{\cos(2\pi b_n h)-1}{h}.\end{aligned}$$

Consequently, $f'(x_0)$ exists iff $f'(0)$ exists, and then

$$f'(x_0)=-2\pi\sum_{n=0}^{m} a_n b_n\sin(2\pi b_n x_0)=-2\pi\sum_{n=0}^{\infty} a_n b_n\sin(2\pi b_n x_0).$$ □

Theorem 3.5.11 (Cf. [Ler88]). *Let* $f:=\boldsymbol{W}_{1,\boldsymbol{a},\boldsymbol{b},0}$, *where:*

- $\boldsymbol{a}=(a_n)_{n=0}^{\infty}\subset\mathbb{R}_{>0}$ *is such that* $\sum_{n=0}^{\infty} a_n<+\infty$,
- $\boldsymbol{b}=(b_n)_{n=0}^{\infty}$, $b_n:=p_0\cdots p_n$, *where* $(p_m)_{m=0}^{\infty}\subset 2\mathbb{N}_0+1$, $b_n\nearrow+\infty$, *and*
- *there exists an* $M>0$ *such that* $a_m b_m-\frac{\pi^2}{b_m}\sum_{n=0}^{m-1} a_n b_n^2\geq M$, $m\in\mathbb{N}$.

Then $f\in\mathbf{ND}(\mathbb{R})$.

Remark 3.5.12. In the original Weierstrass case ($a_n=a^n$, $b_n=b^n$ with $0<a<1$ and $b\in 2\mathbb{N}+1$), the third condition states that

$$(ab)^m-\frac{\pi^2}{b^m}\sum_{n=0}^{m-1}(ab^2)^n\geq M,\quad m\in\mathbb{N}.$$

In the case $ab^2=1$, we get $(ab)^m-\frac{\pi^2}{b^m}m\geq M$, i.e., $\frac{1}{b^m}(1-\pi^2 m)\geq M$, $m\in\mathbb{N}$, which gives a contradiction. Thus $ab^2\neq 1$ and

$$M\leq(ab)^m-\frac{\pi^2}{b^m}\frac{(ab^2)^m-1}{ab^2-1}=(ab)^m\Big(1-\pi^2\frac{1-(ab^2)^{-m}}{ab^2-1}\Big),\quad m\in\mathbb{N},$$

which is equivalent to $ab\geq 1$, $ab^2>1+\pi^2$. Note that this condition is better than the Weierstrass one (cf. Theorem 3.5.1), i.e., $ab>1+\frac{3}{2}\pi$, but the result is weaker.

Proof of Theorem 3.5.11. (Cf. the proof of Theorem 3.5.1.) Fix $x\in\mathbb{R}$ and $m\in\mathbb{N}$. Let $\alpha_m\in\mathbb{Z}$ be such that

$$h_m:=2b_m x-\alpha_m\in(-\tfrac{1}{2},\tfrac{1}{2}].$$

Put $x_m^{\pm}:=\frac{\alpha_m\pm 1}{2b_m}$ and observe that $x_m^{\pm}-x=\frac{\pm 1-h_m}{2b_m}$. In particular, $x_m^-\longrightarrow x-$ and $x_m^+\longrightarrow x+$. Then

$$\boldsymbol{\Delta} f(x, x_m^{\pm}) = \sum_{n=0}^{m-1} a_n \frac{\cos(2\pi b_n x_m^{\pm}) - \cos(2\pi b_n x)}{x_m^{\pm} - x}$$
$$+ \sum_{n=m}^{\infty} a_n \frac{\cos(2\pi b_n x_m^{\pm}) - \cos(2\pi b_n x)}{x_m^{\pm} - x} =: Q'_{m,\pm} + Q''_{m,\pm}.$$

By the mean value theorem, we get

$$Q'_{m,-} - Q'_{m,+} = -\sum_{n=0}^{m-1} a_n \Big(2\pi b_n \sin(2\pi b_n \xi_n^-) - 2\pi b_n \sin(2\pi b_n \xi_n^+)\Big)$$
$$= -4\pi^2 \sum_{n=0}^{m-1} a_n b_n^2 \cos(2\pi b_n \eta_n)(\xi_n^- - \xi_n^+).$$

Thus,

$$|Q'_{m,-} - Q'_{m,+}| \le \frac{4\pi^2}{b_m} \sum_{n=0}^{m-1} a_n b_n^2.$$

For $n \ge m$, we have

$$\cos(2\pi b_n x_m^{\pm}) = \cos(\pi \tfrac{b_n}{b_m}(\alpha_m \pm 1)) = (-1)^{\frac{b_n}{b_m}(\alpha_m \pm 1)} = -(-1)^{\alpha_m},$$
$$\cos(2\pi b_n x) = \cos(\pi \tfrac{b_n}{b_m}(h_m + \alpha_m)) = (-1)^{\alpha_m} \cos(\pi \tfrac{b_n}{b_m} h_m).$$

Hence

$$Q''_{m,\pm} = -(-1)^{\alpha_m} \frac{2b_m}{\pm 1 - h_m} \sum_{n=m}^{\infty} a_n (1 + \cos(\pi \tfrac{b_n}{b_m} h_m)),$$

and therefore

$$Q''_{m,-} - Q''_{m,+} = (-1)^{\alpha_m} \frac{4b_m}{1 - h_m^2} \sum_{n=m}^{\infty} a_n (1 + \cos(\pi \tfrac{b_n}{b_m} h_m)).$$

Observe that

$$\sum_{n=m}^{\infty} a_n (1 + \cos(\pi \tfrac{b_n}{b_m} h_m)) \ge a_m (1 + \cos(\pi h_m)) \ge a_m.$$

Thus

$$|\boldsymbol{\Delta} f(x, x_m^-) - \boldsymbol{\Delta} f(x, x_m^+)| \ge 4\Big(a_m b_m - \frac{\pi^2}{b_m} \sum_{n=0}^{m-1} a_n b_n^2\Big) \ge 4M > 0,$$

which immediately implies that a finite derivative $f'(x)$ does not exist. □

3.5.2 Porter's Results

In the context of Lerch's results (cf. § 3.5.1), another approach was proposed by M.B. Porter in [Por19].

Theorem 3.5.13 (Cf. [Por19]). *Let* $f := \boldsymbol{W}_{1,\boldsymbol{a},\boldsymbol{b},0}$, *where:*

- $\boldsymbol{a} = (a_n)_{n=0}^{\infty} \subset \mathbb{C}$ *is such that* $\sum_{n=0}^{\infty} |a_n| < +\infty$,
- $\boldsymbol{b} = (b_n)_{n=0}^{\infty}$, $b_n := p_0 \cdots p_n$ *with* $(p_m)_{m=0}^{\infty} \subset \mathbb{N}$,
- *the set* $A := \{n \in \mathbb{N} : p_{n+1} \in 4\mathbb{N}\}$ *is infinite,*
- $M_m := |a_m| b_m - \frac{3\pi}{2} \sum_{n=0}^{m-1} |a_n| b_n \underset{A \ni m \to +\infty}{\longrightarrow} +\infty$.

Then $f \in \mathcal{ND}(\mathbb{R})$.

The result remains true for the function $\boldsymbol{W}_{1,\boldsymbol{a},\boldsymbol{b},-\pi/2}$.

Remark 3.5.14 (Cf. [Por19]). Theorem 3.5.13 implies that the following functions are nowhere differentiable (EXERCISE):

(a) $f(x) = \sum_{n=0}^{\infty} a^n \cos(2\pi b^n x)$, $|a| < 1$, $b \in 4\mathbb{N}$, $|a| b > 1 + \frac{3}{2}\pi$.
(b) $f(x) = \sum_{n=0}^{\infty} \frac{a^n}{n!} \cos(2\pi n! x)$, $|a| > 1 + \frac{3}{2}\pi$.
(c) $f(x) = \sum_{n=0}^{\infty} \frac{1}{b^n} \cos(2\pi n! b^n x)$, $b \in \mathbb{N}_2$.

Moreover, in the above examples, one can replace cos by sin.

Proof of Theorem 3.5.13. Fix an $x_0 \in \mathbb{R}$ and an $m \in A$. Let $\frac{b_{m+1}}{b_m} = p_{m+1} = 4q_m$ ($q_m \in \mathbb{N}$) and let $h = h_{m,k} := \frac{k}{b_{m+1}}$ with $k \in \mathbb{Z}$, $|k| \frac{b_m}{b_{m+1}} = \frac{|k|}{4q_m} \leq \frac{3}{4}$. Since $b_m | b_n$ for $n \geq m+1$, we get

$$\boldsymbol{\Delta} f(x_0, x_0 + h) = \Big(\sum_{n=0}^{m-1} a_n \frac{\cos(2\pi b_n (x_0 + h)) - \cos(2\pi b_n x_0)}{h} \Big) + a_m \frac{\cos(2\pi b_m (x_0 + h)) - \cos(2\pi b_m x_0)}{h} = Q'_m + Q''_m.$$

Then $|Q'_{m,\pm}| \leq 2\pi \sum_{n=0}^{m-1} |a_n| b_n$ (cf. Remark 3.2.1(f)). On the other hand, since $|k| \frac{b_m}{b_{m+1}} \leq \frac{3}{4}$, we have

$$|Q''_m| = \left| a_m \frac{2 \sin(\frac{\pi k b_m}{b_{m+1}}) \sin(\pi b_m x_0 + \frac{\pi k b_m}{b_{m+1}})}{\frac{k}{b_{m+1}}} \right| \geq 2\pi |a_m| b_m \frac{1}{\frac{3}{4}\pi\sqrt{2}} \left| \sin\left(\pi b_m x_0 + \frac{\pi k}{4q_m}\right) \right|.$$

Observe (EXERCISE) that there exists an $\ell \in \{-1, 1\}$ such that

$$\left| \sin\left(\pi b_m x_0 + \frac{\pi \ell}{4}\right) \right| \geq \frac{1}{\sqrt{2}}.$$

Consequently, for $h = h_m := h_{m,\ell q_m}$ (note that $0 \neq h_m \longrightarrow 0$), we get

$$|\boldsymbol{\Delta} f(x_0, x_0 + h_m)| \geq 2\pi |a_m| b_m \frac{2}{3\pi} - 2\pi \sum_{n=0}^{m-1} |a_n| b_n = \frac{4}{3} M_n \underset{A \ni m \to +\infty}{\longrightarrow} +\infty,$$

which implies that a finite derivative $f'(x_0)$ does not exist.

The case of the function $\boldsymbol{W}_{1,\boldsymbol{a},\boldsymbol{b},-\pi/2}$ is left to the reader as an EXERCISE. □

3.6 Cellérier's Method

Theorem 3.6.1 (Cf. [Cel90]). *If $b \in 2\mathbb{N}$ and $b \geq 14$, then $\boldsymbol{W}_{1,1/b,b,\boldsymbol{\theta}} \in \mathcal{ND}_{\pm}(\mathbb{R})$ for every $\boldsymbol{\theta}$.*

Remark 3.6.2. (a) In fact, C. Cellérier in [Cel90] considered only the case in which $b \in 2\mathbb{N}$, $b > 1000$, and $\boldsymbol{\theta} = 0$.

(b) We will see (cf. Remark 3.7.2(d)) that in fact, the result is true for all $b \gg 1$.

(c) Notice that in the case $\boldsymbol{\theta} = \theta$, the result is true for all $b \in \mathbb{N}_2$ (Theorem 8.4.1).
[?] It is an open question whether $\boldsymbol{W}_{1,1/b,b,\boldsymbol{\theta}} \in \mathcal{ND}_{\pm}(\mathbb{R})$ for all $b > 1$ and $\boldsymbol{\theta}$ [?]

Proof of Theorem 3.6.1. Put

$$f(x) := \boldsymbol{W}_{1,1/b,b,\boldsymbol{\theta}-\frac{\pi}{2}}(x) = \sum_{n=0}^{\infty} \frac{1}{b^n} \sin(2\pi b^n x + \theta_n), \quad x \in \mathbb{R}.$$

We know that it suffices to prove that $f'_+(0)$ does not exist for every $\boldsymbol{\theta}$ (cf. Remark 3.2.1(k)). Define

$$\Delta_m := \boldsymbol{\Delta} f(0, b^{-m}), \quad \Delta'_m := \boldsymbol{\Delta} f(0, \tfrac{1}{2} b^{-m}), \quad m \in \mathbb{N}.$$

Observe that for $n > m$ and $\eta \in \{\frac{1}{2}, 1\}$, we have

$$\sin(2\pi b^n \eta b^{-m} + \theta_n) = \sin(2\pi \eta b^{n-m} + \theta_n) = \sin \theta_n.$$

Moreover,

$$\sin(2\pi b^m \eta b^{-m} + \theta_m) = \sin(2\pi\eta + \theta_m) = \begin{cases} -\sin\theta_m, & \text{if } \eta = \frac{1}{2} \\ \sin\theta_m, & \text{if } \eta = 1 \end{cases}.$$

Thus for $m \geq 2$, we get

$$\begin{aligned} \Delta_m &= \sum_{n=0}^{m-1} \frac{\sin(2\pi b^{n-m} + \theta_n) - \sin\theta_n}{b^{n-m}} \\ &= \sum_{n=0}^{m-1} \Big(\sin\theta_n \frac{\cos(2\pi b^{n-m}) - 1}{b^{n-m}} + \cos\theta_n \frac{\sin(2\pi b^{n-m})}{b^{n-m}} \Big). \end{aligned}$$

Observe that

$$\left| \frac{\cos(2\pi b^{n-m}) - 1}{b^{n-m}} \right| = \left| \frac{2\sin^2(\pi b^{n-m})}{b^{n-m}} \right| \leq 2\pi^2 b^{n-m},$$

$$\left| 1 - \frac{\sin(2\pi b^{n-m})}{2\pi b^{n-m}} \right| \leq \frac{1}{6}(2\pi b^{n-m})^2.$$

Let

$$c_m := 2\pi \sum_{n=0}^{m-1} \cos\theta_n.$$

Thus

$$|\Delta_m - c_m| \le \sum_{n=0}^{m-1}\left(2\pi^2 b^{n-m} + \frac{2}{3}\pi^2 b^{2(n-m)}\right) < 2\pi^2\frac{1}{b-1} + \frac{2}{3}\pi^2\frac{1}{b^2-1}$$
$$= \frac{2\pi^2}{b^2-1}\left(b+1+\frac{2}{3}\pi\right) =: B.$$

Analogously,

$$|\Delta'_m - c_m + 4\sin\theta_m| < \pi^2\frac{1}{b-1} + 2\pi\frac{1}{6}\pi^2\frac{1}{b^2-1} = \frac{\pi^2}{b^2-1}\left(b+1+\frac{1}{3}\pi\right) =: B'.$$

Suppose $f'_+(0)$ exists and is finite. Then $\Delta_{m+1} - \Delta_m \longrightarrow 0$ and $\Delta_m - \Delta'_m \longrightarrow 0$ when $m \longrightarrow +\infty$. On the other hand,

$$|\Delta_{m+1} - \Delta_m - 2\pi\cos\theta_m| < 2B, \quad |\Delta_m - \Delta'_m - 4\sin\theta_m| < B + B'.$$

Since $\cos^2\theta_m + \sin^2\theta_m = 1$, to get a contradiction we have only to show that

$$\varphi(b) := \left(\frac{B}{\pi}\right)^2 + \left(\frac{B+B'}{4}\right)^2 < 1.$$

We have

$$\varphi(b) = \frac{\pi^2}{144(b^2-1)^2}\Big((576+81\pi^2)b^2 + (1152+768\pi+162\pi^2+90\pi^3)b$$
$$+ 576 + 768\pi + 337\pi^2 + 90\pi^3 + 25\pi^4\Big).$$

Then $\varphi'(b) < 0$ for all $b > 1$ and $\varphi(13) < 1$ (Exercise—use a computer). □

3.7 Dini's Method

The following general method proposed by U. Dini in [Din92] made it possible to solve many new cases. For $p \in \mathbb{N}$, define

$$\delta := \begin{cases} \frac{3}{4}, & \text{if } p \in 2\mathbb{N}_0 + 1 \\ \frac{1}{2}(1-\Theta_p), & \text{if } p \in 2\mathbb{N} \end{cases}, \quad d := \begin{cases} 1, & \text{if } p \in 2\mathbb{N}_0 + 1 \\ \frac{1}{2}, & \text{if } p \in 2\mathbb{N} \end{cases},$$

where $\Theta_p \in (0, \frac{1}{4}]$ is such that $\cos^p(\pi\Theta_p) = \frac{1}{2}$ $(p \in 2\mathbb{N})$. Consider the following two conditions:

$$ab > 1 \text{ and } \frac{p\pi\delta}{ab-1} + \frac{a}{1-a} < \frac{d}{2}, \tag{3.7.1}$$

$$\frac{(p\pi)^2\delta(\delta+d)}{ab^2-1} + \left(1+\frac{\delta}{d}\right)\frac{a}{1-a} < \frac{d}{2}. \tag{3.7.2}$$

Direct calculations give:

- (a,b) satisfies (3.7.1) iff $0 < a < a_1(p)$ and $b > \Psi_1(a)$,
- (a,b) satisfies (3.7.2) iff $0 < a < a_2(p)$ and $b > \Psi_2(a)$,

where

$$\Psi_1(x) := \frac{1}{x}\Big(1 + 2p\pi\delta\frac{1-x}{d-(d+2)x}\Big), \quad 0 < x < \frac{d}{d+2} =: a_1(p),$$

$$\Psi_2(x) := \sqrt{\frac{1}{x}\Big(1 + (p\pi)^2\delta(\delta+d)\frac{2d(1-x)}{d^2-(d^2+2(\delta+d))x}\Big)},$$

$$0 < x < \frac{d^2}{d^2+2(\delta+d)} =: a_2(p) < a_1(p).$$

Theorem 3.7.1 (Cf. [Din92], Chap. 10). *If* ($a < a_1(p)$ *and* $b > \Psi_1(a)$) *or* ($a < a_2(p)$ *and* $b > \Psi_2(a)$), *then* $\boldsymbol{W}_{p,a,b,\boldsymbol{\theta}} \in \mathfrak{ND}_{\pm}(\mathbb{R})$.

Remark 3.7.2. (a) One can easily check (EXERCISE) that $\Psi_1(x) > \Psi_2(x)$ when $x \longrightarrow 0+$, and $\Psi_1(x) < \Psi_2(x)$ when $x \longrightarrow a_2(p)-$. Thus for small a, the estimate $b > \Psi_2(a)$ is better, and for a near $a_2(p)$, the estimate $b > \Psi_1(a)$ is better.

(b) In fact, U. Dini in [Din92] considered only the case that $p = 1$ and $\boldsymbol{\theta} = 0$.

(c) In the case $p = 1$, the theorem states the following:
If ($a < \frac{1}{3}$ *and* $ab > 1 + \frac{3}{2}\pi\frac{1-a}{1-3a}$) *or* ($a < \frac{2}{9}$ *and* $ab^2 > 1 + \frac{21}{4}\pi^2\frac{1-a}{2-9a}$), *then* $\boldsymbol{W}_{1,a,b,\boldsymbol{\theta}} \in \mathfrak{ND}_{\pm}(\mathbb{R})$.

(d) In the case $ab = 1$, condition (3.7.2) states that

$$b > \begin{cases} \frac{21}{8}(p\pi)^2 + \frac{9}{2}, & \text{if } p \in 2\mathbb{N}_0 + 1 \\ (2-\Theta_p)(1-\Theta_p)(p\pi)^2 + 9 - 4\Theta_p, & \text{if } p \in 2\mathbb{N} \end{cases}.$$

In particular,

- if $p = 1$, then $b > \frac{21}{8}\pi^2 + \frac{9}{2} \approx 30.4077$; note that if $b \in 2\mathbb{N}$ and $\boldsymbol{\theta} = 0$, then Theorem 3.6.1 gives a better estimate ($b \geq 14$);
- if $p = 2$, then $b > \frac{21}{4}\pi^2 + 8 \approx 59.8154$.

Proof of Theorem 3.7.1. Put $f := \boldsymbol{W}_{p,a,b,\boldsymbol{\theta}}$. By Remark 3.2.1(k), we have only to prove that a finite $f'_+(0)$ does not exist. Let $\varphi_n(x) := \cos^p(2\pi b^n x + \theta_n)$, $x \in \mathbb{R}$. For $h > 0$, we have

$$|\varphi_n(h) - \varphi_n(0) - \varphi_n'(0)h| = |\tfrac{1}{2}\varphi_n''(\eta)h^2| \leq 2(p\pi b^n h)^2, \tag{3.7.3}$$

where $\eta = \eta(n,h) \in (0,h)$. Fix $m \in \mathbb{N}$, $h > 0$, and write

$$f(h) - f(0) = \Big(\sum_{n=0}^{m-1} a^n(\varphi_n(h) - \varphi_n(0))\Big) + \Big(a^m(\varphi_m(h) - \varphi_m(0))\Big)$$
$$+ \Big(\sum_{n=m+1}^{\infty} a^n(\varphi_n(h) - \varphi_n(0))\Big) =: A_m(h) + B_m(h) + C_m(h).$$

By Remark 3.2.1(f), we obtain $|A_m(h)| < \frac{2p\pi h(ab)^m}{ab-1}$, provided that $ab > 1$. Using (3.7.3), we get

$$\Big|A_m(h) - \sum_{n=0}^{m-1} a^n\varphi_n'(0)h\Big| \leq \sum_{n=0}^{m-1} a^n 2(p\pi b^n h)^2 = 2(p\pi h)^2\frac{(ab^2)^m - 1}{ab^2 - 1}$$
$$< \frac{2(p\pi h)^2(ab^2)^m}{ab^2 - 1},$$

$$|C_m(h)| \le 2 \sum_{n=m+1}^{\infty} a^n = \frac{2a^{m+1}}{1-a}.$$

Observe that there exists an $h_m \in (0, \frac{\delta}{b^m})$ such that $\varphi_m(h_m) - \varphi_m(0) = d\varepsilon_m$, where $\varepsilon_m \in \{-1, +1\}$. Consequently, if $ab > 1$, then we get

$$\boldsymbol{\Delta} f(0, h_m) = \frac{\varepsilon_m d a^m}{h_m} \Big(\alpha_m \frac{2p\pi\delta}{d(ab-1)} + 1 + \gamma_m \frac{2a}{d(1-a)} \Big),$$

where $\alpha_m, \gamma_m \in [-1, 1]$. Note that $\frac{da^m}{h_m} > \frac{d}{\delta}(ab)^m \underset{m\to+\infty}{\longrightarrow} +\infty$. Thus, if $\frac{p\pi\delta}{ab-1} + \frac{a}{1-a} < \frac{d}{2}$, then $|\boldsymbol{\Delta} f(0, h_m)| \underset{m\to+\infty}{\longrightarrow} +\infty$, and therefore a finite $f'_+(0)$ does not exist.

Now let us consider the general case (with $ab \ge 1$). Let $h'_m := \frac{d}{b^m}$. Then $\varphi_m(h'_m) = \varphi_m(0)$. Thus

$$\begin{aligned}
&\boldsymbol{\Delta} f(0, h_m) - \boldsymbol{\Delta} f(0, h'_m) \\
&= \frac{\varepsilon_m d a^m}{h_m} \Big(\alpha'_m \frac{2(p\pi)^2 h_m^2 b^{2m}}{d(ab^2-1)} + \alpha''_m \frac{2(p\pi)^2 h'_m h_m b^{2m}}{d(ab^2-1)} + 1 \\
&\quad + \gamma'_m \frac{2a}{d(1-a)} + \gamma''_m \frac{h_m}{h'_m} \frac{2a}{d(1-a)} \Big) \\
&= \frac{\varepsilon_m d a^m}{h_m} \Big(\alpha_m \frac{2(p\pi)^2 \delta(\delta+d)}{d(ab^2-1)} + 1 + \gamma_m \Big(1 + \frac{\delta}{d}\Big) \frac{2a}{d(1-a)} \Big),
\end{aligned}$$

where $\alpha'_m, \alpha''_m, \alpha_m, \gamma'_m, \gamma''_m, \gamma_m \in [-1, 1]$. Observe that $\frac{da^m}{h_m} > \frac{d}{\delta}(ab)^m \ge \frac{d}{\delta} > 0$. Thus, if $\frac{(p\pi)^2\delta(\delta+d)}{ab^2-1} + (1 + \frac{\delta}{d})\frac{a}{1-a} < \frac{d}{2}$, then a finite $f'_+(0)$ does not exist. □

3.8 Bromwich's Method

Theorem 3.8.1 (Cf. [Bro08]). *Assume that* $b \in 2\mathbb{N} + 1$, $\boldsymbol{\theta} = \theta$ *with* $\theta \in (-\frac{\pi}{2}, \frac{\pi}{2})$, *and* $ab > 1 + \frac{3}{2}\frac{\pi}{\cos\theta}(1-a)$. *Let* $f := \boldsymbol{W}_{1,a,b,\theta}$. *Then*

$$\min\{D_+ f(x), D_- f(x)\} = -\infty, \quad \max\{D^+ f(x), D^- f(x)\} = +\infty.$$

In particular, $f \in \mathbf{ND}^\infty(\mathbb{R})$.

Remark 3.8.2. In fact, T.J.I'A. Bromwich considered only the case $\theta = 0$.

Proof of Theorem 3.8.1. Take $m \in \mathbb{N}$, $p \in \mathbb{Z}$, $q \in \{p+1, p+3\}$, and let $t := \frac{1}{2}pb^{-m}$, $u := \frac{1}{2}qb^{-m}$. Observe that

$$\begin{aligned}
\sum_{n=m}^{\infty} a^n \cos(2\pi b^n t + \theta) = \sum_{n=m}^{\infty} a^n \cos(\pi b^{n-m} p + \theta) &= \sum_{n=m}^{\infty} a^n (-1)^p \cos\theta \\
&= (-1)^p \cos\theta \frac{a^m}{1-a}.
\end{aligned}$$

Hence, in view of Remark 3.2.1(f), we obtain

$$\boldsymbol{\Delta} f(t,u) = \frac{2(ab)^m}{q-p}\Big((-1)^q\frac{2\cos\theta}{1-a} + M(m,p,q)\Big),$$

where $|M(m,p,q)| \le \frac{3\pi}{ab-1}$. The condition $ab > 1 + \frac{3\pi}{2\cos\theta}(1-a)$ implies that $\operatorname{sgn}\boldsymbol{\Delta} f(t,u) = \operatorname{sgn}(-1)^q$.

Now fix $x \in \mathbb{R}$ and $\delta > 0$. Let $m \in \mathbb{N}$ be such that $2b^{-m} < \delta$ and let $r := \lfloor 2xb^m \rfloor$.

If $r < \lfloor 2xb^m \rfloor$, then we put $t_{i,m} = \frac{1}{2}p_{i,m}b^{-m}$, $p_{i,m} := r+1-i$, $u_{i,m} = \frac{1}{2}q_{i,m}b^{-m}$, $i=1,2$, $q_{1,m} := p_{1,m}+1$, $q_{2,m} := p_{2,m}+3$.

If $r = \lfloor 2xb^m \rfloor$, then we take $p_{i,m} := r-i$, $q_{i,m} := p_{i,m}+3$, $i=1,2$.

Then $t_{i,m} \in (x-\delta, x)$, $u_{i,m} \in (x, x+\delta)$, $i=1,2$, and by the first part of the proof, we get

$$\boldsymbol{\Delta} f(t_{i,m}, u_{i,m}) = \frac{2(ab)^m}{q_{i,m}-p_{i,m}}\Big((-1)^{q_{i,m}}\frac{2\cos\theta}{1-a} + M_{i,m}\Big),$$

where $|M_{i,m}| \le \frac{3\pi}{ab-1}$, $i=1,2$. Note that $\operatorname{sgn} q_{1,m} = -\operatorname{sgn} q_{2,m}$, $m \in \mathbb{N}$. Thus, there exist sequences $(t_s^{\pm})_{s=1}^{\infty}$, $(u_s^{\pm})_{s=1}^{\infty}$ with $t_s^{\pm} < x < u_s^{\pm}$, $t_s^{\pm} \longrightarrow x$, $u_s^{\pm} \longrightarrow x$, such that $\boldsymbol{\Delta} f(t_s^{\pm}, u_s^{\pm}) \longrightarrow \pm\infty$. By Remark 2.1.2, we have

$$\min\{\boldsymbol{\Delta} f(x,u_s^{\pm}), \boldsymbol{\Delta} f(x,t_s^{\pm})\} \le \boldsymbol{\Delta} f(t_s^{\pm}, u_s^{\pm}) \le \max\{\boldsymbol{\Delta} f(x,u_s^{\pm}), \boldsymbol{\Delta} f(x,t_s^{\pm})\}.$$

Hence $-\infty = \min\{D_+f(x), D_-f(x)\} < \max\{D^+f(x), D^-f(x)\} = +\infty$. □

3.9 Behrend's Method

This whole section is based on [Beh49]. Let $f := \boldsymbol{C}_{a,b}$.

Theorem 3.9.1. *Assume that* $K > 0$ *is such that* $ab > 1 + \frac{2\pi}{K}$ *and*

$$\forall_{x\in\mathbb{R}}\ \exists_{h\in\mathbb{R}_*} : \boldsymbol{\Delta} f(x, x+h) \ge K \tag{3.9.1}$$

(cf. Remark 3.9.3). Then $f \in \boldsymbol{\mathfrak{N}\mathfrak{D}}^{\infty}(\mathbb{R})$.

It is clear that the result gets better as we increase the constant K.

Proof. Let $A := \frac{1}{1-a}$. Note that $|\boldsymbol{\Delta} f(x,x+h)| \le \frac{2A}{|h|}$. We will prove that

$$\forall_{x\in\mathbb{R},\, C>0}\ \exists_{h=h_{x,C},\, h'=h'_{x,C}\in\mathbb{R}_*} : \boldsymbol{\Delta} f(x,x+h) > C,\ \boldsymbol{\Delta} f(x,x+h') < -C. \tag{3.9.2}$$

Observe that $|h_{x,C}|, |h'_{x,C}| \le \frac{2A}{C}$. In particular, $h(x,C), h'(x,C) \longrightarrow 0$ when $C \longrightarrow +\infty$. Consequently, (3.9.2) implies that $f \in \boldsymbol{\mathfrak{N}\mathfrak{D}}^{\infty}(\mathbb{R})$.

Put

$$W_m(x) := \sum_{n=0}^{m-1} a^n \cos(2\pi b^n x), \quad R_m(x) := a^m f(b^m x), \quad x \in \mathbb{R},\ m \in \mathbb{N}.$$

Then $f = W_m + R_m$. Indeed,

$$\sum_{n=m}^{\infty} a^n \cos(2\pi b^n x) = \sum_{n=0}^{\infty} a^{m+n}\cos(2\pi b^{m+n}x) = a^m f(b^m x).$$

We know that $|\boldsymbol{\Delta} W_m(x, x+h)| < \frac{2\pi(ab)^m}{ab-1}$ (cf. Remark 3.2.1(f)). Since $f(-x) = f(x)$, $x \in \mathbb{R}$, condition (3.9.1) implies that

$$\forall_{x\in\mathbb{R},\ m\in\mathbb{N}}\ \exists_{h=h_{x,m},\ h'=h'_{x,m}\in\mathbb{R}_*} :$$
$$\boldsymbol{\Delta} R_m(x, x+h) \geq K(ab)^m, \quad \boldsymbol{\Delta} R_m(x, x+h') \leq -K(ab)^m.$$

Hence

$$\boldsymbol{\Delta} f(x, x+h_{x,m}) \geq (ab)^m\Big(K - \frac{2\pi}{ab-1}\Big),$$
$$\boldsymbol{\Delta} f(x, x+h'_{x,m}) \leq -(ab)^m\Big(K - \frac{2\pi}{ab-1}\Big),$$

which immediately gives (3.9.2). □

The same proof gives the following general theorem (Exercise).

Theorem 3.9.2. *Let $g : \mathbb{R} \longrightarrow [-1, 1]$ be a function such that*

$$|g(x) - g(x')| \leq M|x - x'| \text{ and } g(-x) = g(x), \quad x, x' \in \mathbb{R}.$$

For $0 < a < 1$ and $b > 1$, define $G(x) := \sum_{n=0}^{\infty} a^n g(b^n x)$, $x \in \mathbb{R}$. Assume that $K > 0$ is such that $ab > 1 + \frac{M}{K}$ and

$$\forall_{x\in\mathbb{R}}\ \exists_{h\in\mathbb{R}_*} : \boldsymbol{\Delta} G(x, x+h) \geq K.$$

Then $G \in \mathbf{ND}^\infty(\mathbb{R})$.

Remark 3.9.3. Let $\varphi : \mathbb{R} \longrightarrow \mathbb{R}$. Fix $x \in \mathbb{R}$, $K > 0$. Observe that the following two conditions are equivalent:

(i) $\exists_{h\in\mathbb{R}_*} : \boldsymbol{\Delta}\varphi(x, x+h) \geq K$;
(ii) $\exists_{x_1\leq x\leq x_2,\ x_1<x_2} : \boldsymbol{\Delta}\varphi(x_1, x_2) \geq K$.

Indeed, the implication (i) $\Longrightarrow$ (ii) is obvious. To prove the opposite implication, observe that using Remark 2.1.2(b), we get

$$K \leq \boldsymbol{\Delta}\varphi(x_1, x_2) \leq \max\{\boldsymbol{\Delta}\varphi(x, x_1), \boldsymbol{\Delta}\varphi(x, x_2)\}.$$

Thus either $\boldsymbol{\Delta}\varphi(x, x_2) \geq K$ or $\boldsymbol{\Delta}\varphi(x, x_1) \geq K$.

To apply Theorem 3.9.1, one should find a constant $K > 0$ such that (3.9.1) is satisfied. We get the following corollaries.

Theorem 3.9.4. *If $b \in \mathbb{N}_2$ and*

$$ab > \begin{cases} 1 + \frac{3\pi}{2}, & \text{if } b \in 2\mathbb{N} \\ 1 + \frac{3\pi}{2}(1-a), & \text{if } b \in 2\mathbb{N}+1 \end{cases},$$

then $f \in \mathbf{ND}^\infty(\mathbb{R})$ (cf. Theorems 3.5.1, 3.8.1).

Proof. Since $f(x) = f(x+1)$, $x \in \mathbb{R}$, it suffices to check (3.9.1) only for $x \in [0, 1)$. As before, let $A := \sum_{n=0}^{\infty} a^n = \frac{1}{1-a}$. In view of Remark 3.9.3, we have only to find $x_1 \leq 0 < 1 \leq x_2$ such that $\boldsymbol{\Delta} f(x_1, x_2) = K > 0$. Take $x_1 := -\frac{1}{2}$, $x_2 := 1$. Then

$$f(x_1) = \sum_{n=0}^{\infty} a^n \cos(\pi b^n) = \begin{cases} -1 + \sum_{n=1}^{\infty} a^n = -2 + A, & \text{if } b \in 2\mathbb{N} \\ -\sum_{n=0}^{\infty} a^n = -A, & \text{if } b \in 2\mathbb{N}+1 \end{cases},$$

$$f(x_2) = \sum_{n=0}^{\infty} a^n \cos(2\pi b^n) = A.$$

Hence

$$K = \boldsymbol{\Delta} f(x_1, x_2) = \begin{cases} \frac{4}{3}, & \text{if } b \in 2\mathbb{N} \\ \frac{4}{3}A, & \text{if } b \in 2\mathbb{N}+1 \end{cases}.$$

It remains to apply Theorem 3.9.1. □

Theorem 3.9.5. *If $b \in 2\mathbb{N} \setminus (3\mathbb{N})$ and $ab > 1 + \frac{16\pi}{9}(1-a)$, then $f \in \mathcal{ND}^\infty(\mathbb{R})$.*

Remark 3.9.6. Observe that $\frac{16\pi}{9}(1-a) < \frac{3\pi}{2}$ for $\frac{5}{32} < a < 1$. Thus for $\frac{5}{32} < a < 1$, Theorem 3.9.5 improves Theorem 3.9.4.

Proof of Theorem 3.9.5. We take $x_1 := -\frac{1}{3}$, $x_2 := 1$. Then $f(x_1) = -\frac{1}{2}A$ and $K = \frac{9}{8}A$. □

Theorem 3.9.7. *Assume that $b \in 2\mathbb{N}$. Let $\mu \in (0, \frac{1}{2})$ be such that*

$$\Big(\frac{3}{2} - \mu\Big)\pi = \frac{1}{\tan(\pi\mu)}$$

($\mu \approx 0.0697$) and let $k = k(b) \in 2\mathbb{N}_0$ be such that $|b\mu - k| \le 1$. If $ab > 1 + \frac{\pi(\frac{3}{2} - \frac{k}{b})}{\cos(\pi\frac{k}{b})} =: C(b)$, then $f \in \mathcal{ND}^\infty(\mathbb{R})$.

Remark 3.9.8. Observe that $\lim_{b\to+\infty} C(b) = 1 + \frac{\pi(\frac{3}{2}-\mu)}{\cos(\pi\mu)} = 1 + \frac{1}{\sin(\pi\mu)} < 1 + \frac{3\pi}{2}$ (because $\frac{3\pi}{2} - \frac{1}{\sin(\pi\mu)} \approx 0.1088$). Thus for $b \gg 1$, Theorem 3.9.7 improves Theorem 3.9.4.

Proof of Theorem 3.9.7. Let $x_1 = x_1(b) := -\frac{1}{2} + \frac{k}{2b}$, $x_2 = x_2(b) := 1 - \frac{k}{2b}$. Then $f(x_1) = -\cos(\pi\frac{k}{b}) + aA$ and $f(x_2) = \cos(\pi\frac{k}{b}) + aA$. Hence $\boldsymbol{\Delta} f(x_1, x_2) = \frac{\cos(\pi\frac{k}{b})}{\frac{3}{2} - \frac{k}{b}}$. The above points x_1, x_2 suffice to prove the nowhere differentiability in $[0, \frac{1}{2}]$. Since $f(x + \frac{1}{2}) = f(x)$ (b is even), we get the nowhere differentiability on $\mathbb{R}$. □

Theorem 3.9.9. *For an arbitrary $b > 3$, if $ab > 1 + \frac{(3+2\varepsilon)\pi}{2\cos(\pi\varepsilon)}(1-a)$, where $\varepsilon := \frac{1}{b-1}$, then $f \in \mathcal{ND}^\infty(\mathbb{R})$.*

Remark 3.9.10. One can prove (Exercise) that the above estimate is better than Dini's (Remark 3.7.2(c)), i.e., if $a < \frac{1}{3}$ and $ab > 1 + \frac{3}{2}\pi\frac{1-a}{1-3a}$, then $b > 3$ and $ab > 1 + \frac{(3+2\varepsilon)\pi}{2\cos(\pi\varepsilon)}(1-a)$ with $\varepsilon := \frac{1}{b-1}$.

Proof of Theorem 3.9.9. We have $\cos(\pi x) \ge \cos(\pi\varepsilon) =: \delta > 0$, $x \in I_p := [2p - \varepsilon, 2p + \varepsilon]$, $p \in \mathbb{Z}$. Take an $m \in \mathbb{Z}$ and let $J_0 := I_m$. Observe that every interval of length $\ge 2 + 2\varepsilon$ contains an interval I_p for some p. Suppose that $J_n := \frac{1}{b^n} I_{p_n}$ (for some p_n). The length of J_n equals

$$\frac{2\varepsilon}{b^n} = \frac{2\varepsilon}{b^{n+1}} b \ge \frac{2\varepsilon}{b^{n+1}}\Big(1 + \frac{1}{\varepsilon}\Big) = \frac{2 + 2\varepsilon}{b^{n+1}}.$$

Thus there exists a p_{n+1} such that $J_{n+1} := \frac{1}{b^{n+1}} I_{p_{n+1}} \subset J_n \subset \dots \subset J_1 \subset J_0$. Let $\bigcap_{n=0}^{\infty} J_n = \{x_{2m}\}$. Then $f(\frac{1}{2}x_{2m}) \ge \frac{\delta}{1-a}$ and $|2m - x_{2m}| \le \varepsilon$.

Analogously, one gets a point x_{2m+1} such that $f(\frac{1}{2}x_{2m+1}) \leq -\frac{\delta}{1-a}$ and $|2m+1-x_{2m+1}| \leq \varepsilon$. We have $x_k \leq k+\varepsilon < k+\frac{1}{2} < k+1-\varepsilon \leq x_{k+1}$. Hence for every $x \in \mathbb{R}$, there exists an $m \in \mathbb{Z}$ such that $2x \in (x_{2m+1}, x_{2m+4})$. Thus

$$\boldsymbol{\Delta} f(\tfrac{1}{2}x_{2m+1}, \tfrac{1}{2}x_{2m+4}) \geq \frac{4\delta}{(1-a)(x_{2m+4}-x_{2m+1})} \geq \frac{4\delta}{(1-a)(3+2\varepsilon)}. \qquad \square$$

To get better characterizations of the case $f \in \mathfrak{N}\mathfrak{D}(\mathbb{R})$, F.A. Behrend proposed the following method.

Theorem 3.9.11. *Assume that $E, D, L > 0$ are such that $ab \geq 1$, $ab^2 > 1 + \frac{2\pi^2}{L}$, and for every $x \in \mathbb{R}$, there exist $x', x'' \in \mathbb{R}$ for which:*

- $|x-x'| + |x-x''| \leq E$,
- $\boldsymbol{\Delta} f(x,x') - \boldsymbol{\Delta} f(x,x'') \geq D$,
- $\boldsymbol{\Delta} f(x,x') - \boldsymbol{\Delta} f(x,x'') \geq L(|x-x'| + |x-x''|)$.

Then $f \in \mathfrak{N}\mathfrak{D}(\mathbb{R})$.

It is clear that the result gets better as we increase the constant L.

Proof. Fix an $x \in \mathbb{R}$ and suppose that a finite $f'(x)$ exists. We use notation from the proof of Theorem 3.9.1. Observe that for the function $\varphi(t) := \cos(2\pi t)$, we have

$$|\boldsymbol{\Delta}\varphi(t,t') - \boldsymbol{\Delta}\varphi(t,t'')| \leq N(|t-t'| + |t-t''|),$$

where $N := 2\pi^2$. Thus we get

$$\begin{aligned} |\boldsymbol{\Delta} W_m(x,x') - \boldsymbol{\Delta} W_m(x,x'')| &\leq \sum_{n=0}^{m-1} (ab^2)^n N(|x-x'| + |x-x''|) \\ &< N\frac{(ab^2)^m}{ab^2-1}(|x-x'| + |x-x''|), \quad m \in \mathbb{N}. \end{aligned}$$

Recall that $R_m(x) = a^m f(b^m x)$. Hence for every $m \in \mathbb{N}$, there exist points x'_m, x''_m such that:

- $|x-x'_m| + |x-x''_m| \leq \frac{E}{b^m}$ (in particular, $x'_m \longrightarrow x$ and $x''_m \longrightarrow x$ as $m \longrightarrow +\infty$),
- $\boldsymbol{\Delta} R_m(x,x'_m) - \boldsymbol{\Delta} R_m(x,x''_m) \geq D(ab)^m$,
- $\boldsymbol{\Delta} R_m(x,x'_m) - \boldsymbol{\Delta} R_m(x,x''_m) \geq L(ab^2)^m(|x-x'_m| + |x-x''_m|)$.

Finally,

$$\boldsymbol{\Delta} f(x,x'_m) - \boldsymbol{\Delta} f(x,x''_m) \geq D(ab)^m\Big(1 - \frac{N}{L}\frac{1}{ab^2-1}\Big), \quad m \in \mathbb{N}.$$

Letting $m \longrightarrow +\infty$, we get a contradiction. $\square$

The same proof gives the following general theorem (Exercise).

Theorem 3.9.12. *Let $g : \mathbb{R} \longrightarrow [-1,1]$ be such that there exists an $N > 0$ with*

$$|\boldsymbol{\Delta} g(x,x') - \boldsymbol{\Delta} g(x,x'')| \leq N(|x-x'| + |x-x''|), \quad x, x', x'' \in \mathbb{R}.$$

For $0 < a < 1$ and $b > 1$, define $G(x) := \sum_{n=0}^{\infty} a^n g(b^n x)$, $x \in \mathbb{R}$. Assume that $E, D, L > 0$ are such that $ab \geq 1$, $ab^2 > 1 + \frac{N}{L}$, and for every $x \in \mathbb{R}$, there exist $x', x'' \in \mathbb{R}$ for which:

- $|x - x'| + |x - x''| \le E$,
- $\boldsymbol{\Delta} G(x, x') - \boldsymbol{\Delta} G(x, x'') \ge D$,
- $\boldsymbol{\Delta} G(x, x') - \boldsymbol{\Delta} G(x, x'') \ge L(|x - x'| + |x - x''|)$.

Then $G \in \mathfrak{ND}(\mathbb{R})$.

Remark 3.9.13. (a) Let $\varphi : \mathbb{R} \longrightarrow \mathbb{R}$. Fix $x \in \mathbb{R}$, $D > 0$. Observe that the following two conditions are equivalent:

(i) $\exists_{x',x'' \in \mathbb{R}} : \boldsymbol{\Delta}\varphi(x, x') - \boldsymbol{\Delta}\varphi(x, x'') \ge D$;

(ii) $\exists_{\substack{x_1' \le x \le x_2',\ x_1' < x_2' \\ x_1'' \le x \le x_2'',\ x_1'' < x_2''}} : \boldsymbol{\Delta}\varphi(x_1', x_2') - \boldsymbol{\Delta}\varphi(x_1'', x_2'') \ge D$.

Indeed, the implication (i) $\Longrightarrow$ (ii) is obvious. The opposite implication follows from Remark 2.1.2(b). We have

$$D \le \boldsymbol{\Delta}\varphi(x, x') - \boldsymbol{\Delta}\varphi(x, x'') \\ \le \max\{\boldsymbol{\Delta}\varphi(x, x_1'), \boldsymbol{\Delta}\varphi(x, x_2')\} - \min\{\boldsymbol{\Delta}\varphi(x, x_1''), \boldsymbol{\Delta}\varphi(x, x_2'')\},$$

which gives (i) with $x' \in \{x_1', x_2'\}$, $x'' \in \{x_1'', x_2''\}$.

(b) Let x_i', x_i'', $i = 1, 2$, be as in (ii) above. Assume that $x_1' < x < x_2'$ and $x_1'' < x < x_2''$. Let $x_0' := \max\{x_1', x_1''\}$, $x_0'' := \min\{x_2', x_2''\}$, $R := [x_0', x_0''] \times \mathbb{R}$, $y_i' := \varphi(x_i')$, $y_i'' := \varphi(x_i'')$, $i = 1, 2$, and define

$$L_i'(x, y) := \frac{y_i' - y}{x_i' - x}, \quad L_i''(x, y) := \frac{y_i'' - y}{x_i'' - x}, \quad (x, y) \in R,\ i = 1, 2.$$

Consider four sets

$$S_{1,1} := \{(x, y) \in R : L_1'(x, y) \le L_2'(x, y),\ L_1''(x, y) \le L_2''(x, y)\},$$
$$S_{1,2} := \{(x, y) \in R : L_1'(x, y) \le L_2'(x, y),\ L_1''(x, y) \ge L_2''(x, y)\},$$
$$S_{2,1} := \{(x, y) \in R : L_1'(x, y) \ge L_2'(x, y),\ L_1''(x, y) \le L_2''(x, y)\},$$
$$S_{2,2} := \{(x, y) \in R : L_1'(x, y) \ge L_2'(x, y),\ L_1''(x, y) \ge L_2''(x, y)\}.$$

Assume that there exists a point $(\sigma, \tau) \in R$ such that:

- $L_1'(\sigma, \tau) = L_2'(\sigma, \tau)$,
- $L_1''(\sigma, \tau) = L_2''(\sigma, \tau)$,
- $(x, y) \in S_{1,2} \Longrightarrow x \ge \sigma$,
- $(x, y) \in S_{2,1} \Longrightarrow x \le \sigma$.

Fix an $x \in [x_0', x_0'']$. Put $\ell(x) := |x - x'| + |x - x''|$. We obtain

$(x, \varphi(x))$ in	x'	x''	$\ell(x)$
$S_{1,1}$	x_2'	x_1''	$x_2' - x_1''$
$S_{1,2}$	x_2'	x_2''	$x_2' + x_2'' - 2x \le x_2' + x_2'' - 2\sigma$
$S_{2,1}$	x_1'	x_1''	$2x - x_1' - x_1'' \le 2\sigma - x_1' - x_1''$
$S_{2,2}$	x_1'	x_2''	$x_2'' - x_1'$

Thus for $x \in [x_0', x_0'']$, we can take

$$E := \max\{x_2' - x_1'', x_2' + x_2'' - 2\sigma, 2\sigma - x_1' - x_1'', x_2'' - x_1'\}, \quad L := \frac{D}{E}.$$

Theorem 3.9.14. *If $b \in \mathbb{N}_2$ and $ab \geq 1$, then $f \in \mathcal{ND}(\mathbb{R})$.*

Proof. Since $f(x+1) = f(x)$ and $f(-x) = f(x)$, $x \in \mathbb{R}$, to apply Theorem 3.9.11 it suffices to perform the construction from Remark 3.9.13(b) for $x \in [0, \frac{1}{2}]$. Take $x_1' := -\frac{1}{2}$, $x_2' := 1$, $x_1'' := 0$, $x_2'' := \frac{1}{2}$. Then

$$f(x_1') = f(x_2'') = \begin{cases} -A, & \text{if } b \in 2\mathbb{N}+1 \\ A-2, & \text{if } b \in 2\mathbb{N} \end{cases}, \quad f(x_2') = f(x_1'') = A,$$

where $A = \frac{1}{1-a}$. Moreover, $(\sigma, \tau) = (\frac{1}{4}, 0)$. Hence

$$D := \boldsymbol{\Delta} f(x_1', x_2') - \boldsymbol{\Delta} f(x_1'', x_2'') = \begin{cases} \frac{16}{3}A, & \text{if } b \in 2\mathbb{N}+1 \\ \frac{16}{3}, & \text{if } b \in 2\mathbb{N} \end{cases},$$

$$E = 1, \quad L = \begin{cases} \frac{16}{3}A, & \text{if } b \in 2\mathbb{N}+1 \\ \frac{16}{3}, & \text{if } b \in 2\mathbb{N} \end{cases}.$$

Thus, if

$$ab \geq 1, \quad ab^2 > 1 + \frac{2\pi^2}{L} = \begin{cases} 1 + \frac{3}{8}\pi^2(1-a), & \text{if } b \in 2\mathbb{N}+1 \\ 1 + \frac{3}{8}\pi^2, & \text{if } b \in 2\mathbb{N} \end{cases},$$

then $f \in \mathcal{ND}(\mathbb{R})$. Observe that if $ab \geq 1$ and $b \geq 5$, then the above condition is automatically satisfied. Indeed, $ab^2 \geq 5 > 1 + \frac{3}{8}\pi^2 > \frac{3}{8}\pi^2(1-a)$.

For $b \in \{2, 3, 4\}$, F.A. Behrend had to use more subtle systems of points x_i', x_i'', $i = 1, 2$ (the reader is asked to check the details).

- $b = 4$. Then $a \geq \frac{1}{4}$ and $A \geq \frac{4}{3}$. It suffices to consider only $x \in [0, \frac{1}{2}]$. We use the following configuration:

x in	x_1'	x_2'	x_1''	x_2''	y_1'	y_2'	y_1''	y_2''	σ	D	E	L
$[0, \frac{1}{6}]$	$-\frac{1}{6}$	$\frac{1}{6}$	0	$\frac{1}{6}$	$1-\frac{1}{2}A$	$1-\frac{1}{2}A$	A	$1-\frac{1}{2}A$	$\frac{1}{6}$	6	$\frac{1}{2}$	12
$[\frac{1}{6}, \frac{1}{4}]$	$\frac{1}{6}$	$\frac{1}{4}$	$\frac{1}{6}$	$\frac{1}{3}$	$1-\frac{1}{2}A$	$A-1$	$1-\frac{1}{2}A$	$-\frac{1}{2}A$	$\frac{1}{6}$	6	$\frac{1}{4}$	24
$[\frac{1}{4}, \frac{1}{3}]$	$\frac{1}{4}$	$\frac{3}{4}$	$\frac{1}{4}$	$\frac{1}{3}$	$A-1$	$A-1$	$A-1$	$-\frac{1}{2}A$	$\frac{1}{4}$	12	$\frac{7}{12}$	$\frac{144}{7}$
$[\frac{1}{3}, \frac{1}{2}]$	$\frac{1}{3}$	$\frac{1}{2}$	$\frac{1}{4}$	$\frac{1}{2}$	$-\frac{1}{2}A$	$A-2$	$A-1$	$A-2$	$\frac{1}{2}$	4	$\frac{5}{12}$	$\frac{48}{5}$

Consequently, we get the condition $a \geq \frac{1}{4}$ and $ab^2 > 1 + \frac{5}{24}\pi^2$, which is always satisfied, because $ab^2 \geq 4 > 1 + \frac{5}{24}\pi^2 \approx 3.0561$.

- $b = 3$. Then $a \geq \frac{1}{3}$, $A \geq \frac{3}{2}$, and $f(\frac{1}{4} + x) = -f(\frac{1}{4} - x)$. Thus it suffices to consider only $x \in [0, \frac{1}{4}]$. We use the following configuration:

x in	x_1'	x_2'	x_1''	x_2''	y_1'	y_2'	y_1''	y_2''	σ	D	E	L
$[0, \frac{1}{6}]$	$-\frac{1}{2}$	$\frac{1}{6}$	0	$\frac{1}{6}$	$-A$	$\frac{3}{2}-A$	A	$\frac{3}{2}-A$	$\frac{1}{6}$	$\frac{45}{4}$	$\frac{5}{6}$	$\frac{27}{2}$
$[\frac{1}{6}, \frac{1}{4}]$	$\frac{1}{6}$	$\frac{1}{4}$	0	$\frac{1}{4}$	$\frac{3}{2}-A$	0	A	0	$\frac{1}{4}$	6	$\frac{1}{3}$	18

Consequently, we get the condition $a \geq \frac{1}{3}$ and $ab^2 > 1 + \frac{4}{27}\pi^2$, which is always satisfied, because $ab^2 \geq 3 > 1 + \frac{4}{27}\pi^2 \approx 2.4621$.

- $b = 2$. Then $a \geq \frac{1}{2}$, $A \geq 2$, $\frac{3}{2}A \geq 2(1+a)$, and $A - 1 - 2a \geq 0$. We use the following configuration:

x in	x'_1	x'_2	x''_1	x''_2	y'_1	y'_2	y''_1	y''_2	σ	D	E	L
$[0, \frac{1}{6}]$	$-\frac{1}{6}$	$\frac{1}{6}$	0	$\frac{1}{6}$	$1 - \frac{1}{2}A$	$1 - \frac{1}{2}A$	A	$1-\frac{1}{2}A$	$\frac{1}{6}$	12	$\frac{1}{2}$	24
$[\frac{1}{6}, \frac{1}{4}]$	$\frac{1}{6}$	$\frac{1}{4}$	$\frac{1}{6}$	$\frac{1}{3}$	$1 - \frac{1}{2}A$	$A-1-2a$	$1 - \frac{1}{2}A$	$-\frac{1}{2}A$	$\frac{1}{6}$	6	$\frac{1}{4}$	24
$[\frac{1}{4}, \frac{1}{3}]$	$\frac{1}{4}$	$\frac{1}{2}$	$\frac{1}{4}$	$\frac{1}{3}$	$A-1-2a$	$A - 2$	$A-1-2a$	$-\frac{1}{2}A$	$\frac{1}{4}$	12	$\frac{1}{3}$	36
$[\frac{1}{3}, \frac{1}{2}]$ $f(x) \geq -\frac{1}{3}$	$\frac{1}{3}$	x	x	$\frac{2}{3}$	$-\frac{1}{2}A$	$f(x)$	$f(x)$	$-\frac{1}{2}A$	x	8	$\frac{1}{3}$	24
$[\frac{1}{3}, \frac{1}{2}]$ $f(x) \leq -\frac{1}{3}$	x	$\frac{1}{2}$	$\frac{1}{4}$	x	$f(x)$	$A - 2$	$A-1-2a$	$f(x)$	x	$\frac{16}{3}$	$\frac{1}{4}$	$\frac{64}{3}$

Consequently, we get the condition $a \geq \frac{1}{2}$ and $ab^2 > 1 + \frac{3}{32}\pi^2$, which is always satisfied, because $ab^2 \geq 2 > 1 + \frac{3}{32}\pi^2 \approx 1.9252$. □

Theorem 3.9.15. *If $b \in \mathbb{R}$, $b > 3$, $ab \geq 1$, and*

$$ab^2 > 1 + \frac{(3+2\varepsilon)(1+2\varepsilon)}{8\cos(\pi\varepsilon)}\pi^2(1-a),$$

where $\varepsilon := \frac{1}{b-1}$, then $f \in \mathbf{ND}(\mathbb{R})$.

Remark 3.9.16. (a) One can prove (EXERCISE) that the above estimate is better than Dini's (Remark 3.7.2(c)), i.e., if $a < \frac{2}{9}$ and $ab^2 > 1 + \frac{21}{4}\pi^2 \frac{1-a}{2-9a}$, then $b > 3$ and $ab^2 > 1 + \frac{(3+2\varepsilon)(1+2\varepsilon)}{8\cos(\pi\varepsilon)}\pi^2(1-a)$ with $\varepsilon := \frac{1}{b-1}$.

(b) The conditions $ab \geq 1$ and $ab^2 > 1 + \varkappa(1-a)$ are always satisfied if $ab \geq 1$ and $b > \varkappa$ (EXERCISE). Hence we conclude that if $ab \geq 1$ and $b > \frac{(3+2\varepsilon)(1+2\varepsilon)}{8\cos(\pi\varepsilon)}\pi^2$ with $\varepsilon := \frac{1}{b-1}$, then $f \in \mathbf{ND}(\mathbb{R})$. One can check (EXERCISE) that the above condition is satisfied if $ab \geq 1$ and $b \geq \frac{20}{3}$ ($b > 6.6208$).

Proof of Theorem 3.9.15. We keep the notation from the proof of Theorem 3.9.9. Recall that $|k - x_k| \leq \varepsilon$, $f(\frac{1}{2}x_{2m}) \geq \delta A$, $f(\frac{1}{2}x_{2m+1}) \leq -\delta A$, where $\delta := \cos(\pi\varepsilon)$. Fix an $m \in \mathbb{Z}$. Suppose that $2x \in [x_{2m}, x_{2m+1}]$. The case $2x \in [x_{2m+1}, x_{2m+2}]$ is left to the reader as an EXERCISE. Put $P := (x, f(x))$. Consider two segments $S' := [Q_{2m}, Q_{2m+1}]$, $S'' := [Q_{2m-1}, Q_{2m+2}]$, where

$$Q_{2m-1} := (m - \tfrac{1}{2} - \tfrac{\varepsilon}{2}, -\delta A), \quad Q_{2m} := (m - \tfrac{\varepsilon}{2}, \delta A),$$
$$Q_{2m+1} := (m + \tfrac{1}{2} + \tfrac{\varepsilon}{2}, -\delta A), \quad Q_{2m+2} := (m + 1 + \tfrac{\varepsilon}{2}, \delta A).$$

One can easily check that $S' \cap S'' = \{R\}$, where $R := (m + \frac{1}{4}, 0)$. Consider the following configurations:

- The point P is below S' and below S''. Then

$$\boldsymbol{\Delta} f(x, \tfrac{1}{2}x_{2m+2}) - \boldsymbol{\Delta} f(x, \tfrac{1}{2}x_{2m}) \geq \operatorname{slope}[Q_{2m+2}, R] - \operatorname{slope}[Q_{2m}, R]$$

$$= \frac{\delta A}{\frac{3}{4} + \frac{\varepsilon}{2}} + \frac{\delta A}{\frac{1}{4} + \frac{\varepsilon}{2}} = \frac{16\delta A(1+\varepsilon)}{(3+2\varepsilon)(1+2\varepsilon)} =: D$$

and

$$|\tfrac{1}{2}x_{2m+2} - x| + |\tfrac{1}{2}x_{2m} - x| = \tfrac{1}{2}x_{2m+2} - x + x - \tfrac{1}{2}x_{2m} \leq 1 + \varepsilon =: E.$$

- The point P is above S' and below S''. Then

$$\boldsymbol{\Delta} f(x, \tfrac{1}{2}x_{2m+2}) - \boldsymbol{\Delta} f(x, \tfrac{1}{2}x_{2m+1}) \geq \operatorname{slope}[Q_{2m+2}, R] - \operatorname{slope}[Q_{2m+1}, R] = D,$$

$$|\tfrac{1}{2}x_{2m+2} - x| + |\tfrac{1}{2}x_{2m+1} - x| \\ \leq m + 1 + \tfrac{\varepsilon}{2} - (m + \tfrac{1}{4}) + (m + \tfrac{1}{2} + \tfrac{\varepsilon}{2}) - (m + \tfrac{1}{4}) = 1 + \varepsilon = E.$$

- The same estimates may be obtained in the remaining two cases—Exercise.

Now we apply Theorem 3.9.11 with $L := \frac{16\delta A}{(3+2\varepsilon)(1+2\varepsilon)}$. □

3.10 Emde Boas's Method

The aim of this section is to present an elementary proof of the following result.

Theorem 3.10.1 (Cf. [Boa69]). $\boldsymbol{S}_{1/2,2} \in \boldsymbol{\mathcal{ND}}(\mathbb{R})$.

Recall that the result was already proved by a different method in Theorem 3.9.14.

Proof. Step 1°. Let

$$g(x) := (2 - \sqrt{2}) \sin(\pi x/2) + \sin(\pi x) + \sin(2\pi x), \quad x \in \mathbb{R}.$$

Then for every $x_0 \in \mathbb{R}$, there exists a $y \in \mathbb{R}$ such that

$$x_0 \in [y, y+1] \text{ and } |g(y)| > \frac{\pi^2}{16} + \frac{1}{20}.$$

Indeed, since $g(x) = g(x+4)$, $x \in \mathbb{R}$, it suffices to find points $0 \leq y_0 < \cdots < y_N \leq 4$ such that $|g(y_i)| > \frac{\pi^2}{16} + \frac{1}{20}$ and $y_i - y_{i-1} < 1$, $i = 0, \dots, N$, with $y_{-1} := y_N - 4$. Using a computer (Exercise), we check that we may take $(y_0, \dots, y_6) = (\frac{1}{8}, \frac{4}{8}, \frac{9}{8}, \frac{13}{8}, \frac{19}{8}, \frac{23}{8}, \frac{28}{8})$.

Step 2°. Let $f : \mathbb{R} \longrightarrow \mathbb{R}$ be continuous and let

$$A(x, h) := \frac{1}{h}\Big(f(x) + f(h) - 2f\Big(x + \frac{h}{2}\Big)\Big).$$

Suppose that

$$\exists_{M>0}\ \forall_{x_0 \in \mathbb{R},\, \delta>0}\ \exists_{x \in \mathbb{R},\, h \in (0,\delta)} : x_0 \in [x, x+h],\ |A(x,h)| \geq M. \tag{3.10.1}$$

Then $f \in \boldsymbol{\mathcal{ND}}(\mathbb{R})$.

Indeed, take $x_0 \in \mathbb{R}$ and $\delta > 0$. Let x, h be as in (3.10.1). Then we have

$$\begin{aligned} A(x,h) &= \boldsymbol{\Delta} f(x, x+h) - \boldsymbol{\Delta} f(x, x+h/2) \\ &= \boldsymbol{\Delta} f(x+h/2, x+h) - \boldsymbol{\Delta} f(x, x+h). \end{aligned} \tag{3.10.2}$$

If $x + h/2 \geq x_0$, then we take $a_1 = a_2 := x$, $b_1 = x+h$, $b_2 := x + h/2$. If $x + h/2 \leq x_0$, then we take $a_1 := x + h/2$, $a_2 := x$, $b_1 = b_2 := x + h$. Using (3.10.2), we get

$$|\boldsymbol{\Delta} f(a_1, b_1) - \boldsymbol{\Delta} f(a_2, b_2)| = |A(x, h)| \geq M.$$

It remains to apply Remark 2.1.4(d).

Step 3°. Now let $f := \boldsymbol{S}_{1/2,2}$. We will check that f satisfies (3.10.1) with $M := \frac{1}{5}$. We have $A(x,h) = \sum_{n=0}^{\infty} A_n(x,h)$, where

$$\begin{aligned} A_n(x,h) &= \frac{1}{h}\frac{1}{2^n}\Big(\sin(\pi 2^{n+1}x) + \sin(\pi 2^{n+1}(x+h)) - 2\sin(\pi 2^{n+1}(x+h/2))\Big) \\ &= \frac{1}{h}\frac{1}{2^n}\Big(2\sin(\pi 2^{n+1}(x+h/2))\cos(\pi 2^{n+1}h/2) - 2\sin(\pi 2^{n+1}(x+h/2))\Big) \\ &= \frac{1}{h}\frac{1}{2^n}2\sin(\pi 2^{n+1}(x+h/2))\Big(\cos(\pi 2^{n+1}h/2) - 1\Big) \\ &= \frac{1}{h}\frac{1}{2^{n-1}}\sin(\pi 2^{n+1}x + \pi 2^n h)(\cos(\pi 2^n h) - 1). \end{aligned}$$

Taking $h = h_m := 1/2^{m+1}$, $m \in \mathbb{N}_2$, we get

$$A_n(x, h_m) = 2^{m-n+2}\sin(\pi 2^{n+1}x + \pi 2^{n-m-1})(\cos(\pi 2^{n-m-1}) - 1).$$

In particular, if $n \geq m+2$, then $A_n(x, h_m) = 0$. Thus

$$A(x, h_m) = R_m(x) + S_m(x),$$

where

$$\begin{aligned} R_m(x) &:= \sum_{n=0}^{m-2} A_n(x, h_m), \\ S_m(x) &:= A_{m-1}(x, h_m) + A_m(x, h_m) + A_{m+1}(x, h_m). \end{aligned}$$

If $n \leq m-2$, then we get

$$|A_n(x, h_m)| \leq 2^{m-n+2}\frac{1}{2}(\pi 2^{n-m-1})^2 = \pi^2 2^{n-m-1}.$$

Hence

$$|R_m(x)| \leq \sum_{n=0}^{m-2} |A_n(x, h_m)| \leq \sum_{n=0}^{m-2} \pi^2 2^{n-m-1} < \frac{\pi^2}{4}.$$

Let $\xi = \xi_{x,m} := 2^{m+1}x + 1/2$. Then we have

$$\begin{aligned} S_m(x) &= 8\sin(\pi 2^m x + \pi/4)(\cos(\pi/4) - 1) \\ &\quad + 4\sin(\pi 2^{m+1}x + \pi/2)(\cos(\pi/2) - 1) + 2\sin(\pi 2^{m+2}x + \pi)(\cos\pi - 1) \\ &= -4\Big((2-\sqrt{2})\sin(\pi\xi/2) + \sin(\pi\xi) + \sin(2\pi\xi)\Big) = -4g(\xi), \end{aligned}$$

where g is as in Step 1°.

Take $x_0 \in \mathbb{R}$, $\delta > 0$. Let $m \in \mathbb{N}_2$ be such that $h_m < \delta$. By Step 1°, there exists a $y \in \mathbb{R}$ such that $2^{m+1}x_0 + 1/2 \in [y, y+1]$ and $|g(y)| > \frac{\pi^2}{16} + \frac{1}{20}$. Define $x := (y - 1/2)/2^{m+1}$. Note that $\xi_{x,m} = y$. Then $x_0 \in [x, x+h_m]$ and $|A(x, h_m)| \geq 4|g(y)| - |R_m(x)| > 4(\frac{\pi^2}{16} + \frac{1}{20}) - \frac{\pi^2}{4} = \frac{1}{5} = M$. □

3.11 The Method of Baouche–Dubuc

In 1992, A. Baouche and S. Dubuc, using elementary means, proved the following strong result.

Theorem 3.11.1 (Cf. [BD92]). *If $b \in 2\mathbb{N}+1$, $ab > 1$, and $\alpha := -\frac{\log a}{\log b} \in (0,1)$, then $\boldsymbol{W}_{1,a,b,0}$ is weakly α-anti-Hölder continuous uniformly with respect to $x \in \mathbb{R}$. In particular, $\boldsymbol{W}_{1,a,b,0} \in \mathcal{ND}(\mathbb{R})$ (cf. Remark 2.5.4*(b)*).*

The result will be generalized in Theorem 8.3.1.

Proof. Put $f := \boldsymbol{W}_{1,a,b,0}$. Fix $x \in \mathbb{R}$ and $m \in \mathbb{N}$. Let $k \in \mathbb{Z}$ be such that $|b^m x - k| \le \frac{1}{2}$. Define $t := \frac{k}{b^m}$, $h := \frac{1}{4b^m}$. Then for $n \ge m$, we get

$$\cos(2\pi b^n(t \pm h)) = \cos\left(2b^{n-m}\left(k \pm \frac{1}{4}\right)\right) = \cos\left(\frac{b^{n-m}}{2}\right) = 0,$$
$$\cos(2\pi b^n t) = \cos(2\pi b^{n-m} k) = 1.$$

Hence,

$$f(t \pm h) = \sum_{n=0}^{m-1} a^n \cos(2\pi b^n (t \pm h)), \quad f(t) = \sum_{n=0}^{m-1} a^n \cos(2\pi b^n t) + \frac{a^m}{1-a}.$$

Note that for $0 \le n \le m-1$, we have

$$\begin{aligned} 2\cos(2\pi b^n t) &- \cos(2\pi b^n(t-h)) - \cos(2\pi b^n(t+h)) \\ &= 2\cos(2\pi b^n t)(1 - \cos(\pi b^n h)) \\ &\ge -2(1-\cos(\pi b^n h)) = -4\sin^2\left(\frac{\pi b^n h}{2}\right) \ge -(\pi b^n h)^2. \end{aligned}$$

Thus

$$\begin{aligned} 2f(t) - f(t-h) - f(t+h) &\ge \frac{2a^m}{1-a} - \sum_{n=0}^{m-1} a^n (\pi b^n h)^2 \\ &= \frac{2a^m}{1-a} - \frac{(ab^2)^m - 1}{ab^2 - 1}(\pi h)^2 > \frac{2a^m}{1-a} - \frac{(ab^2)^m}{ab^2-1}(\pi h)^2 = a^m c, \end{aligned}$$

where

$$\begin{aligned} c :&= \frac{2}{1-a} - \frac{b^{2m}}{ab^2-1}(\pi h)^2 = \frac{2}{1-a} - \frac{\pi^2}{16(ab^2-1)} = \frac{32ab^2 + \pi^2 a - 32 - \pi^2}{16(1-a)(ab^2-1)} \\ &\ge \frac{32b + \frac{\pi^2}{b} - 32 - \pi^2}{16(1-a)(ab^2-1)} = \frac{\frac{1}{b}(32b^2 - (32+\pi^2)b + \pi^2)}{16(1-a)(ab^2-1)} = \frac{2(b-1)(b - \frac{\pi^2}{32})}{b(1-a)(ab^2-1)} \\ &> 0. \end{aligned}$$

On the other hand,

$$\begin{aligned} 2f(t) - f(t-h) &- f(t+h) \\ &= 2(f(t) - f(x)) + (f(x) - f(t-h)) + (f(x) - f(t+h)), \end{aligned}$$

which implies that at least one of the above four summands is greater than $\frac{ca^m}{4}$, i.e., there exists an $x_m \in \{t, t-h, t+h\}$, $|x_m - x| \le \frac{3}{4b^m}$, such that $|f(x_m) - f(x)| > \frac{ca^m}{4}$.

Now put $\varepsilon := \frac{ca}{4}(\frac{4}{3})^\alpha$ and let $\delta \in (0, \frac{3}{4})$. Take an $m \in \mathbb{N}$ such that $\frac{3}{4b^m} \le \delta < \frac{3}{4b^{m-1}}$. Then $\frac{ca^m}{4} = \frac{c}{4b^{m\alpha}} = \frac{c4^\alpha}{4 \cdot 3^\alpha b^\alpha}(\frac{3}{4b^{m-1}})^\alpha > \varepsilon\delta^\alpha$. □

3.12 Summary

In a concentrated tabular form, the best results presented so far (related to nowhere differentiability of the function $\boldsymbol{W}_{p,a,b,\boldsymbol{\theta}}$) can be summarized as follows:

$\mathcal{ND}^\infty$	$\mathcal{ND}_\pm$	$\mathcal{M} \cap \mathcal{ND}^\infty$
$p=1$, $\boldsymbol{\theta}=0$, b odd $ab>1+\frac{3}{2}\pi(1-a)$ Theorem 3.8.1	$p=1$, $\boldsymbol{\theta}$ arbitrary, b even, $b\ge 14$ $a=1/b$ Theorem 3.6.1	p odd, $\boldsymbol{\theta}=0$, b odd $ab>1+\frac{3}{2}p\pi$ Theorem 3.5.1
$p=1$, $\boldsymbol{\theta}=0$, $b\in 2\mathbb{N}\setminus(3\mathbb{N})$ $ab>1+\frac{16\pi}{9}(1-a)$ Theorem 3.9.5	p arbitrary, $\boldsymbol{\theta}$ arbitrary $(a<a_1(p), b>\Psi_1(a))$ or $(a<a_2(p), b>\Psi_2(a))$ Theorem 3.7.1	
$p=1$, $\boldsymbol{\theta}=0$, $b>3$ $ab>1+\frac{(3b-1)\pi}{2(b-1)\cos(\frac{\pi}{b-1})}(1-a)$ Theorem 3.9.9		

Chapter 4
Takagi–van der Waerden-Type Functions I

Summary. The purpose of this chapter is to present basic results related to the nowhere differentiability of the Takagi–van der Waerden function $x \longmapsto \sum_{n=0}^{\infty} a^n \operatorname{dist}(b^n x + \theta_n, \mathbb{Z})$. The discussion will be continued in Chap. 9.

4.1 Introduction

Let $\boldsymbol{\psi} : \mathbb{R} \longrightarrow [0, \frac{1}{2}]$, $\boldsymbol{\psi}(x) := \operatorname{dist}(x, \mathbb{Z})$. Observe that

- $\boldsymbol{\psi}(x+1) = \boldsymbol{\psi}(x)$,
- $\boldsymbol{\psi}(-x) = \boldsymbol{\psi}(x)$, $x \in \mathbb{R}$,
- $|\boldsymbol{\psi}(x) - \boldsymbol{\psi}(y)| \le |x - y|$, $x, y \in \mathbb{R}$, and
- $|\boldsymbol{\psi}'(x)| = 1$, $x \in \mathbb{R} \setminus \frac{1}{2}\mathbb{Z}$.

For $0 < a < 1$, $ab \ge 1$, $\boldsymbol{\theta} = (\theta_n)_{n=0}^{\infty} \subset \mathbb{R}$, define

$$\boldsymbol{T}_{a,b,\boldsymbol{\theta}}(x) := \sum_{n=0}^{\infty} a^n \boldsymbol{\psi}(b^n x + \theta_n), \quad x \in \mathbb{R}.$$

The function $\boldsymbol{T}_{a,b,\boldsymbol{0}}$ is called the *generalized Takagi–van der Waerden function*. The function $\boldsymbol{T} := \boldsymbol{T}_{1/2,2,0}$ is called the *Takagi function*; it is sometimes also called the *blancmange function*. The name "blancmange" comes from the resemblance of the graph of $\boldsymbol{T}$ to a pudding of the same name (cf. Fig. 4.1).

The aim of this section is to present basic properties of the Takagi–van der Waerden function. More developed results (such as Theorem 9.2.1) will be given in Chap. 9.

Remark 4.1.1 (Takagi (blancmange) Function).

(a) T. Takagi in [Tak03] proved that $\boldsymbol{T} \in \mathcal{ND}(\mathbb{R})$ (see also [KV02]).

(b) Independently, T.H. Hildebrandt in [Hil33] proved that $\boldsymbol{T} \in \mathcal{ND}(\mathbb{R})$.

(c) A. Shidfar and K. Sabetfakhri in [SS86] proved that $\boldsymbol{T} \in \mathcal{H}^{\beta}(\mathbb{R})$ for every $\beta \in (0,1)$.

(d) M. Hata in [Hat91] proved the optimal estimate

$$|\boldsymbol{T}(x+h) - \boldsymbol{T}(x)| \le \operatorname{const} |h| \log |h|, \quad x, h \in \mathbb{R}.$$

(e) W.F. Darsow, M.J. Frank, and H.-H. Kairies proved in [KDF88, DFK89] that $\boldsymbol{T} \in \mathcal{ND}_{\pm}(\mathbb{I})$ (Theorem 4.2.1).

M. Jarnicki, P. Pflug, *Continuous Nowhere Differentiable Functions*, Springer Monographs in Mathematics, DOI 10.1007/978-3-319-12670-8_4

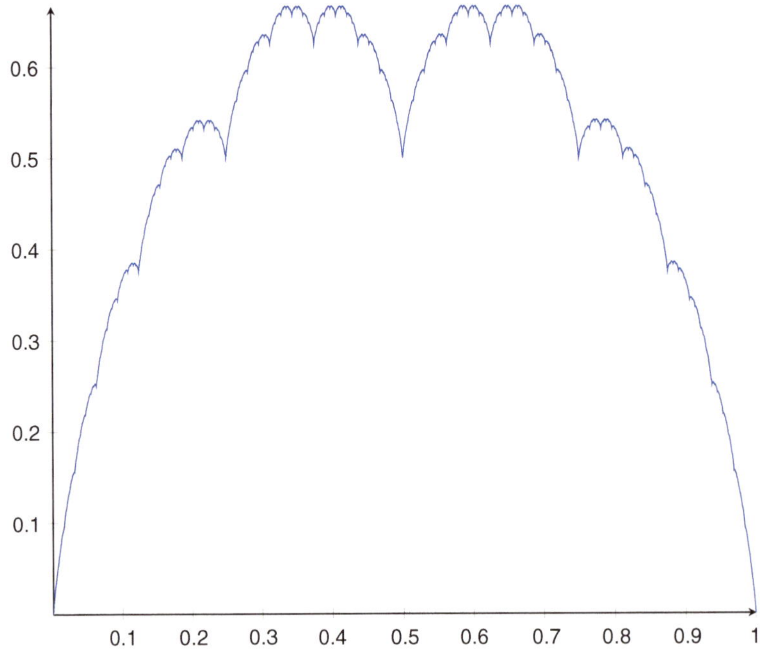

Fig. 4.1 Takagi function $\mathbb{I} \ni x \longmapsto \boldsymbol{T}(x)$

(f) $\boldsymbol{T}$ has several applications in physics (cf. references in [AK06b]). Moreover, $\boldsymbol{T} = \frac{1}{2}T_1$, where $T_n := \frac{1}{n!}\frac{\partial^n L_a}{\partial a^n}|_{a=1/2}$, and L_a stands for the Lebesgue singular function (cf. [AK06b]). One can prove that $T_n \in \mathbf{\mathfrak{ND}}(\mathbb{R})$ (cf. [AK06b], Theorem 5.1).

(g) $\boldsymbol{T}$ has applications in number theory, e.g., we have the following *generalized Trollope's formula* (cf. [AFS09]):

$$S(n) = \frac{1}{2}nm - 2^{m-1}\boldsymbol{T}\Big(\frac{n}{2^m}\Big), \quad 1 \le n \le 2^m, \text{ where}$$

$$S(n) := \sum_{k=0}^{n-1} s(k), \quad s(k) := \sum_{p=0}^{\infty} e_p(k),$$

$$k = \sum_{p=0}^{\infty} e_p(k)2^p \text{ with } e_p(k) \in \{0, 1\}.$$

(h) The Riemann hypothesis can be formulated in terms of the Takagi function [KY00, BKY06]. More precisely, the Riemann hypothesis is equivalent to

$$\sum_{\varrho \in \mathcal{F}_n} \boldsymbol{T}(\varrho) - (\#\mathcal{F}_n)\int_0^1 \boldsymbol{T}(t)dt = O(n^{\frac{1}{2}+\varepsilon}) \text{ when } n \longrightarrow +\infty,$$

where $\mathcal{F}_n$ is the set of all Farey fractions of order n (cf. § A.8).

(i) The function $\boldsymbol{T}$ has been studied in many other papers; see, e.g.,[All13, AK06b, AK10, AK12, Kôn87, Krü07, Krü08, Krü10, Lag12, Vas13].

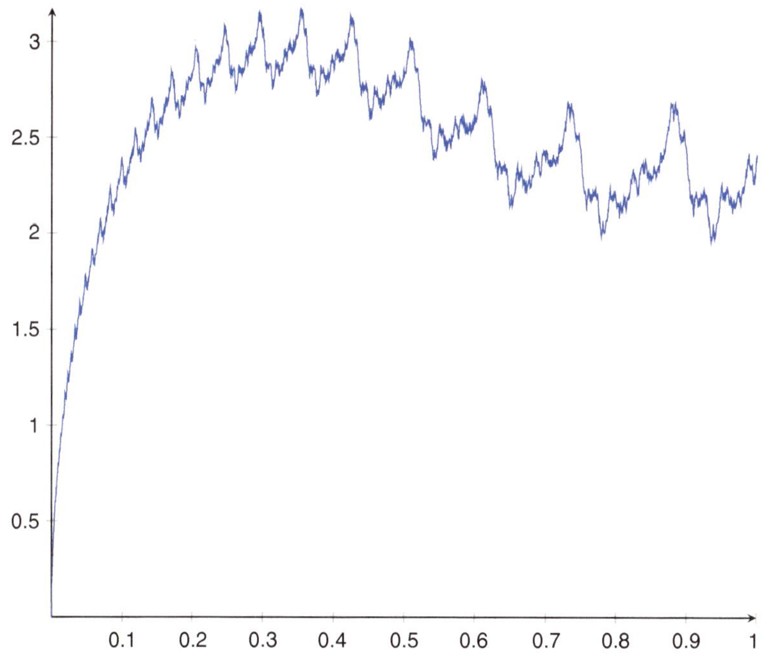

Fig. 4.2 Takagi–van der Waerden type function $\mathbb{I} \ni x \longmapsto \boldsymbol{T}_{0.9,1.2,0}(x)$

Remark 4.1.2 (Takagi–van der Waerden-Type Function; cf. Fig. 4.2).

(a) B.L. van der Waerden in [Wae30] proved that $\boldsymbol{T}_{1/10,10,0} \in \mathcal{ND}(\mathbb{R})$; see also [Hai76].

(b) F.S. Cater in [Cat94] and [Cat03] proved two general results that imply that:
- $\boldsymbol{T}_{1/b,b,\boldsymbol{\theta}} \in \mathcal{ND}_{\pm}(\mathbb{R})$ provided that $b \geq 10$ (Theorem 4.3.1) and
- $\boldsymbol{T}_{a,b,0} \in \mathcal{ND}_{\pm}(\mathbb{R})$ provided that $ab \geq 1$ and $b \in \mathbb{N}_2$ (Theorem 4.3.2); in particular, $\boldsymbol{T}_{1/b,b,0} \in \mathcal{ND}_{\pm}(\mathbb{R})$, provided that $b \in \mathbb{N}_2$.

(c) The function $\boldsymbol{T}_{1/b,b,0}$ with $b \in \mathbb{N}_2$ has been discussed in many other papers, e.g.,[Rha57b, Bab84, Spu04].

(d) K. Knopp in [Kno18] proved that $\boldsymbol{T}_{a,b,0} \in \mathcal{ND}(\mathbb{R})$ for $0 < a < 1$, $ab > 4$, and $b \in 2\mathbb{N}$.

(e) D.P. Minassian and J.W. Gaisser presented in [MG84] an elementary proof showing that $\boldsymbol{T}_{1/2,5,0} \in \mathcal{ND}_{\pm}(\mathbb{R})$.

(f) A. Baouche and S. Dubuc in [BD94] proved that $\boldsymbol{T}_{a,b,\boldsymbol{\theta}} \in \mathcal{M}(\mathbb{R}) \subset \mathcal{ND}_{\pm}(\mathbb{R})$ for $ab > 1$ (Theorem 9.2.1).

(g) The reader may find in [AK12] more historical information on the nowhere differentiability of the generalized Takagi–van der Waerden function.

The Takagi function has the following elementary properties.

Remark 4.1.3 (cf. [Lag12]).

(a) $\boldsymbol{T}(1-x) = \boldsymbol{T}(x)$, $x \in \mathbb{I}$.

(b) $\boldsymbol{T}(x) = 0$ iff $x \in \mathbb{Z}$.

(c) $0 \leq \boldsymbol{T} \leq \frac{2}{3}$.
Indeed, since $\boldsymbol{\psi}(x) + \frac{1}{2}\boldsymbol{\psi}(2x) \leq \frac{1}{2}$, we have $\boldsymbol{T}(x) \leq \frac{1}{2} + \frac{1}{8} + \frac{1}{32} + \cdots = \frac{2}{3}$.

(d) $\boldsymbol{T}(\frac{1}{3}) = \frac{2}{3}$.

(e) $\boldsymbol{T}(\mathbb{Q}) \subset \mathbb{Q}$.

Indeed, fix an $x_0 = \frac{p}{q} \in \mathbb{Q} \cap \mathbb{I}$. Then $r_n := \psi(2^n x_0) = \psi(\{2^n x_0\}) \in \mathbb{Q}$, $n \in \mathbb{N}$, where $\{t\} := t - \lfloor t \rfloor$ denotes the fractional part of t. Since $\{2^n x_0\}$ is one of the numbers $0, \frac{1}{q}, \dots, \frac{q-1}{q}$, there exist $\mu < \nu$ such that $\{2^\mu x_0\} = \{2^\nu x_0\}$. Put $\omega := \nu - \mu$. Then $\{2^{n+\omega} x_0\} = \{2^n x_0\}$ for $n \geq \mu$ (observe that $\{2^{n+1} x_0\} = \{2\{2^n x_0\}\}$ and use induction). Finally,

$$\boldsymbol{T}(x_0) = \Big(\sum_{n=0}^{\mu-1} \frac{r_n}{2^n}\Big) + \sum_{s=0}^{\omega-1} \frac{r_{\mu+s}}{2^{\mu+s}} \sum_{k=0}^{\infty} \frac{1}{2^{k\omega}} \in \mathbb{Q}.$$

Notice that in fact, the original definition of the Takagi function $\boldsymbol{T}$ has been formulated in the language of binary representations of numbers.

Proposition 4.1.4. *Let* $x = \sum_{n=1}^{\infty} \frac{a_n}{2^n} \in \mathbb{I}$ *with* $(a_n)_{n=1}^{\infty} \subset \{0,1\}$ *be a* binary representation *of* x.

(a) (*Cf.* [Tak03]) *Put*

$$\tau_k := \sum_{n=k}^{\infty} \frac{a_n}{2^n}, \quad \tau_k' := \frac{1}{2^{k-1}} - \tau_k, \quad \gamma_k := \begin{cases} \tau_k, & \text{if } a_k = 0 \\ \tau_k', & \text{if } a_k = 1 \end{cases},$$

$$A_k := a_1 + \cdots + a_k,$$

$$b_k := \left.\begin{cases} A_k, & \text{if } a_k = 0 \\ k - A_k, & \text{if } a_k = 1 \end{cases}\right\} = (1 - a_k)A_k + a_k(k - A_k), \quad k \in \mathbb{N}. \tag{4.1.1}$$

Then

$$\boldsymbol{T}(x) = \sum_{k=1}^{\infty} \gamma_k = \sum_{k=1}^{\infty} \frac{b_k}{2^k}.$$

(b) (*Cf.* [Lyc40, AFS09]) *Let*

$$c_k := \left.\begin{cases} 0, & \text{if } a_k = 0 \\ k - 2(A_k - 1), & \text{if } a_k = 1 \end{cases}\right\} = a_k(k - 2(A_k - 1)), \quad k \in \mathbb{N}. \tag{4.1.2}$$

Then $\boldsymbol{T}(x) = \sum_{k=1}^{\infty} \frac{c_k}{2^k}$.

Proof. (a) To get the equality $\boldsymbol{T}(x) = \sum_{k=1}^{\infty} \gamma_k$, it suffices to prove that $\psi(2^k x) = 2^k \gamma_{k+1}$, $k \in \mathbb{N}_0$. Fix a $k \in \mathbb{N}_0$. Then

$$\psi(2^k x) = \psi\Big(2^k \sum_{n=1}^{\infty} \frac{a_n}{2^n}\Big) = \psi\Big(\sum_{n=1}^{\infty} \frac{a_n}{2^{n-k}}\Big) \overset{\psi(t+1)=\psi(t)}{=} \psi\Big(\sum_{n=k+1}^{\infty} \frac{a_n}{2^{n-k}}\Big) = \psi(2^k \tau_{k+1}).$$

It now remains to observe that:

- $2^k \tau_{k+1} \in \mathbb{I}$,
- $a_{k+1} = 0 \Longrightarrow 2^k \tau_{k+1} \in [0, 1/2]$,
- $a_{k+1} = 1 \Longrightarrow 2^k \tau_{k+1} \in [1/2, 1] \Longrightarrow 2^k \tau_{k+1}' \in [0, 1/2]$.

Note that A_k counts the number of 1's in the sequence $(a_1, \dots, a_k)$ and $k - A_k$, the number of 0's. To get the equality $\sum_{k=1}^{\infty} \gamma_k = \sum_{k=1}^{\infty} \frac{b_k}{2^k}$, we proceed as follows:

$$\sum_{k=1}^{\infty}\gamma_k=\sum_{k=1}^{\infty}\Big(a_k\sum_{n=k}^{\infty}\frac{1}{2^n}+(-1)^{a_k}\sum_{n=k}^{\infty}\frac{a_n}{2^n}\Big)$$
$$=\sum_{n=1}^{\infty}\frac{1}{2^n}\Big(\sum_{k=1}^{n}a_k+a_n\sum_{k=1}^{n}(-1)^{a_k}\Big)=\sum_{n=1}^{\infty}\frac{b_n}{2^n}.$$

(b) $$\sum_{k=1}^{\infty}\frac{b_k-c_k}{2^k}=\sum_{k=1}^{\infty}\frac{(1-a_k)A_k+a_k(k-A_k)-a_k(k-2(A_k-1))}{2^k}$$
$$=\sum_{k=1}^{\infty}\frac{A_k-2a_k}{2^k}=\Big(\sum_{k=1}^{\infty}\frac{A_k}{2^k}\Big)-2x=0.$$ □

Remark 4.1.5.
(a) Note that

$$b_1=0,\quad b_{n+1}=\begin{cases}b_n, & \text{if } a_{n+1}=a_n\\ n-b_n, & \text{if } a_{n+1}\neq a_n\end{cases},\quad n\in\mathbb{N}.$$

(b) If $x=\frac{k}{2^m}$, then x has two binary representations:

$$x=\sum_{n=1}^{m}\frac{a_n}{2^n}=\Big(\sum_{n=1}^{m-1}\frac{a_n}{2^n}\Big)+\sum_{n=m+1}^{\infty}\frac{1}{2^n}=:x'$$

with $a_m=1$. Let $(b_n)_{n=1}^{\infty}$, $(c_n)_{n=1}^{\infty}$, $(b'_n)_{n=1}^{\infty}$, $(c'_n)_{n=1}^{\infty}$, be associated to x, x', respectively (using (4.1.1) and (4.1.2)). Proposition 4.1.4 implies that

$$\boldsymbol{T}(x)=\sum_{n=1}^{\infty}\frac{b_n}{2^n}=\sum_{n=1}^{\infty}\frac{b'_n}{2^n}=\sum_{n=1}^{\infty}\frac{c_n}{2^n}=\sum_{n=1}^{\infty}\frac{c'_n}{2^n},$$

i.e., the results are independent of the representation.

The Takagi function may be also defined in an axiomatic way.

Proposition 4.1.6 (cf. [Rha57a]). *The function $f=\boldsymbol{T}$ is the only bounded function on $\mathbb{I}$ satisfying the following equations:*

(1) $f(x)-2f(\frac{x}{2})=-x$, $x\in\mathbb{I}$,
(2) $f(\frac{x+1}{2})-f(\frac{x}{2})=-x+\frac{1}{2}$, $x\in\mathbb{I}$.

Proof. One can easily check that $\boldsymbol{T}$ satisfies (1) and (2).

Let $\mathfrak{E}$ be the Banach space of all bounded functions on $\mathbb{I}$ with the supremum norm $\|F\|:=\sup_{x\in\mathbb{I}}|F(x)|$. Let $L:\mathfrak{E}\longrightarrow\mathfrak{E}$ be given by the formula

$$L(F)(x):=\begin{cases}x+\frac{1}{2}F(2x), & \text{if } 0\le x\le\frac{1}{2}\\ 1-x+\frac{1}{2}F(2x-1), & \text{if } \frac{1}{2}<x\le 1\end{cases},\quad x\in\mathbb{I}.$$

Observe that

$$\|L(F)-L(G)\|\le\frac{1}{2}\|F-G\|,\quad F,G\in\mathfrak{E}.$$

One can easily prove that if $F \in \mathcal{E}$ satisfies (1) and (2) (in particular, if $F = \boldsymbol{T}$), then $L(F) = F$ (EXERCISE), i.e., F is a fixed point of L. Since L is a contraction, the Banach fixed-point theorem implies that $\boldsymbol{T}$ is the only fixed point of L. □

4.2 Kairies's Method

Theorem 4.2.1 (cf. [KDF88, DFK89]). *If a continuous function $f : \mathbb{I} \longrightarrow \mathbb{R}$ satisfies conditions* (1), (2) *from Proposition 4.1.6, and moreover,*
(3) $f(1 - x) = f(x)$, $x \in \mathbb{I}$,
then $f \in \mathcal{ND}_{\pm}(\mathbb{I})$. In particular, $\boldsymbol{T} \in \mathcal{ND}_{\pm}(\mathbb{I})$ (cf. Remark 4.1.3(a)).

Proof. In view of (3), we have only to prove that for every $x_1 \in (0, 1]$, a finite left-sided derivative $f'_-(x_1)$ does not exist. Fix an $x_1 \in (0, 1]$ and suppose that a finite $f'_-(x_1)$ exists.

Iteration of (1) gives

$$f(2^m x) = 2^m f(x) - m2^m x, \quad x \in [0, \tfrac{1}{2^m}], \tag{4.2.1}$$
$$f(\tfrac{x}{2^m}) = \tfrac{1}{2^m}(f(x) + mx), \quad x \in \mathbb{I},\ m \in \mathbb{N}. \tag{4.2.2}$$

We have

$$\begin{aligned} f(x + \tfrac{1}{2^m}) &= f(\tfrac{1}{2^{m-1}}(2^{m-1}x + \tfrac{1}{2})) \\ &\overset{(4.2.2)}{=} \tfrac{1}{2^{m-1}}\Big(f(2^{m-1}x + \tfrac{1}{2}) + (m-1)(2^{m-1}x + \tfrac{1}{2})\Big) \\ &\overset{(2)}{=} \tfrac{1}{2^{m-1}}\Big(f(2^{m-1}x) - 2^m x + \tfrac{1}{2}\Big) + (m-1)(x + \tfrac{1}{2^m}) \\ &\overset{(4.2.1)}{=} f(x) - 2x + \tfrac{m}{2^m}, \quad x \in [0, \tfrac{1}{2^m}]. \end{aligned} \tag{4.2.3}$$

Let $x_1 = \sum_{s=1}^{\infty} \frac{1}{2^{n_s}}$, where $1 \le n_1 < n_2 < \dots$, be the infinite binary representation of x_1. Put $x_k := \sum_{s=k}^{\infty} \frac{1}{2^{n_s}}$. We have $x_k = x_{k+1} + \frac{1}{2^{n_k}}$ and $0 < x_{k+1} \le \frac{1}{2^{n_k}}$. By (4.2.3), we have $f(x_k) = f(x_{k+1}) - 2x_{k+1} + \frac{n_k}{2^{n_k}}$, which gives

$$f(x_1) = f(x_{k+1}) - 2(x_2 + \dots + x_{k+1}) + \frac{n_1}{2^{n_1}} + \dots + \frac{n_k}{2^{n_k}}. \tag{4.2.4}$$

Fix a $k \in \mathbb{N}_2$ and let

$$y_1 := x_1 - \frac{1}{2^{n_k}} = \frac{1}{2^{n_1}} + \dots + \frac{1}{2^{n_{k-1}}} + \frac{1}{2^{n_{k+1}}} + \frac{1}{2^{n_{k+2}}} + \cdots.$$

Let $(y_s)_{s=1}^{\infty}$ be defined for y_1 in the same way as $(x_s)_{s=1}^{\infty}$ for x_1. We have $y_s = x_{s+1}$ for $s \ge k$ and $y_s = x_s - \frac{1}{2^{n_k}}$ for $s \le k - 1$. If we apply (4.2.4) to $(y_1, k-1)$ (instead of (x_1, k)), then we get

$$f(y_1) = f(x_{k+1}) - 2\Big(x_2 - \frac{1}{2^{n_k}} + \dots + x_{k-1} - \frac{1}{2^{n_k}} + x_{k+1}\Big) + \frac{n_1}{2^{n_1}} + \dots + \frac{n_{k-1}}{2^{n_{k-1}}}.$$

Hence

$$\boldsymbol{\Delta} f(x_1, y_1) = 2^{n_k}(f(x_1) - f(y_1)) = 2^{n_k}\Big(-2\frac{k-2}{2^{n_k}} - 2x_k + \frac{n_k}{2^{n_k}}\Big)$$
$$= n_k - 2k + 2(1 - 2^{n_k}x_{k+1}) = n_k - 2k + 2(1 - u_k), \text{ where } u_k := 2^{n_k}x_{k+1}.$$

When x_1 is *dyadic rational*, i.e., $x_1 = \frac{m}{2^q}$ with $1 \leq m \leq 2^q$, we have $(n_1, n_2, \dots) = (n_1, \dots, n_p, n_p + 1, n_p + 2, \dots)$ for some $p \in \mathbb{N}$. Thus for $k \geq p$, we get

$$\boldsymbol{\Delta} f(x_1, y_1) = n_k - 2k + 2(1 - u_k) = k - p + n_p - 2k + 2(1 - u_k)$$
$$= -k + n_p - p + 2(1 - u_k) \leq -k + n_p - p \underset{k \to +\infty}{\longrightarrow} -\infty;$$

a contradiction.

When x_1 is not a dyadic rational, the set

$$S := \{s \in \mathbb{N} : n_{s+1} - n_s \geq 2\}$$

is infinite. Assume that $k \in S$ and let

$$z_1 := x_1 - \frac{1}{2^{n_k+1}} = \frac{1}{2^{n_1}} + \dots + \frac{1}{2^{n_{k-1}}} + \frac{1}{2^{n_k+1}} + \frac{1}{2^{n_{k+1}}} + \frac{1}{2^{n_{k+2}}} + \cdots.$$

We have $z_s = x_s$ for $s \geq k+1$ and $z_s = x_s - \frac{1}{2^{n_k+1}}$ for $s \leq k$. If we apply (4.2.4) to (z_1, k) (instead of (x_1, k)), then we get

$$f(z_1) = f(x_{k+1}) - 2\Big(x_2 - \frac{1}{2^{n_k+1}} + \dots + x_k - \frac{1}{2^{n_k+1}} + x_{k+1}\Big)$$
$$+ \frac{n_1}{2^{n_1}} + \dots + \frac{n_{k-1}}{2^{n_{k-1}}} + \frac{n_k + 1}{2^{n_k+1}}.$$

Consequently,

$$\boldsymbol{\Delta} f(x_1, z_1) = 2^{n_k+1}(f(x_1) - f(z_1)) = 2^{n_k+1}\Big(-2\frac{k-1}{2^{n_k+1}} + \frac{n_k}{2^{n_k}} - \frac{n_k+1}{2^{n_k+1}}\Big)$$
$$= n_k - 2k + 1 \underset{S \ni k \to +\infty}{\longrightarrow} f'_-(x_1).$$

Thus there exist $m, s_0 \in \mathbb{N}$ such that $n_s - 2s + 1 = m$ for $s \in S \cap \mathbb{N}_{s_0}$. In particular, $n_{s'} - n_s = 2(s' - s)$ for $s, s' \in S \cap \mathbb{N}_{s_0}$, $s < s'$. For $k \geq s_0$, we get

$$\boldsymbol{\Delta} f(x_1, y_1) = n_k - 2k + 2(1 - u_k) = m - 1 + 2(1 - u_k) \longrightarrow m,$$

which implies that $u_k \longrightarrow \frac{1}{2}$.

There are two possibilities:

- The set $S_3 := \{s \in S : n_{s+1} - n_s \geq 3\}$ is infinite. Then for $k \in S_3$, we get

$$u_k = 2^{n_k}x_{k+1} \leq 2^{n_k}\Big(\frac{1}{2^{n_k+3}} + \frac{1}{2^{n_k+4}} + \frac{1}{2^{n_k+5}} + \cdots\Big) = \frac{1}{4};$$

a contradiction.

• The set S_3 is finite, i.e., $n_{s+1} - n_s = 2$ for $s \in S \cap \mathbb{N}_{s_1}$ with an $s_1 \geq s_0$. Suppose that there exist $s, s' \in S \cap \mathbb{N}_{s_1}$ such that $s' \geq s + 2$ and $(s, s') \cap S = \varnothing$, i.e., $n_{s'} - n_s = s' - s$. On the other hand, we know that $n_{s'} - n_s = 2(s' - s)$; a contradiction. Thus $n_{s+1} - n_s = 2$ for $s \in \mathbb{N}_{s_2}$ with an $s_2 \geq s_1$. Then

$$u_k = 2^{n_k} x_{k+1} = 2^{n_k}\Big(\frac{1}{2^{n_k+2}} + \frac{1}{2^{n_k+4}} + \frac{1}{2^{n_k+6}} + \cdots\Big) = \frac{1}{3}, \quad k \geq s_2;$$

a contradiction. □

4.3 Cater's Method

We will present two general theorems due to F.S. Cater that give a partial characterization of nowhere differentiability of the function $\boldsymbol{T}_{1/b,b,\boldsymbol{\theta}}$.

Theorem 4.3.1 (cf. [Cat94]). *Let* $\boldsymbol{b} = (b_n)_{n=0}^{\infty} \subset \mathbb{R}_{>0}$, $\boldsymbol{a} = (a_n)_{n=0}^{\infty} \subset \mathbb{R}_*$, $|a_n| = 1/b_n$, $n \in \mathbb{N}_0$, $\boldsymbol{\theta} = (\theta_n)_{n=0}^{\infty} \subset \mathbb{R}$,

$$\boldsymbol{T}_{\boldsymbol{a},\boldsymbol{b},\boldsymbol{\theta}}(x) := \sum_{n=0}^{\infty} a_n \psi(b_n x + \theta_n), \quad x \in \mathbb{R}.$$

If $\frac{b_{n+1}}{b_n} \geq 10$, $n \in \mathbb{N}_0$, *then* $\boldsymbol{T}_{\boldsymbol{a},\boldsymbol{b},\boldsymbol{\theta}} \in \mathcal{ND}_{\pm}(\mathbb{R})$. *In particular, if* $b \geq 10$, *then* $\boldsymbol{T}_{1/b,b,\boldsymbol{\theta}} \in \mathcal{ND}_{\pm}(\mathbb{R})$.

Functions of the class $\boldsymbol{T}_{\boldsymbol{a},2,0}$ have also been studied in [HY84].

Proof of Theorem 4.3.1. Since $\boldsymbol{T}_{\boldsymbol{a},\boldsymbol{b},\boldsymbol{\theta}}(x + x_0) = \boldsymbol{T}_{\boldsymbol{a},\boldsymbol{b},(b_n x_0+\theta_n)_{n=0}^{\infty}}(x)$ and $\boldsymbol{T}_{\boldsymbol{a},\boldsymbol{b},\boldsymbol{\theta}}(-x) = \boldsymbol{T}_{\boldsymbol{a},\boldsymbol{b},-\boldsymbol{\theta}}(x)$, $x, x_0 \in \mathbb{R}$, we have only to prove that for every $\boldsymbol{\theta}$, a finite derivative $(\boldsymbol{T}_{\boldsymbol{a},\boldsymbol{b},\boldsymbol{\theta}})'_+(0)$ does not exist (cf. Remark 3.2.1(k)). Suppose that for some $\boldsymbol{\theta}$, a finite derivative $(\boldsymbol{T}_{\boldsymbol{a},\boldsymbol{b},\boldsymbol{\theta}})'_+(0)$ exists. Put $f := \boldsymbol{T}_{\boldsymbol{a},\boldsymbol{b},\boldsymbol{\theta}}$. Write $f(x) = f(0) + f'_+(0)x + \alpha(x)x$, $x > 0$, where $\lim_{x\to 0+} \alpha(x) = 0$. Let $\varepsilon > 0$ be such that $|\alpha(x)| < 1/100$ for $0 < x < \varepsilon$.

Step 1°. Let $N \in \mathbb{N}$ be such that $5/b_N < \varepsilon$. Then, using Remark 2.1.4(a), for $n \geq N$ and $i \in \{2, 3, 4\}$, we get

$$|\boldsymbol{\Delta} f(\tfrac{i}{b_n}, \tfrac{i+1}{b_n}) - f'_+(0)| = |(i+1)\alpha(\tfrac{i+1}{b_n}) - i\alpha(\tfrac{i}{b_n})| < \tfrac{9}{100} < \tfrac{1}{10}.$$

For some $n \in \mathbb{N}_0$, consider five intervals $I_{n,i} := [\frac{i}{b_n}, \frac{i+1}{b_n})$, $i = 0, \ldots, 4$.

Step 2°. Suppose that the function $x \overset{\psi_j}{\longmapsto} \psi(b_j x + \theta_j)$ is nonlinear on two of the four intervals $I_{n,0} \cup I_{n,1}$, $I_{n,2}$, $I_{n,3}$, $I_{n,4}$. Then $5/b_n > 1/(2b_j)$, which is impossible for $j \in \{0, \ldots, n-1\}$. Thus, for every $j \in \{0, \ldots, n-1\}$, the function ψ_j is nonlinear on at most one of the intervals $I_{n,0} \cup I_{n,1}$, $I_{n,2}$, $I_{n,3}$, $I_{n,4}$. In particular, if ψ_j is nonlinear on one of the intervals $I_{n,2}$, $I_{n,3}$, $I_{n,4}$, then ψ_j is linear on the interval $I_{n,0} \cup I_{n,1}$. Observe that since $\frac{2}{b_n} \geq \frac{5}{b_{n+1}}$, we get

$$\Big[0, \frac{2}{b_n}\Big) = I_{n,0} \cup I_{n,1} \supset \bigcup_{s=1}^{\infty} \Big[\frac{2}{b_{n+s}}, \frac{5}{b_{n+s}}\Big) = \bigcup_{s=1}^{\infty} I_{n+s,2} \cup I_{n+s,3} \cup I_{n+s,4}.$$

Consequently, if ψ_j ($j \in \{0, \ldots, n-1\}$) is nonlinear on one of the intervals $I_{n,2}$, $I_{n,3}$, $I_{n,4}$, then ψ_j is linear on each of the intervals $I_{n+s,2}$, $I_{n+s,3}$, $I_{n+s,4}$ with $s \in \mathbb{N}$.

Step 3°. There are infinitely many even (resp. odd) $n \in \mathbb{N}$ such that there exists $i_n \in \{2,3,4\}$ for which all the functions ψ_j, $j = 0, \dots, n-1$, are linear on I_{n,i_n}.

Suppose that there exists a constant $M \in \mathbb{N}$ such that for every $i \in \{2,3,4\}$ and even (resp. odd) number $n > M$ there exists a $j_{i,n} \in \{0, \dots, n-1\}$ such that $\psi_{j_{i,n}}$ is not linear on $I_{n,i}$. In view of Step 2°, we know that $j_{i,n'} \neq j_{i,n}$ for $n' > n$. Thus for every $q > 3M+8$, we get an injective mapping $A \longrightarrow B$, where

$$A := \{(i,n) : i \in \{2,3,4\},\ M < n \le q,\ n \text{ even (resp. odd)}\},$$
$$B := \{0, \dots, q-1\}.$$

It follows that $q = \#B \ge \#A \ge 3(\frac{1}{2}(q-M)-1)$; a contradiction.

Step 4°. Let N be as in Step 1° and let $n > N$ be even (resp. odd) and such that all functions ψ_j, $j = 0, \dots, n-1$, are linear on I_{n,i_n} (as in Step 3°). Then

$$P_n := b_n \sum_{j=0}^{n-1} a_j \Big(\psi_j\Big(\frac{i_n+1}{b_n}\Big) - \psi_j\Big(\frac{i_n}{b_n}\Big)\Big) = b_n \sum_{j=0}^{n-1} \frac{\varepsilon_j}{b_j} \eta_j \frac{b_j}{b_n} = \sum_{j=0}^{n-1} \varepsilon_j \eta_j,$$

where $\varepsilon_j, \eta_j \in \{-1, +1\}$, $j = 0, \dots, n-1$. Consequently, P_n is even (resp. odd); EXERCISE. Obviously, $\psi_n(\frac{i_n+1}{b_n}) = \psi_n(\frac{i_n}{b_n})$. Moreover,

$$b_n \Big| \sum_{j=n+1}^{\infty} a_j \Big(\psi_j\Big(\frac{i_n+1}{b_n}\Big) - \psi_j\Big(\frac{i_n}{b_n}\Big)\Big)\Big| \le b_n \sum_{j=n+1}^{\infty} |a_j| \frac{1}{2} \le \frac{1}{2} \sum_{n=1}^{\infty} \frac{1}{10^j} = \frac{1}{18}.$$

Hence $|\boldsymbol{\Delta} f(\frac{i_n}{b_n}, \frac{i_n+1}{b_n}) - P_n| < \frac{1}{18}$, and consequently, by Step 1°, $|P_n - f'_+(0)| < \frac{1}{5}$. Taking n even and m odd as in Step 3°, we get $1 \le |P_n - P_m| < \frac{2}{5}$; a contradiction. □

Theorem 4.3.2 (cf. [Cat03]). *Let*

$$T(x) := \sum_{n=0}^{\infty} a_n f_n(b_n x), \quad x \in \mathbb{R},$$

where $\boldsymbol{a} = (a_n)_{n=0}^{\infty} \subset \mathbb{R}_{>0}$, $\sum_{n=0}^{\infty} a_n < +\infty$, $\boldsymbol{b} = (b_n)_{n=0}^{\infty} \subset \mathbb{N}$, $\frac{b_{n+1}}{b_n} \in \mathbb{N}_2$, $\limsup_{n\to+\infty} a_n b_n > 0$, $(f_n)_{n=0}^{\infty} \subset \mathcal{C}(\mathbb{R}, \mathbb{I})$, $f_n(-x) = f_n(x)$, $x \in \mathbb{R}$, $f_n(2k) = 0$, $f_n(2k+1) = 1$, $f_n|_{[2k,2k+2]}$ *is concave*, $k \in \mathbb{Z}$, $n \in \mathbb{N}_0$.

Then $T \in \mathcal{ND}_{\pm}(\mathbb{R})$. In particular, taking $f_n(x) := 2\psi(\frac{x}{2})$, we conclude that if $b \in \mathbb{N}_2$, then $\boldsymbol{T}_{1/b,b,0} \in \mathcal{ND}_{\pm}(\mathbb{R})$.

Remark 4.3.3. An independent proof showing that $\boldsymbol{T} \in \mathcal{ND}_{\pm}(\mathbb{R})$ may be found in [Cat84]. Similar problems have been studied in [Mik56].

Notice that Theorem 4.3.2 is a general tool that may be applied to many other series. For example, one can take $f_n(x) := \varphi_n(2\psi(\frac{x}{2}))$, where $\varphi_n : \mathbb{I} \longrightarrow \mathbb{I}$ is an increasing concave function with $\varphi_n(0) = 0$, $\varphi_n(1) = 1$.

(a) In particular, we may take

$$\varphi_n(t) = t^{1/m}\ (m \ge 1), \quad \varphi_n(t) = \frac{\sin t}{\sin 1}, \quad \varphi_n(t) = \frac{1}{2}(t^{1/3} + t^{1/5});$$

see [Cat03] for more examples.

(b) Taking $\varphi_n := \mathrm{id}$, we easily conclude that the *Faber functions*

$$\sum_{n=0}^{\infty} \frac{1}{10^n} \psi(2^{n!}x), \quad \sum_{n=0}^{\infty} \frac{1}{n!} \psi(2^{n!}x), \quad x \in \mathbb{R},$$

are of the class $\mathfrak{ND}_{\pm}(\mathbb{R})$ (cf. [Fab07, Fab08, Fab10]).

(c) Similarly, the *Lebesgue function*

$$\sum_{n=0}^{\infty} \frac{1}{2^n} \sin(2^{n^2} x), \quad x \in \mathbb{R},$$

is of the class $\mathfrak{ND}_{\pm}(\mathbb{R})$ (cf. [Leb40]).

(d) In an analogous way, one can easily prove (EXERCISE) that the *McCarthy function*

$$\sum_{n=0}^{\infty} \frac{1}{2^n} g(2^{2^n} x), \quad x \in \mathbb{R},$$

where $g(x) := 4\psi(\frac{1}{4}(x+2))-1$, is of the class $\mathfrak{ND}_{\pm}(\mathbb{R})$ (cf. [McC53]). Note that $g(x+4) = g(x)$ and

$$g(x) = \begin{cases} 1+x, & \text{if } x \in [-2,0] \\ 1-x, & \text{if } x \in [0,2] \end{cases}.$$

Proof of Theorem 4.3.2. Obviously, $T \in \mathcal{C}(\mathbb{R})$.

Since $T(-x) = T(x)$, $x \in \mathbb{R}$, we have only to show that $T'_+(x)$ does not exist for every $x \in \mathbb{R}$. Suppose that $T'_+(x_0)$ exists for some $x_0 \in \mathbb{R}$. Write

$$T(x) = T(x_0) + T'_+(x_0)(x - x_0) + \alpha(x)(x - x_0), \quad x > x_0,$$

where $\lim_{x \to x_0+} \alpha(x) = 0$. Let $0 < \varepsilon < \frac{1}{10}L$, where $L := \limsup_{n \to +\infty} a_n b_n$. Take a $\delta > 0$ such that $|\alpha(x)| < \varepsilon$ for $x_0 < x < x_0 + 2\delta$. Fix an $N \in \mathbb{N}$ such that $b_N > \frac{4}{\delta}$ and $a_N b_N > 10\varepsilon$. Take a $k \in \mathbb{Z}$ with $2k - 2 \le b_N x_0 < 2k$ and define $x_i := \frac{2k+i-1}{b_N}$, $i = 1, 2, 3$. Then, by Remark 2.1.4(a), we get

$$\begin{aligned} |\boldsymbol{\Delta} T(x_1, x_2) - \boldsymbol{\Delta} T(x_1, x_3)| &= |b_N(\alpha(x_2)(x_2 - x_0) - \alpha(x_1)(x_1 - x_0)) \\ &\quad - \frac{b_N}{2}(\alpha(x_3)(x_3 - x_0) + \alpha(x_1)(x_1 - x_0))| \\ &\le b_N\Big(\varepsilon\frac{3}{b_N} + \varepsilon\frac{2}{b_N}\Big) + \frac{b_N}{2}\Big(\varepsilon\frac{4}{b_N} + \varepsilon\frac{2}{b_N}\Big) = 8\varepsilon. \end{aligned}$$

If $n > N$ and $m := \frac{b_n}{b_N} \in \mathbb{N}$, then

$$f_n(b_n x_i) = f_n((2k + i - 1)m) = \begin{cases} 0, & \text{if } i \in \{1, 3\} \\ f_n((2k+1)m), & \text{if } i = 2 \end{cases}.$$

Thus

$$\frac{f_n(b_n x_2) - f_n(b_n x_1)}{x_2 - x_1} - \frac{f_n(b_n x_3) - f_n(b_n x_1)}{x_3 - x_1} = \frac{f_n(b_n x_2) - 0}{x_2 - x_1} - \frac{0 - 0}{x_3 - x_1} \ge 0.$$

If $n = N$, then we get

$$\frac{f_N(b_N x_2) - f_N(b_N x_1)}{x_2 - x_1} - \frac{f_N(b_N x_3) - f_N(b_N x_1)}{x_3 - x_1} = \frac{f_N(2k+1) - f_N(2k)}{x_2 - x_1} - \frac{f_N(2k+2) - f_N(2k)}{x_3 - x_1} = b_N.$$

If $n < N$ and $m := \frac{b_N}{b_n} \in \mathbb{N}$, $k = \ell m + r$ ($\ell \in \mathbb{Z}$, $r \in \{0, \dots, m-1\}$), then $2\ell \leq 2\ell + \frac{2r}{m} = b_n x_1 < b_n x_3 = 2\ell + \frac{2r+2}{m} \leq 2\ell + 2$. Thus the function $[x_1, x_3] \ni x \longmapsto f_n(b_n x)$ is concave. Consequently,

$$\frac{f_n(b_n x_2) - f_n(b_n x_1)}{x_2 - x_1} - \frac{f_n(b_n x_3) - f_n(b_n x_1)}{x_3 - x_1} \geq 0.$$

Finally,

$$\boldsymbol{\Delta} T(x_1, x_2) - \boldsymbol{\Delta} T(x_1, x_3) \geq a_N b_N > 10\varepsilon;$$

a contradiction. □

Remark 4.3.4. In [Lan08], G. Landsberg showed that if $(a_n)_{n \in \mathbb{N}_0} \subset \mathbb{R}$, $\sum_{n=0}^{\infty} |a_n| < +\infty$, and if the series $(S) = \sum_{n=0}^{\infty} 2^n a_n$ is divergent, then the function $F : \mathbb{R} \longrightarrow \mathbb{R}$, $F(x) := \sum_{n=0}^{\infty} a_n \psi(2^n x)$, belongs to $\mathcal{ND}(\mathbb{R})$. If the series (S) is absolutely convergent, then F'_+ and F'_- exist as finite values everywhere.

4.4 Differentiability of a Class of Takagi Functions

It is clear that the Takagi function

$$F_a(x) := \boldsymbol{T}_{a,2,0}(x) = \sum_{n=0}^{\infty} a^n \psi(2^n x), \quad x \in \mathbb{R},$$

may be formally defined for all $a \in \mathbb{R}$ with $0 < |a| < \frac{1}{2}$. Obviously, $F_a \in \mathcal{C}(\mathbb{R})$. We are not interested in the nowhere differentiability of F_a, but in its differentiability. The aim of this section is to prove the following Theorem 4.4.2 due to A.L. Thomson and J.N. Hagler.

Remark 4.4.1. Put $f_n(x) := \psi(2^n x)$, $x \in \mathbb{R}$, $n \in \mathbb{N}_0$ ($f_0 \equiv \psi$). Then for every $n \in \mathbb{N}_0$, we have:

(a) f_n is periodic with period $\frac{1}{2^n}$.

(b) $(f_n)'_+(x) = \left\{ \begin{array}{ll} 2^n, & \text{if } 0 \leq x < \frac{1}{2^{n+1}} \\ -2^n, & \text{if } \frac{1}{2^{n+1}} \leq x < \frac{1}{2^n} \end{array} \right\} = 2^n \psi'_+(2^n x).$

(c) $(f_n)'_-(x) = \left\{ \begin{array}{ll} 2^n, & \text{if } 0 < x \leq \frac{1}{2^{n+1}} \\ -2^n, & \text{if } \frac{1}{2^{n+1}} < x \leq \frac{1}{2^n} \end{array} \right\} = 2^n \psi'_-(2^n x).$

(d) $(f_n)'_+$ and $(f_n)'_-$ are periodic with period $\frac{1}{2^n}$.

Theorem 4.4.2 (cf. [TH]). *Let $0 < |a| < \frac{1}{2}$, and for $x \in \mathbb{I}$, let $x = \sum_{n=1}^{\infty} \frac{x_n}{2^n}$ be a binary representation of x. Then:*

(a) $F_{1/4}(x) = 2x(1-x)$.

(b) *If* $x < 1$, *then* $(F_a)'_+(x) = \sum_{n=0}^{\infty} a^n (f_n)'_+(x) \in \mathbb{R}$. *Moreover, if* $\sup\{n \in \mathbb{N} : x_n = 0\} = +\infty$, *then* $(f_n)'_+(x) = 2^n(1 - 2x_{n+1})$, *and consequently,* $(F_a)'_+(x) = \frac{a}{1-2a} - 2\sum_{n=0}^{\infty}(2a)^n x_{n+1}$.

(c) *If* $x > 0$, *then* $(F_a)'_-(x) = \sum_{n=0}^{\infty} a^n (f_n)'_-(x) \in \mathbb{R}$. *Moreover, if* $\sup\{n \in \mathbb{N} : x_n = 1\} = +\infty$, *then* $(f_n)'_-(x) = 2^n(1-2x_{n+1})$, *and consequently,* $(F_a)'_-(x) = \frac{a}{1-2a} - 2\sum_{n=0}^{\infty}(2a)^n x_{n+1}$.

(d) *If* $a \neq \frac{1}{4}$, *then a finite* $F'_a(x)$ *exists iff* x *is not a dyadic rational.*

Proof. (a) Let $\boldsymbol{\mathcal{E}} := \{F \in \mathcal{C}(\mathbb{R}) : \forall_{x\in\mathbb{R}} : F(x+1) = F(x)\}$. Consider $\boldsymbol{\mathcal{E}}$ as a Banach space with the supremum norm $\|F\| := \sup_{x\in\mathbb{R}} |F(x)|$. Define an operator $L : \boldsymbol{\mathcal{E}} \longrightarrow \boldsymbol{\mathcal{E}}$, $L(F)(x) := \boldsymbol{\psi}(x) + \frac{1}{4}F(2x)$, $x \in \mathbb{R}$. Let $H \in \boldsymbol{\mathcal{E}}$ be such that $H(x) = 2x(1-x)$ for $x \in \mathbb{I}$. Observe that for $x \in \mathbb{I}$, we get

$$L(H)(x) = \boldsymbol{\psi}(x) + \frac{1}{4}H(2x) = \begin{Bmatrix} x + x(1-2x), & \text{if } x \le \frac{1}{2} \\ 1 - x + \frac{1}{4}H(2x-1), & \text{if } x \ge \frac{1}{2} \end{Bmatrix} = H(x).$$

One may easily check (EXERCISE) that $\|L^n(\boldsymbol{\psi}) - F_{1/4}\| \underset{n\to+\infty}{\longrightarrow} 0$, where L^n stands for the nth iterate of L.

For every $F, G \in \boldsymbol{\mathcal{E}}$, we have $\|L(F) - L(G)\| = \frac{1}{4}\|F - G\|$. Consequently, by the Banach fixed-point theorem, there exists exactly one $F^* \in \boldsymbol{\mathcal{E}}$ such that $L(F^*) = F^*$, and moreover, for every $F \in \boldsymbol{\mathcal{E}}$, we get $L^n(F) \longrightarrow F^*$.

Thus $F_{1/4} \equiv F^* \equiv H$.

(b) Let $\varepsilon > 0$, and let $k \in \mathbb{N}$ be such that $\frac{2(2|a|)^{k+1}}{1-2|a|} < \varepsilon$. Observe that there exists a $0 < \delta < 1 - x$ such that $f_n(x+h) = f_n(x) + (f_n)'_+(x)h$ for $n = 0, \dots, k$ and $0 < h < \delta$. Consequently, for $0 < h < \delta$, we get

$$\begin{aligned} &\Big|\boldsymbol{\Delta}F_a(x, x+h) - \sum_{n=0}^{\infty} a^n (f_n)'_+(x)\Big| \\ &\le \sum_{n=k+1}^{\infty} |a|^n \Big|\frac{\boldsymbol{\psi}(2^n(x+h)) - \boldsymbol{\psi}(2^n x)}{h}\Big| + \sum_{n=k+1}^{\infty} |a|^n |(f_n)'_+(x)| \\ &\overset{\text{Remark 4.4.1}}{\le} 2\sum_{n=k+1}^{\infty}(2|a|)^n = \frac{2(2|a|)^{k+1}}{1-2|a|} < \varepsilon, \end{aligned}$$

which proves the formula for $(F_a)'_+(x)$.

Let $x' = \sum_{n=1}^{\infty} \frac{x'_n}{2^n}$ with $\sup\{n \in \mathbb{N} : x'_n = 0\} = +\infty$. Then

$$\boldsymbol{\psi}'_+(x') = \begin{Bmatrix} 1, & \text{if } x'_1 = 0 \\ -1, & \text{if } x'_1 = 1 \end{Bmatrix} = 1 - 2x'_1.$$

Take a $k \in \mathbb{N}_0$. Then

$$2^k x = \Big(\sum_{n=1}^{k} 2^{k-n} x_n\Big) + \sum_{n=k+1}^{\infty} \frac{x_n}{2^{n-k}} =: y + x' = y + \sum_{n=1}^{\infty} \frac{x_{k+n}}{2^n} = y + \sum_{n=1}^{\infty} \frac{x'_n}{2^n},$$

where $y \in \mathbb{N}_0$. Hence

$$(f_k)'_+(x') = 2^k \psi'_+(2^k x) = 2^k \psi'_+(x') = 2^k(1 - 2x'_1) = 2^k(1 - 2x_{k+1}).$$

(c) EXERCISE.

(d) If x is not a dyadic rational, then $\sup\{n \in \mathbb{N} : x_n = 0\} = \sup\{n \in \mathbb{N} : x_n = 1\} = +\infty$, and therefore, the result follows from (b) and (c).
If $x = \frac{m}{2^k}$ is a dyadic rational ($m, k \in \mathbb{N}_0$, $m \leq 2^k$, and $(m, 2^k) = 1$), then

$$(f_n)'_+(x) - (f_n)'_-(x) = 2^n(\psi'_+(2^n x) - \psi'_-(2^n x))$$
$$= 2^n\left(\psi'_+\left(\frac{m}{2^{k-n}}\right) - \psi'_-\left(\frac{m}{2^{k-n}}\right)\right) = \begin{cases} 0, & \text{if } n = 0, \ldots, k-2 \\ -2^k, & \text{if } n = k-1 \\ 2^{n+1}, & \text{if } n = k, k+1, \ldots \end{cases},$$

with obvious modifications for $k \in \{0, 1\}$. Hence, using (b) and (c), we get

$$(F_a)'_+(x) - (F_a)'_-(x) = -a^{k-1}2^k + \sum_{n=k}^{\infty} a^n 2^{n+1} = \frac{2(2a)^{k-1}}{1-2a}(4a-1) \neq 0. \qquad \square$$

Chapter 5
Bolzano-Type Functions I

Summary. The goal of this chapter is to prove basic results related to the nowhere differentiability of Bolzano-type functions. Some more advanced properties will be presented in Chap. 10.

5.1 The Bolzano-Type Function

Bolzano-type functions $f : \mathbb{I} \longrightarrow \mathbb{R}$ will be of the form $f(x) = \lim_{n\to+\infty} L_n(x)$, $x \in \mathbb{I}$, where each function $L_n : \mathbb{I} \longrightarrow \mathbb{R}$, $n \in \mathbb{N}_0$, is continuous, piecewise linear, and the convergence is uniform (which guarantees that f is continuous). Moreover, $L_0(x) = x$, $x \in \mathbb{I}$, and L_{n+1} is obtained from L_n via the following geometric procedure (Fig. 5.1).

Fix numbers $N \in \mathbb{N}_2$, $0 = \varphi_0 < \varphi_1 < \cdots < \varphi_N = 1$, $\Phi_1, \dots, \Phi_{N-1} \in \mathbb{R}$. Let $\Phi_0 := 0$, $\Phi_N := 1$.

For two points $P = (a, A), Q = (b, B) \in \mathbb{R}^2$, $a < b$, consider the segment $S := [P, Q] \subset \mathbb{R}^2$ identified with the graph of the affine function

$$J \ni x \longmapsto L_0^{P,Q}(x) := A + \varkappa(x - a),$$

where $J = J(S) := [a, b]$, $\Delta = \Delta(S) := B - A$, $\delta = \delta(J) := b - a$, $\varkappa = \varkappa(S) := \frac{\Delta}{\delta}$. Consider $N + 1$ points

$$R_i = (a_i, A_i) := (a + \varphi_i\delta, A + \Phi_i\Delta), \quad i = 0, \dots, N.$$

Note that $R_0 = P$ and $R_N = Q$. Let $S_i = S_i(P, Q) := [R_{i-1}, R_i]$, $J_i = J_i(P, Q) := [a_{i-1}, a_i]$, $i = 1, \dots, N$, and let $L_1^{P,Q} : J \longrightarrow \mathbb{R}$ be the piecewise affine function corresponding to the union of segments $S_1 \cup \cdots \cup S_N$. We say that the interval J_i *is of type* i (type$(J_i) = i$). We have

$$\Delta(S_i) = (\Phi_i - \Phi_{i-1})\Delta, \quad \delta(J_i) = (\varphi_i - \varphi_{i-1})\delta, \quad \varkappa(S_i) = \frac{\Phi_i - \Phi_{i-1}}{\varphi_i - \varphi_{i-1}}\varkappa, \qquad i = 1, \dots, N.$$

Observe that

$$\max_{1\le i\le N} |\Delta(S_i)| = M_\Delta \cdot |\Delta|, \text{ where } M_\Delta := \max_{1\le i\le N} |\Phi_i - \Phi_{i-1}|,$$
$$\max_{1\le i\le N} \delta(J_i) = M_\delta \cdot \delta, \text{ where } M_\delta := \max_{1\le i\le N} (\varphi_i - \varphi_{i-1}) < 1,$$

M. Jarnicki, P. Pflug, *Continuous Nowhere Differentiable Functions*, Springer Monographs in Mathematics, DOI 10.1007/978-3-319-12670-8_5

$$\max_{1\le i\le N} |\varkappa(S_i)| = M_\varkappa \cdot |\varkappa|, \text{ where } M_\varkappa := \max_{1\le i\le N} \frac{|\Phi_i - \Phi_{i-1}|}{\varphi_i - \varphi_{i-1}},$$
$$|L_1^{P,Q}(x') - L_1^{P,Q}(x'')| \le M_\varkappa |\varkappa| |x' - x''|, \quad x', x'' \in J,$$
$$\max_{x\in J} |L_1^{P,Q}(x) - L_0^{P,Q}(x)| = M_L \cdot |\Delta|, \text{ where } M_L := \max_{1\le i\le N-1} |\Phi_i - \varphi_i|.$$

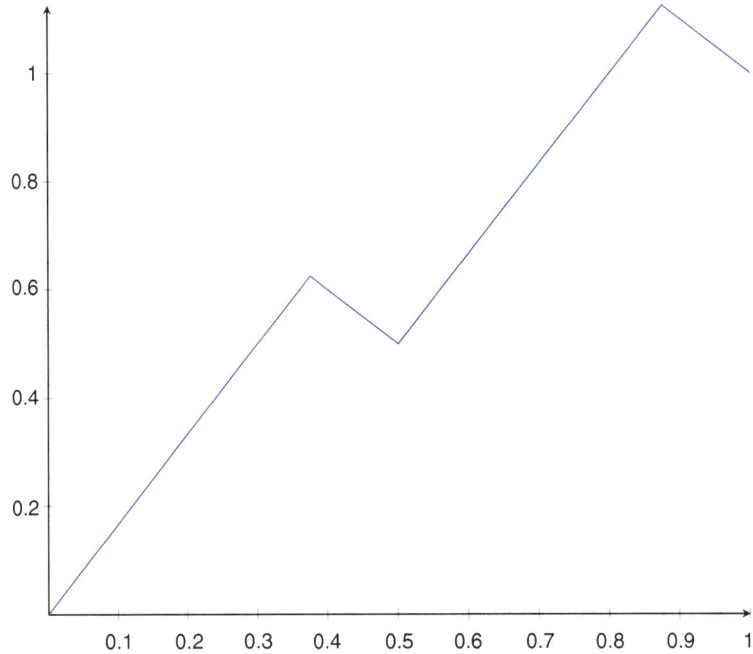

Fig. 5.1 The function L_1 for $N = 4$, $\varphi_1 = \frac{3}{8}$, $\varphi_2 = \frac{1}{2}$, $\varphi_3 = \frac{7}{8}$, $\Phi_1 = \frac{5}{8}$, $\Phi_2 = \frac{1}{2}$, $\Phi_3 = \frac{9}{8}$, $P = (0,0)$, $Q = (1,1)$

Assume additionally that $M_\Delta < 1$.

We are ready to define the *Bolzano-type function* using the following recursive procedure.

We begin with $P = (a, A) := (0,0)$ and $Q = (b, B) := (1,1)$ and continue using the general construction given previously. We get:

- N intervals $J_{1,i} := [a_{1,i-1}, a_{1,i}]$, $i = 1, \dots, N$,
- N segments $S_{1,i} = [(a_{1,i-1}, A_{1,i-1}), (a_{1,i}, A_{1,i})]$, $i = 1, \dots, N$, and
- a continuous piecewise affine function $L_1 : \mathbb{I} \longrightarrow \mathbb{R}$.

We have $\operatorname{type}(J_{1,i}) = i$, $i = 1, \dots, N$. We repeat the above construction for each of the segments $S_{1,i}$, $i = 1, \dots, N$. We get:

- N^2 intervals $J_{2,i} = [a_{2,i-1}, a_{2,i}]$, $i = 1, \dots, N^2$,
- N^2 segments $S_{2,i} = [(a_{2,i-1}, A_{2,i-1}), (a_{2,i}, A_{2,i})]$, $i = 1, \dots, N^2$, and
- a continuous piecewise affine function $L_2 : \mathbb{I} \longrightarrow \mathbb{R}$.

We have $\operatorname{type}(J_{2,sN+i}) = i$, $s = 0, \dots, N-1$, $i = 1, \dots, N$. After n steps, we arrive at:

- N^n intervals $J_{n,i} = [a_{n,i-1}, a_{n,i}]$, $i = 1, \dots, N^n$,
- N^n segments $S_{n,i} = [(a_{n,i-1}, A_{n,i-1}), (a_{n,i}, A_{n,i})]$, $i = 1, \dots, N^n$, and
- a continuous piecewise affine function $L_n : \mathbb{I} \longrightarrow \mathbb{R}$.

We have $\operatorname{type}(J_{n,sN+i}) := i$, $s = 0, \dots, N^{n-1} - 1$, $i = 1, \dots, N$. Set

$$\begin{aligned}\mathfrak{J}_n &:= \{[a_{n,i-1}, a_{n,i}] : i = 1, \dots, N^n\},\\ \mathfrak{S}_n &:= \{[(a_{n,i-1}, A_{n,i-1}), (a_{n,i}, A_{n,i})] : i = 1, \dots, N^n\}.\end{aligned}$$

If $J = [a_{n,i-1}, a_{n,i}] \in \mathfrak{J}_n$, then $S(J) := [(a_{n,i-1}, A_{n,i-1}), (a_{n,i}, A_{n,i})]$. Conversely, if $S = [(a_{n,i-1}, A_{n,i-1}), (a_{n,i}, A_{n,i})] \in \mathfrak{S}_n$, then $J(S) := [a_{n,i-1}, a_{n,i}]$. We have

$$\begin{gathered}|\Delta(S_{n,i})| \le M_\Delta^n, \quad \delta(J_{n,i}) \le M_\delta^n, \quad |\varkappa(S_{n,i})| \le M_\varkappa^n, \quad i = 1, \dots, N^n,\\ |L_n(x') - L_n(x'')| \le M_\varkappa^n |x' - x''|, \quad x', x'' \in \mathbb{I},\\ \max_{x \in \mathbb{I}} |L_{n+1}(x) - L_n(x)| \le M_L M_\Delta^n.\end{gathered}$$

In particular, the series $\sum_{n=1}^\infty (L_{n+1} - L_n)$ is convergent uniformly on $\mathbb{I}$. Consequently, the *Bolzano-type function*

$$f(x) := \lim_{n \to +\infty} L_n(x) = L_1(x) + \sum_{n=1}^\infty (L_{n+1} - L_n(x)), \quad x \in \mathbb{I},$$

is well defined and is continuous on $\mathbb{I}$.

Let $\mathbf{N}_n := \{a_{n,i} : i = 0, \dots, N^n\}$, and let $\mathbf{N} := \bigcup_{n=1}^\infty \mathbf{N}_n$ denote the set of *nodes*.

Remark 5.1.1.

(a) The *classical Bolzano function* $\boldsymbol{B}$ (cf. § 10.1) is the case in which $N = 4$, $\varphi_1 = \frac{3}{8}$, $\varphi_2 = \frac{1}{2}$, $\varphi_3 = \frac{7}{8}$, $\Phi_1 = \frac{5}{8}$, $\Phi_2 = \frac{1}{2}$, $\Phi_3 = \frac{9}{8}$ ($M_\Delta = \frac{5}{8}$) (cf. Fig. 5.2).

(b) $\mathbf{N}$ is countable and dense in $\mathbb{I}$.

(c) If $x_0 \in \mathbf{N}_p$, then $f(x_0) = L_n(x_0)$ for $n \ge p$.

(d) Let $x_0 \in \mathbb{I}$ and let $S_n \in \mathfrak{S}_n$ be such that $x_0 \in J_n := J(S_n) = [a_n, b_n]$. If $x_0 \notin \mathbf{N}_n$, then S_n is uniquely determined, and $a_n < x_0 < b_n$. If p is the minimal number such that $x_0 \in \mathbf{N}_p \setminus \{1\}$, then we choose S_n such that $a_n = x_0$ for all $n \ge p$ (then we say for short that the *sequence* $(S_n)_{n=1}^\infty$ *is of type (L)*). Notice that if p is the minimal number such that $x_0 \in \mathbf{N}_p \setminus \{0\}$, then we may also choose S_n such that $b_n = x_0$ for all $n \ge p$ (then the *sequence* $(S_n)_{n=1}^\infty$ *is of type (R)*). In any case, we have $J_{n+1} \subset J_n$ for all $n \in \mathbb{N}$ and $\{x_0\} = \bigcap_{n=1}^\infty J_n$. We say that $(S_n)_{n=1}^\infty$ is a *determining sequence for* x_0.

(e) Conversely, if $(S_n)_{n=1}^\infty$ is a sequence such that $S_n \in \mathfrak{S}_n$ and $J_{n+1} \subset J_n$ ($J_n := J(S_n)$), then $(S_n)_{n=1}^\infty$ is a determining sequence for x_0, where $\{x_0\} = \bigcap_{n=1}^\infty J_n$.

(f) If $M_\varkappa \le 1$, then f is Lipschitz continuous; in particular, $f'(x)$ exists for a.a. $x \in \mathbb{I}$. Thus, from our "nowhere differentiable" point of view, we have to assume that $M_\varkappa > 1$.

Define

$$\Sigma := \Big\{i \in \{1, \dots, N\} : \frac{|\Phi_i - \Phi_{i-1}|}{\varphi_i - \varphi_{i-1}} > 1\Big\} \quad \Sigma' := \{1, \dots, N\} \setminus \Sigma.$$

Theorem 5.1.2. *Assume that* $M_\Delta < 1$, $\Sigma \ne \varnothing$, *and*

$$\frac{\Phi_i - \Phi_{i-1}}{\varphi_i - \varphi_{i-1}} = -1, \quad i \in \Sigma'. \tag{5.1.1}$$

Then $f \in \mathbf{ND}(\mathbb{I})$.

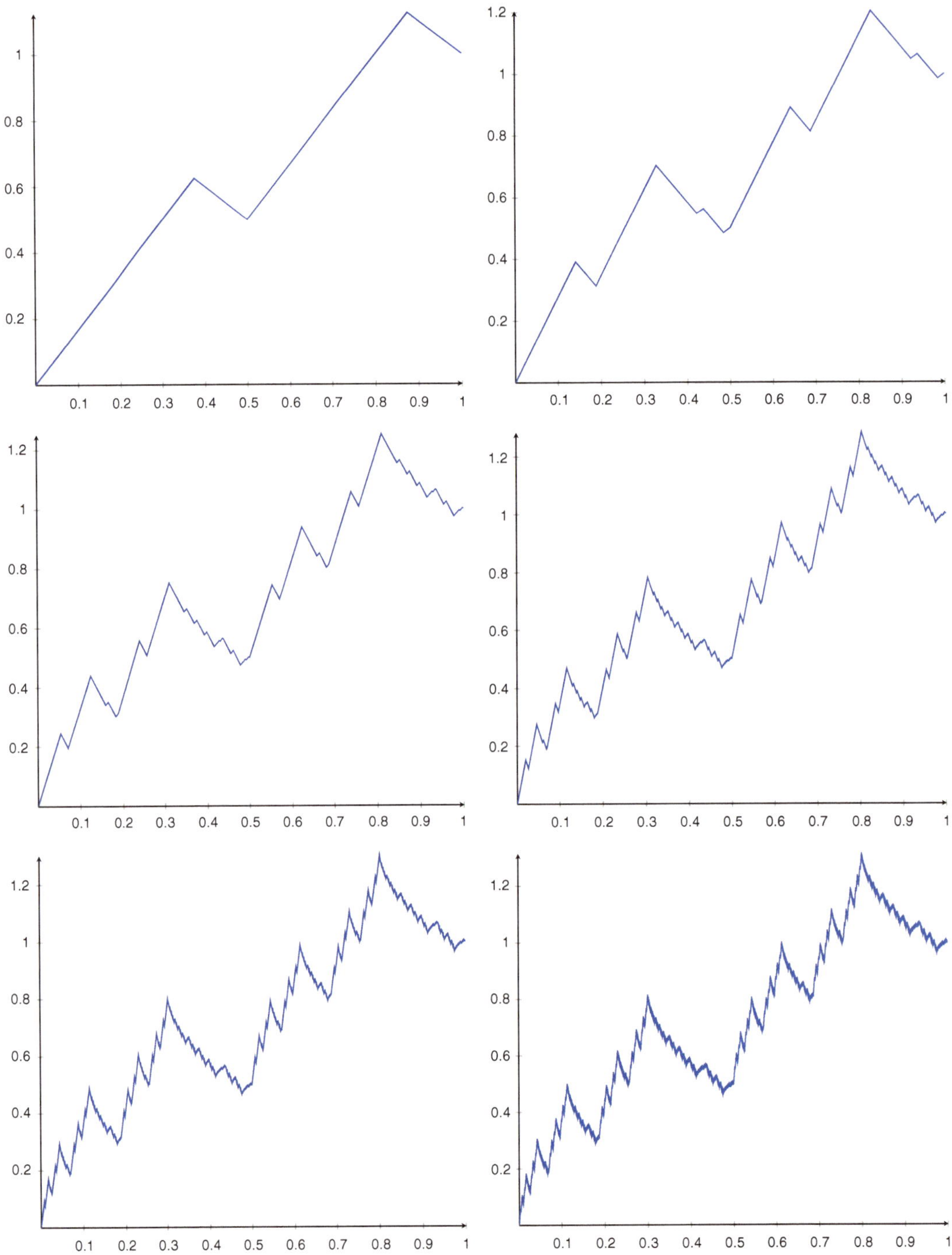

Fig. 5.2 The first six steps of the construction of the classical Bolzano function $\boldsymbol{B}$

Remark 5.1.3.

(a) The above assumptions imply that $|\varkappa(S_{n,i})| \geq 1$, $n \in \mathbb{N}$, $i = 1, \dots, N^n$.

(b) In the Bolzano case (cf. Remark 5.1.1(a)), we have $M_\Delta = \frac{5}{8}$ and $\Sigma = \{1, 3\}$. Moreover, in this case, condition (5.1.1) is also satisfied.

Proof of Theorem 5.1.2. Let

$$M_{\varkappa,0} := \min\left\{\frac{|\Phi_i - \Phi_{i-1}|}{\varphi_i - \varphi_{i-1}} : i \in \Sigma\right\} > 1.$$

Suppose that $f'(x_0) \in \mathbb{R}$ exists. Let $(S_n)_{n=1}^\infty$ be a determining sequence for x_0 (cf. Remark 5.1.1(d)) and let $J_n = J(S_n) = [a_n, b_n]$. Then

$$\varkappa(S_n) = \boldsymbol{\Delta} L_n(a_n, b_n) = \boldsymbol{\Delta} f(a_n, b_n) \underset{n \to +\infty}{\longrightarrow} f'(x_0)$$

(cf. Remark 2.1.2). There are the following two possibilities:

(A) There exists an $n_0 \in \mathbb{N}$ such that $\text{type}(J_n) \in \Sigma$ for $n \geq n_0$. Then $|\varkappa(S_n)| \geq M_{\varkappa,0}^{n-n_0} |\varkappa(S_{n_0})|$ for $n \geq n_0$, and consequently $|\varkappa(S_n)| \longrightarrow +\infty$; a contradiction.

(B) There exists a subsequence $(n_s)_{s=1}^\infty$, $n_1 \geq 2$, such that $\text{type}(J_{n_s}) \in \Sigma'$ for all $s \in \mathbb{N}$. Then $\varkappa(S_{n_s}) = -\varkappa(S_{n_s - 1})$, $s \in \mathbb{N}$. Thus, a finite or infinite limit $\lim_{n \to +\infty} \varkappa(S_n)$ does not exist; a contradiction.

□

Notice that in the case (B), *a finite or infinite derivative $f'(x_0)$ does not exist.*

Exercise 5.1.4. Consider the case $N = 4$, $\varphi_2 = \frac{1}{2} = \Phi_2$.

(a) Prove that $M_\Delta < 1 \iff \Phi_1 \in (-\frac{1}{2}, 1)$ and $\Phi_3 \in (0, \frac{3}{2})$.

(b) Prove that each of the following sets Σ may be realized by a configuration of parameters $0 < \varphi_1 < \frac{1}{2} < \varphi_3 < 1, \Phi_1 \in (-\frac{1}{2}, 1)$, $\Phi_3 \in (0, \frac{3}{2})$:

$$\{1,2,3,4\}, \quad \{2,3,4\}, \quad \{1,3,4\}, \quad \{1,2,4\}, \quad \{1,2,3\}, \quad \{2,4\}, \quad \{1,3\}.$$

5.2 Q-Representation of Numbers

We will describe an alternative tool to define Bolzano-type functions, not via the recursive procedure given above, but via a certain arithmetic representation of real numbers, called *Q-representation* (cf. [PV13]).

We fix $N \in \mathbb{N}_2$ and $\delta_0, \dots, \delta_{N-1} > 0$ such that $\delta_0 + \dots + \delta_{N-1} = 1$. Put $Q := (\delta_0, \dots, \delta_{N-1})$. Define $\varphi_0 := 0$, $\varphi_{i+1} := \varphi_i + \delta_i$, $i = 0, \dots, N-1$ (note that $\varphi_N = 1$). Put $M_\delta := \max\{\delta_0, \dots, \delta_{N-1}\}$ (note that $M_\delta < 1$). For a sequence $\alpha = (\alpha_n)_{n=1}^\infty \subset \{0, \dots, N-1\}$, define

$$x_k(\alpha) := \sum_{n=1}^{k} \varphi_{\alpha_n} d_{n-1}(\alpha), \quad k \in \mathbb{N},$$

where $d_0(\alpha) := 1$ and $d_\ell(\alpha) := \delta_{\alpha_1} \cdots \delta_{\alpha_\ell}$, $\ell \in \mathbb{N}$. Since $d_\ell(\alpha) \le M_\delta^\ell$, the number

$$x(\alpha) := \sum_{n=1}^{\infty} \varphi_{\alpha_n} d_{n-1}(\alpha) = \lim_{k\to+\infty} x_k(\alpha)$$

is well defined.

If $x(\alpha) = x_k(\alpha)$ for some $k \in \mathbb{N}$ (i.e., $\alpha_n = 0$ for $n \ge k+1$), then we say that $x(\alpha)$ is *Q-rational.*

Observe that when $\delta_i := \frac{1}{N}$, $i = 0, \dots, N-1$, we get $\varphi_i = \frac{i}{N}$, $i = 0, \dots, N$, and therefore $x(\alpha) = \sum_{n=1}^{\infty} \frac{\alpha_n}{N^n}$ is an N-adic representation.

The following remark collects basic properties of Q-representations.

Remark 5.2.1 (Details Are Left to the Reader as an EXERCISE).

(a) For every α, we have $x(\alpha) \le x_k(\alpha) + d_k(\alpha) \le 1$, $k \in \mathbb{N}$. In particular, $x(\alpha) \le 1$. Moreover, $x_{k+1}(\alpha) + d_{k+1}(\alpha) \le x_k(\alpha) + d_k(\alpha)$, $k \in \mathbb{N}$.

(b) If $x(\alpha) = x_k(\alpha)$ is Q-rational with $\alpha_k \ge 1$, then $x(\alpha) = x(\beta)$, where $\beta := (\alpha_1, \dots, \alpha_{k-1}, \alpha_k - 1, N-1, N-1, \dots)$.

(c) Every number $x \in \mathbb{I}$ may be written in the form (*Q-representation*) $x = x(\alpha)$ for some α. If x is not Q-rational, then the Q-representation is uniquely determined. If $x = x(\alpha) = x_k(\alpha)$ is Q-rational with $\alpha_k \ge 1$, then x has exactly two different Q-representations (as in (b)).

5.2.1 Continuity of Functions Given via Q-Representation

Let N, δ_i, $i = 0, \dots, N-1$, φ_i, $i = 0, \dots, N$, $d_k(\alpha)$, $x_k(\alpha)$, $x(\alpha)$ be as before. Put $\Xi_\ell := \{0, \dots, N-1\}^\ell$, $\ell \in \mathbb{N}$.

Fix a sequence $(\sigma_n)_{n=1}^{\infty} \subset \mathbb{N}$ with $n \le \sigma_n \le \sigma_{n+1}$. Let $g_n : \Xi_{\sigma(n)} \longrightarrow \mathbb{C}$, $M_n := \max_{\Xi_{\sigma(n)}} |g_n|$, $n \in \mathbb{N}$, and *assume* that $\sum_{n=1}^{\infty} M_n < +\infty$. We assign to each $x = x(\alpha) \in \mathbb{I}$ the value $f(x) := \sum_{n=1}^{\infty} g_n(\alpha_1, \dots, \alpha_{\sigma(n)})$, and we *assume* that $f(x)$ is independent of the different Q-representations of x. More precisely, we assume that

$$\begin{aligned} g_k(\alpha_1, \dots, \alpha_{\sigma(k)}) &+ \sum_{n=k+1}^{\infty} g_n(\alpha_1, \dots, \alpha_k, \underbrace{0, \dots, 0}_{(\sigma(n)-k)\times}) \\ &= g_k(\alpha_1, \dots, \alpha_{k-1}, \alpha_k - 1, \alpha_{k+1}, \dots, \alpha_{\sigma(k)}) \\ &\quad + \sum_{n=k+1}^{\infty} g_n(\alpha_1, \dots, \alpha_{k-1}, \alpha_k - 1, \underbrace{N-1, \dots, N-1}_{(\sigma(n)-k)\times}), \\ &\qquad k \in \mathbb{N},\ (\alpha_1, \dots, \alpha_{\sigma(k)}) \in \Xi_{\sigma(k)},\ \alpha_k > 0. \end{aligned}$$

Lemma 5.2.2. *Under the above assumptions, we have $f \in \mathcal{C}(\mathbb{I})$.*

Proof. Step 1°. Right continuity of f. Fix an $x = x(\alpha) \in [0,1)$ with the Q-representation chosen such that $\sup\{n \in \mathbb{N} : \alpha_n \le N-2\} = +\infty$. For an arbitrary $\varepsilon > 0$, let $p \in \mathbb{N}$ be such that $\sum_{n=p+1}^{\infty} M_n \le \frac{\varepsilon}{2}$. Let $k \ge \sigma(p)$ be such that $\alpha_{k+1} \le N-2$. Take an arbitrary $x' = x(\beta) \in (x, x(\alpha) + d_k(\alpha))$. Then $\alpha_n = \beta_n$, $n = 1, \dots, k$.

Indeed, suppose that $\alpha_1 < \beta_1$. Then $x(\beta) < x_k(\alpha) + d_k(\alpha) \le \varphi_{\alpha_1} + \delta_{\alpha_1} \le \varphi_{\beta_1} \le x(\beta)$; a contradiction. If $\beta_1 < \alpha_1$, then $x(\beta) \le \varphi_{\beta_1} + \delta_{\beta_1} \le \varphi_{\alpha_1} \le x(\alpha)$; a contradiction. Thus $\alpha_1 = \beta_1$.

Consequently, we get $x(\widehat{\alpha}) < x(\widehat{\beta}) < x_{k-1}(\widehat{\alpha}) + d_{k-1}(\widehat{\alpha})$, where $\widehat{\alpha} := (\alpha_2, \alpha_3, \dots)$. A finite induction finishes the proof.

Finally,

$$|f(x') - f(x)| \leq \sum_{n=p+1}^{\infty} |g_n(\beta_1, \dots, \beta_{\sigma(n)}) - g_n(\alpha_1, \dots, \alpha_{\sigma(n)})| \leq 2 \sum_{n=p+1}^{\infty} M_n \leq \varepsilon.$$

Step 2°. Left continuity of f—EXERCISE. □

5.2.2 Bolzano-Type Functions Defined via Q-Representation

Let $N \in \mathbb{N}_2$, φ_i, Φ_i, $i = 0, \dots, N$, be as in § 5.1. Define

$$\delta_i := \varphi_{i+1} - \varphi_i, \quad \Delta_i := \Phi_{i+1} - \Phi_i, \quad i = 0, \dots, N-1.$$

Assume that $M_\Delta := \max\{|\Delta_0|, \dots, |\Delta_{N-1}|\} < 1$.

We are going to define the Bolzano-type function f from § 5.1 via Q-representation (with $Q := (\delta_0, \dots, \delta_{N-1})$) in the sense of § 5.2.1. The definition from § 5.1 immediately implies that

$$f(x(\alpha)) = f\Big(\sum_{n=1}^{\infty} \varphi_{\alpha_n} d_{n-1}(\alpha)\Big) = \Phi_{\alpha_1} + \sum_{n=2}^{\infty} \Phi_{\alpha_n} \Delta_{\alpha_1} \cdots \Delta_{\alpha_{n-1}},$$

$$\alpha = (\alpha_n)_{n=1}^{\infty} \subset \{0, \dots, N-1\};$$

notice that since $M_\Delta < 1$, the series is convergent. On the other hand, we have

$$f(x(\alpha)) = \sum_{n=1}^{\infty} g_n(\alpha_1, \dots, \alpha_n),$$

where

$$g_1(\alpha_1) := \Phi_{\alpha_1}, \quad g_n(\alpha_1, \dots, \alpha_n) := \Phi_{\alpha_n} \Delta_{\alpha_1} \cdots \Delta_{\alpha_{n-1}}, \quad n \in \mathbb{N}_2.$$

Thus we are in the situation of § 5.2.1 with $\sigma(n) := n$ and $M_n \leq M_\Delta^{n-1}$.

Notice that if $0 = \Phi_0 < \Phi_1 < \dots < \Phi_{N-1} < \Phi_N = 1$, then the above series

$$\Phi_{\alpha_1} + \sum_{n=2}^{\infty} \Phi_{\alpha_n} \Delta_{\alpha_1} \cdots \Delta_{\alpha_{n-1}}$$

may be considered a $(\Delta_0, \dots, \Delta_{N-1})$-representation of $f(x(\alpha))$.

Example 5.2.3. Let us illustrate the above procedure with the example of the classical Bolzano function $\boldsymbol{B}$. We have

$$N = 4, \quad \varphi_1 = \frac{3}{8}, \; \varphi_2 = \frac{1}{2}, \; \varphi_3 = \frac{7}{8}, \quad \Phi_1 = \frac{5}{8}, \; \Phi_2 = \frac{1}{2}, \; \Phi_3 = \frac{9}{8}.$$

We get

$$\delta_0 = \frac{3}{8},\ \delta_1 = \frac{1}{8},\ \delta_2 = \frac{3}{8},\ \delta_3 = \frac{1}{8},\quad \Delta_0 = \frac{5}{8},\ \Delta_1 = -\frac{1}{8},\ \Delta_2 = \frac{5}{8},\ \Delta_3 = -\frac{1}{8}.$$

Recall that $M_\Delta = \frac{5}{8} < 1$. Let $\frac{k_i}{8} := \varphi_i$, $\frac{K_i}{8} := \Phi_i$, $i = 0,1,2,3$. Then for $\alpha = (\alpha_n)_{n=1}^\infty \subset \{0,1,2,3\}$, the representation of the point $x = x(\alpha) \in \mathbb{I}$ may be written in the form

$$x = x(\alpha) = \varphi_{\alpha_1} + \sum_{n=2}^\infty \varphi_{\alpha_n}\delta_{\alpha_1}\cdots\delta_{\alpha_{n-1}} = \sum_{n=1}^\infty \frac{1}{8^n}k_{\alpha_n}3^{\beta_n(\alpha)},$$

where $\beta_1(\alpha) := 0$, $\beta_n(\alpha) := \#\{i \in \{1,\dots,n-1\} : \alpha_i \in \{0,2\}\}$, $n \geq 2$. Consequently, the value $\boldsymbol{B}(x)$ may be written in the form

$$\begin{aligned}\boldsymbol{B}(x) = \boldsymbol{B}(x(\alpha)) = \Phi_{\alpha_1} + \sum_{n=2}^\infty \Phi_{\alpha_n}\Delta_{\alpha_1}\cdots\Delta_{\alpha_{n-1}}\\ = \sum_{n=1}^\infty \frac{1}{8^n}K_{\alpha_n}5^{\beta_n(\alpha)}(-1)^{n-1-\beta_n(\alpha)}.\end{aligned}$$

The above formula may easily be used to show that $\boldsymbol{B} \in \mathfrak{ND}(\mathbb{I})$ (cf. Theorem 5.1.3). Indeed, let $\mu : \{0,1,2,3\} \longrightarrow \{0,1,2,3\}$ be given as $\mu(0) := 2$, $\mu(1) := 3$, $\mu(2) := 0$, $\mu(3) := 1$. Observe that $\delta_{\mu(s)} = \delta_s$, $\Delta_{\mu(s)} = \Delta_s$, and $\frac{K_{\mu(s)}-K_s}{k_{\mu(s)}-k_s} = 1$, $s = 0,1,2,3$.

Take a point $x = x(\alpha) \in \mathbb{I}$. For $m \in \mathbb{N}_2$, define $\alpha^m = (\alpha_n^m)_{n=1}^\infty$:

$$\alpha_n^m = \alpha_n, \quad n \neq m, \qquad \alpha_m^m := \mu(\alpha_m).$$

Obviously, $\beta_n(\alpha^m) = \beta_n(\alpha)$, $n \in \mathbb{N}$. Put $x^m := x(\alpha^m)$. Then

$$x^m - x = \frac{1}{8^m}(k_{\mu(\alpha_m)} - k_{\alpha_m})3^{\beta_m(\alpha)}.$$

In particular, $x^m \longrightarrow x$ when $m \longrightarrow +\infty$. Moreover,

$$\boldsymbol{B}(x^m) - \boldsymbol{B}(x) = \frac{1}{8^n}(K_{\mu(\alpha_m)} - K_{\alpha_m})5^{\beta_m(\alpha)}(-1)^{m-1-\beta_m(\alpha)}.$$

Thus

$$\begin{aligned}\boldsymbol{\Delta B}(x,x^m) = \frac{(K_{\mu(\alpha_m)} - K_{\alpha_m})5^{\beta_m(\alpha)}(-1)^{m-1-\beta_m(\alpha)}}{(k_{\mu(\alpha_m)} - k_{\alpha_m})3^{\beta_m(\alpha)}}\\ = \left(\frac{5}{3}\right)^{\beta_m(\alpha)}(-1)^{m-1-\beta_m(\alpha)}.\end{aligned}$$

There are the following two possibilities:

- $\beta_m(\alpha) \nearrow +\infty$. Then $|\boldsymbol{\Delta B}(x,x^m)| \longrightarrow +\infty$, and consequently, a finite derivative $\boldsymbol{B}'(x)$ does not exist.
- The sequence $(\beta_m(\alpha))_{m=1}^\infty$ is bounded, i.e., $\beta_m(\alpha) = \beta_0 = \text{const}$ for $m \geq m_0$. Then

$$\boldsymbol{\Delta B}(x,x^m) = \left(\frac{5}{3}\right)^{\beta_0}(-1)^{m-1-\beta_0} =: (-1)^m c_0, \quad m \geq m_0$$

($c_0 \neq 0$), which easily implies that a finite or infinite derivative $\boldsymbol{B}'(x)$ does not exist.

Notice that the above example may be generalized to other Bolzano-type functions (cf., e.g., [Sin28]).

5.3 Examples of Bolzano-Type Functions

5.3.1 The Hahn Function

In 1899, E. Steinitz (cf. [Ste99]) considered the case $\varphi_j = \frac{j}{N}$, $j = 1, \dots, N$, $M_\Delta < 1$, and $\Sigma = \{1, \dots, N\}$ (i.e., $|\Phi_i - \Phi_{i-1}| > \frac{1}{N}$, $i = 1, \dots, N$). Based on this construction, H. Hahn studied the case $N = 6$, $\Phi_1 = 1/2, \Phi_2 = 0, \Phi_3 = 1/2, \Phi_4 = 1, \Phi_5 = 1/2$ (see also [Hah17]; cf. Fig. 5.3). In virtue of Theorem 5.1.2 it follows that $f \in \mathfrak{ND}(\mathbb{I})$, where f is the Bolzano type function associated to the above data.

In his book [Pas14], M. Pasch asked the following question: is there a function $g \in \mathfrak{ND}(\mathbb{I})$ such that for every $x \in (0,1)$, the finite limit $\lim_{h\to 0} |\boldsymbol{\Delta} g(x, x+h)|$ exists? A first positive answer was found by W. Sierpiński (see [Sie14b]), but his construction was very sophisticated. H. Hahn used his example from above to give a short proof.

Theorem 5.3.1. *If f denotes the Hahn function, then $f \in \mathfrak{ND}(\mathbb{I})$, and for every $x \in (0,1)$, the finite limit $\lim_{h\to 0} |\boldsymbol{\Delta} g(x, x+h)|$ exists.*

Proof. Fix an $x \in (0,1)$ and write $x = \sum_{j=1}^{\infty} \frac{c_j}{6^j}$, where infinitely many of the c_j are different from 5. Fix an $n \in \mathbb{N}$ and put $x_n := \sum_{j=1}^{n} \frac{c_j}{6^j}$ and $x_n' := x_n + \frac{1}{6^n}$. Then $x_n \le x \le x_n'$. We may assume that $f(x_n) \neq f(x_n')$, and we discuss only the case $f(x_n) < f(x_n')$ (the inverse case is done is the same way). Then by construction of f, we have

$$f(x_n) = f(x_n + \tfrac{2}{6^{n+1}}), \quad f(x_n') = f(x_n + \tfrac{4}{6^{n+1}}),$$
$$f(x_n + \tfrac{1}{6^{n+1}}) = f(x_n + \tfrac{3}{6^{n+1}}) = f(x_n + \tfrac{5}{6^{n+1}}) = \tfrac{1}{2}(f(x_n) + f(x_n')).$$

Moreover,

$$f(x_n) \le f(x) \le \tfrac{1}{2}(f(x_n) + f(x_n')), \quad x_n \le x \le x_n + \tfrac{3}{6^{n+1}},$$
$$\tfrac{1}{2}(f(x_n) + f(x_n')) \le f(x) \le f(x_n'), \quad x_n + \tfrac{3}{6^{n+1}} \le x \le x_n'.$$

Hence, if $y \in (f(x_n), f(x_n'))$, then $f^{-1}(y)$ has at least two elements in $[x_n, x_n + \frac{3}{6^{n+1}}]$ or in $[x_n + \frac{3}{6^{n+1}}, x_n']$. In particular, there exists a point $\xi_n \in [x_n, x_n']$ with $f(\xi_n) = f(x)$ and $0 < |\xi_n - x| < \frac{1}{6^n}$. □

Corollary 5.3.2. *Let f be as above. Then there is no $x \in (0,1)$ such that both one-sided derivatives $f_+'(x)$ and $f_-'(x)$ exist and satisfy $|f_+'(x)| = |f_-'(x)| = +\infty$.*

Remark 5.3.3. Note that for the classical Bolzano function $\boldsymbol{B}$, there are points $x \in (0,1)$ such that $\boldsymbol{B}_-'(x) = -\boldsymbol{B}_+'(x) = +\infty$ (cf. Theorem 10.1.1).

5.3.2 The Kiesswetter Function

In 1966, K. Kiesswetter introduced the following function (see [Kie66]), which belongs to the class $\mathfrak{ND}(\mathbb{I})$. Let $M := \{0, 1, 2, 3\}$ and let $X : M \longrightarrow \mathbb{R}$ be given by

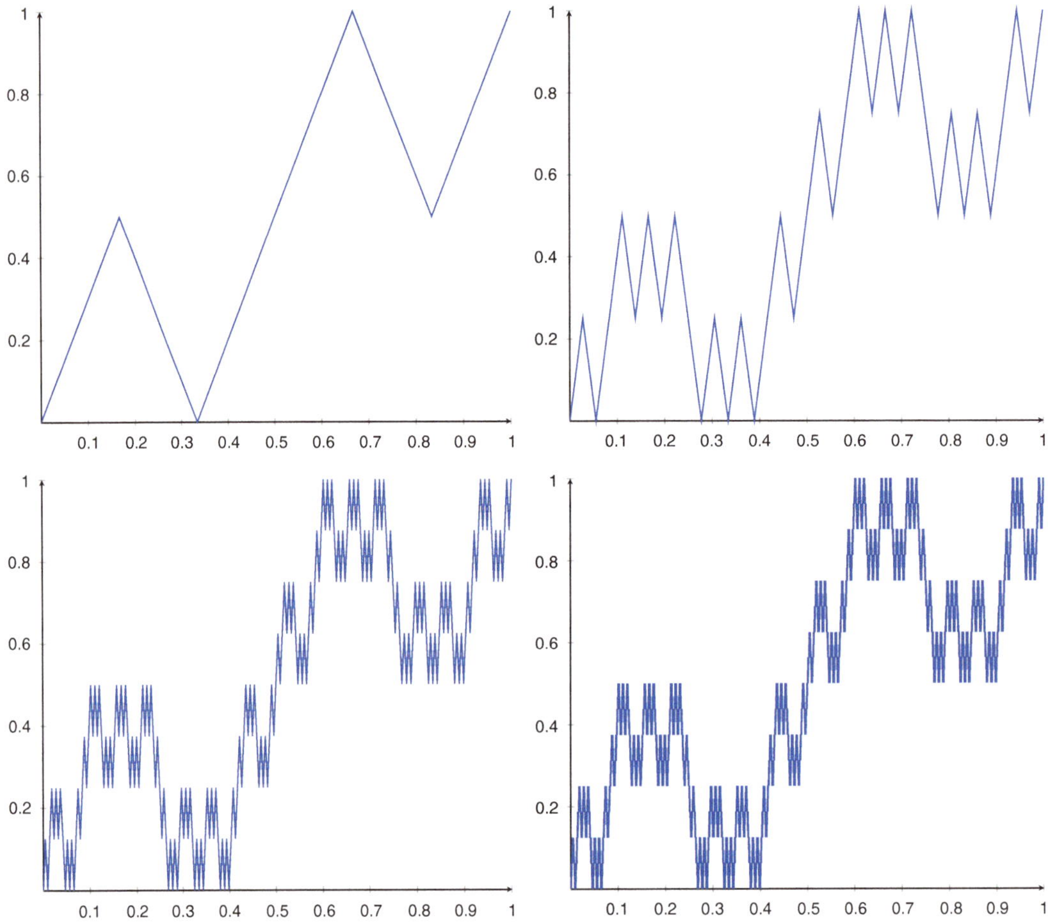

Fig. 5.3 The first four steps of the construction of the Hahn function

$$X(j) := \begin{cases} j-2, & \text{if } j \neq 0 \\ 0, & \text{if } j = 0 \end{cases}.$$

If $x \in \mathbb{I}$ is given as $x = \sum_{j\in\mathbb{N}} \frac{x_j}{4^j}$ with $x_j \in M$, then put

$$\boldsymbol{K}(x) := \sum_{n=1}^{\infty} (-1)^{s_n} \frac{X(x_n)}{2^n},$$

where $s_n := \#\{k \in \mathbb{N} : k < n,\ x_k = 0\}$. Note that $\boldsymbol{K}$ is well defined (Exercise), i.e., is independent of the representation of x.

Looking at its graph, we see that $\boldsymbol{K}$ is a Bolzano-type function with the following data: $N = 4$, $\varphi_j = j/4$, $\Phi_1 = -1/2$, $\Phi_2 = 0$, and $\Phi_3 = 1/2$ (cf. Fig. 5.4). Then $M_\Delta < 1$ and $\Sigma = \{1, 2, 3, 4\}$. Hence applying Theorem 5.1.2, we get

Proposition 5.3.4. *The Kiesswetter function belongs to* $\mathfrak{ND}(\mathbb{I})$.

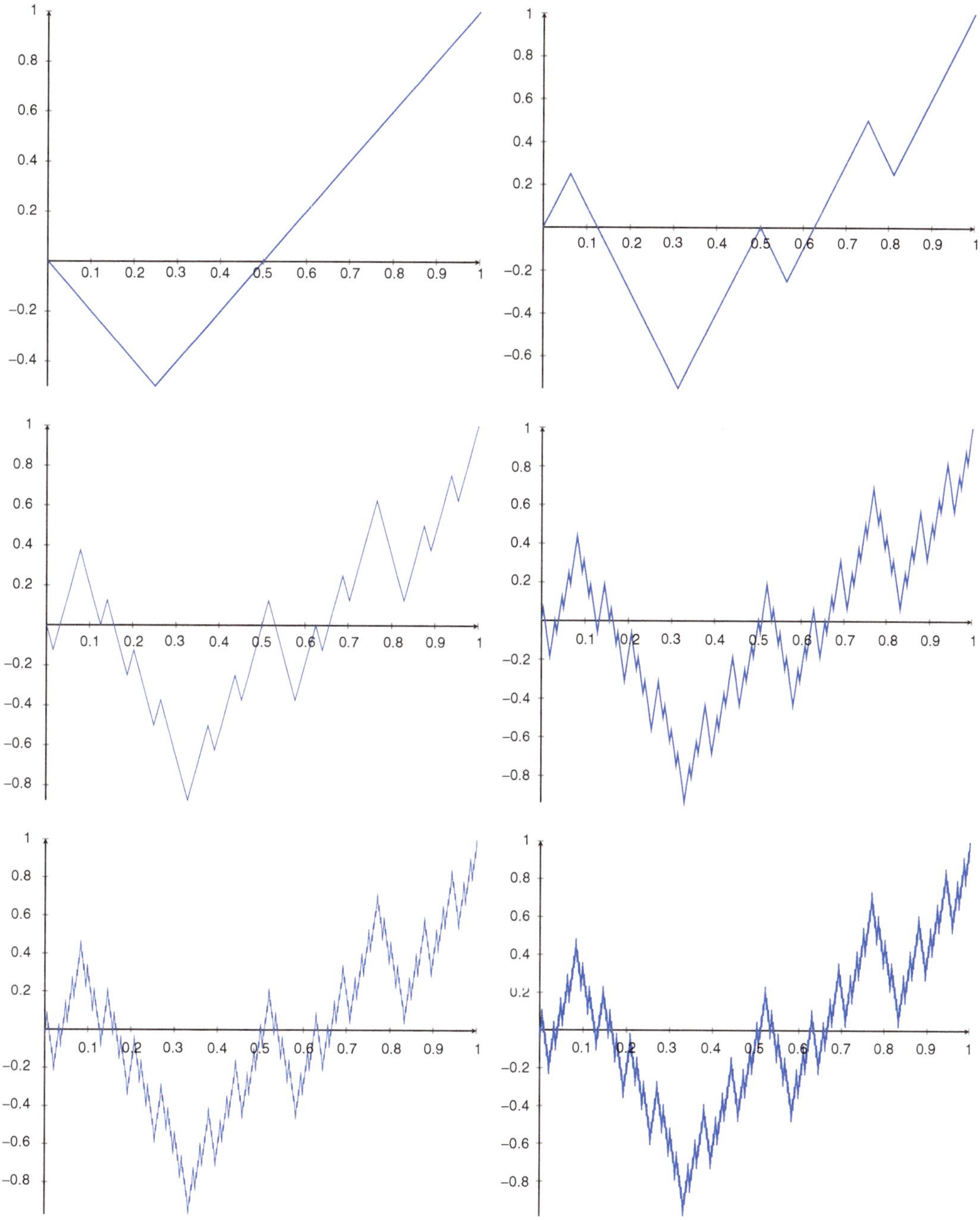

Fig. 5.4 The first six steps of the construction of the Kiesswetter function $\boldsymbol{K}$

Exercise 5.3.5. Prove the proposition using the arithmetic definition of $\boldsymbol{K}$ (this is the proof usually presented in the literature).

Remark 5.3.6. Note that $\boldsymbol{K}$ is not α-Hölder continuous if $\alpha > 1/2$.
Indeed, take $x_k = 1/4^k$ and $y_k = 2/4^k$, $k \in \mathbb{N}_2$. Then

$$\boldsymbol{K}(x_k) - \boldsymbol{K}(y_k) = 1/2^k = 2^{(2\alpha-1)k}|x_k - y_k|^\alpha,$$

where the constant $2^{(2\alpha-1)k}$ is unbounded if $k \longrightarrow \infty$.

One may try to generalize the Kiesswetter function in a straightforward way substituting the base 4 (resp. 2) by a base $a \in \mathbb{N}_2$ (resp. $b \in \mathbb{N}_2$) and defining

$$\mathbb{I} \ni x = \sum_{j=1}^{\infty} \frac{x_j}{a^j} \overset{f}{\longmapsto} \sum_{j=1}^{\infty} (-1)^{s_j(x)} \frac{X(x_j)}{b^j},$$

where $s_1(x) = 0$, $s_j(x) := \#\{k \in \{1, \ldots, j-1\} : x_k = 0\}$, $j \in \mathbb{N}_2$, and $X : M_a \longrightarrow \mathbb{R}$ is a bounded function ($M_a := \{0, 1 \ldots, a-1\}$). Note that because of the nonuniqueness of the a-adic representation, the first thing to do is to prove under what conditions on X the function f is well defined.

Lemma 5.3.7. *Let a, b, and f be as above. Then the following statements are equivalent:*

(i) *f is well defined;*
(ii) *X fulfills the following conditions:*

$$X(1) = -\frac{X(a-1)}{b-1} + \frac{bX(0)}{b+1},$$
$$X(j) = X(j-1) + \frac{X(a-1)}{b-1} - \frac{X(0)}{b+1}, \quad j = 2, \ldots, a-1. \tag{5.3.1}$$

Proof. Recall that the only case in which the a-adic representation is not unique is that in which x is a-rational, i.e., in which x has the following two representations:

$$x = \sum_{j=1}^{n} \frac{x_j}{a^j} = \sum_{j=1}^{n-1} \frac{x_j}{a^j} + \frac{x_n - 1}{a^n} + \sum_{j=n+1}^{\infty} \frac{a-1}{a^j},$$

where $n \in \mathbb{N}$ and $x_n \geq 1$. The fact that f is well defined is equivalent to

$$\sum_{j=1}^{n-1} \frac{(-1)^{s_j} X(x_j)}{b^j} + \frac{(-1)^{s_n} X(x_n)}{b^n} + \sum_{j=n+1}^{\infty} \frac{(-1)^{s_j} X(x_j)}{b^j}$$
$$= \sum_{j=1}^{n-1} \frac{(-1)^{s'_j} X(x_j)}{b^j} + \frac{(-1)^{s'_n} X(x_n - 1)}{b^n} + \sum_{j=n+1}^{\infty} \frac{(-1)^{s'_j} X(a-1)}{b^j},$$

where the s_j's (resp. s'_j's) correspond to the first (resp. second) representation of x, $j \in \mathbb{N}$. Note that $s_j = s'_j$ if $j \leq n$.

Case 1°. Let $x_n = 1$. Then $s_j = s_n + j - n - 1$ and $s'_j = s_n + 1$, $j > n$. Therefore, the fact that $f(x)$ is well defined is equivalent to

$$\frac{(-1)^{s_n}X(1)}{b^n} + X(0)\sum_{j=n+1}^{\infty}\frac{(-1)^{s_n+j-n-1}}{b^j}$$
$$= \frac{(-1)^{s_n}X(0)}{b^n} + X(a-1)\sum_{j=n+1}^{\infty}\frac{(-1)^{s_n+1}}{b^j}.$$

Exploiting the sums in the last equality leads to $X(1) = -\frac{X(a-1)}{b-1} + \frac{bX(0)}{b+1}$.

Case 2°. Let $x_n > 1$. Then $s_j = s_n + j - n - 1$ and $s'_j = s_n$, $j > n$. As before, we obtain that $f(x)$ is well defined if and only if

$$X(x_n) = X(x_n - 1) + \frac{X(a-1)}{b-1} - \frac{X(0)}{b+1}.$$

It follows that for all a-rational x with $x_n > 1$, the value $f(x)$ is well defined if and only if all the second equations in (5.3.1) are satisfied. □

Corollary 5.3.8. *If the two equations in* (5.3.1) *are satisfied, then*

$$\frac{b-a+2}{b-1}X(a-1) = \frac{b-a+2}{b+1}X(0).$$

Since we are interested in the functions $f \in \mathcal{ND}(\mathbb{I})$, the constant functions f have to be excluded.

Lemma 5.3.9. *Let a, b, and f be as above and let X satisfy the equations* (5.3.1) *from Lemma 5.3.7, i.e., f is well defined. Then f is identically constant on $\mathbb{I}$ if and only if* $\frac{X(a-1)}{b-1} = \frac{X(0)}{b+1}$.

Proof. Step 1°. Let f be a constant function. Then $f(0) = f(1)$. Using the definition of f gives

$$\frac{X(0)}{b+1} = \sum_{j=1}^{\infty}(-1)^{j-1}\frac{X(0)}{b^j} = f(0) = f(1) = \sum_{j=1}^{\infty}\frac{X(a-1)}{b^j} = \frac{X(a-1)}{b-1}.$$

Step 2°. Assume that the equality of Lemma 5.3.9 is true. Then (5.3.1) implies that

$$X(1) = X(2) = \cdots = X(a-1) = \frac{b-1}{b+1}X(0).$$

Take an $x \in \mathbb{I}$ with $x = \sum_{j=1}^{\infty}\frac{x_j}{a^j}$. Put $c_j := \frac{(-1)^{s_j}X(x_j)}{b^j}$. Then $c_j = \frac{X(0)}{b+1}\left(\frac{(-1)^{s_j}}{b^{j-1}} - \frac{(-1)^{s_{j+1}}}{b^j}\right)$. Finally, the definition of f leads to

$$f(x) = \lim_{n\to\infty}\sum_{j=1}^{n} c_j = \lim_{n\to\infty}\frac{X(0)}{b+1}\left(1 - \frac{(-1)^{s_{n+1}}}{b^n}\right) = \frac{X(0)}{b+1}.$$

Since x was arbitrarily chosen from $\mathbb{I}$, we see that f is identically constant on $\mathbb{I}$. □

Remark 5.3.10. Let a, b, f, and X be as in Lemma 5.3.7.

- If $b \neq a - 2$, then f is identically constant.
- If $b = a - 2$, then $a \geq 4$. Moreover, the condition for $j = a - 1$ in the second equation in (5.3.1) is an automatic consequence of the equations for $j = 2, \ldots, a-2$.

Theorem 5.3.11 ([LM14]). *Let $a \geq 4$, $b = a - 2$. Assume that X satisfies all the equations from (5.3.1) together with $\frac{X(a-1)}{b-1} \neq \frac{X(0)}{b+1}$. If f is the associated well-defined function from above, then:*

(a) $|f(x) - f(y)| \leq C|x-y|^{\frac{\log b}{\log a}}$, *in particular f is continuous;*

(b) $f \in \mathcal{ND}(\mathbb{I})$.

Proof. (a) Take two distinct points x, y in $\mathbb{I}$ with $x < y$. Then there exists an $n \in \mathbb{N}$ such that $\frac{1}{a^{n+1}} \leq y - x \leq \frac{1}{a^n}$. Let $t_j := \frac{j}{a^n}$ and put $J_j := [t_{j-1}, t_j]$, $j = 1, \dots, a^n$, where $t_0 = 0$. Then both points x and y lie either in the same interval J_m or in adjacent intervals. Suppose first that $x, y \in J_m$ for some m, $1 \leq m \leq a^n$. Then $x = t_{m-1} + \sum_{j=n+1}^{\infty} \frac{x_j}{a^j}$ and $y = t_{m-1} + \sum_{j=n+1}^{\infty} \frac{y_j}{a^j}$. Hence

$$
\begin{aligned}
|f(x) - f(y)| &= \Big| \sum_{j=n+1}^{\infty} \frac{(-1)^{s_j} X(x_j)}{b^j} - \sum_{j=n+1}^{\infty} \frac{(-1)^{s'_j} X(y_j)}{b^j} \Big| \\
&\leq \frac{2M}{b^n(b-1)} \leq \frac{2Mb}{b-1} |x-y|^{\frac{\log b}{\log a}},
\end{aligned}
$$

where M denotes the maximum of the function $|X|$.
The remaining case is a simple consequence of the triangle inequality and the former reasoning.

(b) Fix a non-a-rational point $x = \sum_{j=1}^{\infty} \frac{x_j}{a^j} \in \mathbb{I}$. Define

$$
a_n := \sum_{j=1}^{n} \frac{x_j}{a^j}, \quad b_n := \sum_{j=1}^{n} \frac{x_j}{a^j} + \sum_{j=n+1}^{\infty} \frac{a-1}{a^j}.
$$

Note that $a_n < x < b_n$ and $b_n - a_n = \frac{1}{a^n}$. Therefore, $a_n \longrightarrow x$ and $b_n \longrightarrow x$. Then

$$
|f(b_n) - f(a_n)| = \frac{1}{b^n} \Big| \frac{X(a-1)}{b-1} - \frac{X(0)}{b+1} \Big| = \frac{C}{b^n}
$$

with $C > 0$. Hence, $|\boldsymbol{\Delta} f(a_n, b_n)| = C(\frac{a}{b})^n \longrightarrow \infty$. Applying Remark 2.1.2(a) gives that f is not differentiable at x. The case in which x is an a-rational point is left as an EXERCISE. □

Summarizing, the above theorem describes all Kiesswetter functions in $\mathcal{ND}(\mathbb{I})$ that are built using the bases $a, b \in \mathbb{N}_2$ and the functions $s_j : M_a \longrightarrow \mathbb{R}$ from above.

Recently, other types of functions in $\mathcal{ND}(\mathbb{I})$ similar to the Kiesswetter function have been studied in [YG04] and [Yon10]. We give only the main definitions and results; details are left to the reader.

Remark 5.3.12. (a) A Kiesswetter-type function defined via the quinary representation (see [YG04]).

Let $M := \{0, 1, 2, 3, 4\}$, $U, \alpha : M \longrightarrow \mathbb{R}$ be such that $|U(j)| \leq 2$ and $\alpha(j) \in \{0, 1\}$, $j \in M$. Then if $x \in \mathbb{I}$ is written as $x = \sum_{n=1}^{\infty} \frac{x_n}{5^n}$, then $f(x)$ is defined as

$$
f(x) := \sum_{n=1}^{\infty} (-1)^{\alpha(x_1) + \cdots \alpha(x_{n-1})} \frac{U(x_n)}{3^n}.
$$

Obviously, this series is absolutely convergent. It is easily seen that under each of the following conditions, the definition of f is independent of the quinary representation:

- $U(j) := \begin{cases} 2-j, & \text{if } j \in M \setminus \{4\} \\ 0, & \text{if } j = 4 \end{cases}, \quad \alpha(j) := \begin{cases} 1, & \text{if } j = 4 \\ 0, & \text{if } j \neq 4 \end{cases}.$
- $U(j) := \begin{cases} j-2, & \text{if } j \in M \setminus \{0\} \\ 0, & \text{if } j = 0 \end{cases}, \quad \alpha(j) := \begin{cases} 1, & \text{if } j = 0 \\ 0, & \text{if } j \neq 0 \end{cases}.$
- $U(j) := \begin{cases} 2-j, & \text{if } j \in M \setminus \{0\} \\ 0, & \text{if } j = 0 \end{cases}, \quad \alpha(j) := \begin{cases} 1, & \text{if } j = 0 \\ 0, & \text{if } j \neq 0 \end{cases}.$
- $U(j) := \begin{cases} j-2, & \text{if } j \in M \setminus \{4\} \\ 0, & \text{if } j = 4 \end{cases}, \quad \alpha(j) := \begin{cases} 1, & \text{if } j = 4 \\ 0, & \text{if } j \neq 4 \end{cases}.$

Let f be the function defined via a pair U, α from above. Then $f \in \mathbf{ND}(\mathbb{I})$, and if $x, y \in \mathbb{I}$, $x \leq y$, then

$$\frac{1}{c}|x-y|^{\frac{\log 3}{\log 5}} \leq \max_{x \leq \xi \leq \eta \leq y} |f(\xi) - f(\eta)| \leq c|x-y|^{\frac{\log 3}{\log 5}},$$

where $c \geq 1$.

(b) Another Kiesswetter-type function can be defined via the septenary representation of x (see [Yon10]).

Let $M := \{0, 1, \ldots, 6\}$, $U, \alpha : M \longrightarrow \mathbb{R}$ be functions satisfying $\alpha(M) \subset \{0, 1\}$ and $|U| \leq 2$. If $\mathbb{I} \ni x = \sum_{n=1}^{\infty} \frac{x_n}{7^n}$, put

$$f(x) := \sum_{n=1}^{\infty} (-1)^{\alpha(x_1)+\cdots+\alpha(x_{n-1})} \frac{U(x_n)}{3^n}.$$

If the functions U and α satisfy certain conditions (we skip details here), then f is well defined and a continuous function on $\mathbb{I}$. Here we give only one example, namely

$$U(j) := \begin{cases} 2-j, & \text{if } j \in M \setminus \{5, 6\} \\ j-6, & \text{if } j = 5, 6 \end{cases}, \quad \alpha(j) := \begin{cases} 1, & \text{if } j = 5 \\ 0, & \text{if } j \neq 5 \end{cases}.$$

In this case, we have $f \in \mathbf{ND}(\mathbb{I})$; moreover, if $0 \leq x < y \leq 1$, then

$$\frac{1}{c}|x-y|^{\frac{\log 3}{\log 7}} \leq \max_{x \leq \xi \leq \eta \leq y} |f(\xi) - f(\eta)| \leq c|x-y|^{\frac{\log 3}{\log 7}}$$

for a suitable constant $c > 1$.

5.3.3 The Okamoto Function

Fix an $\alpha \in (0, 1)$ and define a Bolzano-type function F_α with respect to the data $N = 3$, $\varphi_j := j/3$, and $\Phi_1 := \alpha$, $\Phi_2 := 1 - \alpha$; cf. Fig. 5.5. The function F_α is studied by H. Okamoto in [Oka05] (see also [Kob09, OW07]). We call it the *Okamoto function.*

Remark 5.3.13. Note that:

- $F_{2/3}$ is the function studied in N. Bourbaki (see [Bou04], page 35, Problem 1-2) and in [Kat91];
- $F_{5/6}$ is studied by F.W. Perkins in [Per27];
- $F_{1/3} = \mathrm{id}|_{\mathbb{I}}$ (in particular, it is everywhere differentiable).

Observe that $M_\Delta = \max\{\alpha, |1-2\alpha|\} < 1$. Thus, F_α is a continuous function on $\mathbb{I}$. Moreover, we have:

- if $\alpha \in (\frac{2}{3}, 1)$, then $M_\Delta = \alpha$ and $\Sigma = \{1,2,3\}$;
- if $\alpha = \frac{2}{3}$, then $M_\Delta = \frac{2}{3}$ and $\Sigma = \{1,3\}$.

Note that condition (5.1.1) is also satisfied. Therefore, applying Theorem 5.1.2, we have the following result.

Proposition 5.3.14. *If $2/3 \leq \alpha < 1$, then $F_\alpha \in \mathfrak{N}\mathfrak{D}(\mathbb{I})$.*

Moreover, we have the following behavior of F_α (Fig. 5.6).

Lemma 5.3.15. *If $\alpha \in (0, 1/2]$, then F_α is nondecreasing; in particular, it is almost everywhere differentiable.*

Proof. The proof of being nondecreasing is left as an EXERCISE (use induction). □

Let $p(t) := 54t^3 - 27t^2 - 1$ and denote by α_0 the uniquely defined real zero of this polynomial. Note that $\alpha_0 \approx 0.5592 < 2/3$. Then we have the following result.

Proposition 5.3.16. *Let $\alpha \in [\alpha_0, 2/3)$. Then F_α has no finite derivative at almost all $x \in (0,1)$.*

Note that this result was proved in [Oka05] if $\alpha \in (\alpha_0, 2/3)$. The case $\alpha = \alpha_0$ is due to K. Kobayashi (see [Kob09]). The proof is based on § 5.2.2 and on a refinement of the Borel result on normal numbers (cf. Definition A.9.1), the law of iterated logarithms. Let $x \in (0,1)$ with the following triadic representation:

$$x = \sum_{n=1}^{\infty} \frac{\xi_n(x)}{3^n}, \quad \text{where } \xi_n(x) \in \{0,1,2\}.$$

If

$$c(t) := \begin{cases} 1, & \text{if } t \in \{0,2\} \\ -2, & \text{if } t = 1 \end{cases},$$

then put for $n \in \mathbb{N}$

$$S_n(x) := \sum_{j=1}^{n} c(\xi_j(x)).$$

Lemma 5.3.17 (Law of iterated logarithms). *For almost all $x \in (0,1)$ the following is true:*

$$\limsup_{n\to\infty} \frac{S_n(x)}{\sqrt{4n\log(\log n)}} = 1, \quad \liminf_{n\to\infty} \frac{S_n(x)}{\sqrt{4n\log(\log n)}} = -1.$$

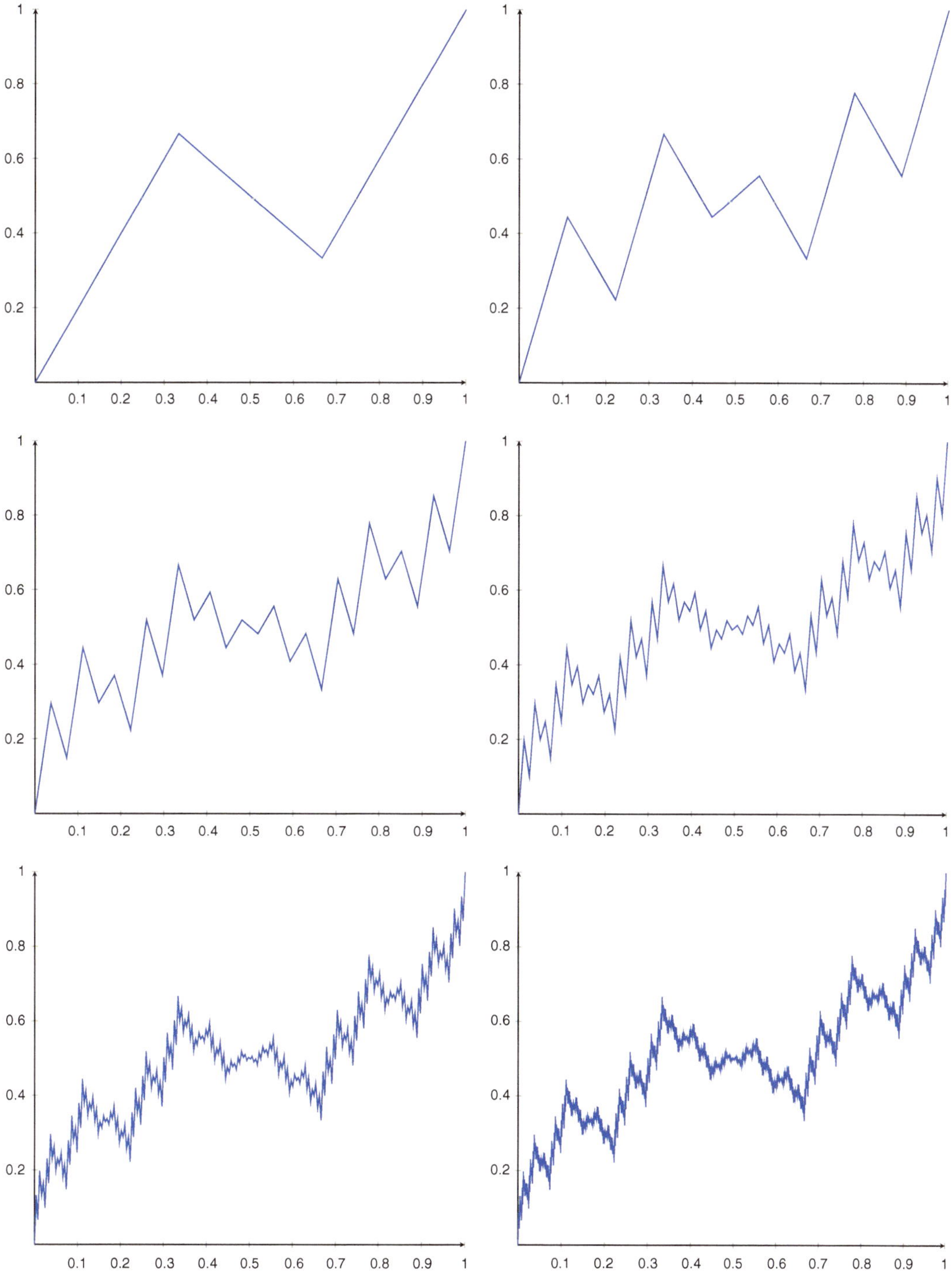

Fig. 5.5 The first six steps of the construction of the Okamoto function with $\alpha = 2/3$

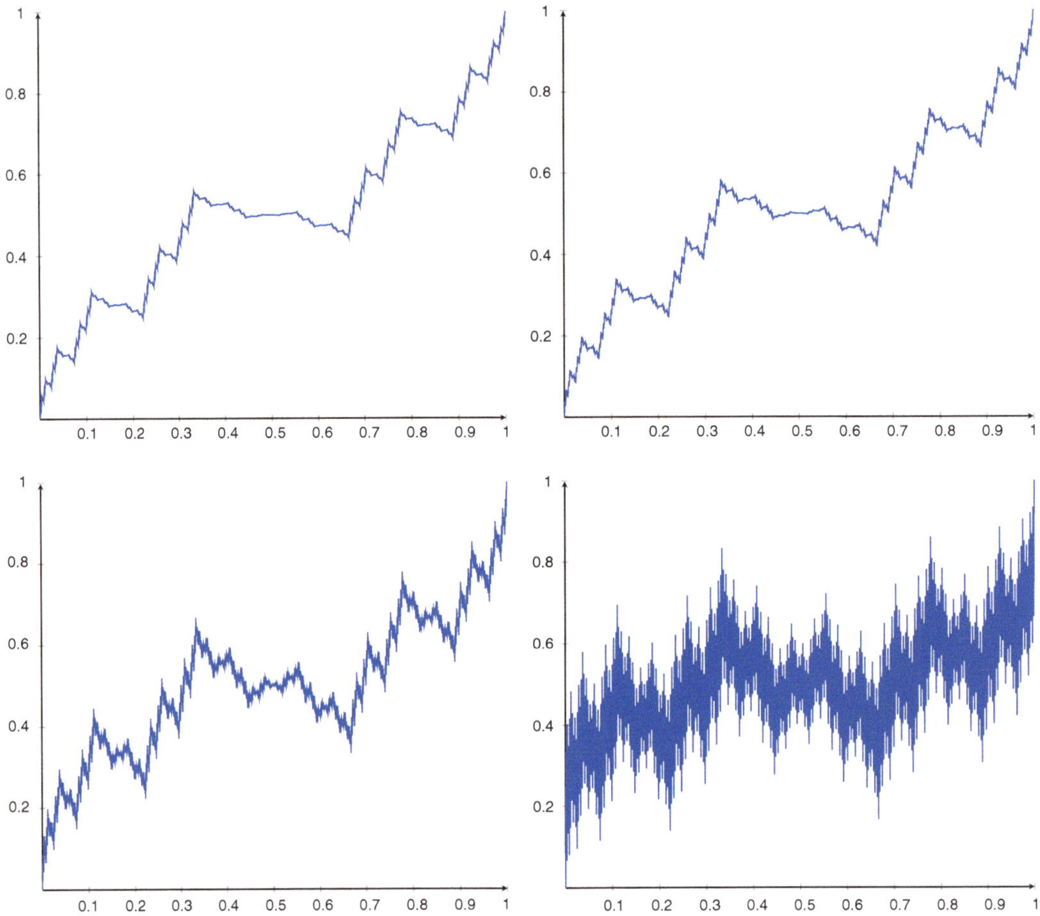

Fig. 5.6 Okamoto functions with $\alpha = \alpha_0 \approx 0.5592, 7/12, 2/3, 5/6$, respectively

Proof. Since the proof is based on probabilistic methods, it is skipped. For a proof, see, for example, [HW41].[1] □

Proof of Proposition 5.3.16. Step 1°. Fix a point x for which the former lemma holds. Then it follows that there are infinitely many n's (resp. m's) such that $\frac{S_n(x)}{\sqrt{n}} > 1$ (resp. $\frac{S_m(x)}{\sqrt{m}} < -1$). Therefore, we obtain a strictly increasing sequence $(r_n)_{n\in\mathbb{N}} \subset \mathbb{N}$ such that

$$S_{r_n}(x) \geq \sqrt{r_n} \quad \text{and} \quad \xi_{r_n+1}(x) = 1, \quad n \in \mathbb{N}.$$

Step 2°. Put

$$y_n = \sum_{j=1}^{r_n} \frac{\xi_j(x)}{3^j}.$$

Then

$$\frac{1}{3^{r_n+1}} \leq x - y_n \leq \frac{1}{3^{r_n}}.$$

[1] In fact, a weaker result is sufficient to prove Proposition 5.3.16, namely, that the set $\{x \in (0,1) :$ there are infinitely many n such that $S_n(x)/\sqrt{n} > 1$ and $\xi_{n+1}(x) = 1\}$ has full measure 1. This result can be proved also via probabilistic methods. Nevertheless, it would be interesting to have a direct proof.

Applying now Sect. 5.2.2, we get

$$F_\alpha(x) = \Phi_{x_1} + \sum_{j=2}^{\infty} \Phi_{x_j} \Delta_{x_1} \cdots \Delta_{x_{j-1}}, \quad F_\alpha(y_n) = \Phi_{x_1} + \sum_{j=2}^{r_n} \Phi_{x_j} \Delta_{x_1} \cdots \Delta_{x_{j-1}},$$

where $x_j := \xi_j(x)$,

$$\Phi_s := \begin{cases} 0, & \text{if } s = 0 \\ \alpha, & \text{if } s = 1 \\ 1-\alpha, & \text{if } s = 2 \end{cases}, \quad \Delta_s := \begin{cases} \alpha, & \text{if } s = 0 \\ 1-2\alpha, & \text{if } s = 1 \\ \alpha, & \text{if } s = 2 \end{cases}.$$

Hence,

$$\begin{aligned}
|\boldsymbol{\Delta} F_\alpha(x, y_n)| &\geq 3^{r_n} \Big| \sum_{j=r_n+1}^{\infty} \Phi_{x_j} \Delta_{x_1} \cdots \Delta_{x_{j-1}} \Big| \\
&\geq 3^{r_n} \Big(\Phi_{x_{r_n+1}} |\Delta_{x_1} \cdots \Delta_{x_{r_n}}| - \sum_{j=r_n+2}^{\infty} |\Phi_{x_1} \Delta_{x_1} \cdots \Delta_{x_{j-1}}| \Big) \\
&\overset{(*)}{\geq} 3^{r_n} |\Delta_{x_1} \cdots \Delta_{x_{r_n}}| \Big(\alpha - (2\alpha - 1) \sum_{j=1}^{\infty} \alpha^j \Big) \\
&= \frac{\alpha(2-3\alpha)}{1-\alpha} 3^{r_n} |\Delta_{x_1} \cdots \Delta_{x_{r_n}}| = \frac{\alpha(2-3\alpha)}{1-\alpha} \exp\Big(\sum_{j=1}^{r_n} \log |3\Delta_{x_j}| \Big),
\end{aligned}$$

where in $(*)$, the fact that $x_{r_n+1} = 1$ has been used.

Evaluating now the Δ's leads to

$$\begin{aligned}
&\log |3\Delta_0| = \log |3\Delta_2| = \log(3\alpha), \\
&\log |3\Delta_1| = \log(3(2\alpha - 1)) = \log \tfrac{54\alpha^3 - 27\alpha^2}{9\alpha^2} \geq -2 \log(3\alpha);
\end{aligned}$$

use that α lies on the right side of the zero of the polynomial p. Hence,

$$\begin{aligned}
|\boldsymbol{\Delta} F_\alpha(x, y_n)| \geq \frac{\alpha(2-3\alpha)}{1-\alpha} \exp\Big(\log(3\alpha) S_{r_n}(x) \Big) \\
\geq \frac{\alpha(2-3\alpha)}{1-\alpha} (3\alpha)^{\sqrt{r_n}} \underset{n\to\infty}{\longrightarrow} +\infty,
\end{aligned}$$

giving that F_α has no finite derivative at the point x. □

Remark 5.3.18.

(a) Using again § 5.2.2, it is easy to see that if $\alpha \in [1/2, \alpha_0)$, then F_α is differentiable at every point $x = (2k+1)/3^N$, $k = 0, \ldots, 3^N - 1$, with $F_\alpha(x) = 0$ (Exercise; use that in the triadic representation of x we have $\xi_j(x) = 1$, $j \in \mathbb{N}_N$). This example shows that the assumptions on Σ in Theorem 5.1.2 cannot be dropped.

(b) Even more is true, namely, if $\alpha < \alpha_0$, then F_α is differentiable almost everywhere on $\mathbb{I}$ (see [Oka05]).

(c) In [OW07], one finds the following result. Suppose that $\alpha \in (0, 1/2)$, $\alpha \neq 1/3$. Then F_α is a continuous, strictly increasing, and singular function (i.e., F_α' is zero almost everywhere).

Remark 5.3.19. Using ideas underlying the nowhere differentiability of functions of Bolzano type, F.W. Perkins proved in [Per29] the following result.

Let $f \in \mathcal{C}(\mathbb{I})$ be such that

$$\exists_{\delta>0}\ \forall_{0\le x_0<x_3\le 1: f(x_0)\neq f(x_3)}\ \exists_{x_0<x_1<x_2<x_3} : \frac{f(x_1)-f(x_2)}{f(x_3)-f(x_0)} > \delta.$$

Then there exists a continuous increasing function $\Phi : \mathbb{I} \longrightarrow \mathbb{I}$ such that $f \circ \Phi \in \mathcal{ND}(\mathbb{I})$.

5.4 Continuity of Functions Given by Arithmetic Formulas

Parallel to the Q-representation, we have another tool to define Bolzano-type functions, namely the Cantor representation of real numbers; cf. § A.1.

Fix a sequence $(\mathfrak{q}_n)_{n=1}^{\infty} \subset \mathbb{N}_2$ and put

$$\Xi_\ell := \{(a_1, \dots, a_\ell) : a_j \in \{0, \dots, \mathfrak{q}_j - 1\},\ j = 1, \dots, \ell\}, \quad \ell \in \mathbb{N}.$$

For a sequence $(\sigma_n)_{n=1}^{\infty} \subset \mathbb{N}$ with $n \le \sigma_n \le \sigma_{n+1}$, let $\varphi_n : \Xi_{\sigma(n)} \longrightarrow \mathbb{C}$, $M_n := \max_{\Xi_{\sigma(n)}} |\varphi_n|$, $n \in \mathbb{N}$. *Assume* that $\sum_{n=1}^{\infty} M_n < +\infty$. We assign to each $x = \sum_{n=1}^{\infty} \frac{a_n}{\mathfrak{q}_1\cdots\mathfrak{q}_n} \in \mathbb{I}$ (with $a_n \in \{0, \dots, \mathfrak{q}_n - 1\}$) the value $f(x) := \sum_{n=1}^{\infty} \varphi_n(a_1, \dots, a_{\sigma(n)})$, and we *assume* that $f(x)$ is independent of the different representations of x, i.e., that

$$\begin{aligned} &\varphi_k(a_1, \dots, a_{\sigma(k)}) + \sum_{n=k+1}^{\infty} \varphi_n(a_1, \dots, a_k, \underbrace{0, \dots, 0}_{(\sigma(n)-k)\times}) \\ &\quad = \varphi_k(a_1, \dots, a_{k-1}, a_k - 1, a_{k+1}, \dots, a_{\sigma(k)}) \\ &\qquad + \sum_{n=k+1}^{\infty} \varphi_n(a_1, \dots, a_{k-1}, a_k - 1, \mathfrak{q}_{k+1} - 1, \dots, \mathfrak{q}_{\sigma(n)} - 1), \\ &\qquad\qquad k \in \mathbb{N},\ (a_1, \dots, a_{\sigma(k)}) \in \Xi_{\sigma(k)},\ a_k > 0. \end{aligned}$$

Lemma 5.4.1. *Under the above assumptions, we have $f \in \mathcal{C}(\mathbb{I})$.*

Proof. Step 1°. Right continuity of f. Fix an $x = \sum_{n=1}^{\infty} \frac{a_n}{\mathfrak{q}_1\cdots\mathfrak{q}_n} \in [0, 1)$ with the representation chosen such that $\sup\{n \in \mathbb{N} : a_n \le \mathfrak{q}_n - 2\} = +\infty$. Take an $\varepsilon > 0$, and let $p \in \mathbb{N}$ be such that $\sum_{n=p+1}^{\infty} M_n \le \frac{\varepsilon}{2}$. Let $k \ge \sigma(p)$ be such that $a_{k+1} \le \mathfrak{q}_{k+1} - 2$. We will use notation from Proposition A.1.1. Put

$$x^* := S_k(x) + \sum_{n=k+1}^{\infty} \frac{\mathfrak{q}_n - 1}{\mathfrak{q}_1 \cdots \mathfrak{q}_n} = S_k(x) + \frac{1}{\mathfrak{q}_1 \cdots \mathfrak{q}_k} > x.$$

Take an arbitrary $x' = \sum_{n=1}^{\infty} \frac{a'_n}{\mathfrak{q}_1\cdots\mathfrak{q}_n} \in (x, x^*)$. Then $a'_n = a_n$ for $n = 1, \dots, k$.

Indeed, it suffices to prove that $S_k(x) = S_k(x')$. Suppose that $S_k(x') < S_k(x)$. Then $x' \le S_k(x') + \frac{1}{\mathfrak{q}_1\cdots\mathfrak{q}_k} \le S_k(x) \le x$; a contradiction. If $S_k(x') > S_k(x)$, then $x' \ge S_k(x') \ge S_k(x) + \frac{1}{\mathfrak{q}_1\cdots\mathfrak{q}_k} = x^*$; a contradiction.

Consequently,

$$|f(x') - f(x)| \le \sum_{n=p+1}^{\infty} |\varphi_n(a'_1, \dots, a'_{\sigma(n)}) - \varphi_n(a_1, \dots, a_{\sigma(n)})| \le 2 \sum_{n=p+1}^{\infty} M_n \le \varepsilon.$$

Step 2°. Left continuity of f—EXERCISE. □

5.5 Sierpiński Function

For $x = \sum_{n=1}^{\infty} \frac{a_n}{5^n} \in \mathbb{I}$, where $a_n \in \{0, \dots, 4\}$, define the *Sierpiński function* $\boldsymbol{S} : \mathbb{I} \longrightarrow \mathbb{R}$,

$$\boldsymbol{S}(x) := \sum_{n=1}^{\infty} \frac{\varepsilon_n b_n}{3^n},$$

where

$$b_n := a_n - 2\Big\lfloor \frac{a_n}{3} \Big\rfloor,$$

$$\varepsilon_1 := 1,\ \varepsilon_n := \begin{cases} 1, & \text{if } \#\{j \in \{1, \dots, n-1\} : a_j = 2\} \in 2\mathbb{N}_0 \\ -1, & \text{if } \#\{j \in \{1, \dots, n-1\} : a_j = 2\} \in 2\mathbb{N}_0 + 1 \end{cases}.$$

Remark 5.5.1. The Sierpiński function is well defined, i.e., $\boldsymbol{S}(x)$ is independent of the representation of x.

Indeed, for

$$x = \sum_{n=1}^{k} \frac{a_n}{5^n} = \sum_{n=1}^{k-1} \frac{a_n}{5^n} + \frac{a_k - 1}{5^k} + \sum_{n=k+1}^{\infty} \frac{4}{5^n} =: x'$$

with $a_k > 0$, we have

$$\boldsymbol{S}(x) = \Big(\sum_{n=1}^{k-1} \frac{\varepsilon_n b_n}{3^n} \Big) + \frac{\varepsilon_k b_k}{3^k},$$

$$\boldsymbol{S}(x') = \Big(\sum_{n=1}^{k-1} \frac{\varepsilon_n b_n}{3^n} \Big) + \frac{\varepsilon_k b'_k}{3^k} + \sum_{n=k+1}^{\infty} \frac{\varepsilon'_n b'_n}{3^n} = \Big(\sum_{n=1}^{k-1} \frac{\varepsilon_n b_n}{3^n} \Big) + \frac{\varepsilon_k b'_k}{3^k} + \frac{\varepsilon'_{k+1}}{3^k},$$

where $b'_k = a_k - 1 - 2\lfloor \frac{a_k - 1}{3} \rfloor$, $b'_{k+m} = 2$, and $\varepsilon'_{k+m} = \begin{cases} \varepsilon_k, & \text{if } a_k - 1 \neq 2 \\ -\varepsilon_k, & \text{if } a_k - 1 = 2 \end{cases}$, $m \in \mathbb{N}$. It remains to verify that $2\lfloor \frac{a_k}{3} \rfloor = 1 + 2\lfloor \frac{a_k - 1}{3} \rfloor - \varepsilon_k \varepsilon'_{k+1}$, which is easily seen by discussing the concrete cases $a_k = 1, 2, 3, 4$.

Theorem 5.5.2 (cf. [Sie14a]). $\boldsymbol{S} \in \mathcal{ND}_{\pm}(\mathbb{I})$.

Proof. Put $f := \boldsymbol{S}$.

Step 1°. Continuity of f follows from Lemma 5.4.1 and Remark 5.5.1 with $\mathfrak{q}_n := 5$, $\sigma_n := n$, $\varphi_n(a_1, \dots, a_n) := \frac{\varepsilon_n b_n}{3^n}$ $(M_n = \frac{2}{3^n})$.

Step 2°. Nowhere differentiability of f.

Let $x_0 \in (0, 1]$ be given as $x_0 = \sum_{j=1}^{\infty} \frac{a_j}{5^j}$, where $a_j \neq 0$ for infinitely many j's. Fix an $n \in \mathbb{N}_2$ with $a_n \neq 0$. Put $x_n := \sum_{j=1}^{n-1} \frac{a_j}{5^j}$. Then

$$0 \le x_0 - x_n \le 1/5^{n-1} \quad \text{and} \quad |f(x_n) - f(x_0)| = \Big| \sum_{j=n}^{\infty} \frac{\varepsilon_j b_j}{3^j} \Big|.$$

If $a_n = 1$ or $a_n = 3$, then $b_n = 1$ and $\varepsilon_n = \varepsilon_{n+1}$. Thus

$$|\boldsymbol{\Delta} f(x_0, x_n)| \ge 5^{n-1} \Big(\frac{b_n}{3^n} + \frac{b_{n+1}}{3^{n+1}} - \sum_{j=n+2}^{\infty} \frac{2}{3^j} \Big) \ge \frac{2}{9} \Big(\frac{5}{3} \Big)^{n-1}.$$

If $a_n = 2$ or $a_n = 4$, then $b_n = 2$. Therefore,

$$|\boldsymbol{\Delta} f(x_0, x_n)| \geq 5^{n-1}\Big(\frac{2}{3^n} - \sum_{j=n+1}^{\infty} \frac{2}{3^j}\Big) = \frac{1}{3}\Big(\frac{5}{3}\Big)^{n-1}.$$

Hence f has no finite left-sided derivative at the point x_0.

Let $x_0 \in [0, 1)$ be given as $x_0 = \sum_{j=1}^{\infty} \frac{a_j}{5^j}$. Assume that there are infinitely many j's with $a_j \in \{0, 1, 3\}$. Note that this assumption is fulfilled for $x_0 = 0$. Fix an n with $a_n \in \{0, 1, 3\}$ and put

$$y_n := \sum_{j=1}^{\infty} \frac{a'_j}{5^j}, \quad \text{where } a'_j := \begin{cases} 4, & \text{if } j = n \\ a_j, & \text{if } j \neq n \end{cases}.$$

Then $0 < y_n - x_0 \leq 4/5^n$ and $|f(y_n) - f(x_0)| \geq 1/3^n$. Thus $|\boldsymbol{\Delta} f(x_0, y_n)| \geq (1/4)(5/3)^n$, implying that f has no finite right-sided derivative at the point x_0. Now assume that there is a j_0 such that for all $j \geq j_0$, we have $a_j \in \{2, 4\}$, but there are infinitely many j's with $a_j < 4$. Take an $n > j_0$ with $a_n = 2$ and $\varepsilon_n = 1$ (note that there are infinitely many n's with this property). Put

$$y_n := \sum_{j=1}^{\infty} \frac{a'_j}{5^j}, \quad \text{where } a'_j := \begin{cases} 4, & \text{if } j \geq n \\ a_j, & \text{if } j < n \end{cases}.$$

Then $0 < y_n - x_0 \leq 4/5^n$ and $b_j = b'_j = 2$ for $j \geq j_0$, $-\varepsilon_{n+1} = \varepsilon_n = \varepsilon'_{n+k}$, $k \geq 1$. Thus,

$$f(y_n) - f(x_0) = 2\frac{2\varepsilon_n}{3^{n+1}} + 2\sum_{j=n+2}^{\infty} \frac{\varepsilon_n - \varepsilon_j}{3^j} \geq \frac{4}{3^{n+1}}.$$

Hence, $\boldsymbol{\Delta} f(x_0, y_n) \geq (1/3)(5/3)^n$. Therefore, f has no finite right-sided derivative at x_0. □

5.6 The Pratsiovytyi–Vasylenko Functions

Let $N \in \mathbb{N}_2$, δ_i, $i = 1, \ldots, N-1$, Q, and $N' \in \mathbb{N}_2$, δ'_i, $i = 1, \ldots, N'-1$, Q' be two systems as in § 5.2. We write $x_Q(\alpha)$ (resp. $x_{Q'}(\alpha)$) for the Q- (resp. Q'-) representation. Suppose that we are given a function that assigns to each sequence $\alpha = (\alpha_n)_{n=1}^{\infty} \subset \{0, \ldots, N-1\}$ a sequence $\beta(\alpha) = (\beta_n)_{n=1}^{\infty} \subset \{0, \ldots, N'-1\}$. Then we may define the function $f : \mathbb{I} \longrightarrow \mathbb{I}$, $f(x_Q(\alpha)) := x_{Q'}(\beta(\alpha))$. Of course, one must guarantee that f is well defined, i.e., that $f(x)$ is independent of the particular Q-representation of x.

The following function constructed in [PV13] may be thought of as a generalization of the Sierpiński function from § 5.5.

Assume that N is odd, $N \geq 5$, $N' = 3$, and let $\gamma : \{0 \ldots, N-1\} \longrightarrow \{0, 1, 2\}$,

$$\gamma(i) := \begin{cases} 0, & \text{if } i = 0 \\ 1, & \text{if } 1 \leq i \leq N-2 \\ 2, & \text{if } i = N-1 \end{cases}.$$

For $\alpha = (\alpha_n)_{n=1}^{\infty} \subset \{0, \ldots, N-1\}$, put

$$c_1(\alpha) := 0,\ c_{k+1}(\alpha) := \begin{cases} c_k, & \text{if } \alpha_k \in \{0, 1, 3, \ldots, N-2, N-1\} \\ 1 - c_k, & \text{if } \alpha_k \in \{2, 4, \ldots, N-3\} \end{cases},\ k \in \mathbb{N},$$

$$\beta(\alpha) = (\beta_n)_{n=1}^{\infty} \subset \{0, 1, 2\},$$

$$\beta_1 := \gamma(\alpha_1), \quad \beta_k := \begin{cases} \gamma(\alpha_k), & \text{if } c_k(\alpha) = 0 \\ 2 - \gamma(\alpha_k), & \text{if } c_k(\alpha) \neq 0 \end{cases},$$
$$f : \mathbb{I} \longrightarrow \mathbb{I}, \quad f(x_Q(\alpha)) := x_{Q'}(\beta(\alpha)).$$

Then (see [PV13]):

- f is well defined and continuous;
- if $\min\{\delta'_0, \delta'_2\} \geq \max\{\delta_0, \delta_{N-1}\}$, then f has no finite derivative at Q-rational points;
- if $\min\{\delta'_0, \delta'_1, \delta'_2\} \geq \max\{\delta_0, \ldots, \delta_{N-1}\}$, then f has no finite derivative at Q-irrational points;
- if $N = 5$, $\delta_0 = \cdots = \delta_4 = 1/5$, and $\delta'_0 = \delta'_1 = \delta'_2 = 1/3$, then f coincides with the Sierpiński function, and the former conditions are fulfilled.

5.7 Petr Function

For $x = \sum_{n=1}^{\infty} \frac{a_n}{10^n} \in \mathbb{I}$, where $a_n \in \{0, \ldots, 9\}$, define the *Petr function* $\boldsymbol{P} : \mathbb{I} \longrightarrow \mathbb{R}$,

$$\boldsymbol{P}(x) := \sum_{n=1}^{\infty} \frac{\varepsilon_n b_n}{2^n},$$

where

$$b_n \in \{0, 1\}, \ b_n \equiv a_n (\operatorname{mod} 2),$$
$$\varepsilon_1 := 1, \ \varepsilon_n := \begin{cases} -\varepsilon_{n-1}, & \text{if } a_{n-1} \in \{1, 3, 5, 7\} \\ \varepsilon_{n-1}, & \text{otherwise} \end{cases}, \quad n \geq 2.$$

Remark 5.7.1.

(a) The Petr function is well defined.
Indeed, for

$$x = \sum_{n=1}^{k} \frac{a_n}{10^n} = \sum_{n=1}^{k-1} \frac{a_n}{10^n} + \frac{a_k - 1}{10^k} + \sum_{n=k+1}^{\infty} \frac{9}{10^n} =: x'$$

with $a_k > 0$, we have

$$\boldsymbol{P}(x) = \Big(\sum_{n=1}^{k-1} \frac{\varepsilon_n b_n}{2^n}\Big) + \frac{\varepsilon_k b_k}{2^k},$$
$$\boldsymbol{P}(x') = \Big(\sum_{n=1}^{k-1} \frac{\varepsilon_n b_n}{2^n}\Big) + \frac{\varepsilon_k (1 - b_k)}{2^k} + \sum_{n=k+1}^{\infty} \frac{\varepsilon_k}{2^n} = \Big(\sum_{n=1}^{k-1} \frac{\varepsilon_n b_n}{2^n}\Big) + \frac{\varepsilon_k b_k}{2^k}.$$

(b) For $n \in \mathbb{N}_2$, we have $\varepsilon_n = \varphi(a_{n-1}) \cdots \varphi(a_1)$, where $\varphi : \{0, \ldots, 9\} \longrightarrow \{-1, +1\}$,

$$\varphi(p) := \begin{cases} -1, & \text{if } p \in \{1, 3, 5, 7\} \\ +1, & \text{otherwise} \end{cases}.$$

Theorem 5.7.2 (cf. [Pet20]). $\boldsymbol{P} \in \mathcal{ND}^{\infty}((0, 1))$.

Proof. Put $f := \boldsymbol{P}$.

Step 1°. Continuity of f follows from Lemma 5.4.1 and Remark 5.7.1(a) with $\mathfrak{q}_n := 10$, $\sigma_n := n$, $\varphi_n(a_1, \dots, a_n) := \frac{\varepsilon_n b_n}{2^n}$ ($M_n = \frac{1}{2^n}$).

Step 2°. Nowhere differentiability of f.

Fix an $x = \sum_{n=1}^{\infty} \frac{a_n}{10^n} \in (0, 1)$ with $\sup\{n \in \mathbb{N} : a_n \leq 8\} = +\infty$. We are going to prove that for arbitrary $\ell \in \mathbb{N}$, there exist $h, h' \in \mathbb{R}$ such that $|h|, |h'| \leq \frac{2}{10^\ell}$ and

$$|\boldsymbol{\Delta} f(x, x + h)| \geq \frac{5^\ell}{4}, \quad \boldsymbol{\Delta} f(x, x + h') = 0,$$

which immediately implies that a finite or infinite derivative $f'(x)$ does not exist.

Fix an $\ell \in \mathbb{N}$ and let $k \geq \ell$ be such that $a_k \leq 8$. Define $h := \frac{\mu_k}{10^k} + \frac{\mu_{k+1}}{10^{k+1}}$, where $\mu_k, \mu_{k+1} \in \{-1, 0, +1\}$ and the pair (μ_k, μ_{k+1}) is chosen according to the following table:

$a_k \backslash a_{k+1}$	$0, \dots, 7$	8	9
$0, \dots, 7$	$(1, 1)$	$(1, -1)$	$(0, -1)$
8	$(-1, 1)$	$(-1, -1)$	$(1, -1)$

Observe that $a_n + \mu_n \in \{0, \dots, 9\}$, $n \in \{k, k + 1\}$, and

$$\varphi(a_k)\varphi(a_{k+1}) = \varphi(a_k + \mu_k)\varphi(a_{k+1} + \mu_{k+1}).$$

Put $x' := x + h = \sum_{n=1}^{\infty} \frac{a'_n}{10^n} \in (0, 1)$. Let $f(x') = \sum_{n=1}^{\infty} \frac{\varepsilon'_n b'_n}{2^n}$. We get

$a'_n = a_n$ for $n = 1, \dots, k - 1$, $\qquad \varepsilon'_n = \varepsilon_n$ for $n = 1, \dots, k$,

$a'_n = a_n$ for $n \geq k + 2$, $\qquad \varepsilon'_n = \varepsilon_n$ for $n \geq k + 2$.

Thus, either

- $|f(x + h) - f(x)| = |\frac{1}{2^k} \pm \frac{1}{2^{k+1}}| \geq \frac{1}{2^{k+1}}$ (if $\mu_k \neq 0$) or
- $|f(x + h) - f(x)| = \frac{1}{2^{k+1}}$ (if $\mu_k = 0$).

Now we define h':

- if $a_k \leq 6$, then $h' := \frac{2}{10^k}$;
- if $a_k \in \{7, 8\}$, then $h' := -\frac{2}{10^k}$.

Then in both cases, we get $f(x + h') = f(x)$. □

Remark 5.7.3. Using analogous ideas, K. Rychlik constructed in [Ryc23] an example of a continuous nowhere differentiable function in the field of p-adic numbers.

Exercise 5.7.4. Prove the following version of the Petr theorem (cf. [Pet20]; see also [Sin35], p. 51).

For $x = \sum_{n=1}^{\infty} \frac{a_n}{10^n} \in \mathbb{I}$, *where* $a_n \in \{0, \dots, 9\}$, *define*

$$f(x) := \sum_{n=1}^{\infty} \frac{\varepsilon_n b_n}{4^n},$$

where b_n is chosen according to the table

a_n	0	1	2	3	4	5	6	7	8	9
b_n	0	1	2	1	0	1	2	1	2	3

and

$$\varepsilon_1 := 1, \ \varepsilon_n := \begin{cases} -\varepsilon_{n-1}, & \textit{if } a_{n-1} \in \{2,3,6\} \\ \varepsilon_{n-1}, & \textit{otherwise} \end{cases}, \quad n \geq 2.$$

Then $f \in \mathcal{ND}^\infty((0,1))$.

5.8 Wunderlich–Bush–Wen Function

Fix $\mathfrak{b} \in \mathbb{N}_2$, $\lambda > 1$, and $c \in \mathbb{R}$. Define $\varphi(u) := (1-\lambda)(u-c)$, $u \in \mathbb{R}$. For $x = \sum_{n=1}^{\infty} \frac{a_n}{\mathfrak{b}^n} \in \mathbb{I}$, where $a_n \in \{0, \ldots, \mathfrak{b}-1\}$, define the *Wunderlich–Bush–Wen function* $\boldsymbol{U} : \mathbb{I} \longrightarrow \mathbb{R}$,

$$\boldsymbol{U}(x) := \sum_{n=1}^{\infty} \frac{b_n}{\lambda^n},$$

where

$$b_1 := 1, \quad b_{n+1} := \begin{cases} b_n, & \text{if } a_n = a_{n+1} \\ \varphi(b_n), & \text{if } a_n \neq a_{n+1} \end{cases}, \quad n \in \mathbb{N}.$$

Remark 5.8.1.

(a) The function $\boldsymbol{U}$ is well defined.
Indeed, if

$$x = \sum_{n=1}^{k} \frac{a_n}{\mathfrak{b}^n} = \Big(\sum_{n=1}^{k-1} \frac{a_n}{\mathfrak{b}^n}\Big) + \frac{a_k - 1}{\mathfrak{b}^k} + \sum_{n=k+1}^{\infty} \frac{\mathfrak{b}-1}{\mathfrak{b}^n} =: x'$$

with $a_k > 0$, then

$$\boldsymbol{U}(x) - \sum_{n=1}^{k-1} \frac{b_n}{\lambda^n} = \frac{b_k}{\lambda^k} + \sum_{n=k+1}^{\infty} \frac{\varphi(b_k)}{\lambda^n} = \frac{b_k}{\lambda^k} + \frac{\varphi(b_k)}{\lambda^k(\lambda-1)} = \frac{c}{\lambda^k},$$

$$\boldsymbol{U}(x') - \sum_{n=1}^{k-1} \frac{b_n}{\lambda^n} = \frac{b'_k}{\lambda^k} + \sum_{n=k+1}^{\infty} \frac{\varphi(b'_k)}{\lambda^n} = \frac{b'_k}{\lambda^k} + \frac{\varphi(b'_k)}{\lambda^k(\lambda-1)} = \frac{c}{\lambda^k}.$$

Notice that the function $\varphi(u) := (1-\lambda)(u-c)$ is the most natural function for which the function $\boldsymbol{U}$ is well defined.

(b) $\boldsymbol{U}(0) = \boldsymbol{U}(1) = \frac{1}{\lambda-1}$.

(c) Define

$$s_n(x) := \begin{cases} 0, & \text{if } n = 1 \\ \#\{i \in \{2, \ldots, n\} : a_{i-1} \neq a_i\}, & \text{if } n \in \mathbb{N}_2 \end{cases}.$$

Then (EXERCISE)

$$b_n = \frac{1}{\lambda}\Big((-1)^{s_n(x)}(\lambda-1)^{s_n(x)}(\lambda - c(\lambda-1)) + c(\lambda-1)\Big), \quad n \in \mathbb{N}.$$

(d) Since $s_n(x) \le n-1$, we have

$$|b_n| \le \begin{cases} (\lambda-1)^n(1+|c|)+|c|, & \text{if } \lambda \ge 2 \\ 1+2|c|, & \text{if } 1 < \lambda \le 2 \end{cases}, \quad n \in \mathbb{N}.$$

Consequently, there exists a $C = C(\lambda, c) > 0$ such that

$$\frac{|b_n|}{\lambda^n} \le C\theta^n, \quad n \in \mathbb{N},$$

where

$$\theta = \theta(\lambda, c) := \begin{cases} 1-\frac{1}{\lambda}, & \text{if } \lambda \ge 2 \\ \frac{1}{\lambda}, & \text{if } 1 < \lambda \le 2 \end{cases}.$$

Theorem 5.8.2 (cf. [Wen00]). *If* $\mathfrak{b} \in \mathbb{N}_3$, $\lambda \in (\frac{\mathfrak{b}}{\mathfrak{b}-1}, \mathfrak{b})$, *and* $c \neq \frac{\lambda}{\lambda-1}$, *then* $\boldsymbol{U} \in \boldsymbol{\mathcal{ND}}_{\pm}(\mathbb{I})$.

Remark 5.8.3. (a) The case in which $\mathfrak{b} = 3$, $\lambda = 2$, and $c = 1$ was discussed by W. Wunderlich in [Wun52], where he proved that $\boldsymbol{U} \in \boldsymbol{\mathcal{ND}}(\mathbb{I})$ (see also [Swi61]).

(b) The case in which $\mathfrak{b} \ge 3$, $\lambda = 2$, and $c = 1$ was discussed by K.A. Bush in [Bus52], where he proved that $\boldsymbol{U} \in \boldsymbol{\mathcal{ND}}(\mathbb{I})$.

(c) Assume that $\mathfrak{b} = 3$, $\lambda = 2$, and $c = 1$. Then $\boldsymbol{U}'_+(0) = -\infty$.
Indeed, take a $k \in \mathbb{N}$ and let $1/3^{k+1} \le x < 1/3^k$. Then $x = \sum_{n=k+1}^{\infty} \frac{a_n}{3^n}$ with $a_{k+1} \neq 0$, and hence

$$\boldsymbol{U}(x) = \sum_{n=1}^{k} \frac{1}{2^n} + \sum_{n=k+2} \frac{b_n}{2^n} \le 1 - \frac{1}{2^k} + \frac{1}{2^{k+1}} = 1 - \frac{1}{2^{k+1}}.$$

Consequently,

$$\boldsymbol{\Delta U}(0,x) \le -\frac{\frac{1}{2^{k+1}}}{\frac{1}{3^k}} = -\frac{1}{2}\Big(\frac{3}{2}\Big)^k \underset{k\to+\infty}{\longrightarrow} -\infty.$$

Proof of Theorem 5.8.2. Put $f := \boldsymbol{U}$.

Step 1°. Continuity of f follows from Lemma 5.4.1 and Remark 5.8.1(a) with $\mathfrak{q}_n := \mathfrak{b}$, $\sigma_n := n$, $\varphi_n(a_1, \dots, a_n) := \frac{b_n}{\lambda^n}$ $(M_n \le C\theta^n)$.

Step 2°. Nondifferentiability of f.

We are going to prove that for every x, a finite right-sided derivative $f'_+(x)$ does not exist. The case of the left-sided derivative is left for the reader as an EXERCISE.

Fix an $x \in [0,1)$ and suppose that $f'_+(x) \in \mathbb{R}$ exists. We may assume that for every $\ell \in \mathbb{N}$, there exists a $k = k(\ell) \ge \ell$ such that $a_{k+1} \le \mathfrak{b} - 2$ and $k(\ell+1) > k(\ell)$. Put

$$A_k := \Big(\sum_{n=1}^{k} \frac{a_n}{\mathfrak{b}^n}\Big) + \frac{\mathfrak{b}-1}{\mathfrak{b}^{k+1}}, \quad B_k := A_k + \frac{\mathfrak{b}-1}{\mathfrak{b}^{k+2}}.$$

Then $x < A_k < B_k$, $B_k - x < \frac{1}{\mathfrak{b}^k}$, and $\frac{1}{\mathfrak{b}^{k+2}} < B_k - A_k = \frac{\mathfrak{b}-1}{\mathfrak{b}^{k+2}} < \frac{1}{\mathfrak{b}^{k+1}}$. Hence $B_k - A_k > \frac{1}{\mathfrak{b}^2}(B_k - x)$. Thus (by Remark 2.1.4)

$$\lim_{\ell\to+\infty} \boldsymbol{\Delta} f(A_{k(\ell)}, B_{k(\ell)}) = f'_+(x).$$

On the other hand, we have

$$f(A_k) = \Big(\sum_{n=1}^{k} \frac{b_n}{\lambda^n}\Big) + \frac{b'_{k+1}}{\lambda^{k+1}} + \sum_{n=k+2}^{\infty} \frac{\varphi(b'_{k+1})}{\lambda^n}$$
$$= \Big(\sum_{n=1}^{k} \frac{b_n}{\lambda^n}\Big) + \frac{b'_{k+1}}{\lambda^{k+1}} + \varphi(b'_{k+1})\frac{1}{(\lambda-1)\lambda^{k+1}} = \Big(\sum_{n=1}^{k} \frac{b_n}{\lambda^n}\Big) + \frac{c}{\lambda^{k+1}},$$
$$f(B_k) = \Big(\sum_{n=1}^{k} \frac{b_n}{\lambda^n}\Big) + \frac{b'_{k+1}}{\lambda^{k+1}} + \frac{b'_{k+1}}{\lambda^{k+2}} + \sum_{n=k+3}^{\infty} \frac{\varphi(b'_{k+1})}{\lambda^n}$$
$$= \Big(\sum_{n=1}^{k} \frac{b_n}{\lambda^n}\Big) + \frac{b'_{k+1}}{\lambda^{k+1}} + \frac{b'_{k+1}}{\lambda^{k+2}} + \varphi(b'_{k+1})\frac{1}{(\lambda-1)\lambda^{k+2}}$$
$$= \Big(\sum_{n=1}^{k} \frac{b_n}{\lambda^n}\Big) + \frac{b'_{k+1}}{\lambda^{k+1}} + \frac{c}{\lambda^{k+2}}.$$

Hence, using Remark 5.8.1(c), we get

$$f(B_k) - f(A_k) = \frac{b'_{k+1}}{\lambda^{k+1}} + \frac{c}{\lambda^{k+2}} - \frac{c}{\lambda^{k+1}}$$
$$= \frac{1}{\lambda^{k+2}}(-1)^{s_{k+1}(A_k)}(\lambda-1)^{s_{k+1}(A_k)}(\lambda - c(\lambda-1)).$$

Finally,

$$|\boldsymbol{\Delta} f(A_k, B_k)| \geq \frac{1}{\mathfrak{b}}\Big(\frac{\mathfrak{b}}{\lambda}\Big)^{k+2}(\lambda-1)^{s_{k+1}(A_k)}|\lambda - c(\lambda-1)|$$
$$\geq \frac{1}{\mathfrak{b}}\Big(\frac{\mathfrak{b}}{\lambda}\Big)^{k+2} \cdot \begin{Bmatrix} (\lambda-1)^{k+2}|\lambda - c(\lambda-1)|, & \text{if } \lambda \leq 2 \\ |\lambda - c(\lambda-1)|, & \text{if } \lambda \geq 2 \end{Bmatrix} \underset{k\to+\infty}{\longrightarrow} +\infty;$$

a contradiction. □

5.9 Wen Function

L. Wen in [Wen01] proposed another type of a nowhere differentiable function based on Cantor series (see also [Sin27]).

Fix a sequence $(\mathfrak{q}_n)_{n=1}^{\infty} \subset \mathbb{N}_2$. For $x = \sum_{n=1}^{\infty} \frac{a_n}{\mathfrak{q}_1 \cdots \mathfrak{q}_n} \in \mathbb{I}$, where $a_n \in \{0, \dots, \mathfrak{q}_n - 1\}$, define the *Wen function* $\boldsymbol{W}_1 : \mathbb{I} \longrightarrow \mathbb{R}$,

$$\boldsymbol{W}_1(x) := \sum_{n=1}^{\infty} \frac{b_n}{n(n+1)},$$

where $b_1 := 1$ and

$$b_{n+1} := \begin{cases} -\frac{b_n}{n}, & \text{if } (a_n > 0,\ a_{n+1} = 0) \text{ or } (a_n < \mathfrak{q}_n - 1,\ a_{n+1} = \mathfrak{q}_{n+1} - 1) \\ b_n, & \text{otherwise} \end{cases}.$$

Remark 5.9.1. (a) The Wen function is well defined.

Indeed, for

$$x = \sum_{n=1}^{k} \frac{a_n}{\mathfrak{q}_1 \cdots \mathfrak{q}_n} = \Big(\sum_{n=1}^{k-1} \frac{a_n}{\mathfrak{q}_1 \cdots \mathfrak{q}_n}\Big) + \frac{a_k - 1}{\mathfrak{q}_1 \cdots \mathfrak{q}_k} + \sum_{n=k+1}^{\infty} \frac{\mathfrak{q}_n - 1}{\mathfrak{q}_1 \cdots \mathfrak{q}_n} =: x'$$

with $a_k > 0$, we have

$$\boldsymbol{W}_1(x) - \sum_{n=1}^{k-1} \frac{b_n}{n(n+1)} = \frac{b_k}{k(k+1)} - \frac{b_k}{k} \sum_{n=k+1}^{\infty} \frac{1}{n(n+1)} = 0,$$

$$\boldsymbol{W}_1(x') - \sum_{n=1}^{k-1} \frac{b_n}{n(n+1)} = \frac{b'_k}{k(k+1)} - \frac{b'_k}{k} \sum_{n=k+1}^{\infty} \frac{1}{n(n+1)} = 0.$$

(b) $|b_n| \geq \frac{1}{(n-1)!}$, $n \in \mathbb{N}$.

Theorem 5.9.2 (cf. [Wen01]). *If* $(\mathfrak{q}_n)_{n=1}^{\infty} \subset \mathbb{N}_3$ *and* $\frac{\mathfrak{q}_1 \cdots \mathfrak{q}_n}{n!} \longrightarrow +\infty$, *then* $\boldsymbol{W}_1 \in \mathcal{ND}_{\pm}(\mathbb{I})$.

Proof. Put $f := \boldsymbol{W}_1$.

Step 1°. Continuity of f follows from Lemma 5.4.1 and Remark 5.9.1(a) with $\sigma_n := n$, $\varphi_n(a_1, \dots, a_n) := \frac{b_n}{n(n+1)}$ $(M_n \leq \frac{1}{n(n+1)})$.

Step 2°. Nowhere differentiability of f.

We are going to prove that for every x, a finite right-sided derivative $f'_+(x)$ does not exist. The case of the left-sided derivative is left for the reader as an EXERCISE.

Fix an $x \in [0, 1)$ and suppose that $f'_+(x) \in \mathbb{R}$ exists. We may assume that for every $\ell \in \mathbb{N}$, there exists a $k = k(\ell) \geq \ell$ such that $a_k < \mathfrak{q}_k - 1$. We may assume that $k(\ell + 1) > k(\ell)$. Put

$$A_k := \Big(\sum_{n=1}^{k-1} \frac{a_n}{\mathfrak{q}_1 \cdots \mathfrak{q}_n}\Big) + \frac{a_k + 1}{\mathfrak{q}_1 \cdots \mathfrak{q}_k}, \quad B_k := A_k + \sum_{n=k+1}^{\infty} \frac{\mathfrak{q}_n - 2}{\mathfrak{q}_1 \cdots \mathfrak{q}_n}.$$

Then $x < A_k < B_k$,

$$B_k - A_k < \frac{1}{\mathfrak{q}_1 \cdots \mathfrak{q}_k},$$

$$B_k - x < \frac{1}{\mathfrak{q}_1 \cdots \mathfrak{q}_k} + \sum_{n=k+1}^{\infty} \frac{\mathfrak{q}_n - 2}{\mathfrak{q}_1 \cdots \mathfrak{q}_n} < \frac{2}{\mathfrak{q}_1 \cdots \mathfrak{q}_k},$$

$$B_k - A_k = \sum_{n=k+1}^{\infty} \frac{\mathfrak{q}_n - 2}{\mathfrak{q}_1 \cdots \mathfrak{q}_n} = \frac{1}{2} \sum_{n=k+1}^{\infty} \frac{2\mathfrak{q}_n - 4}{\mathfrak{q}_1 \cdots \mathfrak{q}_n}$$
$$\geq \frac{1}{2} \sum_{n=k+1}^{\infty} \frac{\mathfrak{q}_n - 1}{\mathfrak{q}_1 \cdots \mathfrak{q}_n} = \frac{1}{2\mathfrak{q}_1 \cdots \mathfrak{q}_k}.$$

Hence $B_k - A_k > \frac{1}{4}(B_k - x)$. Thus (by Remark 2.1.4)

$$\lim_{\ell \to +\infty} \boldsymbol{\Delta} f(A_{k(\ell)}, B_{k(\ell)}) = f'_+(x).$$

On the other hand, we have

$$f(A_k) = \Big(\sum_{n=1}^{k} \frac{b_n}{n(n+1)}\Big) - \frac{b'_k}{k} \sum_{n=k+1}^{\infty} \frac{1}{n(n+1)} = \Big(\sum_{n=1}^{k} \frac{b_n}{n(n+1)}\Big) - \frac{b'_k}{k(k+1)},$$

$$f(B_k) = \Big(\sum_{n=1}^{k} \frac{b_n}{n(n+1)}\Big) + b'_k \sum_{n=k+1}^{\infty} \frac{1}{n(n+1)} = \Big(\sum_{n=1}^{k} \frac{b_n}{n(n+1)}\Big) + \frac{b'_k}{k+1}.$$

Hence, $|f(B_k) - f(A_k)| = \frac{|b'_k|}{k} \geq \frac{1}{k!}$. Finally,

$$|\boldsymbol{\Delta} f(A_k, B_k)| \geq \frac{\mathfrak{q}_1 \cdots \mathfrak{q}_k}{k!} \underset{k\to+\infty}{\longrightarrow} +\infty;$$

a contradiction. □

5.10 Singh Functions

We present three interesting examples (Theorems 5.10.2, 5.10.4, 5.10.6) of nowhere differentiable functions due to A.N. Singh [Sin30, Sin35].

Fix $p \in 2\mathbb{N}+1$, $m \in \mathbb{N}_2$, $r \in \mathbb{N}$, $r \leq m$. For $x = \sum_{n=1}^{\infty} \frac{a_n}{p^n} \in \mathbb{I}$, where $a_n \in \{0, \dots, p-1\}$, define the *Singh function* $\boldsymbol{S}_1 : \mathbb{I} \longrightarrow \mathbb{R}$,

$$\boldsymbol{S}_1(x) := \sum_{n=1}^{\infty} \frac{b_n}{p^n},$$

where

$$b_n := \begin{cases} a_{(n-1)m+r}, & \text{if } A_n \in 2\mathbb{N}_0 \\ p-1-a_{(n-1)m+r}, & \text{if } A_n \in 2\mathbb{N}_0+1 \end{cases}$$

with

$$\begin{aligned} A_n := {} & (a_1 + \cdots + a_{r-1}) + (a_{r+1} + \cdots + a_{m+r-1}) \\ & + (a_{m+r+1} + \cdots + a_{2m+r-1}) + \cdots + (a_{(n-2)m+r+1} + \cdots + a_{(n-1)m+r-1}); \end{aligned}$$

if $r = 1$, we put $b_1 := a_1$, define A_n only for $n \geq 2$, and skip the first group $(a_1 + \cdots + a_{r-1})$ in the definition of A_n.

Remark 5.10.1. The function $\boldsymbol{S}_1$ is well defined.

Let

$$x = \sum_{n=1}^{k} \frac{a_n}{p^n} \text{ with } a_k \geq 1, \quad x' := \Big(\sum_{n=1}^{k-1} \frac{a_n}{p^n}\Big) + \frac{a_k - 1}{p^k} + \sum_{n=k+1}^{\infty} \frac{p-1}{p^n}.$$

Let A'_n, b'_n be the sequences constructed for x'. There are the following two cases:

- $\exists_{s\in\mathbb{N}_0} : k = sm + r$. Then $(-1)^{A_n} = (-1)^{A'_n}$ for all $n \in \mathbb{N}$. If A_{s+1} is even, then

$$b = (b_1, \dots, b_s, a_k, 0, 0, \dots), \quad b' = (b_1, \dots, b_s, a_k - 1, p-1, p-1, \dots).$$

If A_{s+1} is odd, then

$$b = (b_1, \dots, b_s, p-1-a_k, p-1, p-1, \dots), \quad b' = (b_1, \dots, b_s, p-a_k, 0, 0, \dots).$$

It is clear that in both cases, we have $\boldsymbol{S}_1(x) = \boldsymbol{S}_1(x')$.

- $\forall_{s \in \mathbb{N}_0} : k \neq sm + r$. Write $k = sm + r + t$ with $s \in \mathbb{N}_0$ and $t \in \{1, \dots, m-1\}$. Then $(-1)^{A_n} = (-1)^{A'_n}$ for $n = 1, \dots, s+1$, and $(-1)^{A_n} = -(-1)^{A'_n}$ for all $n \geq s+2$. If A_s is even, then $b = b' = (b_1, \dots, b_{s+1}, 0, 0, \dots)$. If A_s is odd, then $b = b' = (b_1, \dots, b_{s+1}, p-1, p-1, \dots)$. So in both cases, we have $\boldsymbol{S}_1(x) = \boldsymbol{S}_1(x')$.

Theorem 5.10.2 (cf. [Sin35]). $\boldsymbol{S}_1 \in \mathcal{ND}^\infty((0,1))$.

Proof. Put $f := \boldsymbol{S}_1$.

Step 1°. Continuity of f follows from Lemma 5.4.1 and Remark 5.10.1 with $\mathfrak{q}_n := p$, $\sigma_n := (n-1)m + r$, $\varphi_n(a_1, \dots, a_{\sigma_n}) := \frac{b_n}{p^n}$ $(M_n \leq \frac{p-1}{p^n})$.

Step 2°. Nowhere differentiability of f. Fix an $x = \sum_{n=1}^\infty \frac{a_n}{p^n}$ with $\sup\{n \in \mathbb{N} : a_n \leq p-2\} = +\infty$. Consider the following cases:

- The set $S := \{s \in \mathbb{N}_0 : a_{sm+r} \leq p-2\}$ is infinite. For $s \in S$, take $x' := x + \frac{1}{p^{sm+r}}$. Then $b_n = b'_n$ for $n \neq s+1$ and

$$b'_{s+1} = \begin{cases} a_{sm+r} + 1, & \text{if } A_{s+1} \in 2\mathbb{N}_0 \\ (p-1) - (a_{sm+r} + 1), & \text{if } A_{s+1} \in 2\mathbb{N}_0 + 1 \end{cases}.$$

Hence

$$|\boldsymbol{\Delta} f(x, x')| = \frac{\frac{1}{p^{s+1}}}{\frac{1}{p^{sm+r}}} = p^{s(m-1)-1+r} \underset{S \ni s \to +\infty}{\longrightarrow} +\infty.$$

- The set $S := \{s \in \mathbb{N}_0 : a_{sm+r} \geq 1\}$ is infinite. Then for $s \in S$, we take $x' := x - \frac{1}{p^{sm+r}}$ and argue as above (Exercise).

Observe that at least one of the above two possibilities holds. Thus a finite derivative $f'(x)$ does not exist.

- There exists a $t \in \{1, \dots, m-1\}$ such that the set $S := \{s \in \mathbb{N}_0 : a_{sm+r+t} \leq p-3\}$ is infinite. For $s \in S$ take $x' := x + \frac{2}{p^{sm+r+t}}$ and then $f(x) = f(x')$.
- There exists a $t \in \{1, \dots, m-1\}$ such that the set $S := \{s \in \mathbb{N}_0 : a_{sm+r+t} \geq 2\}$ is infinite. For $s \in S$, take $x' := x - \frac{2}{p^{sm+r+t}}$ and then $f(x) = f(x')$.

Observe that if at least one of the above two possibilities holds (e.g., $p \geq 5$), then an infinite derivative $f'(x)$ does not exist.

- It remains to consider the case $p = 3$ and $a_n = 1$ for $n \gg 1$. Then we take $x' := x + \frac{1}{p^{sm+r+1}} - \frac{1}{p^{sm+r+2}}$, $s \gg 1$, and we get $f(x) = f(x')$. □

Let $(\mathfrak{q}_n)_{n=1}^\infty = (3, 5, 3, 5, \dots)$. Observe that $\mathfrak{q}_1 \cdots \mathfrak{q}_{2r} = 15^r$, $\mathfrak{q}_1 \cdots \mathfrak{q}_{2r+1} = 3 \cdot 15^r$. For $x = \sum_{n=1}^\infty \frac{a_n}{\mathfrak{q}_1 \cdots \mathfrak{q}_n}$ with $a_n \in \{0, \dots, \mathfrak{q}_n - 1\}$, define the *Singh function* $\boldsymbol{S}_2 : \mathbb{I} \longrightarrow \mathbb{R}$,

$$\boldsymbol{S}_2(x) := \sum_{n=1}^\infty \frac{b_n}{3^n},$$

where

$$b_1 := a_1, \quad b_n := \begin{cases} a_{2n+1}, & \text{if } A_n \in 2\mathbb{N}_0 \\ 2 - a_{2n+1}, & \text{if } A_n \in 2\mathbb{N}_0 + 1 \end{cases}, \quad n \in \mathbb{N}_2,$$

with $A_n := a_2 + a_4 + \cdots + a_{2n}$.

Remark 5.10.3. The function $\boldsymbol{S}_2$ is well defined.

Let

$$x = \sum_{n=1}^{k} \frac{a_n}{\mathfrak{q}_1 \cdots \mathfrak{q}_n} \text{ with } a_k \geq 1,$$

$$x' := \Big(\sum_{n=1}^{k-1} \frac{a_n}{\mathfrak{q}_1 \cdots \mathfrak{q}_n}\Big) + \frac{a_k - 1}{\mathfrak{q}_1 \cdots \mathfrak{q}_k} + \sum_{n=k+1}^{\infty} \frac{\mathfrak{q}_n - 1}{\mathfrak{q}_1 \cdots \mathfrak{q}_n}.$$

Let A'_n, b'_n be the sequences constructed for x'. There are the following two cases:

- $k = 2s + 1$ is odd. Then $(-1)^{A'_n} = (-1)^{A_n}$ for all $n \in \mathbb{N}_2$.

If A_s is even, then

$$b = (b_1, \dots, b_{s-1}, a_k, 0, 0, \dots), \quad b' = (b_1, \dots, b_{s-1}, a_k - 1, 2, 2, \dots),$$

which implies that $\boldsymbol{S}_2(x) = \boldsymbol{S}_2(x')$.

If A_s is odd, then

$$b = (b_1, \dots, b_{s-1}, 2 - a_k, 2, 2, \dots), \quad b' = (b_1, \dots, b_{s-1}, 3 - a_k, 0, 0, \dots),$$

which also implies that $\boldsymbol{S}_2(x) = \boldsymbol{S}_2(x')$.

- $k = 2s$ is even. Then $A_n = A'_n$ for $n = 1, \dots, s-1$, and $(-1)^{A_n} = -(-1)^{A'_n}$ for $n \geq s$.

If A_s is even, then $b' = b = (b_1, \dots, b_{s-1}, 0, 0, \dots)$, which gives $\boldsymbol{S}_2(x) = \boldsymbol{S}_2(x')$.

If A_s is odd, then $b' = b = (b_1, \dots, b_{s-1}, 2, 2, \dots)$. Thus once again, $\boldsymbol{S}_2(x) = \boldsymbol{S}_2(x')$.

Theorem 5.10.4 (cf. [Sin35]). $\boldsymbol{S}_2 \in \boldsymbol{\mathcal{ND}}^\infty((0,1))$.

Proof. Put $f := \boldsymbol{S}_2$.

Step 1°. Continuity of f follows from Lemma 5.4.1 and Remark 5.10.3 with $(\mathfrak{q}_n)_{n=1}^\infty = (3, 5, 3, 5, \dots)$, $\sigma_n := 2n + 1$, $\varphi_n(a_1, \dots, a_{\sigma_n}) := \frac{b_n}{3^n}$ $(M_n \leq \frac{4}{3^n})$.

Step 2°. Nowhere differentiability of f. Fix an $x = \sum_{n=1}^\infty \frac{a_n}{\mathfrak{q}_1 \cdots \mathfrak{q}_n}$ with $\sup\{n \in \mathbb{N} : a_n \leq \mathfrak{q}_n - 2\} = +\infty$. Consider the following cases:

- The set $S := \{s \in \mathbb{N} : a_{2s+1} \leq 1\}$ is infinite. For $s \in S$, take $x' := x + \frac{1}{\mathfrak{q}_1 \cdots \mathfrak{q}_{2s+1}} = x + \frac{1}{3 \cdot 15^s}$. Then $b_n = b'_n$ for $n \neq s$ and

$$b'_s = \begin{cases} a_{2s+1} + 1, & \text{if } A_s \in 2\mathbb{N}_0 \\ 1 - a_{2s+1}, & \text{if } A_s \in 2\mathbb{N}_0 + 1 \end{cases}.$$

Hence

$$|\boldsymbol{\Delta} f(x, x')| = \frac{\frac{1}{3^s}}{\frac{1}{3 \cdot 15^s}} = 3 \cdot 5^s \underset{S \ni s \to +\infty}{\longrightarrow} +\infty.$$

- The set $S := \{s \in \mathbb{N} : a_{2s+1} \geq 1\}$ is infinite. Then for $s \in S$, we take $x' := x - \frac{1}{\mathfrak{q}_1 \cdots \mathfrak{q}_{2s+1}}$ and argue as above (Exercise).

Observe that at least one of the above two possibilities holds. Thus a finite derivative $f'(x)$ does not exist.

- The set $S := \{s \in \mathbb{N} : a_{2s} \leq 1\}$ is infinite. For $s \in S$, take $x' := x + \frac{2}{q_1 \cdots q_{2s}}$ and then $f(x) = f(x')$.
- The set $S := \{s \in \mathbb{N} : a_{2s} \geq 2\}$ is infinite. For $s \in S$, take $x' := x - \frac{2}{q_1 \cdots q_{2s}}$ and then $f(x) = f(x')$.

Observe that if at least one of the above two possibilities holds, and therefore, an infinite derivative $f'(x)$ does not exist. □

Let $R := 2r$, $r \in 2\mathbb{N}_2 + 1$ (e.g., $r = 5$). For $x = \sum_{n=1}^{\infty} \frac{a_n}{R^n}$ with $a_n \in \{0, \dots, R-1\}$, define the *Singh function* $\boldsymbol{S}_3 : \mathbb{I} \longrightarrow \mathbb{R}$,

$$\boldsymbol{S}_3(x) := \sum_{n=1}^{\infty} \frac{\varepsilon_n b_n}{(2R)^n},$$

where

$$b_n := 2a_{2n-1} + \psi(a_{2n}),$$

$$\psi(t) := \begin{cases} 0, & \text{if } t \in \{0, 2, 4, \dots, r-1\} \\ 1, & \text{if } t \in \{1, 3, 5, \dots, 2r-1\} \\ 2, & \text{if } t \in \{r+1, r+3, \dots, 2r-2\} \end{cases}, \quad n \in \mathbb{N},$$

$$\varepsilon_1 := 1, \quad \varepsilon_n := \varepsilon_{n-1}\varphi(a_{2n-2}),$$

$$\varphi(t) := \begin{cases} 1, & \text{if } t \in \{0, 2, 4, \dots, r-1, r, r+2, r+4, \dots, 2r-1\} \\ -1, & \text{if } t \in \{1, 3, 5, \dots, r-2, r+1, r+3, \dots, 2r-2\} \end{cases}, \quad n \in \mathbb{N}_2.$$

Notice that $\varepsilon_n b_n$ depends on a_{2n-2}, a_{2n-1}, and a_{2n} $(n \in \mathbb{N}_2)$.

Remark 5.10.5. The function $\boldsymbol{S}_3$ is well defined.

Let

$$x = \sum_{n=1}^{k} \frac{a_n}{R^n} \text{ with } a_k \geq 1, \quad x' := \Big(\sum_{n=1}^{k-1} \frac{a_n}{R^n}\Big) + \frac{a_k - 1}{R^k} + \sum_{n=k+1}^{\infty} \frac{R-1}{R^n}.$$

Let ε'_n, b'_n be the sequences constructed for x'. There are the following two cases:

- $k = 2s - 1$. Then

$$\varepsilon' = \varepsilon = (\varepsilon_1, \dots, \varepsilon_{s-1}, \varepsilon_s, \varepsilon_s, \dots),$$
$$b = (b_1, \dots, b_{s-1}, 2a_k, 0, 0, \dots),$$
$$b' = (b_1, \dots, b_{s-1}, 2(a_k - 1) + 1, 2R - 1, 2R - 1, \dots),$$

which immediately gives $\boldsymbol{S}_3(x) = \boldsymbol{S}_3(x')$.

- $k = 2s$. Then

$$\varepsilon = (\varepsilon_1, \dots, \varepsilon_s, \varepsilon_{s+1}, \varepsilon_{s+1}, \dots), \quad \varepsilon' = (\varepsilon_1, \dots, \varepsilon_s, \varepsilon'_{s+1}, \varepsilon'_{s+1}, \dots),$$
$$\varepsilon'_{s+1} = \varepsilon_s \varphi(a_k - 1),$$
$$b = (b_1, \dots, b_{s-1}, 2a_{2s-1} + \psi(a_k), 0, 0, \dots),$$
$$b' = (b_1, \dots, b_{s-1}, 2a_{2s-1} + \psi(a_k - 1), 2R - 1, 2R - 1, \dots).$$

Consequently, the equality $\boldsymbol{S}_3(x) = \boldsymbol{S}_3(x')$ reduces to the identity $\psi(t+1) = \psi(t) + \varphi(t)$, $t \in \{0, \dots, R-2\}$, which may be easily verified (EXERCISE).

Theorem 5.10.6 (cf. [Sin30, Sin35]). $\boldsymbol{S}_3 \in \mathcal{ND}^\infty((0,1))$.

Proof. Put $f := \boldsymbol{S}_3$.

Step 1°. Continuity of f follows from Lemma 5.4.1 and Remark 5.10.5 with $\mathfrak{q}_n = R$, $\sigma_n := 2n$, $\varphi_n(a_1, \dots, a_{2n}) := \frac{\varepsilon_n b_n}{(2R)^n}$ $(M_n \leq \frac{1}{(2R)^{n-1}})$.

Step 2°. Nowhere differentiability of f. Fix an $x = \sum_{n=1}^\infty \frac{a_n}{R^n}$ with $\sup\{n \in \mathbb{N} : a_n \leq R - 2\} = +\infty$. We are going to prove that for infinitely many $n \in \mathbb{N}$, there exist $\varepsilon'_n, \varepsilon''_n \in \{-1, +1\}$ such that for $x'_n := x + \varepsilon'_n \frac{2}{R^{2n}}$, $x''_n := x + \varepsilon''_n \frac{r}{R^{2n}}$, we have $a_{2n} + 2\varepsilon'_n$, $a_{2n} + r\varepsilon''_n \in \{0, \dots, R-1\}$ and

$$\psi(a_{2n} + 2\varepsilon'_n) = \psi(a_{2n}), \qquad \varphi(a_{2n} + 2\varepsilon'_n) = \varphi(a_{2n}),$$
$$\psi(a_{2n} + r\varepsilon'_n) = \psi(a_{2n}) + 1, \qquad \varphi(a_{2n} + r\varepsilon''_n) = \varphi(a_{2n}).$$

Consequently,

$$f(x'_n) = f(x), \quad |\boldsymbol{\Delta} f(x, x''_n)| = r^{n-1},$$

which obviously will imply that a finite or infinite $f'(x)$ does not exist.

Consider the four sets

$$\begin{aligned} S_1 &:= \{n \in \mathbb{N} : a_{2n} \in \{0, \dots, r-3\}\}, \\ S_2 &:= \{n \in \mathbb{N} : a_{2n} \in \{r-2, r-1\}\}, \\ S_3 &:= \{n \in \mathbb{N} : a_{2n} \in \{r, \dots, 2r-3\}\}, \\ S_4 &:= \{n \in \mathbb{N} : a_{2n} \in \{2r-2, 2r-1\}\}, \end{aligned}$$

and note that at least one of them is infinite. For $n \in S_i$, define the numbers ε'_n, ε''_n according to the following table:

n	ε'_n	ε''_n
$n \in S_1$	$+1$	$+1$
$n \in S_2$	-1	$+1$
$n \in S_3$	$+1$	-1
$n \in S_4$	-1	-1

It remains to check (EXERCISE) that ε'_n, ε''_n (defined above) fulfill our requirements. □

Remark 5.10.7. ? Based on the above ideas, the reader may try to create his or her own nowhere differentiable functions ?

Chapter 6
Other Examples

Summary. It is not surprising that there are many examples of nowhere differentiable functions that are outside the above three main types discussed so far. We will present only two of them.

6.1 Schoenberg Functions

During the discussion of so-called space-filling curves, other examples of nowhere differentiable functions occurred. Here we restrict ourselves to presenting Schoenberg's curve. A few more details will be given in the remark at the end.

Theorem 6.1.1. *Let*

$$\Phi(x) := \frac{1}{2}\sum_{n=0}^{\infty} \frac{1}{2^n} p(3^{2n}x), \quad \Psi(x) := \Phi(3x), \quad x \in \mathbb{R},$$

where

$$p(x) := \begin{cases} 0, & \text{if } x \in [0, 1/3] \cup [5/3, 2] \\ 3x - 1, & \text{if } x \in [1/3, 2/3] \\ 1, & \text{if } x \in [2/3, 4/3] \\ 5 - 3x, & \text{if } x \in [4/3, 5/3] \end{cases}, \quad p(x+2) = p(x), \quad x \in \mathbb{R}.$$

Then $\Phi, \Psi \in \mathcal{ND}(\mathbb{R})$ *(cf. Fig. 6.1).*

The first proof of Theorem 6.1.1 was given in [Als81]. The one presented here can be found in [Sag92].

Proof of Theorem 6.1.1. We have only to show that $\Phi \in \mathcal{ND}(\mathbb{R})$. Observe that $\Phi(x + 2) = \Phi(x)$. Thus, it suffices to prove that $\Phi'(x)$ does not exist for $x \in [0, 2)$. Suppose that a finite derivative $\Phi'(x_0)$ exists for some $x_0 \in [0, 2)$.

- If $x_0 = 0$, then we have

$$\boldsymbol{\Delta}\Phi(0, 1/9^k) = 9^k \frac{1}{2} \sum_{n=0}^{\infty} \frac{1}{2^n} p(9^{n-k}).$$

M. Jarnicki, P. Pflug, *Continuous Nowhere Differentiable Functions*, Springer Monographs in Mathematics, DOI 10.1007/978-3-319-12670-8_6

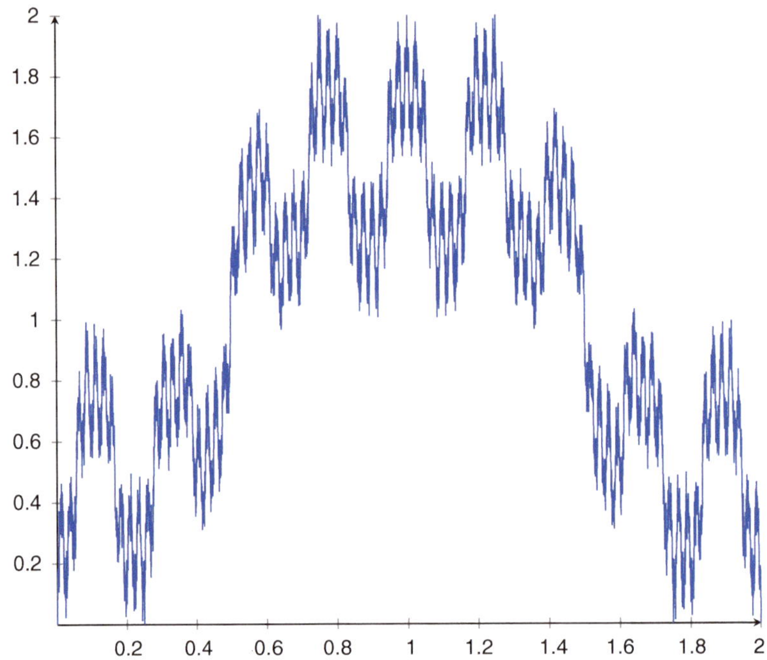

Fig. 6.1 Schoenberg function $\mathbb{I} \ni x \longmapsto \frac{1}{2}\sum_{n=0}^{\infty} \frac{1}{2^n} p(3^{2n}x)$

Note that

$$p(9^{n-k}) = \begin{cases} 0, & \text{if } n < k \\ 1, & \text{if } n \geq k \end{cases}.$$

Hence

$$\boldsymbol{\Delta}\Phi(0, 1/9^k) = 9^k \frac{1}{2} \sum_{n=k}^{\infty} \frac{1}{2^n} = \left(\frac{9}{2}\right)^k \longrightarrow +\infty;$$

a contradiction.

- If $x_0 \in (0, 2)$, then let $N_k \in \mathbb{N}_0$ be such that $9^k x_0 \in [N_k, N_k + 1)$. Put $a_k := \frac{N_k}{9^k}$, $b_k := \frac{N_k+1}{9^k}$, $k \in \mathbb{N}$. Then $\boldsymbol{\Delta}\Phi(a_k, b_k) \longrightarrow \Phi'(x_0) \in \mathbb{R}$ (cf. Remark 2.1.2). Let $A := \{k \in \mathbb{N} : N_k \in 2\mathbb{N}_0\}$, $B := \{k \in \mathbb{N} : N_k \in 2\mathbb{N}_0 + 1\}$. Of course, at least one of the sets A, B is infinite.

– If A is infinite, then for $k \in A$, we get

$$\begin{aligned}
\Phi(b_k) - \Phi(a_k) &= \frac{1}{2} \sum_{n=0}^{\infty} \frac{1}{2^n} p(9^{n-k}N_k + 9^{n-k}) - \frac{1}{2} \sum_{n=0}^{\infty} \frac{1}{2^n} p(9^{n-k}N_k) \\
&= \frac{1}{2} \sum_{n=0}^{k-1} \frac{1}{2^n} \Big(p(9^{n-k}N_k + 9^{n-k}) - p(9^{n-k}N_k) \Big) + \frac{1}{2} \sum_{n=k}^{\infty} \frac{1}{2^n} \\
&\geq -\frac{1}{2} \sum_{n=0}^{k-1} \frac{1}{2^n} 3 \cdot 9^{n-k} + \frac{1}{2^k} = -\frac{3}{2 \cdot 9^k} \sum_{n=0}^{k-1} \left(\frac{9}{2}\right)^n + \frac{1}{2^k} \\
&= -\frac{3}{7 \cdot 9^k} \left(\left(\frac{9}{2}\right)^k - 1 \right) + \frac{1}{2^k}.
\end{aligned}$$

Hence

$$\boldsymbol{\Delta}\Phi(a_k, b_k) \geq -\frac{3}{7}\Big(\Big(\frac{9}{2}\Big)^k - 1\Big) + \Big(\frac{9}{2}\Big)^k = \frac{3}{7} + \frac{4}{7}\Big(\frac{9}{2}\Big)^k \underset{A\ni k\to+\infty}{\longrightarrow} +\infty;$$

a contradiction.

– If B is infinite, then for $k \in B$, an analogous argument gives (Exercise)

$$\boldsymbol{\Delta}\Phi(a_k, b_k) \leq -\frac{3}{7} - \frac{4}{7}\Big(\frac{9}{2}\Big)^k \underset{B\ni k\to+\infty}{\longrightarrow} -\infty;$$

a contradiction.

□

Notice that the *Schoenberg curve*

$$\mathbb{I} \ni t \overset{\gamma}{\longmapsto} (\Phi(t), \Psi(t)) \in \mathbb{I} \times \mathbb{I}$$

is a so-called *space-filling curve*, i.e., $\gamma(\mathbb{I}) = \mathbb{I} \times \mathbb{I}$. More precisely, we have the following result.

Proposition 6.1.2 (cf. [Sch38]). *Let* $\mathfrak{C} \subset \mathbb{I}$ *stand for the standard* Cantor ternary set. *Then* $\gamma(\mathfrak{C}) = \mathbb{I} \times \mathbb{I}$.

Proof. Every $t \in \mathfrak{C}$ has a ternary representation of the form $t = \sum_{k=1}^{\infty} \frac{2t_k}{3^k}$, where $t_k \in \{0, 1\}$. We have

$$3^{2n}t = \Big(\sum_{k=1}^{2n} 3^{3n-k} 2t_k\Big) + \sum_{s=1}^{\infty} \frac{2t_{2n+s}}{3^s} =: A_n(t) + B_n(t),$$

$$3^{2n+1}t = \Big(\sum_{k=1}^{2n+1} 3^{3n+1-k} 2t_k\Big) + \sum_{s=1}^{\infty} \frac{2t_{2n+1+s}}{3^s} =: C_n(t) + D_n(t),$$

where $A_n(t), C_n(t) \in 2\mathbb{N}_0$. In particular, $p(3^{2n}t) = p(B_n(t))$ and $p(3^{2n+1}t) = p(D_n(t))$. Observe that

$$\max\Big\{\sum_{s=2}^{\infty} \frac{2t_{2n+s}}{3^s}, \sum_{s=2}^{\infty} \frac{2t_{2n+1+s}}{3^s}\Big\} \leq \sum_{s=2}^{\infty} \frac{2}{3^s} = \frac{1}{3}.$$

Consequently, $p(B_n(t)) = t_{2n+1}$ and $p(D_n(t)) = t_{2n+2}$. Thus,

$$\gamma(t) = \Big(\frac{1}{2}\sum_{n=0}^{\infty} \frac{p(3^{2n}t)}{2^n}, \frac{1}{2}\sum_{n=0}^{\infty} \frac{p(3^{2n+1}t)}{2^n}\Big) = \Big(\sum_{n=1}^{\infty} \frac{t_{2n-1}}{2^n}, \sum_{n=1}^{\infty} \frac{t_{2n}}{2^n}\Big).$$

This shows that for every point $P_0 = (\xi_0, \eta_0) \in \mathbb{I}\times\mathbb{I}$ with binary representations $\xi_0 = \sum_{n=1}^{\infty} \frac{\xi_n}{2^n}$, $\eta_0 = \sum_{n=1}^{\infty} \frac{\eta_n}{2^n}$, we have $P_0 = \gamma(t_0)$, where $t_0 := \frac{2\xi_1}{3} + \frac{2\eta_1}{3^2} + \frac{2\xi_2}{3^3} + \frac{2\eta_2}{3^4} + \cdots \in \mathfrak{C}$. □

Remark 6.1.3. (a) The Schoenberg curve $\gamma = (\Phi, \Psi) : \mathbb{I} \longrightarrow \mathbb{I} \times \mathbb{I}$ belongs to a large class of space-filling curves whose coordinate functions Φ, Ψ are nowhere differentiable; cf. [Sag94].

(b) Notice the following surprising result.

Theorem 6.1.4 ([Mor87]). *The following statements are equivalent:*

(i) *there exists a mapping* $f = (f_1, f_2) : \mathbb{R} \longrightarrow \mathbb{R}^2$ *such that* $f(\mathbb{R}) = \mathbb{R}^2$, *and for each* $x \in \mathbb{R}$, *at least one of the finite derivatives* $f_1'(x)$, $f_2'(x)$ *exists;*

(ii) *the continuum hypothesis is true.*

One can also prove (cf. [Mor87]) that if f is as in (i), then f_1 and f_2 are not Lebesgue measurable.

6.2 Second Wen Function

Parallel to nowhere differentiable functions given by series, there are also functions given by infinite products. We present an example of such a function due to L. Wen.

Theorem 6.2.1 (cf. [Wen02]). *Let*

$$\boldsymbol{W}_2(x) := \prod_{n=1}^{\infty}(1 + a_n \sin(\pi b_n x)), \quad x \in \mathbb{R},$$

where $0 < a_n < 1$, $\sum_{n=1}^{\infty} a_n < +\infty$, $p_k \in 2\mathbb{N}$, $b_n := p_1 \cdots p_n$, $\lim_{n\to+\infty} \frac{2^n}{a_n p_n} = 0$. *Then* $\boldsymbol{W}_2 \in \mathfrak{ND}_{\pm}(\mathbb{R})$ *(Fig. 6.2).*

Proof. Put $f := \boldsymbol{W}_2$. Since $\sum_{n=1}^{\infty} \sup_{x\in\mathbb{R}} |a_n \sin(\pi b_n x)| \le \sum_{n=1}^{\infty} a_n < +\infty$, we easily conclude that the product $\prod_{n=1}^{\infty}(1 + a_n \sin(\pi b_n x))$ is uniformly convergent to a

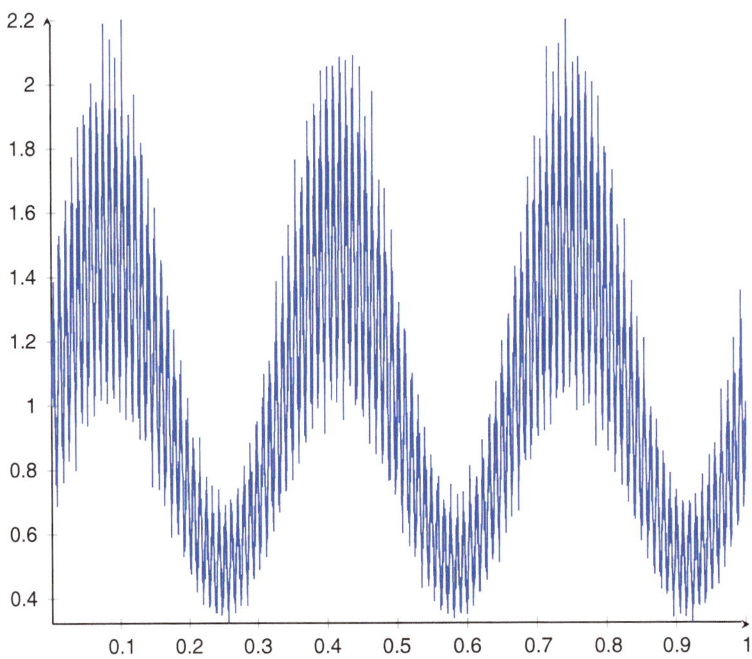

Fig. 6.2 Wen function $\mathbb{I} \ni x \longmapsto \prod_{n=1}^{\infty}(1 + a_n \sin(\pi b_n x))$ with $a_n := 2^n$, $b_n := \sqrt{6}^{\,n(n+1)}$

continuous function. Fix an $x_0 \in \mathbb{R}$. We are going to show that a finite $f'_+(x_0)$ does not exist. A proof for $f'_-(x_0)$ is left for the reader as an EXERCISE. Let $N_n \in \mathbb{Z}$ be such that $b_n x_0 \in [N_n, N_n + 1)$, $n \in \mathbb{N}$. Define $x_n := \frac{N_n+1}{b_n}$, $x'_n := \frac{N_n+3/2}{b_n}$, $n \in \mathbb{N}$. Then

$$x_0 < x_n < x'_n, \quad x'_n - x_0 \leq \frac{3}{2b_n}, \quad \frac{1}{2b_n} = x'_n - x_n \geq \frac{1}{3}(x'_n - x_0) > \frac{1}{3}(x_n - x_0).$$

Define

$$A := \prod_{n=1}^{\infty}(1 - a_n), \quad B := \prod_{n=1}^{\infty}(1 + a_n), \quad I_k(x) := \prod_{n=1}^{k}(1 + a_n \sin(\pi b_n x)), \quad x \in \mathbb{R},\ k \in \mathbb{N}.$$

Observe that for $n > k$, we get

$$\begin{aligned}\sin(\pi b_n x_k) &= \sin(\pi p_n \cdots p_{k+1}(N_k + 1)) = 0,\\ \sin(\pi b_n x'_k) &= \sin(\pi p_n \cdots p_{k+1}(N_k + 3/2)) = 0.\end{aligned}$$

If $n = k$, then

$$\begin{aligned}\sin(\pi b_k x_k) &= \sin(\pi(N_k + 1)) = 0,\\ \sin(\pi b_k x'_k) &= \sin(\pi(N_k + 3/2)) = (-1)^{N_k+1}.\end{aligned}$$

Consequently,

$$\begin{aligned}f(x'_k) - f(x_k) &= I_{k-1}(x'_k)(1 + a_k(-1)^{N_k+1}) - I_{k-1}(x_k)\\ &= I_{k-1}(x'_k) - I_{k-1}(x_k) + a_k(-1)^{N_k+1} I_{k-1}(x'_k).\end{aligned}$$

If $n < k$, then

$$|\sin(\pi b_n x'_k) - \sin(\pi b_n x_k)| \leq \pi b_n (x'_k - x_k) = \pi b_n \frac{1}{2b_k} \leq \frac{\pi}{2p_k}.$$

Thus $a_n \sin(\pi b_n x'_k) = a_n \sin(\pi b_n x_k) + \sigma_{k,n}$, where $|\sigma_{k,n}| < \frac{\pi}{2p_k}$. Then

$$\begin{aligned}&|I_{k-1}(x'_k) - I_{k-1}(x_k)|\\ &\quad= \Big|\prod_{n=1}^{k-1}(1 + a_n \sin(\pi b_n x_k) + \sigma_{k,n}) - \prod_{n=1}^{k-1}(1 + a_n \sin(\pi b_n x_k))\Big|\\ &\quad= \Big|\sum_{\substack{Q,Q' \subset \{1,\dots,k-1\} = Q \cup Q'\\ Q \cap Q' = \varnothing,\ \#Q' \geq 1}} \prod_{n \in Q}(1 + a_n \sin(\pi b_n x_k)) \prod_{n \in Q'} \sigma_{k,n}\Big|\\ &\quad\leq (2^{k-1} - 1)B\frac{\pi}{2p_k} < \frac{\pi B 2^{k-1}}{p_k}.\end{aligned}$$

Hence

$$|f(x'_k) - f(x_k)| \geq |a_k I_{k-1}(x'_k)| - \frac{\pi B 2^{k-1}}{p_k} \geq a_k A - \frac{\pi B 2^{k-1}}{p_k}.$$

Thus,

$$|\boldsymbol{\Delta} f(x_k, x_k')| \geq 2a_k b_k \Big(A - \frac{\pi B}{2}\frac{2^k}{a_k p_k}\Big) \underset{k\to+\infty}{\longrightarrow} +\infty.$$

Finally, since

$$\begin{aligned}|\boldsymbol{\Delta} f(x_k, x_k')| &\leq \frac{|f(x_k') - f(x_0)|}{x_k' - x_k} + \frac{|f(x_k) - f(x_0)|}{x_k' - x_k} \\ &\leq 3(\boldsymbol{\Delta} f(x_0, x_k') + \boldsymbol{\Delta} f(x_0, x_k)),\end{aligned}$$

we easily conclude that a finite $f'_+(x_0)$ does not exist. □

Part II
Topological Methods

Chapter 7
Baire Category Approach

Summary. While in Part I, some concrete functions were discussed, this chapter shows how Baire category methods lead to a description of typical continuous functions on the interval $\mathbb{I} = [0, 1]$. In Sect. 7.2, we prove that most (in the categorial sense) of the continuous functions on $\mathbb{I}$ belong to $\mathcal{ND}_{\pm}(\mathbb{I})$, while in Sect. 7.5, it is shown that the set $\mathcal{ND}_{\pm}^{\infty}(\mathbb{I})$ of all continuous functions on $\mathbb{I}$ having nowhere a unilateral (finite or infinite) derivative is a thin set (in the categorial sense). Nevertheless, later, in Sect. 11.1, we will see that $\mathcal{ND}_{\pm}^{\infty}(\mathbb{I})$ is not empty.

7.1 Metric Spaces and First Baire Category

The idea of this section is to collect and recall some information that will be used in this chapter.

Definition 7.1.1. A *metric space* is a pair (X, d), where X is a nonempty set and the function $d : X \times X \longrightarrow \mathbb{R}_+$ is a *metric*, i.e., d is symmetric, satisfies the triangle inequality (i.e., $d(x, y) \leq d(x, z) + d(z, y)$, $x, y, z \in X$), and is positive definite (i.e., $d(x, y) = 0$ if and only if $x = y$).

The metric space mainly discussed in this section is the set $\mathcal{C}(\mathbb{I}, \mathbb{C})$ together with its standard metric $d(f, g) := \|f - g\|_{\mathbb{I}}$.

Let X be a metric space (in case there is no confusion, we will always omit noting the metric d). Then d induces on X the structure of a topological space specifying the open sets of X. Recall that a subset $M \subset X$ is called *open* if for every $x \in M$, there exists a positive $r \in \mathbb{R}$ such that the ball $\mathbb{B}_d(x, r) := \{y \in X : d(x, y) < r\}$ is a subset of M. A set M is said to be *closed* if $X \setminus M$ is open. A point $a \in M$ is an *interior point* of M if some ball $\mathbb{B}_d(a, r)$ with center a is contained in M.

A subset $M \subset X$ is called *dense* in X if $\overline{M} = X$, where $\overline{M}$ is the closure of M, i.e., $\overline{M} = \{x \in X : \forall_{r \in (0,\infty)} : \mathbb{B}_d(x, r) \cap M \neq \varnothing\}$. On the other hand, M is called *nowhere dense* if $\operatorname{int} \overline{M} = \varnothing$, where $\operatorname{int} L := \{x \in L : \exists_{r \in (0,\infty)} : \mathbb{B}_d(x, r) \subset L\}$ denotes the *interior* of L, $L \subset X$.

Recall that the set of (real-valued) polynomials is dense in $\mathcal{C}(\mathbb{I})$ (Exercise; use Bernstein polynomials).

M. Jarnicki, P. Pflug, *Continuous Nowhere Differentiable Functions*, Springer Monographs in Mathematics, DOI 10.1007/978-3-319-12670-8_7

Definition 7.1.2. Let X be a metric space. A subset $M \subset X$ is said to be *of the first (Baire) category* (or *meagre*) if $M = \bigcup_{j=1}^{\infty} M_j$, where M_j are nowhere dense subsets of X. A set M is said to be *of second (Baire) category* if it is not of first category. Moreover, M is said to be *residual* in X if $X \setminus M$ is of first category.

Sets of first Baire category may be thought as small sets.

Theorem 7.1.3 (The Baire Theorem). *If $X = (X, d)$ is a complete metric space (i.e., every Cauchy sequence with respect to d converges to a point of X), then X is of second Baire category.*

Note that $\mathcal{C}(\mathbb{I}, \mathbb{C})$ is a complete metric space with the metric from above (Exercise). In particular, if $M \subset \mathcal{C}(\mathbb{I}, \mathbb{C})$ is a set of first Baire category, then $\mathcal{C}(\mathbb{I}, \mathbb{C}) \setminus M \neq \varnothing$.

We say that a function $f \in \mathcal{C}(\mathbb{I}, \mathbb{C})$ is *typical (with respect to a certain property (P))* if f satisfies (P) and if the set $\{g \in \mathcal{C}(\mathbb{I}, \mathbb{C}) : g \text{ does not fulfill } (P)\}$ is of first Baire category in $\mathcal{C}(\mathbb{I}, \mathbb{C})$.

7.2 The Banach–Jarnik–Mazurkiewicz Theorem

The main result in this section is the following: the typical continuous function on $\mathbb{I}$ has everywhere on $\mathbb{I}$ an infinite upper or lower right Dini derivative; in particular, it is nowhere differentiable on $\mathbb{I}$. To be more precise we have the following result.

Theorem 7.2.1 (The Banach–Jarnik–Mazurkiewicz Theorem). *There exists a subset $S \subset \mathcal{C}(\mathbb{I})$ of first category such that if $f \in \mathcal{C}(\mathbb{I}) \setminus S$, then the following properties hold:*

(a) *$f \in \mathfrak{M}(\mathbb{I})$; in particular, f has nowhere on $[0,1)$ (resp. on $(0,1]$) a finite right (resp. left) derivative;*

(b) *there exists a set $E \subset \mathbb{I}$ with $\mathcal{L}(E) = 1$ such that*

$$D^+f(x) = D^-f(x) = +\infty, \quad D_+f(x) = D_-f(x) = -\infty, \quad x \in E;$$

(c) *for every $x \in (0,1)$ and $\alpha \in \overline{\mathbb{R}}$, there exists a sequence $(h_j)_{j=1}^{\infty} \subset \mathbb{R}_*$ with $\lim_{j\to+\infty} h_j = 0$ such that*

$$\lim_{j\to+\infty} \frac{f(x+h_j) - f(x)}{h_j} = \alpha,$$

i.e., every number in $\overline{\mathbb{R}}$ is a derived number *of f at x. In particular, nowhere on $(0,1)$ does the two-sided derivative of f, finite or infinite, exist.*

Remark 7.2.2. (i) The first statement is due to [Ban31]; a weaker form may be found in [Maz31]. The remaining facts are contained in [Jar33].

(ii) Recall that a point $x \in \mathbb{I}$ with the property in (b) is called a *knot point* of f. Thus (b) implies that a *typical function in $\mathcal{C}(\mathbb{I})$ has almost everywhere a knot point.*

(iii) While the theorem states that the typical function $f \in \mathcal{C}(\mathbb{I})$ has nowhere a finite unilateral derivative, nothing is said about infinite one-sided derivatives. We will discuss such functions later, beginning with Sect. 7.5.

Proof of Theorem 7.2.1. (a) Put

$$S_1^+ := \{f \in \mathcal{C}(\mathbb{I}) : \exists_{x\in[0,1)} : \max\{|D^+f(x)|, |D_+f(x)|\} < +\infty\}.$$

We will prove that S_1^+ is of first category in $\mathcal{C}(\mathbb{I})$. Define

$$E_k^+ := \{f \in \mathcal{C}(\mathbb{I}) : \exists_{a\in[0,1-1/k]} : |\boldsymbol{\Delta} f(a, a+t)| \le k, \quad 0 < t \le 1-a\}, \quad k \in \mathbb{N}_2.$$

Obviously, E_k^+ is closed in $\mathcal{C}(\mathbb{I})$, and $S_1^+ = \bigcup_{k=1}^{\infty} E_k^+$ (Exercise). It remains to verify that the sets E_k^+ are without interior points.

Indeed, fix a k_0 and assume that $E_{k_0}^+ \supset B(f, 2r)$, where $f \in E_{k_0}^+$, $r > 0$, and $B(f, 2r) := \{g \in \mathcal{C}(\mathbb{I}) : \|g - f\|_{\mathbb{I}} < 2r\}$. Since the real-valued polynomials are dense in $\mathcal{C}(\mathbb{I})$, we may take a polynomial $\widetilde{p}$ with $\|f - \widetilde{p}\|_{\mathbb{I}} < r$. Then $B(p, r) \subset E_{k_0}^+$, where $p := \widetilde{p}|_{\mathbb{I}}$.

Let $g \in \mathcal{C}(\mathbb{I})$ be such that g has finite right-sided derivatives on $[0, 1)$ and

$$\|g\|_{\mathbb{I}} < r, \quad |g'_+(x)| > \|p'\|_{\mathbb{I}} + k_0, \quad x \in [0, 1).$$

Note that such a g always exists (Exercise; use continuous piecewise linear functions).

Then $h := g + p \in B(p, r) \subset E_{k_0}^+$ and

$$|h'_+(x)| \ge |g'_+(x)| - |p'(x)| > \|p'\|_{\mathbb{I}} + k_0 - \|p'\|_{\mathbb{I}} = k_0, \quad x \in [0, 1).$$

Hence, $h \notin E_{k_0}^+$; a contradiction.

A similar argument shows that the set

$$S_1^- := \{f \in \mathcal{C}(\mathbb{I}) : \exists_{x\in(0,1]} : \max\{|D^- f(x)|, |D_- f(x)|\} < +\infty\}$$

is also of first category in $\mathcal{C}(\mathbb{I})$ (Exercise). Thus, $S_1 := S_1^+ \cup S_1^-$ is of first category.

Before starting to prove the remaining two statements let us introduce a special "zigzag" function $\boldsymbol{z}_{r,s}$ on $\mathbb{I}$, where $s \in (0, 1/2)$ and $r > 0$. Define

$$\boldsymbol{z}_{r,s}(t) := \begin{cases} 2r\boldsymbol{\psi}(\frac{t}{2s}), & \text{if } 0 \le t \le 2Ns \\ 0, & \text{if } 2Ns < t \le 1 \end{cases},$$

where $N := \lfloor \frac{1}{2s} \rfloor$ and $\boldsymbol{\psi}(x) = \operatorname{dist}(x, \mathbb{Z})$.

(b) Put

$$E^+(f) := \{x \in [0, 1) : D^+ f(x) < +\infty\}, \quad S_2^+ := \{f \in \mathcal{C}(\mathbb{I}) : \mathcal{L}(E^+(f)) > 0\}.$$

We will show that S_2^+ is of first category.

For $n \in \mathbb{N}_3$, put

$$E_n^+(f) := \Big\{x \in \Big[0, 1 - \frac{1}{n}\Big] : \boldsymbol{\Delta} f(x, x+h) \le n, \ 0 < h \le \frac{1}{n}\Big\}.$$

Then $E^+(f) = \bigcup_{n\in\mathbb{N}_3} E_n^+(f)$. Hence, $\mathcal{L}(E^+(f)) > 0$ if and only if there exist an $n \in \mathbb{N}_3$ and a $k \in \mathbb{N}$ with $\mathcal{L}(E_n^+(f)) \ge 1/k$. Put

$$V_{n,k} := \{f \in \mathcal{C}(\mathbb{I}) : \mathcal{L}(E_n^+(f)) \ge 1/k\}.$$

Then $S_2^+ = \bigcup_{n\in\mathbb{N}_3} \bigcup_{k\in\mathbb{N}} V_{n,k}$. Observe that the sets $V_{n,k}$ are closed in $\mathcal{C}(\mathbb{I})$. Indeed, let $(f_j)_{j\in\mathbb{N}} \subset V_{n,k}$ with $f_j \Longrightarrow f \in \mathcal{C}(\mathbb{I})$. By assumption, $\mathcal{L}(E_n^+(f_j)) \ge 1/k$, $j \in \mathbb{N}$. If $x \in E_n^+(f_j)$ for infinitely many j's, then $x \in E_n^+(f)$. Therefore,

$$\bigcap_{s=1}^{\infty} \bigcup_{j=s}^{\infty} E_n^+(f_j) \subset E_n^+(f).$$

Hence we have $\mathcal{L}(E_n^+(f)) \geq 1/k$, i.e. $f \in V_{n,k}$.

It remains to verify that each of the sets $V_{n,k}$ has no interior point. Assume the contrary. Then one $V_{n,k}$ contains a ball $B(f_0, 2r)$, and therefore (as in (a)) there exists a polynomial $\widetilde{p}$ such that $B(p,r) \subset V_{n,k}$, where $p := \widetilde{p}|_{\mathbb{I}}$. Put

$$q := \sup\Big\{\Big|\frac{p(x+h)-p(x)}{h}\Big| : x \in \mathbb{I},\ h \in \mathbb{R}_*,\ 0 \leq x+h \leq 1\Big\}$$

and note that $q < +\infty$. Now fix an $s \in (0, \min\{\frac{1}{4n}, \frac{r}{8qk}\})$ with $n < \frac{r}{8ks}$. Let $\boldsymbol{z}_{r,s}$ be the corresponding "zigzag" function. Then $g := p + \boldsymbol{z}_{r,s} \in B(p,r) \subset V_{n,k}$.

Take a $t \in [0, 1-\frac{1}{n}]$ such that $\boldsymbol{z}_{r,s}(t) < r(1-\frac{1}{2k})$. Then one finds a point $h' \in (0, 2s]$ with $\boldsymbol{z}_{r,s}(t+h') = r$. Note that $0 < h' \leq \frac{1}{n}$. Moreover,

$$\begin{aligned}\boldsymbol{\Delta} g(t, t+h') &= \boldsymbol{\Delta} p(t, t+h') + \boldsymbol{\Delta} \boldsymbol{z}_{r,s}(t, t+h') \\ &\geq \frac{1}{h'}\Big(r - r\Big(1-\frac{1}{2k}\Big) - qh'\Big) \geq \frac{r}{8sk} > n.\end{aligned}$$

In particular, $t \notin E_n^+(g)$.

By the geometry of the "zigzag" function, it is easily seen that

$$\mathcal{L}(E_n^+(g)) \leq \mathcal{L}\Big(\Big\{t \in \Big[0, 1-\frac{1}{n}\Big] : \boldsymbol{z}_{r,s}(t) \geq r\Big(1-\frac{1}{2k}\Big)\Big\}\Big) \leq \frac{1}{2k} < \frac{1}{k}.$$

Thus $g \notin V_{n,k}$; a contradiction.

(c) Put

$$\begin{aligned}S_3 := \Big\{f \in \mathcal{C}(\mathbb{I}) : \exists_{\alpha_0 \in \overline{\mathbb{R}}},\ \exists_{a \in (0,1)},\ \forall_{(h_j)_{j=1}^{\infty} \subset \mathbb{R}_*,\ \lim_{j\to\infty} h_j = 0} : \\ \boldsymbol{\Delta} f(a, a+h_j) \not\longrightarrow \alpha_0\Big\}.\end{aligned}$$

We want to verify that S_3 is a set of first category in $\mathcal{C}(\mathbb{I})$. To this end, define

$$\begin{aligned}S_{3,n,\alpha,\beta} := \{f \in \mathcal{C}(\mathbb{I}) : \exists_{a \in (1/n, 1-1/n)},\ \forall_{h \in \mathbb{R}_*,\ |h| < 1/n} : \\ \boldsymbol{\Delta} f(a, a+h) \leq \alpha \quad \text{or} \quad \boldsymbol{\Delta} f(a, a+h) \geq \beta\},\end{aligned}$$

where $n \in \mathbb{N}_3$, $\alpha, \beta \in \mathbb{Q}$, $\alpha < \beta$.

Note that $S_{3,n,\alpha,\beta} \subset S_3$. Indeed, let $f \in S_{3,n,\alpha,\beta}$ and choose $a \in (1/n, 1-1/n)$ according to the above definition. Suppose that $f \notin S_3$. Then there exists a sequence $(h_j)_{j=1}^{\infty} \subset \mathbb{R}_*$, $h_j \longrightarrow 0$, such that $\boldsymbol{\Delta} f(a, a+h_j) \longrightarrow \frac{\alpha+\beta}{2}$. Therefore, for all large j, we have that $|h_j| < \frac{1}{n}$ and $\alpha < \boldsymbol{\Delta} f(a, a+h_j) < \beta$; a contradiction.

Next we observe that the sets $S_{3,n,\alpha,\beta}$ are closed in $\mathcal{C}(\mathbb{I})$. To see this, take a sequence $(f_j)_{j=1}^{\infty} \subset S_{3,n,\alpha,\beta}$ with $f_j \Longrightarrow f \in \mathcal{C}(\mathbb{I})$. Choose points $a_j \in [\frac{1}{n}, 1-\frac{1}{n}]$ such that for all $h \in \mathbb{R}_*$, $|h| < \frac{1}{n}$, one of the following inequalities holds:

$$\boldsymbol{\Delta} f_j(a_j, a_j+h) \leq \alpha \quad \text{or} \quad \boldsymbol{\Delta} f_j(a_j, a_j+h) \geq \beta.$$

We may assume that $a_j \longrightarrow a \in [1/n, 1 - 1/n]$ by taking an appropriate subsequence. Fix an $h \in \mathbb{R}_*$, $|h| < 1/n$. Without loss of generality, we may even assume that the first of the above inequalities holds for all j. By uniform convergence and continuity, we conclude that this kind of inequality remains true for the limit function f. Hence, $f \in S_{3,n,\alpha,\beta}$, proving that $S_{3,n,\alpha,\beta}$ is closed.

Finally, we have that

$$S_3 = \bigcup_{n\in\mathbb{N}_3} \bigcup_{\alpha\in\mathbb{Q}} \bigcup_{\mathbb{Q}\ni\beta>\alpha} S_{3,n,\alpha,\beta}.$$

According to the remark above, it suffices to verify only the inclusion $\subset$. Indeed, let $f \in S_3$. Then we find a point $a \in (0,1)$ and a value $\alpha \in \overline{\mathbb{R}}$ such that if $(h_j)_{j=1}^\infty \subset \mathbb{R}_*$ with $h_j \longrightarrow 0$, then the sequence $(\boldsymbol{\Delta} f(a, a+h_j))_{j=1}^\infty$ does not converge to α.

Let α be a real number. Suppose that f does not belong to the set on the right-hand side. Then choose a strictly increasing sequence $(n_j)_{j=1}^\infty \subset \mathbb{N}_3$ with $a \in [\frac{1}{n_1}, 1 - \frac{1}{n_1}]$ and sequences of rational numbers $(\alpha_j)_{j=1}^\infty$, $(\beta_j)_{j=1}^\infty$ such that $\alpha_j \nearrow \alpha$ and $\beta_j \searrow \alpha$. By assumption, there are $h_j \in \mathbb{R}_*$, $|h_j| < \frac{1}{j}$, such that

$$\alpha_j \leq \boldsymbol{\Delta} f(a, a+h_j) \leq \beta_j, \quad j \in \mathbb{N},$$

which immediately implies that $\boldsymbol{\Delta} f(a, a+h_j) \longrightarrow \alpha$; a contradiction. The cases $\alpha = \pm\infty$ are left as an EXERCISE.

To complete the proof, we show that the sets $S_{3,n,\alpha,\beta}$ have no interior points. Indeed, suppose that $B(f, 2r) \subset S_{3,n,\alpha,\beta}$. Then by the Weierstrass approximation theorem, one finds a polynomial $\widetilde{p}$ such that $p := \widetilde{p}|_{\mathbb{I}} \in B(f, r)$, implying that $B(p, r) \subset S_{3,n,\alpha,\beta}$. Let $q := \|p\|_{\mathbb{I}}$. Next, we choose a positive number s such that

(i) $s < \min\{\frac{1}{4n}, \frac{r}{4q+2|\alpha+\beta|}\}$,

(ii) $|\boldsymbol{\Delta} p(t, t+h) - p'(t)| < \frac{\beta-\alpha}{2}$ for all $0 \leq t \leq 1$ and $0 < |h| \leq 2s$.

Let $\boldsymbol{z}_{r,s}$ be the "zigzag" function from above. Fix a point $t \in [\frac{1}{n}, 1-\frac{1}{n}]$. Then $p + \boldsymbol{z}_{r,s} \subset B(p, r)$. Note that $1 - \frac{1}{n} + 2s < 2s(\frac{1}{2s} - 1) \leq 2Ns$. Therefore,

$$\{\boldsymbol{z}_{r,s}(t+h) - \boldsymbol{z}_{r,s}(t) : 0 < h \leq 2s\} = [-\boldsymbol{z}_{r,s}(t), r - \boldsymbol{z}_{r,s}(t)].$$

Taking into account the special form of our "zigzag" function, we conclude that

$$\Big[\frac{-\boldsymbol{z}_{r,s}(t)}{2s}, \frac{r - \boldsymbol{z}_{r,s}(t)}{2s}\Big] \subset \Big\{\boldsymbol{\Delta} \boldsymbol{z}_{r,s}(t, t+h) : 0 < h \leq 2s\Big\}.$$

By a similar argument, we obtain

$$\Big[-\frac{r - \boldsymbol{z}_{r,s}(t)}{2s}, \frac{\boldsymbol{z}_{r,s}(t)}{2s}\Big] \subset \Big\{\boldsymbol{\Delta} \boldsymbol{z}_{r,s}(t, t+h) : 0 > h \geq -2s\Big\}.$$

Using the inequality $\max\{\boldsymbol{z}_{r,s}(t), r - \boldsymbol{z}_{r,s}(t)\} \geq r/2$, it follows that

$$\Big[-\frac{r}{4s}, \frac{r}{4s}\Big] \subset \Big\{\boldsymbol{\Delta} \boldsymbol{z}_{r,s}(t, t+h) : 0 < |h| \leq 2s\Big\} =: M.$$

Finally, the choice of s implies that $[-q - \frac{|\alpha+\beta|}{2}, q + \frac{|\alpha+\beta|}{2}] \subset M$. Therefore, we obtain an h', $0 < |h'| \leq 2s < 1/n$, such that

$$\boldsymbol{\Delta} \boldsymbol{z}_{r,s}(t, t+h') = -p'(t) + \frac{\alpha+\beta}{2}.$$

Finally, applying property (ii) yields

$$\left|\boldsymbol{\Delta}(p+\boldsymbol{z}_{r,s})(t,t+h') - \frac{\alpha+\beta}{2}\right| < \frac{\beta-\alpha}{2}.$$

Hence,

$$\alpha < \boldsymbol{\Delta}(p+\boldsymbol{z}_{r,s})(t,t+h') < \beta,$$

meaning that $p+\boldsymbol{z}_{r,s} \notin S_{3,n,\alpha,\beta}$; a contradiction. □

Remark 7.2.3. Observe that the set S in Theorem 7.2.1 may contain a function, say F, with $F'_+(a) = \pm\infty$ for some $a \in [0,1)$, i.e., functions in S may have infinite right derivatives. So it remains open, at least for the moment, whether there exists a function $f \in \mathcal{C}(\mathbb{I})$ that has nowhere a one-sided derivative, finite or infinite.

Remark 7.2.4. (a) In [Jar33], also a more general differentiation is discussed. Let $\varphi : \mathbb{R} \longrightarrow \mathbb{R}$ with $h\varphi(h) > 0$ for $h \in \mathbb{R}_*$ and $\lim_{\mathbb{R}_* \ni h \to 0} \varphi(h) = 0$, e.g., $\varphi(t) = t$. For an $f \in \mathcal{C}(\mathbb{I})$ and $x \in (0,1)$, put

$$D^{\varphi}(x) := \limsup_{\mathbb{R}_* \ni h \to 0} \frac{f(x+h)-f(x)}{\varphi(h)}, \quad D_{\varphi}(x) := \liminf_{\mathbb{R}_* \ni h \to 0} \frac{f(x+h)-f(x)}{\varphi(h)}.$$

For example, it is shown that the set

$$\{f \in \mathcal{C}(\mathbb{I}) : \ D^{\varphi}(x) = +\infty = -D_{\varphi}f(x), \ x \in (0,1)\}$$

is residual in $\mathcal{C}(\mathbb{I})$.

(b) In [Kos72], P. Kostyrko studied symmetric derivatives. Let $g \in \mathcal{C}(\mathbb{I})$ and $x \in (0,1)$. Then the *symmetric differential quotient* at x is given by

$$\boldsymbol{\Delta}_s g(x,h) := \frac{g(x+h)-g(x-h)}{2h}, \ x \in (0,1), \ x \pm h \in \mathbb{I}.$$

Then the set

$$M := \{f \in \mathcal{C}(\mathbb{I}) : \ \limsup_{\mathbb{R}_* \ni h \to 0} \boldsymbol{\Delta}_s f(x,h) = +\infty = -\liminf_{\mathbb{R}_* \ni h \to 0} \boldsymbol{\Delta}_s f(x,h) \quad \text{for all } x \in (0,1)\}$$

is residual in $\mathcal{C}(\mathbb{I})$. The proof is based on an example, due to L. Filipczak (see [Fil69]), of a function that belongs to M. Later, other concrete functions belonging to M were constructed by P. Kalášek (see [Kal73]).

(c) In [Pet58], V. Petrův generalized the symmetric derivatives. Let $g \in \mathcal{C}(\mathbb{I})$ and $\varphi : (0,\infty) \longrightarrow (0,\infty)$ with $\lim_{h\to 0+} \varphi(h) = 0$. For an $x \in (0,1)$, define

$$D^{\varphi}_{+}g(x) := \limsup_{h \to 0+} \frac{g(x+h)-g(x-h)}{\varphi(h)},$$

$$D^{+}_{\varphi}g(x) := \liminf_{h \to 0+} \frac{g(x+h)-g(x-h)}{\varphi(h)}.$$

Then the set

$$M := \{f \in \mathcal{C}(\mathbb{I}) : D^{\varphi}_{+} f(x) = +\infty = -D^{+}_{\varphi} f(x) \text{ for all } x \in (0,1)\}$$

is residual in $\mathcal{C}(\mathbb{I})$.

7.3 Typical Functions in the Disk Algebra

In this section, we will discuss recent results for the disk algebra $\mathcal{A}(\mathbb{D})$ (see [Esk14, EM14]) that are in the spirit of the discussions above.

Recall that the disk algebra $\mathcal{A}(\mathbb{D})$ is given by

$$\mathcal{A}(\mathbb{D}) := \{f \in \mathcal{C}(\overline{\mathbb{D}}, \mathbb{C}) : f|_{\mathbb{D}} \text{ is holomorphic}\},$$

where $\mathbb{D}$ denotes the open unit disk in the complex plane, i.e., $\mathbb{D} := \{z \in \mathbb{C} : |z| < 1\}$. For more details, see Appendix A.2.

Now we can formulate a recent result on typical functions of the disk algebra (see [Esk14, EM14]).

Theorem 7.3.1. *Let E denote the set of all functions $f \in \mathcal{A}(\mathbb{D})$ such that the functions u_f, $u_f(\theta) := \operatorname{Re} f(e^{i\theta})$, and v_f, $v_f(\theta) := \operatorname{Im} f(e^{i\theta})$ are nowhere differentiable on $\mathbb{R}$. Then E is residual in the Banach space $\mathcal{A}(\mathbb{D})$ equipped with the supremum norm.*

Before beginning the proof, let us introduce a notational convention. If $f \in \mathcal{C}(\mathbb{T}, \mathbb{C})$, then we set $u_f(\theta) := \operatorname{Re} f(e^{i\theta})$ and $v_f(\theta) := \operatorname{Im} f(e^{i\theta})$, $\theta \in \mathbb{R}$. Obviously, these functions are 2π-periodic continuous functions. Conversely, if u is a 2π-periodic continuous function ($u \in \mathcal{C}_{2\pi}(\mathbb{R}, \mathbb{R})$), then we set $h_u : \overline{\mathbb{D}} \longrightarrow \mathbb{R}$, the continuous function on $\overline{\mathbb{D}}$ that coincides on $\mathbb{T}$ with $\mathbb{T} \ni e^{i\theta} \longmapsto u(\theta)$ and is harmonic in $\mathbb{D}$ (see Proposition A.2.3). Moreover, let $\widetilde{h}_u$ be the conjugate harmonic function to h_u with $\widetilde{h}_u(0) = 0$ (see Proposition A.2.4). When $\widetilde{h}_u$ extends to a continuous function on $\overline{\mathbb{D}}$, we arrive at the function $\widetilde{f}_u := h_u + i\widetilde{h}_u \in \mathcal{A}(\mathbb{D})$ with $u_{\widetilde{f}_u} = u$.

Proof of Theorem 7.3.1. Step 1°. It suffices to prove that the set

$$E^+ := \{f \in \mathcal{A}(\mathbb{D}) : u_f \in \mathbf{\mathfrak{N}\mathfrak{D}}(\mathbb{R})\}$$

is residual in $\mathcal{A}(\mathbb{D})$. Indeed, suppose that E^+ is residual in $\mathcal{A}(\mathbb{D})$, i.e., $\mathcal{A}(\mathbb{D}) \setminus E^+ = \bigcup_{n=1}^{\infty} F_n$, where the closure of F_n has empty interior, $n \in \mathbb{N}$. Then $E = E^+ \cap (iE^+)$, and since $\mathcal{A}(\mathbb{D}) \ni g \longmapsto ig \in \mathcal{A}(\mathbb{D})$ is a homeomorphism, iE^+ is also residual, which immediately shows that E itself is residual.

Step 2°. We say that a function $u \in \mathcal{C}_{2\pi}(\mathbb{R}, \mathbb{R})$ satisfies the condition (D_n) if for every $\theta \in \mathbb{R}$, there exists a $y \in (\theta, \theta + \frac{1}{n})$ such that $|\boldsymbol{\Delta} u(\theta, y)| > n$, $n \in \mathbb{N}$. Moreover, put

$$E_n := \{f \in \mathcal{A}(\mathbb{D}) : u_f \text{ satisfies condition } (D_n)\}, \quad n \in \mathbb{N}.$$

We claim that the set E_n is open in $\mathcal{A}(\mathbb{D})$. Indeed, let $(f_j)_{j=1}^{\infty} \subset \mathcal{A}(\mathbb{D}) \setminus E_n$ with $f_j \longrightarrow f \in \mathcal{A}(\mathbb{D})$. Put $u_j := u_{f_j}$ and $u := u_f$. Then $(u_j)_{j=1}^{\infty}$ is uniformly convergent to u. Moreover, using the periodicity of the functions u_j, we see that there are points $\theta_j \in [0, 2\pi]$ such that

$$|u_j(y) - u_j(\theta_j)| \le n|y - \theta_j|, \quad y \in (\theta_j, \theta_j + \frac{1}{n}).$$

By a compactness argument, we may assume (extracting a subsequence if necessary) that $\theta_j \longrightarrow \theta \in [0, 2\pi]$. If $\theta < y < \theta + \frac{1}{n}$, then for all sufficiently large j, we have $\theta_j < y < \theta_j + 1/n$ and thus $|u_j(y) - u_j(\theta_j)| \leq n|y - \theta_j|$. Hence, the uniform convergence leads to $|u(y) - u(\theta)| \leq n|y - \theta|$, showing that u does not satisfy condition (D_n) or that $f \notin E_n$.

Step 3°. Fix an n and let $f \in \mathcal{A}(\mathbb{D})$. Define $f_0(z) := \sum_{n=0}^{\infty} (\frac{1}{2})^n z^{99^n}$, $z \in \overline{\mathbb{D}}$. Obviously, $f_0 \in \mathcal{A}(\mathbb{D})$ and $u_{f_0}(x) = \sum_{n=0}^{\infty} (\frac{1}{2})^n \cos(99^n x)$ for $\in \mathbb{R}$. Then $f - f_0 \in \mathcal{A}(\mathbb{D})$, and therefore, if $\varepsilon > 0$ is given, then one finds a complex polynomial p such that $\|f - f_0 - p\|_{\overline{\mathbb{D}}} < \varepsilon$ (Exercise). If we fix a $\theta \in \mathbb{R}$, then $|\boldsymbol{\Delta} u_p(\theta, y)| \leq M$ on $(\theta, \theta + \frac{1}{n})$ for a sufficiently large M. Now recall that $\max\{|D^+ u_{f_0}(x)|, |D_+ u_{f_0}(x)|\} = +\infty$, $x \in \mathbb{R}$ (see Theorem 3.5.1). Thus there exists a $y_n \in (\theta, \theta + \frac{1}{n})$ with $\boldsymbol{\Delta} u_{f_0}(\theta, y_n) > n + M$, which finally gives that $f_0 + p \in E_n$. Hence $\mathcal{A}(\mathbb{D}) \setminus E_n$ is nowhere dense in $\mathcal{A}(\mathbb{D})$.

Step 4°. To conclude the proof, it remains only to mention that $E \supset \bigcap_{n \in \mathbb{N}} E_n$. □

Remark 7.3.2. (a) The use of the existence of the concrete function $f_0 \in E$ in the proof above led to a simple proof of Theorem 7.3.1. In [Esk14], another proof is given that is not based on the Weierstrass function but on some kind of "zigzag" functions.

(b) Similar results (with respect to directional derivatives) are also known for the boundary values of functions in $\mathcal{A}(\mathbb{D}^n)$.

The following result may be understood in contrast to Theorem 7.3.1, namely that the set of 2π-periodic continuous nowhere differentiable functions u with a "good" associated harmonic conjugate $\widetilde{h}_u$ is small in $\mathcal{C}_{2\pi}(\mathbb{R})$.

Proposition 7.3.3. *The set*

$$L := \{u \in \mathcal{C}_{2\pi}(\mathbb{R}, \mathbb{R}) : \widetilde{h}_u \in \mathcal{C}(\overline{\mathbb{D}}, \mathbb{C}),\ u, v_{\widetilde{f}} \in \mathbf{ND}(\mathbb{R})\},$$

where $\widetilde{f} = \widetilde{f}_u = h_u + i\widetilde{h}_u$, *is dense and of first category in* $\mathcal{C}_{2\pi}(\mathbb{R}, \mathbb{R})$ *(equipped with the supremum norm).*

Proof. Step 1°. To prove that L is dense in $\mathcal{C}_{2\pi}(\mathbb{R}, \mathbb{R})$, fix a function $u \in \mathcal{C}_{2\pi}(\mathbb{R})$ and a positive ε. Since h_u is continuous on $\overline{\mathbb{D}}$, there exists an $r \in (0, 1)$ such that $|h_u(z) - h_u(rz)| < \varepsilon/2$, $z \in \overline{\mathbb{D}}$. Put $u_r(\theta) := \operatorname{Re} h_r(e^{i\theta})$.

Denote by $\widetilde{h}_u$ the harmonic conjugate to h_u. Then the harmonic conjugate to h_r, $h_r(z) := h_u(rz)$, is given by $\mathbb{D} \ni z \overset{\widetilde{h}_r}{\longmapsto} \widetilde{h}_u(rz)$, which obviously extends to a continuous function on $\overline{\mathbb{D}}$. Put $f := h_r + i\widetilde{h}_r$. Then $f \in \mathcal{A}(\mathbb{D})$ with $u_f(\theta) = u_r(\theta)$, $\theta \in \mathbb{R}$. Applying Theorem 7.3.1, we obtain a function $g \in \mathcal{A}(\mathbb{D})$ such that

$$\|f - g\|_{\overline{\mathbb{D}}} < \varepsilon/2 \quad \text{and} \quad u_g, v_g \in \mathbf{ND}(\mathbb{R}).$$

Moreover, we have

$$\|u - u_g\|_{\mathbb{R}} \leq \|u - u_r\|_{\mathbb{R}} + \|u_r - u_g\|_{\mathbb{R}} \leq \varepsilon/2 + \|f - g\|_{\overline{\mathbb{D}}} < \varepsilon.$$

Hence, L is dense in $\mathcal{C}_{2\pi}(\mathbb{R})$.

Step 2°. For $n \in \mathbb{N}$ put $L_n := \{u \in \mathcal{C}_{2\pi}(\mathbb{R}) : \|\widetilde{h}_u\|_{\mathbb{D}} \leq n\}$. Obviously, $L \subset \bigcup_{n=1}^{\infty} L_n$. Note that L_n is a closed subset of $\mathcal{C}_{2\pi}(\mathbb{R})$. Indeed, let $(u_j)_{j=1}^{\infty} \subset L_n$ with $u_j \longrightarrow u$ in $\mathcal{C}_{2\pi}(\mathbb{R})$. Then $(h_j)_{j=1}^{\infty}$, $h_j := h_{u_j}$, converges uniformly to h_u on $\mathbb{T}$ and thus on $\overline{\mathbb{D}}$ (use the maximum principle for harmonic functions). Moreover, $\|\widetilde{h}_j\|_{\mathbb{D}} \leq n$, where $\widetilde{h}_j := \widetilde{h}_{u_j}$, $j \in \mathbb{N}$. By virtue of the Carathéodory inequality (see Proposition A.2.6), we have

$$|\widetilde{h}_u(z) - \widetilde{h}_j(z)| \le 2\frac{1+|z|}{1-|z|}\|h_j - h_u\|_{\overline{\mathbb{D}}} \underset{j\to\infty}{\longrightarrow} 0,$$

implying that $\|\widetilde{h}_u\|_{\mathbb{D}} \le n$, i.e., $u \in L_n$.

Step 3°. It remains to show that L_n has no interior points, or equivalently, that $\mathcal{C}_{2\pi}(\mathbb{R}) \setminus L_n$ is dense in $\mathcal{C}_{2\pi}(\mathbb{R})$, $n \in \mathbb{N}$. Indeed, fix an $n \in \mathbb{N}$, a function $u \in \mathcal{C}_{2\pi}(\mathbb{R})$, and a positive ε. Then $u - u^* \in \mathcal{C}_{2\pi}(\mathbb{R})$, where u^* is the function from Example A.2.5. By virtue of the theorem of Fejér (see Proposition A.2.7), we obtain a complex polynomial $p \in \mathbb{C}[z]$ such that $|u(\theta) - u^*(\theta) - \operatorname{Re} p(e^{i\theta})| < \varepsilon$, $\theta \in \mathbb{R}$. Moreover, $\|\widetilde{h}_{u^*} + \operatorname{Im} p\|_{\mathbb{D}} = \infty$. Thus, the function $\mathbb{R} \ni \theta \longmapsto u^* + \operatorname{Re} p(e^{i\theta})$ does not belong to L_n. □

7.4 The Jarnik–Marcinkiewicz Theorems

The main result here deals with derived numbers instead of differentiability.

Theorem 7.4.1 (cf. [Mar35]). *Let $(h_j)_{j=1}^\infty \subset \mathbb{R}_*$ be a sequence converging to 0. Denote by M the set of all $f \in \mathcal{C}(\mathbb{I})$ such that for every measurable almost everywhere finite function g on $\mathbb{I}$, there exist a subsequence $(h_{j_k})_{k=1}^\infty$ of $(h_j)_{j=1}^\infty$ and a full-measure set $E_g \subset (0,1)$ such that*

$$\lim_{k\to\infty} \boldsymbol{\Delta} f(x, x + h_{j_k}) = g(x), \quad x \in E_g.$$

Then M is residual in $\mathcal{C}(\mathbb{I})$.

Remark 7.4.2. (a) If we take g to be the constant function $\lambda \in \mathbb{R}$, then λ is a derived number of f for almost all $a \in \mathbb{I}$. Hence, one may read this result as follows: the typical function in $\mathcal{C}(\mathbb{I})$, although very wildly behaved, possesses some kind of regularity, i.e., for each real λ, its behavior near every point outside of some null set is similar to its behavior near every other point relative to the sequence $(h_k)_{k=1}^\infty$.

(b) We may also think of f as a universal primitive for all measurable, almost everywhere finite functions g on $\mathbb{I}$. Note that according to a result of W. Sierpiński (see [Sie35]), we have for every measurable, almost everywhere finite function on $\mathbb{I}$ and for every sequence $(h_j)_{j=1}^\infty \subset \mathbb{R}_*$ with $\lim_{j\to\infty} h_j = 0$, a function f on $\mathbb{I}$ such that $\boldsymbol{\Delta} f(x, x+h_j) \underset{j\to\infty}{\longrightarrow} g(x)$, $x \in (0,1)$. Note that this result is true without some exceptional set.

(c) According to the footnote in [Mar35], Theorem 7.4.1 is due to S. Saks.

The proof of Theorem 7.4.1 requires some preparation.

Lemma 7.4.3. *Let $E_k \subset \mathbb{I}$, $k \in \mathbb{N}$, be measurable sets with $\sum_{k=1}^\infty \mathcal{L}(E_k) < +\infty$. Then there exists a subset $E \subset \mathbb{I}$, $\mathcal{L}(E) = 0$, such that if $x \in \mathbb{I} \setminus E$, then there exists an index j_x with $x \notin \bigcup_{j\ge j_x} E_j$.*

Lemma 7.4.3 is a standard result from measure theory; nevertheless, we repeat its simple proof.

Proof. Put $A := \{x \in \mathbb{I} : x \in E_k$ for infinitely many $k\}$. We have to show that $\mathcal{L}(E) = 0$. Put $g(x) := \sum_{k=1}^\infty \chi_{E_k}(x)$, $x \in \mathbb{I}$, where χ_{E_k} denotes the characteristic function of the set E_k, i.e., $\chi_{E_k} = 1$ on E_k, while $\chi_{E_k} = 0$ on $\mathbb{I} \setminus E_k$. Note that g is the limit function of an increasing sequence of integrable functions with convergent integrals. Then, applying the theorem on monotone convergence leads to the fact that g is an integrable function, i.e., g is almost everywhere finite, which immediately implies that $\mathcal{L}(A) = 0$ (observe that $x \in A$ iff $g(x) = +\infty$). □

Moreover, we quote the following results, which will be used in the proof. For their proofs, the reader is asked to consult the corresponding books on measure theory, such as [Rud74].

Theorem 7.4.4 (Lusin's Theorem). *Suppose f is a measurable real-valued function on $\mathbb{I}$. Then there exists an $g \in \mathcal{C}(\mathbb{I})$ such that $\mathcal{L}(\{x \in \mathbb{I} : f(x) \neq g(x)\}) < \varepsilon$.*

Theorem 7.4.5 (Egorov's Theorem). *Let $(f_j)_{j=1}^\infty$ be a sequence of measurable functions on $\mathbb{I}$ that converges at every point in $\mathbb{I}$ and let $\varepsilon > 0$. Then there is a measurable set $E \subset \mathbb{I}$ with $\mathcal{L}(\mathbb{I} \setminus E) < \varepsilon$ such that $(f_j)_{j=1}^\infty$ converges uniformly on E.*

Let $\mathcal{P}$ denote the set of all real-valued polynomials with rational coefficients. Obviously, $\mathcal{P}$ is a countable set. So we fix an enumeration of $\mathcal{P}$, i.e., $\mathcal{P} = \{p_j \in \mathcal{P} : j \in \mathbb{N}\}$, where $p_j \neq p_k$ if $j \neq k$.

Lemma 7.4.6. *Let $(h_j)_{j=1}^\infty \subset \mathbb{R}_*$ be a sequence with $\lim h_j = 0$ and let $f \in \mathcal{C}(\mathbb{I})$. Then the following statements are equivalent:*

(i) *for every measurable, almost everywhere finite function g on $\mathbb{I}$ there exist a subsequence $(h_{j_k})_{k=1}^\infty \subset (h_j)_{j=1}^\infty$ and a set $E_g \subset (0,1)$ with $\mathcal{L}(E_g) = 1$ such that*

$$\lim_{k\to\infty} \boldsymbol{\Delta} f(x, x + h_{j_k}) = g(x), \quad x \in E_g;$$

(ii) *for every pair $(k,n) \in \mathbb{N}^2$, there exist infinitely many indices $s > n$ such that*

$$\mathcal{L}\Big(\Big\{x \in (0,1) : \ x + h_s \in \mathbb{I}, \ \Big|\boldsymbol{\Delta} f(x, x+h_s) - p_k(x)\Big| \geq \frac{1}{n}\Big\}\Big) < \frac{1}{n};$$

(iii) *for every pair $(k,n) \in \mathbb{N}^2$, there exists an index $s > n$ such that*

$$\mathcal{L}\Big(\Big\{x \in (0,1) : \ x + h_s \in \mathbb{I}, \ \Big|\boldsymbol{\Delta} f(x, x+h_s) - p_k(x)\Big| \geq \frac{1}{n}\Big\}\Big) < \frac{1}{n}.$$

Proof. (i)$\Longrightarrow$(ii): Fix $k, n \in \mathbb{N}$. Since p_k is a measurable function on $\mathbb{I}$, we may choose a subsequence $(h_{j_s})_{s=1}^\infty \subset (h_j)_{j=1}^\infty$ such that

$$\boldsymbol{\Delta} f(\cdot, \cdot + h_{j_s}) \underset{s\to\infty}{\longrightarrow} p_k \text{ almost everywhere on } (0,1).$$

Now fix an index $s'_n \in \mathbb{N}$ with $n < s'_n$ such that $|h_{j_s}| < \frac{1}{4n}$ for all $s \geq s'_n$. Then the sequence $(\boldsymbol{\Delta} f(x, x+h_{j_s}))_{s \geq s'_n}$ is defined on $(\frac{1}{4n}, 1 - \frac{1}{4n}) =: J_n$ and converges there almost everywhere to p_k.

By virtue of Egorov's theorem (see Theorem 7.4.5), there exists a set $E_{k,n} \subset J_n$, $\mathcal{L}(E_{k,n}) < \frac{1}{2n}$, such that the former sequence converges uniformly on $J_n \setminus E_{k,n}$ to p_k. In particular, we find an index $s_n > s'_n$ such that for all $s \geq s_n$, one has

$$\Big|\boldsymbol{\Delta} f(x, x + h_{j_s}) - p_k(x)\Big| < \frac{1}{n}, \quad x \in J_n \setminus E_{k,n}, \ s \geq s_n.$$

It remains to note that $j_s > n$ for all $s \geq s_n$.

(ii)$\Longrightarrow$(iii): There is nothing to prove.

(iii)$\Longrightarrow$(i): We begin with a measurable, almost everywhere finite function g on J. Applying Lusin's theorem (see Theorem 7.4.4), we see that there exists a sequence $(g_j)_{j=1}^\infty \subset \mathcal{C}(\mathbb{I})$ such that

$$\mathcal{L}(E_j) < \frac{1}{2^j}, \text{ where } E_j := \{x \in \mathbb{I} : g(x) \neq g_j(x)\}, \ j \in \mathbb{N}.$$

Then according to Lemma 7.4.3, one may find a set $E_0 \subset \mathbb{I}$, $\mathcal{L}(E_0) = 0$, such that every $x \in \mathbb{I} \setminus E_0$ lies in only a finite number of the sets E_j. In other words, for every $x \in \mathbb{I} \setminus E_0$, there exists an index j'_x such that $x \notin \bigcup_{j \geq j'_x} E_j$, i.e., $g(x) = g_j(x)$ if $j \geq j'_x$.

Applying the Weierstrass approximation theorem, we obtain a sequence of polynomials $(q_j)_{j=1}^{\infty}$ with $\|g_j - q_j\|_{\mathbb{I}} < \frac{1}{j}$, $j \in \mathbb{N}$. Then it is easy to construct a subsequence $(p_{k_j})_{j=1}^{\infty} \subset (p_k)_{k=1}^{\infty}$ such that $\|p_{k_j} - q_j\|_{\mathbb{I}} < \frac{1}{j}$, $j \in \mathbb{N}$. Hence, for every $x \in \mathbb{I} \setminus E_0$, we have that if $j \geq j'_x$, then $|g(x) - p_{k_j}(x)| < \frac{2}{j}$. Thus the sequence $(p_{k_j})_{j=1}^{\infty}$ converges on $\mathbb{I} \setminus E_0$ pointwise to g.

Using the assumption in (iii), there are strictly increasing sequences $(m_j)_{j=1}^{\infty}$ and $(s_j)_{j=1}^{\infty}$ of natural numbers with $s_{j+1} > 2^{m_{j+1}} > s_j > 2^{m_j}$ such that

$$\mathcal{L}\Big(\Big\{x \in (0,1) : \, x + h_{s_j} \in \mathbb{I}, \, \Big|\boldsymbol{\Delta} f(x, x + h_{s_j}) - p_{k_j}(x)\Big| \geq \frac{1}{2^{m_j}}\Big\}\Big) < \frac{1}{2^{m_j}}.$$

Note that $(h_{s_j})_{j=1}^{\infty}$ is a subsequence of $(h_j)_{j=1}^{\infty}$.

Put

$$S_j := \{x \in (0,1) : \, x + h_{s_j} \in \mathbb{I}, \, |\boldsymbol{\Delta} f(x, x + h_{s_j}) - p_{k_j}(x)| < \frac{1}{2^{m_j}}\}.$$

As above, using Lemma 7.4.3, we see that there exists a set $S_0 \subset (0,1)$, $\mathcal{L}(S_0) = 0$, such that if $x \in (0,1) \setminus S_0$, then there exists a $j''_x \in \mathbb{N}$ such that $x \in \bigcap_{j \geq j''_x} S_j$.

Fix a point $x \in (0,1) \setminus (E_0 \cup S_0)$. Then there exists an index $j_x > \max\{j'_x, j''_x\}$ such that $x + h_{s_j} \in (0,1)$, $j \geq j_x$. Thus if $j \geq j_x$, then

$$\begin{aligned}|\boldsymbol{\Delta} f(x, x + h_{s_j}) - g(x)| &\leq |\boldsymbol{\Delta} f(x, x + h_{s_j}) - p_{k_j}(x)| + |p_{k_j}(x) - g(x)| \\ &< \frac{1}{2^{m_j}} + \frac{2}{j} \underset{j \to \infty}{\longrightarrow} 0.\end{aligned}$$

Hence, the sequence $(\boldsymbol{\Delta} f(x, x + h_{s_j}))_{j \in \mathbb{N}}$ converges to $g(x)$ for almost every $x \in (0,1)$. □

Lemma 7.4.7. *Let $f_1, f_2 \in \mathcal{C}(\mathbb{I})$. Moreover, it is assumed that f_2 is almost everywhere differentiable on $\mathbb{I}$. If $\varepsilon > 0$, then there exists a continuous function h on $\mathbb{I}$, almost everywhere differentiable on $\mathbb{I}$, such that $\|f_1 - h\|_{\mathbb{I}} < \varepsilon$ and $h' = f'_2$ almost everywhere on $\mathbb{I}$.*

Remark 7.4.8. (a) This lemma was used by M. Lusin to prove that a finite measurable function is almost everywhere the primitive of a continuous function (see the footnote in [Mar35]).

(b) Even more is true: every measurable finite function f on $\mathbb{I}$ possesses a uniformly smooth primitive F, i.e., there exists a function F on $\mathbb{I}$, uniformly smooth, such that $F' = f$ almost everywhere. Recall that F is said to be *uniformly smooth on* $\mathbb{I}$ if F is continuous and

$$\frac{F(x+h) + F(x-h) - 2F(x)}{h} = o(1) \text{ as } h \longrightarrow 0,$$

uniformly in $x \in (0,1)$ (see [Hov09]).

Proof of Lemma 7.4.7. Take a partition $t_0 = 0 < t_1 < \cdots < t_N = 1$ of $\mathbb{I}$ such that on each of the intervals $J_k := [t_k, t_{k+1}]$, one has $\sup\{|f(x) - f(y)| : x, y \in J_k\} < \frac{\varepsilon}{2}$, where $f := f_1 - f_2$. Fix a $k \in \{0, \ldots, N-1\}$. On J_k choose a continuous monotone function g_k with $g_k(t_k) = f(t_k)$, $g_k(t_{k+1}) = f(t_{k+1})$, and $g'_k = 0$ almost everywhere on J_k (use, for example, the Cantor ternary function; see [KK96]). Put $g(x) := g_k(x)$ if $x \in J_k$. Then g is a continuous function on $\mathbb{I}$, and if $x \in J_k$, then

$$|f(x) - g(x)| \leq |f(x) - f(t_{s_k})| + |f(t_{s_k}) - g(x)| \leq \varepsilon,$$

where $t_{s_k} \in \{t_k, t_{k+1}\}$ is the endpoint of J_k with $f(t_{s_k}) = \max\{f(t_k), f(t_{k+1})\}$. Hence, $h := f_2 + g$ satisfies the properties claimed in the lemma. □

Proof of Theorem 7.4.1. For $(k,n) \in \mathbb{N}^2$, put

$$M_{k,n} := \Big\{ f \in \mathcal{C}(\mathbb{I}) : \forall_{s>n} : \\ \mathcal{L}\Big(\Big\{x \in (0,1) : x + h_s \in \mathbb{I},\ |\boldsymbol{\Delta} f(x, x+h_s) - p_k(x)| \geq \frac{1}{n}\Big\}\Big) \geq \frac{1}{n}\Big\}.$$

Note that $\mathcal{C}(\mathbb{I}) \setminus M = \bigcup_{k,n\in\mathbb{N}} M_{k,n}$. So we have to show that each of the sets $M_{k,n}$ is closed and without interior points.

Fix a pair (k,n).

Step 1°. Take a sequence $(f_j)_{j=1}^{\infty} \in M_{k,n}$ such that $f_j \underset{j\to\infty}{\Longrightarrow} f_0 \in \mathcal{C}(\mathbb{I})$ and fix an $s > n$. Put

$$S_j := \Big\{x \in (0,1) :\ x + h_s \in \mathbb{I},\ |\boldsymbol{\Delta} f_j(x, x+h_s) - p_k(x)| \geq \frac{1}{n}\Big\}, \quad j \in \mathbb{N}_0.$$

We wish to show that $S^* := \bigcap_{m=1}^{\infty} \bigcup_{\ell \geq m} S_\ell \subset S_0$. Indeed, let $x \in S^*$. Then for every $m \in \mathbb{N}$, we find an index $\ell_m \geq m$ such that $x \in S_{\ell_m}$. Without loss of generality, we may assume that $\ell_{m+1} > \ell_m$ (otherwise, take a suitable subsequence). Then $x + h_s \in \mathbb{I}$ and $|\boldsymbol{\Delta} f_{\ell_m}(x, x+h_s) - p_k(x)| \geq \frac{1}{n}$, $m \in \mathbb{N}$. Passing to the limit gives $x \in S_0$.

What remains is to observe that $\mathcal{L}(S^*) \geq \frac{1}{n}$ (Exercise). Hence, $f_0 \in M_{k,n}$.

Step 2°. Take a function $f \in \mathcal{C}(\mathbb{I})$ and a positive ε. Applying Lemma 7.4.7 for the data p_k, f, ε, one obtains a continuous function h on $\mathbb{I}$, almost everywhere differentiable, such that $\|f - h\|_{\mathbb{I}} < \varepsilon$ and $h' = p_k$ almost everywhere. It remains to observe that $h \in M_{k,n}$. □

A generalization of Theorem 7.4.1 is due to V. Jarnik. To be able to state his result, let us first recall that a measurable set $E \subset [0,1)$ has *right upper density α, $\alpha \in [0,1]$, at a point $x \in [0,1)$ if*

$$\limsup_{h\to 0} \frac{\mathcal{L}(E \cap (x, x+h))}{h} = \alpha.$$

Theorem 7.4.9 (cf. [Jar34]). *Let M be the set of all $f \in \mathcal{C}(\mathbb{I})$ for which there exists a measurable set $E \subset [0,1)$, $\mathcal{L}(E) = 1$, satisfying the following property: if $x \in E$ and $\lambda \in \overline{\mathbb{R}}$, then there is a measurable set $E_x \subset \mathbb{I}$ with right upper density 1 at x such that*

$$\lim_{E_x \setminus \{x\} \ni y \to x+} \frac{f(y) - f(x)}{y - x} = \lambda.$$

Then M is residual in $\mathcal{C}(\mathbb{I})$.

Remark 7.4.10. λ is called an *essential right derivative of f at x.*

Proof of Theorem 7.4.9. Step 1°. Let $f \in \mathcal{C}(\mathbb{I})$ and let (a,b) be an open interval in $\mathbb{R}$. We say that the interval is *essential for f* if there exists a set $B_f \subset \mathbb{I}$, $\mathcal{L}(B_f) = 0$, such that if $x \in [0,1) \setminus B_f$, then we can find a measurable set $E_x \subset \mathbb{I}$ with right upper density 1 at x such that

$$\boldsymbol{\Delta} f(x,y) \subset (a,b) \quad \text{for all} \quad y \in E_x,\ x < y.$$

We enumerate all the intervals with rational endpoints, i.e., one gets a sequence of intervals $\big((a_k,b_k)\big)_{k=1}^{\infty}$. Put

$$M_k=\{f\in\mathcal{C}(\mathbb{I}):(a_k,b_k)\text{ is essential for }f\}.$$

We claim that $\bigcap_{k=1}^{\infty}M_k\subset M$. Indeed, fix a function $f\in\bigcap_{k=1}^{\infty}M_k$. Using the fact that $I_k:=(a_k,b_k)$ is essential for f, we see that there exist sets $B_k:=B_{f,k}\subset\mathbb{I}$, $\mathcal{L}(B_k)=0$, such that for every k and $x\in[0,1)\setminus B_k$, we obtain a set $E_{k,x}\subset\mathbb{I}$ of right upper density 1 at x such that $\boldsymbol{\Delta}f(x,y)\subset I_k$ whenever $y\in E_{k,x}$, $x<y$.

Put $B_f:=\bigcup_{k=1}^{\infty}B_{f,k}\subset\mathbb{I}$. Obviously, B_f is, as a countable union of Lebesgue zero sets, again a Lebesgue zero set. Now fix a point $x\in[0,1)\setminus B_f$ and a $\lambda\in\overline{\mathbb{R}}$. Choose a subsequence $\big((a_{k_n},b_{k_n})\big)_{n\in\mathbb{N}}$ of the above intervals such that $\lim_{n\to\infty}a_{k_n}=\lim_{n\to\infty}b_{k_n}=\lambda$. Since $x\notin B_{f,k_n}$, we have the measurable set $E_n:=E_{k_n,x}\subset\mathbb{I}$ of right upper density 1 at x such that $\boldsymbol{\Delta}f(x,y)\subset I_{k_n}$ whenever $y\in E_n$, $x<y$.

Recall that if $n\in\mathbb{N}$, then there is a sequence $(s_{n,\ell})_{\ell=1}^{\infty}\subset(0,1)$ with $\lim_{\ell\to\infty}s_{n,\ell}=0$ such that

$$\frac{\mathcal{L}(E_n\cap(x,x+s_{n,\ell}))}{s_{n,\ell}}=1.$$

Let $h_1:=s_{1,1}$. Obviously, $\mathcal{L}(E_1\cap(x,x+h_1))>0=(1-\frac{1}{1})h_1$. Assume that a positive number h_n, $n\geq 1$, with $\mathcal{L}(E_n\cap(x,x+h_n))>(1-\frac{1}{n})h_n$ has been constructed. Then choose an ℓ_n so large that

$$s_{n+1,\ell_n}<\frac{h_n}{n}\quad\text{and}\quad\mathcal{L}(E_{n+1}\cap(x,x+s_{n+1,\ell_n}))>\Big(1-\frac{1}{n+1}\Big)s_{n+1,\ell_n}.$$

Put $h_{n+1}:=s_{n+1,\ell_n}$.

Finally, we introduce the following measurable set:

$$E=E_x:=\bigcup_{n=1}^{\infty}(x+h_{n+1},x+h_n)\cap E_n.$$

Using the estimate

$$\begin{aligned}1&\geq\frac{\mathcal{L}(E\cap(x,x+h_n))}{h_n}\\&\geq\frac{\mathcal{L}(E_n\cap(x,x+h_n))}{h_n}-\frac{\mathcal{L}(E_n\cap(x,x+h_{n+1}))}{h_n}\geq\Big(1-\frac{2}{n}\Big)\underset{n\to\infty}{\longrightarrow}1,\end{aligned}$$

it immediately follows that E has right upper density 1 at x.

Recall that $a_{k_n}<\boldsymbol{\Delta}f(x,y)<b_{k_n}$ if $y\in E_n\cap(x+h_{n+1},x+h_n)$. Thus, $\lim_{E\setminus\{x\}\ni y\to x+}\boldsymbol{\Delta}f(x,y)=\lambda$. Since $x\in[0,1)\setminus B_f$ has been arbitrarily chosen, we get $f\in M$.

Thus it remains to show that all the sets M_k are residual in $\mathcal{C}(\mathbb{I})$ (Exercise).

Step 2°. Fix a $k\in\mathbb{N}$. Let $f\in\mathcal{C}(\mathbb{I})$. Put

$$A_k(f):=\Big\{x\in[0,1):\{y\in\mathbb{I}:y>x,\ \boldsymbol{\Delta}f(x,y)\subset I_k\}\ \text{has right upper density less than 1 at }x\Big\}.$$

Note that $f\notin M_k$ if and only if $\mathcal{L}(A_k(f))>0$.

Moreover, for $s \in \mathbb{N}$, set

$$B_{k,s}(f) := \Big\{x \in [0,1) : \text{ if } 0 < u \le \min\{1/s, 1-x\}, \\ \text{then } \mathcal{L}(\{y \in (x, x+u] : \boldsymbol{\Delta} f(x,y) \subset I_k\}) \le (1-1/s)u\Big\}.$$

Then $A_k(f) = \bigcup_{\ell,s\in\mathbb{N}} B_{k,s}(f)$ (Exercise). Note that $\mathcal{L}(A_k(f)) > 0$ if and only if $\mathcal{L}(B_{k,s}(f)) > 0$ for some s.

Finally, put

$$C_k(\ell, s) := \{f \in \mathcal{C}(\mathbb{I}) : \mathcal{L}(B_{k,s}(f)) \ge 1/\ell\}, \quad \ell, s \in \mathbb{N}.$$

Then $\mathcal{C}(\mathbb{I}) \setminus M_k = \bigcup_{\ell,s\in\mathbb{N}} C_k(\ell, s)$. Thus it remains to show that the sets $C_k(\ell, s)$ are nowhere dense in $\mathcal{C}(\mathbb{I})$.

Step 3°. We will prove that each of the sets $C_k(\ell, s)$ is closed in $\mathcal{C}(\mathbb{I})$. Indeed, fix ℓ, s and take a sequence $(f_j)_{j=1}^\infty \subset C_k(\ell, s)$ with $f_j \Longrightarrow f \in \mathcal{C}(\mathbb{I})$. It suffices to show that $\mathcal{L}(B_{k,s}(f)) \ge 1/\ell$.

Put $F := \bigcap_{m=1}^\infty \bigcup_{j\ge m} B_{k,s}(f_j)$. Since $\mathcal{L}(b_{k,s}(f_j) \ge 1/\ell$, it follows that $\mathcal{L}(F) \ge 1/\ell$. So it remains to verify that $F \subset B_{k,s}(f)$. To do so, fix a point $x \in F$. Then there exists a subsequence $(f_{j_m})_{m=1}^\infty$ such that $x \in B_{k,s}(f_{j_m})$ for all $m \in \mathbb{N}$.

Fix a $u \in (0, \frac{1}{s}]$, $x + u \le 1$. Then

$$\mathcal{L}(\{y \in (x, x+u] : \boldsymbol{\Delta} f_{k_m}(x,y) \not\subset I_k\}) \ge \frac{u}{s}.$$

Put

$$G(x,u) := \bigcap_{m=1}^\infty \bigcup_{\varkappa \ge m} \{y \in (x, x+u] : \boldsymbol{\Delta} f_{k_\varkappa}(x,y) \not\subset I_k\}.$$

Then $\mathcal{L}(G(x,u)) \ge \frac{u}{s}$.

Moreover, using the uniform convergence and the fact that $\mathbb{R} \setminus I_k$ is closed, we see that

$$G(x,u) \subset \{y \in (x, x+u] : \boldsymbol{\Delta} f(x, x+y) \not\subset I_k\}.$$

Therefore, the measure of the set on the right is greater than or equal to $\frac{u}{s}$. Recall that u was arbitrarily chosen with the above conditions. Therefore, $x \in B_{k,s}(f)$.

Step 4°. It remains to prove that all the sets $C_k(\ell, s)$ do not have interior points. Indeed, fix $\ell, s \in \mathbb{N}$ and assume that the function $f \in \mathcal{C}(\mathbb{I})$ is an interior point of $C_k(\ell, s)$, i.e., that there exists a positive r such that the open ball $B(f, 2r)$ with center f and radius $2r$ is contained in $C_k(\ell, s)$. Choose a real-valued polynomial p such that $\|p - f\|_{\mathbb{I}} < r$. Then $B(p, r) \subset C_k(\ell, s)$.

Take a large positive number t such that

$$|\boldsymbol{\Delta} p(x,y)| < t - |c_k| \quad \text{and} \quad \|p'\|_{\mathbb{I}} < t - |c_k|,$$

where $c_k := \frac{a_k + b_k}{2}$. Moreover, fix a natural number m such that $t < mr$ and if $0 < |y - x| \le \frac{1}{m}$, then $\max\{|\boldsymbol{\Delta} p(x,y) - p'(x)|, |p'(x) - p'(y)|\} < \frac{b_k - a_k}{4}$.

We now introduce the following function $g \in \mathcal{C}(\mathbb{I})$:

$$g(x) := \begin{cases} 0, & \text{if } x = \frac{\sigma}{m}, \sigma = 0, 1, \dots, m \\ \widetilde{g}(x), & \text{if } \frac{\sigma}{m} \le x \le \frac{\sigma+1}{m} - \frac{1}{2\ell m}, \ \sigma = 0, \dots, m-1 \\ \widehat{g}(x), & \text{if } \frac{\sigma+1}{m} - \frac{1}{2\ell m} \le x \le \frac{\sigma+1}{m}, \ \sigma = 0, \dots, m-1 \end{cases},$$

where

$$\widetilde{g}(x) := \Big(-p'(\frac{\sigma}{m}) + c_k\Big)\Big(x - \frac{\sigma}{m}\Big),$$
$$\widehat{g}(x) := 2\ell\Big(-p'(\frac{\sigma}{m}) + c_k\Big)\Big(1 - \frac{1}{2\ell}\Big)\Big(\frac{\sigma+1}{m} - x\Big).$$

Then

$$\|g\|_{\mathbb{I}} \le \frac{1}{m}(\|p'\|_{\mathbb{I}} + |c_k|)(1 - \frac{1}{2\ell}) < \frac{t}{m} < r.$$

Thus $p + g \in C_k(\ell, s)$, and so $\mathcal{L}(B_{k,s}(p+g)) \ge \frac{1}{\ell}$.

Now fix a point $x \in [\frac{\sigma}{m}, \frac{\sigma+1}{m} - \frac{1}{2\ell m})$ for an arbitrary $\sigma \in \{0,\dots,m-1\}$ and take a point u with $0 < u \le \min\{\frac{1}{s}, \frac{\sigma+1}{m} - \frac{1}{2\ell m} - x\}$. If $x < y \le x+u$, then

$$|\boldsymbol{\Delta}(p+g)(x,y) - c_k| \le |\boldsymbol{\Delta}p(x,y) - p'(x)| + |p'(x) - p'(\frac{\sigma}{m})| < \frac{b_k - a_k}{2}.$$

In other words, $\boldsymbol{\Delta}(p+g)(x,y) \in I_k$, which implies that

$$\mathcal{L}(\{y \in (x, x+u) : \boldsymbol{\Delta}(p+g)(x,y) \subset I_k\}) \ge u.$$

Hence, $x \notin B_{k,s}(p+g)$.

So we end with the following inclusion:

$$B_{k,s}(p+g) \subset \bigcup_{\sigma=0}^{m-1} \Big(\frac{\sigma+1}{m} - \frac{1}{2\ell m}, \frac{\sigma+1}{m}\Big),$$

which immediately leads to $\frac{1}{\ell} \le \mathcal{L}(B_{k,s}(p+q)) \le \frac{1}{2\ell}$; a contradiction. □

7.5 The Saks Theorem

So far, we have seen that the typical continuous function on $\mathbb{I}$ has neither an infinite two-sided derivative nor a finite one-sided derivative on $\mathbb{I}$. To be more precise, the complement of this set of functions is of first category in $\mathcal{C}(\mathbb{I})$. So the question remains whether there exists a function $f \in \mathcal{C}(\mathbb{I})$ that has nowhere a unilateral derivative, finite or infinite. Functions with this property are called *Besicovitch functions*, because the first example of such a function was found by A.S. Besicovitch in [Bes24]. For a while, there was hope of proving the existence of such a function by showing that the set $\mathfrak{B}(\mathbb{I})$ of Besicovitch functions is a residual one in $\mathcal{C}(\mathbb{I})$. But in [Sak32], S. Saks showed that $\mathfrak{B}(\mathbb{I})$ is a set of first category. Unfortunately, his proof used very advanced tools, and therefore we will omit it. The proof we are going to present is due to F.S. Cater (see [Cat86]); it uses only simple tools, as we will see. Other proofs have been given by D. Preiss (see [Bru84]) and by K.M. Garg (see [Gar70]). Note that this result suggests that a priori there is no direct way to apply Baire category theory to prove the existence of a Besicovitch function. We will return to this problem in Chap. 11.

Theorem 7.5.1 (The Saks Theorem)**.** *The set*

$$\mathfrak{B}(\mathbb{I}) = \{f \in \mathcal{C}(\mathbb{I}) : f \textit{ has nowhere a finite or infinite unilateral derivative}\}$$

of all Besicovitch functions is of first category in $\mathcal{C}(\mathbb{I})$.

Obviously, it is enough to study the functions with right-sided derivatives $-\infty$. The proof of Theorem 7.5.1 is by the following lemmas.

Lemma 7.5.2. *The set*

$$\mathfrak{F}_1 := \{f \in \mathcal{C}(\mathbb{I}) : \mathcal{L}(\{x \in (0,1) : D^+ f(x) < 0\}) > 0\}$$

is of first category in $\mathcal{C}(\mathbb{I})$.

Remark 7.5.3. Note that by Remark 2.1.7, the set

$$A_f := \{x \in (0,1) : D^+ f(x) < 0\}$$

is Borel measurable.

Proof of Lemma 7.5.2. Let

$$S := \{(a,b,d) \in \mathbb{Q}^3 : 0 < a < b < b+d < 1\}.$$

For $(a,b,d) \in S$, put

$$F(a,b,d) := \Big\{g \in \mathcal{C}(\mathbb{I}) : \mathcal{L}(\{x \in (a,b) : g(x) \geq g(\xi),\ x < \xi \leq x+d\}) > \frac{b-a}{2}\Big\}.$$

It suffices to prove that

(a) $\mathfrak{F}_1 \subset \bigcup_{(a,b,d)\in S} F(a,b,d)$,
(b) $F(a,b,d)$ is closed and without interior points.

Indeed, to verify (a), fix an $f \in \mathfrak{F}_1$ and a point $c \in A_f$. Then there exists a positive h_0 with $c + h_0 \leq 1$ such that

$$\sup\Big\{\boldsymbol{\Delta} f(c, c+h) :\ 0 < h \leq h_0\Big\} < \frac{1}{2} D^+ f(c) < 0.$$

Thus $f(c) > f(\xi)$ for $c < \xi \leq c + h_0$.

Let

$$A_k := \Big\{x \in (0,1) :\ x + \frac{1}{k} \leq 1 \text{ and } f(x) \geq f(\xi),\ x \leq \xi \leq x + \frac{1}{k}\Big\}, \quad k \in \mathbb{N}_2.$$

Then $A_k \subset A_{k+1}$ and $A_f \subset \bigcup_{k\in\mathbb{N}_2} A_k$. Since $\mathcal{L}(A_f) > 0$, there exists a $k_0 \in \mathbb{N}_2$ with $\mathcal{L}(A_{k_0}) > 0$. For simplicity, we write $A = A_{k_0}$. Take an open set $U \subset (0,1)$, $A \subset U$, such that $\mathcal{L}(U \setminus A) < \mathcal{L}(A)/3$. Then $\mathcal{L}(U) < 4\mathcal{L}(A)/3$.

Now write $U = \bigcup_{j \in M \subset \mathbb{N}} J_j$, where J_j are the pairwise disjoint connected components of U. Note that J_j are open intervals. Therefore,

$$\mathcal{L}(U) = \sum_{j\in M} \mathcal{L}(J_j) < \frac{4\mathcal{L}(A)}{3} = \frac{4}{3}\sum_{j\in M} \mathcal{L}(A \cap J_j).$$

Thus one of the intervals J_j, say $J_{j_0} = J = (a_0, b_0)$, is such that

$$b_0 - a_0 = \mathcal{L}(J) < 4\mathcal{L}(A \cap J)/3.$$

Finally, one may choose $a, b \in \mathbb{Q}$, $a_0 < a < b < b_0$, such that $\frac{b-a}{2} < \mathcal{L}(A \cap (a,b))$. Hence, $f \in F(a,b,d)$ with $d := \min\{\frac{1}{k_0}, 1-b\}$.

(b) Step 1°. We will show that the set $F(a,b,d)$ is closed in $\mathcal{C}(\mathbb{I})$. Indeed, let $(g_j)_{j=1}^{\infty} \subset F(a,b,d)$ and assume that this sequence is uniformly convergent to a $g \in \mathcal{C}(\mathbb{I})$. Put

$$E_j := \{x \in (a,b) : g_j(x) \geq g_j(\xi),\ x \leq \xi \leq x+d\}.$$

Then $\mathcal{L}(E_j) \geq \frac{b-a}{2}$. Define $E := \bigcap_{k=1}^{\infty} \bigcup_{j=k}^{\infty} E_j$. Then

$$\frac{b-a}{2} \leq \mathcal{L}(E_k) \leq \mathcal{L}\Big(\bigcup_{j=k}^{\infty} E_j\Big) \underset{k\to\infty}{\longrightarrow} \mathcal{L}(E).$$

Let $x \in E$. Then $x \in E_{j_k}$ for a strictly increasing sequence $(j_k)_{k=1}^{\infty} \subset \mathbb{N}$. Thus $g_{j_k}(x) \geq g_{j_k}(\xi)$, $x \leq \xi \leq x+d$, which immediately implies that $g(x) \geq g(\xi)$ for all $\xi \in [x, x+d]$. Hence, $g \in F(a,b,d)$.

Step 2°. It remains to verify that $F(a,b,d)$ has no interior points. Let us assume the contrary, i.e., that $B(f,\varepsilon) \subset F(a,b,d)$ for some $f \in F(a,b,d)$ and $\varepsilon > 0$. Choose a partition $a = t_0 < t_1 < \cdots < t_n = b$ such that $t_j - t_{j-1} < d$ and $\sup\{f(t) : t \in [t_{j-1}, t_j]\} - \inf\{f(t) : t \in [t_{j-1}, t_j]\} < \varepsilon/4$. Then there exists a "zigzag" function $h \in \mathcal{C}(\mathbb{I})$, $0 \leq h \leq \frac{\varepsilon}{2}$, such that $h(t_j) = \varepsilon/2$, $j = 0, \dots, n$, and $\mathcal{L}(N(h) \cap (a,b)) > \frac{b-a}{2}$, where $N(h)$ denotes the zero set of h. Then $g := f + h \in B(f,\varepsilon) \subset F(a,b,d)$.

Let $x \in (t_{j-1}, t_j)$ with $h(x) = 0$. Then $t_j \in (x, x+d)$, and therefore,

$$(f+h)(x) = f(x) < f(t_j) + \varepsilon/4 < (f+h)(t_j).$$

Hence,

$$N(h) \cap (a,b) \subset \{x \in (a,b) :\ (f+h)(x) < \sup\{(f+h)(\xi) : x \leq \xi \leq x+d\}\}.$$

On the other hand, we know that

$$\mathcal{L}(\{x \in (a,b) :\ g(x) \geq g(\xi),\ x \leq \xi \leq x+d\}) \geq \frac{b-a}{2},$$

which implies that $\mathcal{L}((a,b)) > (b-a)$; a contradiction. □

Lemma 7.5.4. *The set*

$$\mathfrak{F}_2 := \{f \in \mathcal{C}(\mathbb{I}) : f|_J \text{ is monotone for some open subinterval } J \subset \mathbb{I}\}$$

is of first category in $\mathcal{C}(\mathbb{I})$.

Proof. We only mention that the sets $\{f \in \mathcal{C}(\mathbb{I}) :\ f|_{(a,b)} \text{ is monotone}\}$, where $0 < a < b < 1$ are rational numbers, are closed sets in $\mathcal{C}(\mathbb{I})$ without interior points. Details are left for the reader (EXERCISE). □

Corollary 7.5.5. *The set* $\mathfrak{F}_1 \cup \mathfrak{F}_2$ *is of first category in* $\mathcal{C}(\mathbb{I})$.

So far we know that "most" of the functions $f \in \mathcal{C}(\mathbb{I})$ are nowhere monotone and A_f is of zero measure.

Lemma 7.5.6. *Let* $f \in \mathcal{C}(\mathbb{I}) \setminus (\mathfrak{F}_1 \cup \mathfrak{F}_2)$. *Then there exists a set* $M \subset \mathbb{I}$ *such that*

(a) $f'_+(x) = -\infty$ *for all* $x \in M$;
(b) $M \cap J$ *has the cardinality of the continuum for every open subinterval* $J \subset \mathbb{I}$.

Proof. Choose open sets $U_n \supset A_f$, $U_n \subset \mathbb{I}$, with $\mathcal{L}(U_n) < 1/2^n$, $n \in \mathbb{N}$, and put

$$F(x) := x + \sum_{n=1}^{\infty} \int_0^x \chi_{U_n}(t)dt, \quad x \in \mathbb{I},\ n \in \mathbb{N},$$

where χ_{U_n} means the characteristic function of U_n. Note that the series is a uniformly convergent series of continuous functions. Hence, $F \in \mathcal{C}(\mathbb{I})$. By a simple calculation, one sees that $D_+F(x) \geq 1$, $x \in \mathbb{I}$, and $F'(x) = +\infty$ for $x \in A_f$.

Take an open interval $J \subset \mathbb{I}$. Since f is not monotone on J, there are points $r_J = r, s_J = s \in J$ with $r < s$ and $f(r) > f(s)$. Then there exists a positive number ε such that $(f+\varepsilon F)(r) > (f+\varepsilon F)(s)$. Let y be between these two numbers and set $x_y := \sup\{t \in (r,s) : (f+\varepsilon F)(t) = y\}$. Obviously, $r < x_y < s$. In particular, $(f+\varepsilon F)(t) < y$ whenever $x_y < t < s$. Therefore, $D^+(f+\varepsilon F)(x_y) \leq 0$. By virtue of $D_+F(x_y) \geq 1$, we get $D^+f(x_y) \leq -\varepsilon$, i.e., $x_y \in A_f$. Thus, $F'_+(x_y) = +\infty$. Using $D^+(f+\varepsilon F)(x_y) \leq 0$ leads to $D^+f(x_y) = -\infty$. Hence, $f'_+(x_y) = -\infty$.

Let $M_J := \{x_y \in (r_y, s_y) : y \in \big((f+\varepsilon F)(s), (f+\varepsilon F)(r)\big)\}$ and $M := \bigcup_J M_J$. According to the construction above, each of the sets M_J is uncountable. Thus its cardinality is that of the continuum. Hence, M satisfies the properties stated in the lemma. □

What we have proved shows that the set of functions without infinite unilateral derivatives is contained in a set of first category, i.e., most of the functions in $\mathcal{C}(\mathbb{I})$ are not Besicovitch functions. Therefore, it seems difficult to find a concrete Besicovitch function or to show at least, by an abstract argument, their existence. Nevertheless, a clever modification of the categorial argument, as will be done in Chap. 11, will prove the existence of Besicovitch functions.

Remark 7.5.7. From the reasoning above, we see that the set of $f \in \mathcal{C}(\mathbb{I})$ that have a dense set of knot points is residual in $\mathcal{C}(\mathbb{I})$; see Theorem 7.2.1(b). One has only to observe that the sets

$$\{f \in \mathcal{C}(\mathbb{I}) : \mathcal{L}(\{x \in (0,1) : D^+f(x) < m\}) > 0\}, \quad m \in \mathbb{N},$$

are of first category. Hence, the set

$$\{f \in \mathcal{C}(\mathbb{I}) : D^+f(x) = +\infty \text{ for almost all } x \in [0,1)\}$$

is a residual one.

7.6 The Banach–Mazurkiewicz Theorem Revisited

Recall that $\mathfrak{M}(\mathbb{I})$ is a residual set in $\mathcal{C}(\mathbb{I})$, i.e., $\mathcal{C}(\mathbb{I}) \setminus \mathfrak{M}(\mathbb{I})$ is small in the categorial sense. There exists a stronger notion of being small, which even has a nice geometric interpretation. It concerns the sizes of pores in $\mathcal{C}(\mathbb{I}) \setminus \mathfrak{M}(\mathbb{I})$ near a function $f \in \mathcal{C}(\mathbb{I}) \setminus \mathfrak{M}(\mathbb{I})$.

Definition 7.6.1. Let $X = (X,d)$ be a metric space, $A \subset X$, and $x \in X$.

(a) A is said to be *porous at* x if

$$p(x,A) := \limsup_{R \to 0+} \frac{\gamma(x,R,A)}{R} > 0,$$

where

$$\gamma(x, R, A) := \sup\{r > 0 : \exists_{z \in X} : \mathbb{B}_d(z, r) \subset \mathbb{B}_d(x, R) \setminus A\}$$

($p(x, A) := 0$ if there is no positive r as above for certain R).

(b) A is said to be *porous* if A is porous at every point $a \in A$.

(c) A is called *σ-porous* if A can be written as $A = \bigcup_{j=1}^{\infty} A_j$, where all the sets A_j are porous.

Remark 7.6.2. (a) There are, in fact, two notions of porosity at a point, namely that of the previous definition and so-called *lower porosity*, where the lim sup in the above definition is substituted by lim inf. Therefore, porosity in the sense of the above definition is also called *upper porosity*. Since we are dealing only with the notion defined above, we will simply speak of porous sets.

(b) Porosity in $\mathbb{R}$ was used by A. Denjoy in 1920 (under a different notation). The theory of σ-porous sets began with investigations of the boundary behavior of functions by E.P. Dolženko (see [Dol67]). It seems that he was the first to use the term "porous."

(c) Let $A \subset X$ be a porous set. Then A is nowhere dense in X. Therefore, every σ-porous set is meagre or of first category (EXERCISE).

(d) If $A \subset \mathbb{R}^n$ is porous, then its Lebesgue measure is zero (use the Lebesgue density theorem). But there exists a closed nowhere dense set $A \subset \mathbb{R}^n$ of Lebesgue measure zero that is not σ-porous. For a proof see [Zaj87].

(e) Let X be a complete metric space without isolated points. Then there exists a closed nowhere dense set $A \subset X$ that is not σ-porous (see [Zaj87]). Therefore, saying that a set is σ-porous is, in general, a stronger statement than claiming that it is of first category.

(f) For more information on porous sets, see [Zaj87] and [Zaj05].

(g) Moreover, a set $A \subset X$ is porous at a point x if and only if there exists a positive ϱ such that for every $\varepsilon > 0$, there are $R \in (0, \varepsilon]$ and $z \in X$ such that $\mathbb{B}_d(z, \varrho R) \subset \mathbb{B}_d(x, R) \setminus A$ (EXERCISE).

Theorem 7.6.3 (cf. [Ani93]). *Let*

$$M := \{f \in \mathcal{C}(\mathbb{I}) : f \textit{ without a finite or infinite derivative in } (0, 1)\}.$$

Then $\mathfrak{M}(\mathbb{I}) \cap M$ *has a σ-porous complement in* $\mathcal{C}(\mathbb{I})$.

Proof. Step 1°. Put

$$\mathfrak{M}^{\pm}(\mathbb{I}) := \Big\{f \in \mathcal{C}(\mathbb{I}) : \max\{|D^{\pm}f(x)|, |D_{\pm}f(x)|\} = +\infty \quad \text{for all } x \in \mathbb{I} \setminus \Big\{\frac{1 \pm 1}{2}\Big\}\Big\}.$$

Then $\mathfrak{M}(\mathbb{I}) = \mathfrak{M}^{+}(\mathbb{I}) \cup \mathfrak{M}^{-}(\mathbb{I})$.

It suffices to prove that $\mathfrak{M}^{+}(\mathbb{I})$ has a σ-porous complement in $\mathcal{C}(\mathbb{I})$. For $n \in \mathbb{N}_2$, put

$$A_n := \Big\{f \in \mathcal{C}(\mathbb{I}) : \exists_{x \in [0, 1 - \frac{1}{n}]} : |\boldsymbol{\Delta} f(x, x + t)| \le nt,\ t \in \Big(0, \frac{1}{n}\Big]\Big\}.$$

Note that $\bigcup_{n \in \mathbb{N}_2} A_n = \mathcal{C}(\mathbb{I}) \setminus \mathfrak{M}^{+}(\mathbb{I})$.

Fix an $n \in \mathbb{N}_2$. It remains to prove that A_n is porous at each of its functions. Fix a function $f \in A_n$ and a number $\varepsilon \in (0, 1)$. Then there exists a positive δ such that $|f(x') - f(x'')| < \varepsilon$ for all $x', x'' \in \mathbb{I}$ with $|x' - x''| < \delta$.

Take a positive a with $a < \min\{\frac{\varepsilon}{n}, \delta\}$. For $x \in (0, 2a]$, put $u_a(x) := 2\psi(\frac{x}{2a})$, where $\psi(t) = \operatorname{dist}(t, \mathbb{Z})$, and extend this function to the whole of $\mathbb{R}$ with a period of $2a$. For simplicity, denote the extension again by u_a. If $x \in \mathbb{R}$, then there exists an interval $I_a(x) \subset [\frac{a}{8}, a]$ of length $\frac{a}{8}$ such that $|u_a(x+t) - u_a(x)| \geq \frac{t}{3a}$ for all $t \in I_a(x)$. Indeed, we have only to study points $x \in [0, 2a]$. If $x \in [0, \frac{3a}{4}) \cup [a, \frac{7a}{4})$, then take $I_a(X) = [\frac{a}{8}, \frac{a}{4}]$. If $x \in [\frac{3a}{4}, a) \cup [\frac{7a}{4}, 2a]$, then $I_a(x) = [\frac{7a}{8}, a]$ does the job.

In particular, $u_a|_{\mathbb{I}} \in \mathcal{C}(\mathbb{I})$, $\|u_a\|_{\mathbb{I}} = 1$, and $|\boldsymbol{\Delta} u_a(x, x+t)| \geq \frac{1}{3a}$ whenever $x \in [0, 1 - \frac{1}{n}]$ and $t \in I_a(x)$. Put $g := 75\varepsilon u_a$ and $f^* := g + f$. Then $B(f^*, \varepsilon) \cap A_n = \varnothing$. In fact, for $h \in B(f^*, \varepsilon)$, $x \in [0, 1 - \frac{1}{n}]$, and $t \in I_a(x)$ we have

$$|h(x+t) - h(x)| \geq |g(x+t) - g(x)| - |(h-g)(x+t) - (h-g)(x)|$$
$$- |(h - f^*)(x+t) - (h - f^*)(x)| \geq \frac{75\varepsilon t}{3a} - \varepsilon - 2\varepsilon \geq \frac{\varepsilon t}{8} > nt.$$

Hence, $h \notin A_n$. Moreover, recall that $\|f^* - f\|_{\mathbb{I}} = \|g\|_{\mathbb{I}} = 75\varepsilon$. Therefore, $B(f^*, \varepsilon) \subset B(f, 76\varepsilon) \setminus A_n$, which implies that A_n is porous at f and, since f was arbitrarily chosen, that A_n is porous.

Step 2°. Let

$$S := \{f \in \mathcal{C}(\mathbb{I}) : \liminf_{t \to 0+} \boldsymbol{\Delta}_s f(x, t) = -\infty,\ \limsup_{t \to 0+} \boldsymbol{\Delta}_s f(x, t) = +\infty,$$
$$x \in (0, 1)\}.$$

Obviously, $S \subset M$.

For $n \in \mathbb{N}_3$ put

$$A_n^{\pm} := \Big\{ f \in \mathcal{C}(\mathbb{I}) : \exists_{x \in [\frac{1}{n}, 1 - \frac{1}{n}]} : \pm \boldsymbol{\Delta}_s f(x, t) \leq n,\ t \in \Big(0, \frac{1}{n}\Big] \Big\}.$$

Then $\mathcal{C}(\mathbb{I}) \setminus S = \bigcup_{n \in \mathbb{N}_3} (A_n^+ \cup A_n^-)$.

Fix an $n \in \mathbb{N}_3$. Then A_n^+ is porous. Indeed, take a function $f \in A_n^+$ and a positive ε. Choose $\delta > 0$ such that $|f(x') - f(x'')| < \varepsilon$ for all $x', x'' \in \mathbb{I}$ with $|x' - x''| < \delta$. Moreover, fix a positive a with $a < \min\{\frac{1}{6n}, \frac{\delta}{12}, \frac{\varepsilon}{2n}\}$.

Similarly to what was done in Step 1°, we find a continuous function u_a on $\mathbb{R}$ with period $7a$, $|u_a| \leq 1 = |u_a(a)|$ on $\mathbb{R}$, and for every $x \in \mathbb{R}$, an interval $I_a(x) \subset [a, 6a]$ of length equal to $\frac{a}{10}$ such that $u_a(x+t) - u_a(x-t) \geq \frac{t}{100a}$, $t \in I_a(x)$. In fact, u_a is given on $[0, 7a]$ by the piecewise linear function whose graph connects the points $(0,0)$, $(a, -1)$, $(3a, 1)$, and $(7a, 0)$ by segments and that is extended to $\mathbb{R}$ by $u_a(x + 7a) = u_a(x)$, $x \in [0, 7a]$ (Exercise).

Put $f^* := 400\varepsilon u_a$ and $g := f + f^*$. Then $\mathbb{B}(g, \varepsilon) \subset \mathbb{B}(f, 401\varepsilon) \setminus A_n^+$ (Exercise), giving that A_n^+ is porous.

In a similar way, it is shown that also A_n^- is porous. Therefore, the complement of S in $\mathcal{C}(\mathbb{I})$ is σ-porous. □

Remark 7.6.4. (a) Looking at the proof of the former theorem shows that even the set S (see Step 2° of the proof) has a σ-porous complement in $\mathcal{C}(\mathbb{I})$.

(b) V. Aniusu also proved that the set of continuous functions having nowhere on $\mathbb{I}$ a finite one-sided approximate derivative has a porous complement in $\mathcal{C}(\mathbb{I})$.

Remark 7.6.5. The complement of the set $\mathfrak{ND}_{\pm}^{\infty}$ cannot be σ-porous because of Theorem 7.5.1.

7.7 The Structure of $\mathcal{ND}(\mathbb{I})$

In the Scottish book, S. Banach asked for a description of the structure of the set D of all differentiable functions on $\mathbb{I}$. In [Maz31], S. Mazurkiewicz proved that the complement of D in $\mathcal{C}(\mathbb{I})$ is analytic, but D itself is not. Recall that a set is called *analytic* (or a *Souslin set*) if it is the continuous image of a Borel set. On the other hand, due to R.D. Mauldin (see [Mau79]), the set $\mathcal{ND}(\mathbb{I})$ also has an analytic complement in $\mathcal{C}(\mathbb{I})$, but it is not a Borel sct in $\mathcal{C}(\mathbb{I})$.

Theorem 7.7.1. (a) *$\mathcal{C}(\mathbb{I}) \setminus \mathcal{ND}(\mathbb{I})$ is an analytic set, i.e., it is the continuous image of some Borel set.*
(b) *$\mathcal{ND}(\mathbb{I})$ is not a Borel set in $\mathcal{C}(\mathbb{I})$.*

Proof. (a) Recall that a function $f \in \mathcal{C}(\mathbb{I})$ has a finite derivative at a point $x \in \mathbb{I}$ if and only if for every $n \in \mathbb{N}$, there exists an $m \in \mathbb{N}$ such that if $h_1, h_2 \in \mathbb{R}_*$ with $\max\{|h_1|, |h_2|\} < \frac{1}{m}$ and $x + h_1, x + h_2 \in \mathbb{I}$, then

$$|\boldsymbol{\Delta} f(x, x + h_1) - \boldsymbol{\Delta} f(x, x + h_2)| \leq \frac{1}{n}.$$

Let $E_{n,m}$ denote the set of all pairs $(f, x) \in \mathcal{C}(\mathbb{I}) \times \mathbb{I}$ satisfying the above condition. Note that $E_{n,m}$ is a closed subset of $\mathcal{C}(\mathbb{I}) \times \mathbb{I}$. Since $\mathcal{C}(\mathbb{I}) \setminus \mathcal{ND}(\mathbb{I})$ is the projection of $\bigcap_{n=1}^{\infty} \bigcup_{m=1}^{\infty} E_{n,m}$, it is an analytic set.

(b) The proof of the fact that $\mathcal{ND}(\mathbb{I})$ is not a Borel set is significantly more complicated. It is based on the fact that there exists an analytic subset of the Cantor ternary set that is not a Borel set. For further details, the reader is asked to consult [Mau79].

□

Thus far, it is known that the set $\mathcal{ND}(\mathbb{I})$ is residual in $\mathcal{C}(\mathbb{I})$. There were attempts to reformulate this statement to say that "almost" all functions from $\mathcal{C}(\mathbb{I})$ are nowhere differentiable, which finally led to the notion of *prevalence* (see [HSY92]). We will not give full details here but only a bit of an idea of what is going on in this direction.

Theorem 7.7.2 (cf. [Hun94]). *There exist two functions $g, h \in \mathcal{C}(\mathbb{I})$ such that for every function $f \in \mathcal{C}(\mathbb{I})$, the set*

$$M_f := \{(\lambda, \mu) \in \mathbb{R}^2 : f + \lambda g + \mu h \in \mathcal{ND}(\mathbb{I})\} \subset \mathbb{R}^2$$

is of full Lebesgue measure.

In other words, the space $\mathcal{C}(\mathbb{I})$ can be partitioned into parallel planes such that in each plane, almost all (in the sense of Lebesgue) functions are nowhere differentiable. In fact, the plane is generated by the following two functions:

$$g(x) := \sum_{n=1}^{\infty} \frac{1}{n^2} \cos(2^n \pi x), \quad h(x) := \sum_{n=1}^{\infty} \frac{1}{n^2} \sin(2^n \pi x), \quad x \in \mathbb{I}.$$

Note that g, h are exactly the functions whose existence is claimed in Theorem 7.7.2.

Lemma 7.7.3. *There exists a positive number c such that if $\alpha, \beta \in \mathbb{R}$ and $J \subset \mathbb{I}$ is a closed interval of length $\varepsilon \leq \frac{1}{2}$, then*

$$\max_J\{\alpha g + \beta h\} - \min_J\{\alpha g + \beta h\} \geq \frac{c\sqrt{\alpha^2 + \beta^2}}{(\log \varepsilon)^2}.$$

Proof. Put $F := \alpha g + \beta h$. Then

$$F(x) = \sum_{n=1}^{\infty} \frac{1}{n^2}\Big(\alpha \cos(2^n \pi x) + \beta \sin(2^n \pi x)\Big) = \sqrt{\alpha^2+\beta^2} \sum_{n=1}^{\infty} \frac{1}{n^2} \cos(2^n \pi x + \theta),$$

where $\theta \in [0, 2\pi]$ is correctly chosen. Note that without loss of generality, we may assume that $\alpha^2 + \beta^2 = 1$.

Let $J \subset \mathbb{I}$ be a closed interval of length $\frac{1}{2^m}$. If $k \in \mathcal{C}(\mathbb{I})$, then

$$\max_J k - \min_J k \geq 2^m \pi \int_J k(x) \cos(2^{m+j}\pi x + \theta) dx, \quad j \in \mathbb{N}. \tag{7.7.1}$$

Indeed, in order to prove (7.7.1), one may assume that

$$-\min_J k = \max_J k =: K \geq 0$$

(add to both sides of (7.7.1) a suitable constant and use the fact that $\int_J \cos(2^{n+j}\pi x)dx = 0$, $j \in \mathbb{N}$). Then $|k| \leq K$. Therefore,

$$\int_J k(x) \cos(2^{m+j}\pi x + \theta) dx \leq K \int_J |\cos(2^{m+j}\pi x + \theta)| dx = K \frac{2}{2^m \pi},$$

which gives (7.7.1).

Continuing with the function F, (7.7.1) leads to

$$\begin{aligned}
\max_J F - \min_J F &\geq 2^m \pi \int_J \sum_{n=1}^{\infty} \frac{1}{n^2} \cos(2^n \pi x + \theta) \cos(2^{m+j}\pi x + \theta) dx \\
&= \sum_{n=1}^{\infty} \frac{2^m \pi}{n^2} \frac{1}{2} \int_J \Big(\cos\big((2^{m+j} - 2^n)\pi x\big) + \cos\big((2^{m+j} + 2^n)\pi x + 2\theta\big)\Big) dx.
\end{aligned}$$

Since the length of J is $\frac{1}{2^m}$, the integral of $\cos\big((2^{m+j} \pm 2^n)\pi x + \theta\big)$ over J is zero for $n > m$ unless $n = m + j$ and the minus sign is chosen. Let $n \leq m$ and $J = [a, a + \frac{1}{2^m}]$. Then

$$\begin{aligned}
&\int_J \cos\big((2^{m+j} \pm 2^n)\pi x\big) dx \\
&= \frac{\sin\big((2^{m+j} \pm 2^n)\pi(a + 2^{-m}) + \theta\big) - \sin\big((2^{m+j} \pm 2^n)\pi a + \theta\big)}{(2^{m+j} \pm 2^n)\pi} \\
&\geq -\frac{2^n}{2^m(2^{m+j} \pm 2^n)}.
\end{aligned}$$

Hence we obtain

$$\begin{aligned}
\max_J F - \min_J F &\geq \frac{\pi}{2(m+j)^2} - \sum_{n=1}^{m} \frac{\pi}{2n^2} \left(\frac{2^n}{2^{m+j} - 2^n} + \frac{2^n}{2^{m+j} + 2^n} \right) \\
&\geq \frac{\pi}{2(m+j)^2} - \frac{\pi}{2^m(2^j - 1)} \sum_{n=1}^{m} \frac{2^n}{n^2}.
\end{aligned}$$

To continue, one needs the inequality $\sum_{n=1}^{m} \frac{2^n}{n^2} \leq 5\frac{2^m}{m^2}$, which can be verified directly for $m = 1, 2, 3, 4$ and then by induction for $m \geq 4$ (Exercise). Therefore, the former inequality may be rewritten as

$$\max_J F - \min_J F \geq \frac{\pi}{2(m+j)^2} - \frac{5\pi}{m^2(2^j - 1)}.$$

In particular, for $j = 10$ and $m \geq 2$, one is led to

$$\max_J F - \min_J F \geq \frac{\pi}{2(m+10)^2} - \frac{5\pi}{m^2(2^{10} - 1)} \geq \frac{\pi}{2(6m)^2} - \frac{\pi}{200m^2} = \frac{2\pi}{225m^2}.$$

Finally, let J be an interval of length $\varepsilon \leq \frac{1}{2}$. Choose $m \in \mathbb{N}_2$ such that $\frac{1}{2^m} < \varepsilon \leq \frac{1}{2^{m-1}}$. Let $J_m \subset J$ be an interval of length 2^{-m}. Then

$$\max_J F - \min_J F \geq \max_{J_m} F - \min_{J_m} F \geq \frac{2\pi}{225m^2} \geq \frac{\pi}{450(m-1)^2} \geq \frac{(\log 2)^2\pi}{450(\log\varepsilon)^2},$$

proving the lemma. □

The main step in verifying Theorem 7.7.2 is contained in the next result, dealing with nowhere Lipschitz functions. Recall that the set of all nowhere Lipschitz functions is a Borel set (see Remark 2.5.2(d)), while the set $\mathcal{ND}(\mathbb{I})$ is not.

Proposition 7.7.4. *Let g, h be the functions from above. Then for every $f \in \mathcal{C}(\mathbb{I})$, the set*

$$\{(\lambda, \mu) \in \mathbb{R}^2 : f + \lambda g + \mu h \text{ is nowhere Lipschitz on } \mathbb{I}\} \subset \mathbb{R}^2$$

is of full Lebesgue measure.

Proof. Obviously, it suffices to show that the set of (λ, μ) for which $f + \lambda g + \mu h$ is M-Lipschitz at some point of $\mathbb{I}$ has measure zero. Denote this set by S_M, i.e.,

$$S_M = \{(\lambda, \mu) \in \mathbb{R}^2 : f + \lambda g + \mu h \text{ is } M\text{-Lipschitz at some } x \in \mathbb{I}\}.$$

Fix an $N \in \mathbb{N}_2$ and cover $\mathbb{I}$ with N closed intervals $J_1, \ldots, J_N$ of lengths $\varepsilon = \varepsilon_N := \frac{1}{N}$. Fix such an interval $J_k = J$ and denote by $S_{M,J}$ the subset of parameters $(\lambda, \mu) \in S_M$ for which the function $f + \lambda g + \mu h$ is M-Lipschitz at a point of J. We will discuss the diameter of $S_{M,J}$.

Let $(\lambda_j, \mu_j) \in S_{M,J}$, put $f_j = f + \lambda_j g + \mu_j h$, and assume that f_j is M-Lipschitz at the point $x_j \in J$, $j = 1, 2$. If $x \in \mathbb{I}$, then

$$|f_j(x) - f_j(x_j)| \leq M|x - x_j| \leq M\varepsilon.$$

Thus, $|f_1(x) - f_2(x) - (f_1(x_1) - f_2(x_2))| \leq 2M\varepsilon$, $x \in \mathbb{I}$, and therefore, $\max_J(f_1 - f_2) - \min_J(f_1 - f_2) \leq 4M\varepsilon$. Since $f_1 - f_2 = (\lambda_1 - \lambda_2)g + (\mu_1 - \mu_2)h$, Lemma 7.7.3 yields

$$\frac{c\sqrt{(\lambda_1 - \lambda_2)^2 + (\mu_1 - \mu_2)^2}}{(\log\varepsilon)^2} \leq 4M\varepsilon,$$

where c does not depend on J. Then

$$\operatorname{diam} S_{M,J} \leq \frac{4M\varepsilon(\log\varepsilon)^2}{c}.$$

Thus $S_{M,J}$ sits in a ball of radius $\frac{4M\varepsilon(\log\varepsilon)^2}{c}$. Therefore, the measure of $S_{M,J}$ is bounded from above by $\pi\frac{16M^2}{c^2}\varepsilon^2(\log\varepsilon)^4$. Since $S_M = \cup_{k=1}^N S_{M,J_k}$, it follows that the measure of S_M is bounded from above by $N\pi\frac{16M^2}{c^2}\varepsilon_N^2(\log\varepsilon_N)^4$. Letting $N \longrightarrow \infty$ finally shows that S_M is of measure zero. □

Proof of Theorem 7.7.2. It remains to recall that a function $f \in \mathcal{C}(\mathbb{I})$ that is nowhere Lipschitz on $\mathbb{I}$ belongs to $\mathfrak{ND}(\mathbb{I})$. □

Part III
Modern Approach

Chapter 8
Weierstrass-Type Functions II

Summary. In this chapter, using more advanced tools, we extend results stated in Chap. 3.

8.1 Introduction

Recall (§ 3.1) that

$$\boldsymbol{W}_{p,a,b,\boldsymbol{\theta}}(x) := \sum_{n=0}^{\infty} a^n \cos^p(2\pi b^n x + \theta_n), \quad x \in \mathbb{R},$$

where $p \in \mathbb{N}$, $0 < a < 1$, $ab \geq 1$, $\boldsymbol{\theta} := (\theta_n)_{n=0}^{\infty} \subset \mathbb{R}$. The reader is asked to recall the list of all partial results related to the function $\boldsymbol{W}_{p,a,b,\boldsymbol{\theta}}$ presented in Remark 3.1.1. Now our aim is to prove the following general theorems:

(9) If $ab \geq 1$, then $\boldsymbol{C}_{a,b}, \boldsymbol{S}_{a,b} \in \mathfrak{ND}(\mathbb{R})$ (Theorems 8.2.1 and 8.2.12).
(10) If $ab > 1$, then $\boldsymbol{W}_{1,a,b,\theta} \in \mathfrak{M}(\mathbb{R}) \subset \mathfrak{ND}_{\pm}(\mathbb{R})$ (Theorem 8.3.1). Using different tools, an analogous result will be proved in Theorem 8.7.3.
(11) If $b \in \mathbb{N}_2$ and $ab \geq 1$, then $\boldsymbol{W}_{1,a,b,\theta} \in \mathfrak{ND}_{\pm}(\mathbb{R})$ (Theorem 8.4.1).
(12) If ($p \in 2\mathbb{N}_0 + 1$ and $b > p$) or ($p \in 2\mathbb{N}$ and $b > \frac{p}{2}$), then $\boldsymbol{W}_{p,a,b,\boldsymbol{\theta}} \in \mathfrak{ND}(\mathbb{R})$ (Theorem 8.6.7).
(13) If $ab > 1$, then there exists a zero-measure set $\Xi \subset \mathbb{R}$ such that x is a knot point of the function $f := \boldsymbol{W}_{1,a,b,\theta}$ (i.e., $D^+f(x) = D^-f(x) = +\infty$, $D_+f(x) = D_-f(x) = -\infty$) for arbitrary $\theta \in \mathbb{R}$ and $x \in \mathbb{R} \setminus \Xi$ (Theorem 8.7.4).
(14) Let $\psi^* \approx 1.3518 \in (0, \frac{\pi}{2})$ be such that $\tan\psi^* = \pi + \psi^*$. If $ab \geq H := 1 + \frac{1}{\cos\psi^*} \approx 5.6034$, then $\boldsymbol{W}_{1,a,b,\theta} \in \mathfrak{M}(\mathbb{R}) \cap \mathfrak{ND}^\infty(\mathbb{R}) \subset \mathfrak{ND}_{\pm}(\mathbb{R}) \cap \mathfrak{ND}^\infty(\mathbb{R})$ (Theorem 8.7.6). Observe that the constant H is better than the original Weierstrass constant $1 + \frac{3}{2}\pi \approx 5.7123$ (cf. Theorem 3.5.1).

It is clear that many cases remain undecided. The most important open problems are the following:

[?] Is it true that if ($p \in 2\mathbb{N}_0 + 1$ and $b > p$) or ($p \in 2\mathbb{N}$ and $b > \frac{p}{2}$), then $\boldsymbol{W}_{p,a,b,\boldsymbol{\theta}} \in \mathfrak{ND}_{\pm}(\mathbb{R})$ [?] This would be a simultaneous generalization of (10), (11), and (12).

[?] Characterize the set of all $p, a, b, \boldsymbol{\theta}$ such that $\boldsymbol{W}_{p,a,b,\boldsymbol{\theta}} \in \mathfrak{ND}^\infty(\mathbb{R})$ [?] (cf. (14)).

M. Jarnicki, P. Pflug, *Continuous Nowhere Differentiable Functions*, Springer Monographs in Mathematics, DOI 10.1007/978-3-319-12670-8_8

8.2 Hardy's Method

This entire section is based on [Har16]. We wish to point out that nowadays, the main results presented in this section, Theorems 8.2.1 and 8.2.12, are direct consequences of more general results, e.g., of Theorem 8.6.7 or Theorem 8.7.3. Nevertheless, innovative for his time, Hardy's methods are in our opinion worth being presented.

Theorem 8.2.1 (cf. [Har16]). *Assume that* $0 < a < 1$, $ab > 1$, $\alpha := -\frac{\log a}{\log b}$. *Then it is impossible for the function* $t \longmapsto \boldsymbol{C}_{a,b}(\frac{t}{2\pi})$ *to be* $o(|t - t_0|^\alpha)$ *when* $t \longrightarrow t_0$ *for some* $t_0 \in \mathbb{R}$. *Consequently,*

- *the constant* α *is the maximal number such that* $\boldsymbol{C}_{a,b}$ *is* α*-Hölder continuous at some point* $t_0 \in \mathbb{R}$ *(cf. Remark 3.2.1*(g)*);*
- a finite $\boldsymbol{C}'_{a,b}(t_0)$ does not exist for every $t_0 \in \mathbb{R}$, and therefore, $\boldsymbol{C}_{a,b} \in \mathcal{ND}(\mathbb{R})$.

An analogous result is true for the function $\boldsymbol{S}_{a,b}$.

We will consider only the case of $\boldsymbol{C}_{a,b}$. The case of $\boldsymbol{S}_{a,b}$ is left to the reader as an EXERCISE. Put $\boldsymbol{H}_+ := \{s = \sigma + it \in \mathbb{C} : \sigma > 0\}$ and define

$$F(s) := \sum_{n=0}^{\infty} a^n e^{-sb^n}, \quad s \in \overline{\boldsymbol{H}}_+.$$

Remark 8.2.2. (a) Since

$$|a^n e^{-sb^n}| = a^n e^{-\sigma b^n} \le a^n, \quad s = \sigma + it \in \overline{\boldsymbol{H}}_+, \quad n \in \mathbb{N}_0,$$

the function F is well defined, $F \in \mathcal{O}(\boldsymbol{H}_+) \cap \mathcal{C}(\overline{\boldsymbol{H}}_+, \mathbb{C})$, and $|F(s)| \le A := \frac{1}{1-a}$, $s \in \overline{\boldsymbol{H}}_+$. We have

$$F(s) = \sum_{n=0}^{\infty} a^n e^{-\sigma b^n} \cos(tb^n) - i \sum_{n=0}^{\infty} a^n e^{-\sigma b^n} \sin(tb^n)$$
$$=: G(s) + iH(s), \quad s = \sigma + it \in \overline{\boldsymbol{H}}_+.$$

Let $g(t) := G(it) = \boldsymbol{C}_{a,b}(\frac{t}{2\pi})$, $t \in \mathbb{R}$.

(b) Using Remark 2.4.1, we see that

$$\frac{\partial G}{\partial t}(s) = -\sum_{n=1}^{\infty} (ab)^n e^{-\sigma b^n} \sin(tb^n),$$
$$\frac{\partial G}{\partial \sigma}(s) = -\sum_{n=1}^{\infty} (ab)^n e^{-\sigma b^n} \cos(tb^n), \quad s = \sigma + it \in \boldsymbol{H}_+.$$

For $\varrho > 0$, define

$$\varphi_\varrho(s) := \sum_{n=0}^{\infty} b^{n\varrho} e^{-sb^n}, \quad s \in \boldsymbol{H}_+.$$

Remark 8.2.3. (a) For each $\sigma_0 > 0$, we have $|b^{n\varrho}e^{-(\sigma+it)b^n}| \le |b^{n\varrho}e^{-\sigma_0 b^n}|$ for $\sigma \ge \sigma_0$ and $t \in \mathbb{R}$. Since

$$|b^{n\varrho}e^{-\sigma_0 b^n}|^{1/n} = b^{\varrho}e^{-\frac{1}{n}\sigma_0 b^n} \underset{n\longrightarrow+\infty}{\longrightarrow} 0,$$

we conclude that the function φ_ϱ is well defined, $\varphi_\varrho \in \mathcal{O}(\boldsymbol{H}_+)$, and $|\varphi_\varrho(\sigma+it)| \le \varphi_\varrho(\sigma_0)$ for $\sigma \ge \sigma_0$, $t \in \mathbb{R}$.

(b) By the Weierstrass theorem for holomorphic functions, we get

$$\varphi_\varrho^{(p)}(s) = (-1)^p \sum_{n=0}^{\infty} b^{n(\varrho+p)}e^{-sb^n} = (-1)^p\varphi_{\varrho+p}(s), \quad s \in \boldsymbol{H}_+,\ p \in \mathbb{N}. \tag{8.2.1}$$

(c) We have

$$\varphi_{1-\alpha}(\sigma+it_0) = \sum_{n=0}^{\infty} b^{n(1-\alpha)}e^{-(\sigma+it_0)b^n} = \sum_{n=0}^{\infty}(ab)^n e^{-(\sigma+it_0)b^n}.$$

Before going into detail, we will describe the main proof. G.H. Hardy mainly used results on the solution of the Dirichlet problem for boundary values on the real axis; see Lemma 8.2.4.

Proof of Theorem 8.2.1. Suppose that $\boldsymbol{C}_{a,b}(\frac{t}{2\pi}) = g(t) = o(|t-t_0|^\alpha)$ when $t \longrightarrow t_0$.

(1) First we obtain an integral representation for the function G (Lemma 8.2.4), which gives us formulas for $\frac{\partial G}{\partial \sigma}$ and $\frac{\partial G}{\partial t}$ (Remark 8.2.5).
(2) Using these formulas, we prove that $\frac{\partial G}{\partial t}(\sigma+it_0) = o(\sigma^{\alpha-1})$ and $\frac{\partial G}{\partial \sigma}(\sigma+it_0) = o(\sigma^{\alpha-1})$ when $\sigma \longrightarrow 0+$ (Lemma 8.2.6).
(3) Hence, by Remarks 8.2.2(b) and 8.2.3(c), we conclude that $\varphi_{1-\alpha}(\sigma+it_0) = o(\sigma^{\alpha-1})$ when $\sigma \longrightarrow 0+$.
(4) On the other hand, we will prove that $|\varphi_\varrho(\frac{\varrho}{b^m}+it)| \ge \frac{1}{2}(\frac{\varrho}{b^m})^{-\varrho}$, $t \in \mathbb{R}$, $m \in \mathbb{N}$, $\varrho \ge \varrho_0 \gg 1$ (Lemma 8.2.8).
(5) Lemma 8.2.9 will show that we always have $\varphi_\varrho(\sigma+it) = O(1/\sigma^\varrho)$ when $\sigma \longrightarrow 0+$. In particular, by Remark 8.2.3(b), $\varphi_{1-\alpha}^{(p)}(\sigma+it) = O(1/\sigma^{1-\alpha+p})$ when $\sigma \longrightarrow 0+$ ($p \in \mathbb{N}$).
(6) Now, by (3) and Lemma 8.2.11, we get $\varphi_{1-\alpha}^{(p)}(\sigma+it_0) = o(1/\sigma^{1-\alpha+p})$ when $\sigma \longrightarrow 0+$ ($p \in \mathbb{N}$).
(7) Using once again Remark 8.2.3(b), we see that $\varphi_{1-\alpha+p}(\sigma+it_0) = o(1/\sigma^{1-\alpha+p})$ when $\sigma \longrightarrow 0+$ ($p \in \mathbb{N}$).
(8) Taking $p \in \mathbb{N}$ such that $1-\alpha+p \ge \varrho_0$, where ϱ_0 is as in (4), we get

$$\Big(\frac{1-\alpha+p}{b^m}\Big)^{1-\alpha+p}\Big|\varphi_{1-\alpha+p}\Big(\frac{1-\alpha+p}{b^m}+it_0\Big)\Big| \ge \frac{1}{2}, \quad m \in \mathbb{N};$$

a contradiction. □

Lemma 8.2.4 (Schwarz Integral Formula). *Let $P = Q + iR \in \mathcal{O}(\boldsymbol{H}_+) \cap \mathcal{C}(\overline{\boldsymbol{H}}_+, \mathbb{C})$, $|P| \le A$, $q(t) := Q(it)$, $t \in \mathbb{R}$. Then*

$$Q(s) = \frac{\sigma}{\pi}\int_{-\infty}^{+\infty}\frac{q(u)du}{\sigma^2+(u-t)^2}, \quad s = \sigma+it \in \boldsymbol{H}_+.$$

Proof. Fix an $s = \sigma + it \in \boldsymbol{H}_+$. For $r > |s|$, let C_r denote the closed contour of the form $[ir, -ir] \cup \Gamma_r$, where $\Gamma_r(\theta) = re^{i\theta}$, $\theta \in [-\pi/2, \pi/2]$. By the Cauchy formula, we obtain

$$
\begin{aligned}
P(s) &= \frac{1}{2\pi i}\int_{C_r}\frac{P(\zeta)}{\zeta - s}d\zeta = \frac{1}{2\pi i}\int_{C_r}\frac{P(\zeta)}{\zeta - s}d\zeta - \frac{1}{2\pi i}\int_{C_r}\frac{P(\zeta)}{\zeta + \overline{s}}d\zeta \\
&= \frac{1}{\pi i}\int_{C_r}\frac{\sigma P(\zeta)}{(\zeta - s)(\zeta + \overline{s})}d\zeta \\
&= \frac{1}{\pi i}\int_{[ir,-ir]}\frac{\sigma P(\zeta)d\zeta}{(\zeta - s)(\zeta + \overline{s})} + \frac{1}{\pi i}\int_{\Gamma_r}\frac{\sigma P(\zeta)d\zeta}{(\zeta - s)(\zeta + \overline{s})} =: I_r' + I_r''.
\end{aligned}
$$

For $r > 2|s|$, we have

$$
|I_r''| \le \frac{1}{\pi}\int_{-\pi/2}^{\pi/2}\frac{\sigma A}{(\frac{1}{2}r)^2}r\,du = \frac{4\sigma A}{r} \underset{r\longrightarrow+\infty}{\longrightarrow} 0.
$$

Consequently,

$$
\begin{aligned}
Q(s) &= \lim_{r\to+\infty}\operatorname{Re}(I_r') = -\lim_{r\to+\infty}\frac{\sigma}{\pi}\int_{-r}^{r}\operatorname{Re}\Big(\frac{q(u)+iR(iu)}{(iu - s)(iu+\overline{s})}\Big)du \\
&= \lim_{r\to+\infty}\frac{\sigma}{\pi}\int_{-r}^{r}\operatorname{Re}\Big(\frac{q(u)+iR(iu)}{|iu - s|^2}\Big)du = \frac{\sigma}{\pi}\int_{-\infty}^{+\infty}\frac{q(u)du}{\sigma^2 + (u-t)^2}.
\end{aligned}
$$

□

Remark 8.2.5. (a) Taking $P := 1$, we get

$$
1 = \frac{\sigma}{\pi}\int_{-\infty}^{+\infty}\frac{du}{\sigma^2 + (u-t)^2}, \quad \sigma + it \in \boldsymbol{H}_+.
$$

(b) Let

$$
h(\sigma, t, u) := \frac{\sigma q(u)}{\sigma^2 + (u-t)^2}, \quad (\sigma, t, u) \in \boldsymbol{H}_+ \times \mathbb{R}.
$$

Observe that

$$
\frac{\partial h}{\partial \sigma}(\sigma, t, u) = q(u)\frac{(u-t)^2 - \sigma^2}{(\sigma^2 + (u-t)^2)^2}
$$

and

$$
\Big|\frac{\partial h}{\partial \sigma}(\sigma, t, u)\Big| \le A\frac{|(u-t)^2 - \sigma^2|}{(\sigma^2 + (u-t)^2)^2}.
$$

Moreover, for each $(\sigma, t) \in \boldsymbol{H}_+$, the last function is integrable with respect to u on $\mathbb{R}$ (Exercise). This permits us to calculate $\frac{\partial Q}{\partial \sigma}$ by differentiating under the integral sign, i.e.,

$$
\frac{\partial Q}{\partial \sigma}(s) = \frac{1}{\pi}\int_{-\infty}^{+\infty}\frac{(u-t)^2 - \sigma^2}{(\sigma^2 + (u-t)^2)^2}q(u)du, \quad s = \sigma + it \in \boldsymbol{H}_+.
$$

(c) Analogously,

$$
\frac{\partial h}{\partial t}(\sigma, t, u) = q(u)\frac{2\sigma(u-t)}{(\sigma^2 + (u-t)^2)^2}
$$

and

$$
\Big|\frac{\partial h}{\partial t}(\sigma, t, u)\Big| \le 2A\sigma\frac{|u-t|}{(\sigma^2 + (u-t)^2)^2}.
$$

Thus,

$$\frac{\partial Q}{\partial t}(s) = \frac{2\sigma}{\pi}\int_{-\infty}^{+\infty} \frac{u-t}{(\sigma^2+(u-t)^2)^2} q(u)du, \quad s=\sigma+it\in \boldsymbol{H}_+.$$

(d) In view of (a), by differentiating with respect to σ, resp. t, we get

$$\frac{1}{\pi}\int_{-\infty}^{+\infty} \frac{(u-t)^2-\sigma^2}{(\sigma^2+(u-t)^2)^2} du = 0,$$
$$\frac{2\sigma}{\pi}\int_{-\infty}^{+\infty} \frac{u-t}{(\sigma^2+(u-t)^2)^2} du = 0, \quad \sigma+it\in \boldsymbol{H}_+.$$

Lemma 8.2.6. *If $g(t)-g(t_0)=o(|t-t_0|^\alpha)$ when $t\longrightarrow t_0$, then*

$$\frac{\partial G}{\partial t}(\sigma+it_0) = o(\sigma^{\alpha-1}), \quad \frac{\partial G}{\partial \sigma}(\sigma+it_0)=o(\sigma^{\alpha-1}) \textit{ when } \sigma\longrightarrow 0+.$$

Proof. Let $\gamma := g-g(t_0)$. Note that $|\gamma|\le 2A$. Take an $\varepsilon>0$ and let $\delta>0$ be such that $|\gamma(u)|\le\varepsilon|u-t_0|^\alpha$ for $u\in[t_0-\delta,t_0+\delta]$. Using Lemma 8.2.4 and Remark 8.2.5, we get

$$\begin{aligned}\frac{\partial G}{\partial t}(\sigma+it_0) &= \frac{2\sigma}{\pi}\int_{-\infty}^{+\infty} \frac{u-t_0}{(\sigma^2+(u-t_0)^2)^2} g(u)du\\ &= \frac{2\sigma}{\pi}\int_{-\infty}^{+\infty} \frac{u-t_0}{(\sigma^2+(u-t_0)^2)^2} \gamma(u)du.\end{aligned}$$

Hence

$$\begin{aligned}\frac{\partial G}{\partial t}(\sigma+it_0) = \frac{2\sigma}{\pi}\Big(\int_{|u-t_0|\le\delta}+\int_{|u-t_0|>\delta}\Big)\frac{u-t_0}{(\sigma^2+(u-t_0)^2)^2}\gamma(u)du\\ =: I'_\sigma + I''_\sigma.\end{aligned}$$

For $|u-t_0|>\delta$, we have

$$\Big|\frac{u-t_0}{(\sigma^2+(u-t_0)^2)^2}\gamma(u)\Big| \le \frac{2A}{|u-t_0|^3},$$

and the right-hand-side function is integrable on $|u-t_0|>\delta$. Hence, $I''_\sigma = O(\sigma)$ when $\sigma\longrightarrow 0+$. On the other hand,

$$\begin{aligned}|I'_\sigma| &\le \frac{2\sigma}{\pi}\int_{|u-t_0|\le\delta}\frac{|u-t_0|\varepsilon|u-t_0|^\alpha}{(\sigma^2+(u-t_0)^2)^2}du \le \varepsilon\frac{2\sigma}{\pi}\int_{-\infty}^{+\infty}\frac{|u|^{\alpha+1}}{(\sigma^2+u^2)^2}du\\ &= \varepsilon\sigma^{\alpha-1}\frac{2}{\pi}\int_{-\infty}^{+\infty}\frac{|v|^{\alpha+1}}{(1+v^2)^2}dv =: \varepsilon\sigma^{\alpha-1}C.\end{aligned}$$

Finally,

$$\sigma^{1-\alpha}\Big|\frac{\partial G}{\partial t}(\sigma+it_0)\Big| \le \varepsilon C + \operatorname{const}\sigma^{2-\alpha}\le 2\varepsilon C \text{ for } 0<\sigma\ll 1.$$

The proof for $\frac{\partial G}{\partial \sigma}(\sigma + it_0)$ is analogous:

$$\begin{aligned}\frac{\partial G}{\partial \sigma}(\sigma + it_0) &= \frac{1}{\pi}\int_{-\infty}^{+\infty} \frac{(u-t_0)^2-\sigma^2}{(\sigma^2+(u-t_0)^2)^2} g(u)du \\ &= \frac{1}{\pi}\int_{-\infty}^{+\infty} \frac{(u-t_0)^2-\sigma^2}{(\sigma^2+(u-t_0)^2)^2} \gamma(u)du.\end{aligned}$$

Hence

$$\frac{\partial G}{\partial \sigma}(\sigma + it_0) = \frac{1}{\pi}\Big(\int_{|u-t_0|\le\delta} + \int_{|u-t_0|>\delta}\Big)\frac{(u-t_0)^2-\sigma^2}{(\sigma^2+(u-t_0)^2)^2}\gamma(u)du$$
$$=: I'_\sigma + I''_\sigma.$$

For $|u-t_0| > \delta > \sigma$, we have

$$\Big|\frac{(u-t_0)^2-\sigma^2}{(\sigma^2+(u-t_0)^2)^2}\gamma(u)\Big| \le \frac{2A}{|u-t_0|^2},$$

and the right-hand-side function is integrable on $|u-t_0| > \delta$. Hence, $I''_\sigma = O(1)$ when $\sigma \longrightarrow 0+$. On the other hand,

$$\begin{aligned}|I'_\sigma| &\le \frac{1}{\pi}\int_{|u-t_0|\le\delta} \frac{|(u-t_0)^2-\sigma^2|\varepsilon|u-t_0|^\alpha}{(\sigma^2+(u-t_0)^2)^2}du \\ &\le \varepsilon\frac{1}{\pi}\int_{-\infty}^{+\infty}\frac{|u^2-\sigma^2||u|^\alpha}{(\sigma^2+u^2)^2}du = \varepsilon\sigma^{\alpha-1}\frac{1}{\pi}\int_{-\infty}^{+\infty}\frac{|v^2-1||v|^\alpha}{(1+v^2)^2}dv =: \varepsilon\sigma^{\alpha-1}C.\end{aligned}$$

Finally,

$$\sigma^{1-\alpha}\Big|\frac{\partial G}{\partial \sigma}(\sigma+it_0)\Big| \le \varepsilon C + \operatorname{const}\sigma^{1-\alpha} \le 2\varepsilon C \text{ for } 0<\sigma\ll 1.$$ □

In view of Remark 8.2.2(b), the above lemma implies the following result.

Corollary 8.2.7. *If $g(t) - g(t_0) = o(|t-t_0|^\alpha)$ when $t \longrightarrow t_0$, then*

$$\sum_{n=0}^{\infty}(ab)^n e^{-\sigma b^n}\sin(t_0 b^n) = o(\sigma^{\alpha-1}),$$
$$\sum_{n=0}^{\infty}(ab)^n e^{-\sigma b^n}\cos(t_0 b^n) = o(\sigma^{\alpha-1}) \text{ when } \sigma \longrightarrow 0+.$$

Consequently,

$$\sum_{n=0}^{\infty}(ab)^n e^{-(\sigma+it_0)b^n} = o(\sigma^{\alpha-1}) \text{ when } \sigma \longrightarrow 0+. \tag{8.2.2}$$

Lemma 8.2.8. *There exists a $\varrho_0 > 0$ such that for every $\varrho \ge \varrho_0$, we have*

$$\Big|\varphi_\varrho\Big(\frac{\varrho}{b^m}+it\Big)\Big| \ge \frac{1}{2}\Big(\frac{\varrho}{b^m}\Big)^{-\varrho}, \quad t\in\mathbb{R},\ m\in\mathbb{N}.$$

Proof. For $m \in \mathbb{N}$ and $s = \frac{\varrho}{b^m} + it$, write

$$\varphi_\varrho(s) = \Big(\sum_{n=0}^{m-1} b^{n\varrho} e^{-sb^n}\Big) + b^{m\varrho}e^{-sb^m} + \sum_{n=m+1}^{\infty} b^{n\varrho}e^{-sb^n} =: f_1(s) + f_2(s) + f_3(s).$$

Obviously,

$$|\varphi_\varrho(s)| \geq |f_2(s)| - |f_1(s)| - |f_3(s)| = b^{m\varrho}e^{-\varrho}\Big(1 - b^{-m\varrho}e^{\varrho}|f_1(s)| - b^{-m\varrho}e^{\varrho}|f_3(s)|\Big).$$

We are going to show that

$$b^{-m\varrho}e^{\varrho}|f_k(s)| \leq \frac{e^{-B_k\varrho}}{1 - e^{-B_k\varrho}}, \quad k = 1, 3, \tag{8.2.3}$$

where

$$B_1 := \log b - 1 + \tfrac{1}{b}, \quad B_3 := -\log b - 1 + b.$$

One can easily check (EXERCISE) that $B_1, B_3 > 0$. Consequently, for $\varrho \geq \varrho_0 \gg 1$, we get (EXERCISE)

$$|\varphi_\varrho(s)| \geq \tfrac{1}{2}b^{m\varrho}e^{-\varrho} \geq \tfrac{1}{2}(\tfrac{\varrho}{b^m})^{-\varrho},$$

which will complete the proof. To prove (8.2.3), we proceed as follows:

$$b^{-m\varrho}e^{\varrho}|f_1(s)| \leq \sum_{n=0}^{m-1} b^{(n-m)\varrho}e^{(1-b^{n-m})\varrho} = \sum_{n=0}^{m-1} e^{-\varrho((m-n)\log b - 1 + b^{n-m})}$$

$$= \sum_{k=1}^{m} e^{-\varrho(k\log b - 1 + b^{-k})} \overset{\text{EXERCISE}}{\leq} \sum_{k=1}^{m} e^{-\varrho k(\log b - 1 + b^{-1})} < \frac{e^{-B_1\varrho}}{1 - e^{-B_1\varrho}};$$

$$b^{-m\varrho}e^{\varrho}|f_3(s)| \leq \sum_{n=m+1}^{\infty} b^{(n-m)\varrho}e^{(1-b^{n-m})\varrho} = \sum_{n=m+1}^{\infty} e^{-\varrho((m-n)\log b - 1 + b^{n-m})}$$

$$= \sum_{k=1}^{\infty} e^{-\varrho(-k\log b - 1 + b^{k})} \overset{\text{EXERCISE}}{\leq} \sum_{k=1}^{\infty} e^{-\varrho k(-\log b - 1 + b)} = \frac{e^{-B_3\varrho}}{1 - e^{-B_3\varrho}}.$$

□

Lemma 8.2.9. *For every* $\varrho > 0$, *we have*

$$|\varphi_\varrho(\sigma + it)| \leq \varphi_\varrho(\sigma) = O(1/\sigma^\varrho) \text{ when } \sigma \longrightarrow 0+, \quad t \in \mathbb{R}.$$

Consequently, by (8.2.1), *for all* $\varrho > 0$ *and* $p \in \mathbb{N}_0$ *we have*

$$|\varphi_\varrho^{(p)}(\sigma + it)| \leq \varphi_{\varrho+p}(\sigma) = O(1/\sigma^{\varrho+p}) \text{ when } \sigma \longrightarrow 0+, \quad t \in \mathbb{R}.$$

Proof. [1] Fix a $\sigma \in (0,1)$ and let $m \in \mathbb{N}$ be such that $\tau := \sigma b^m \in [1, b)$. Then we get

$$\sigma^\varrho\varphi_\varrho(\sigma) = \sum_{n=0}^{\infty}(\sigma b^n)^\varrho e^{-\sigma b^n} = \Big(\sum_{n=0}^{m-1}(\sigma b^n)^\varrho e^{-\sigma b^n}\Big) + \sum_{n=m}^{\infty}(\sigma b^n)^\varrho e^{-\sigma b^n}$$

$$\leq \Big(\sum_{k=1}^{m}(\sigma b^{m-k})^\varrho\Big) + \sum_{n=0}^{\infty}(\sigma b^{m+n})^\varrho e^{-\sigma b^{m+n}}$$

---

[1] The authors would like to thank W. Jarnicki for helpful remarks related to the proofs of Lemmas 8.2.9 and 8.2.10.

$$\leq \Big(\sum_{k=1}^{m}(\tau b^{-k})^{\varrho}\Big) + \sum_{n=0}^{\infty}(\tau b^{n})^{\varrho}e^{-\tau b^{n}}$$

$$\leq \Big(\sum_{k=1}^{\infty}(b^{1-k})^{\varrho}\Big) + \tau^{\varrho}\varphi_{\varrho}(\tau) \leq \frac{1}{1-b^{-\varrho}} + b^{\varrho}\varphi_{\varrho}(1) = \text{const}\,. \qquad \square$$

Lemma 8.2.10. *Let $\Phi : (0,+\infty) \longrightarrow \mathbb{C}$ be a $\mathcal{C}^2$ function such that*

$$\Phi(\sigma) = o(1), \quad \Phi'(\sigma) = O(1/\sigma), \quad \Phi''(\sigma) = O(1/\sigma^2) \text{ when } \sigma \longrightarrow 0+.$$

Then $\Phi'(\sigma) = o(1/\sigma)$ when $\sigma \longrightarrow 0+$.

Proof. We may assume that $\Phi : (0,+\infty) \longrightarrow \mathbb{R}$. Suppose that $\sigma\Phi'(\sigma) \not\longrightarrow 0$ when $\sigma \longrightarrow 0+$. Since the function $\sigma \longmapsto \sigma\Phi'(\sigma)$ is bounded near zero, there exists a sequence $1 \geq \sigma_n \searrow 0$ such that $\sigma_n\Phi'(\sigma_n) \longrightarrow 4c \neq 0$. We may assume that $\sigma_n\Phi'(\sigma_n) \geq 2c > 0$, $n \in \mathbb{N}$. Let $M > 1$ be such that $\sigma^2\Phi''(\sigma) \leq M$, $\sigma \in (0, 1+c]$. Put $\delta_n := \frac{c\sigma_n}{M}$, $\tau_n := \sigma_n + \delta_n \in (0, 1+c]$, $n \in \mathbb{N}$. Obviously, $\tau_n \searrow 0$. By the mean value theorem, for $\eta \in (\sigma_n, \tau_n)$, we get

$$|\Phi'(\eta) - \Phi'(\sigma_n)| = (\eta - \sigma_n)|\Phi''(\xi_n)| \leq \delta_n\frac{M}{\xi_n^2} \leq \delta_n\frac{M}{\sigma_n^2} = \frac{c}{\sigma_n}.$$

Hence $\Phi'(\eta) \geq \Phi'(\sigma_n) - \frac{c}{\sigma_n} \geq \frac{2c}{\sigma_n} - \frac{c}{\sigma_n} = \frac{c}{\sigma_n}$. Using once again the mean value theorem, we get

$$\Phi(\tau_n) - \Phi(\sigma_n) = \delta_n\Phi'(\eta_n) \geq \delta_n\frac{c}{\sigma_n} = \frac{c^2}{M} > 0;$$

a contradiction. $\square$

Lemma 8.2.11. *Let $\varrho > 0$ and let $\Psi : (0,+\infty) \longrightarrow \mathbb{C}$ be a $\mathcal{C}^\infty$ function such that*

$$\Psi(\sigma) = o(1/\sigma^{\varrho}), \quad \Psi^{(p)}(\sigma) = O(1/\sigma^{\varrho+p}) \text{ when } \sigma \longrightarrow 0+, \quad p \in \mathbb{N}.$$

Then for each $p \in \mathbb{N}$, we get $\Psi^{(p)}(\sigma) = o(1/\sigma^{\varrho+p})$.

Proof. It suffices to prove that $\Psi'(\sigma) = o(1/\sigma^{\varrho+1})$ when $\sigma \longrightarrow 0+$, and then replace Ψ by Ψ'. Define $\Phi(\sigma) := \sigma^{\varrho}\Psi(\sigma)$ and observe that Φ satisfies the assumptions of Lemma 8.2.10. Indeed,

$$\begin{aligned}\sigma\Phi'(\sigma) &= \sigma(\varrho\sigma^{\varrho-1}\Psi(\sigma) + \sigma^{\varrho}\Psi'(\sigma)) = \varrho\sigma^{\varrho}\Psi(\sigma) + \sigma^{\varrho+1}\Psi'(\sigma),\\ \sigma^2\Phi''(\sigma) &= \sigma^2(\varrho(\varrho-1)\sigma^{\varrho-2}\Psi(\sigma) + 2\varrho\sigma^{\varrho-1}\Psi'(\sigma) + \sigma^{\varrho}\Psi''(\sigma))\\ &= \varrho(\varrho-1)\sigma^{\varrho}\Psi(\sigma) + 2\varrho\sigma^{\varrho+1}\Psi'(\sigma) + \sigma^{\varrho+2}\Psi''(\sigma).\end{aligned}$$

Hence, using Lemma 8.2.10, we conclude that

$$0 = \lim_{\sigma\to 0+}(\varrho\sigma^{\varrho}\Psi(\sigma) + \sigma^{\varrho+1}\Psi'(\sigma)) = \lim_{\sigma\to 0+}\sigma^{\varrho+1}\Psi'(\sigma). \qquad \square$$

Now we move to the more difficult case $ab = 1$.

Theorem 8.2.12 (cf. [Har16]). *Assume that $b > 1$, $a := 1/b$. Then $\boldsymbol{C}_{1/b,b}$, $\boldsymbol{S}_{1/b,b} \in \mathfrak{ND}(\mathbb{R})$.*

We will consider only the case of $\boldsymbol{C}_{1/b,b}$. The case of $\boldsymbol{S}_{1/b,b}$ is left to the reader as an EXERCISE. The idea of the proof is similar to that of the proof of Theorem 8.2.1.

We keep previous notation.

Lemma 8.2.13. *If a finite derivative $g'(t_0)$ exists, then*

$$\frac{\partial G}{\partial t}(\sigma+it_0)\longrightarrow g'(t_0),\quad \frac{\partial^2 G}{\partial t^2}(\sigma+it_0)=o(1/\sigma)\ \textit{when}\ \sigma\longrightarrow 0+.$$

Proof. Write $g(u)=g(t_0)+g'(t_0)(u-t_0)+\mu(u)(u-t_0)$, where $\mu(u)\longrightarrow 0$ when $u\longrightarrow t_0$. Note that the function μ is bounded. Take an $\varepsilon>0$ and let $\delta>0$ be such that $|\mu(u)|\le\varepsilon$, $u\in[t_0-\delta,t_0+\delta]$. Then (using Remark 8.2.5) we get

$$\begin{aligned}\frac{\partial G}{\partial t}(\sigma+it_0)&=\frac{2\sigma}{\pi}\int_{-\infty}^{+\infty}\frac{u-t_0}{(\sigma^2+(u-t_0)^2)^2}g(u)du\\ &=\frac{2\sigma}{\pi}\int_{-\infty}^{+\infty}\frac{u-t_0}{(\sigma^2+(u-t_0)^2)^2}\Big(g(t_0)+g'(t_0)(u-t_0)+\mu(u)(u-t_0)\Big)du\\ &=g'(t_0)\frac{2\sigma}{\pi}\int_{-\infty}^{+\infty}\frac{(u-t_0)^2du}{(\sigma^2+(u-t_0)^2)^2}+\frac{2\sigma}{\pi}\int_{-\infty}^{+\infty}\frac{(u-t_0)^2\mu(u)du}{(\sigma^2+(u-t_0)^2)^2}=:I'_\sigma+I''_\sigma.\end{aligned}$$

Moreover,

$$\begin{aligned}\frac{2\sigma}{\pi}\int_{-\infty}^{+\infty}\frac{(u-t_0)^2du}{(\sigma^2+(u-t_0)^2)^2}&=\frac{2}{\pi}\int_{-\infty}^{+\infty}\frac{v^2dv}{(1+v^2)^2}\\ &=\frac{1}{\pi}\Big(\arctan v-\frac{v}{1+v^2}\Big)\Big|_{-\infty}^{+\infty}=1.\end{aligned}\tag{8.2.4}$$

Hence $I'_\sigma=f'(t_0)$. On the other hand,

$$I''_\sigma=\frac{2\sigma}{\pi}\Big(\int_{|u-t_0|\le\delta}+\int_{|u-t_0|>\delta}\Big)\frac{(u-t_0)^2\mu(u)du}{(\sigma^2+(u-t_0)^2)^2}=:J'_\sigma+J''_\sigma.$$

Similarly as in the proof of Lemma 8.2.6, one can easily prove that $J''_\sigma=O(\sigma)$ when $\sigma\longrightarrow 0+$ (Exercise). It remains to estimate J'_σ. In view of (8.2.4), we obtain

$$|J'_\sigma|\le\varepsilon\frac{2\sigma}{\pi}\int_{|u-t_0|\le\delta}\frac{(u-t_0)^2du}{(\sigma^2+(u-t_0)^2)^2}\le\varepsilon.$$

Finally,

$$\Big|\frac{\partial G}{\partial t}(\sigma+it_0)-g'(t_0)\Big|\le\operatorname{const}\sigma+\varepsilon\le 2\varepsilon\ \text{for}\ 0<\sigma\ll 1,$$

which proves that $\frac{\partial G}{\partial t}(\sigma+it_0)\longrightarrow g'(t_0)$ when $\sigma\longrightarrow 0+$.

Analogously as in Remark 8.2.5, one can easily prove that $\frac{\partial^2 G}{\partial t^2}$ may be calculated by differentiating under the integral sign (Exercise). Hence

$$\begin{aligned}\frac{\partial^2 G}{\partial t^2}(\sigma+it_0)&=\frac{2\sigma}{\pi}\int_{-\infty}^{+\infty}\frac{3(u-t_0)^2-\sigma^2}{(\sigma^2+(u-t_0)^2)^3}g(u)du\\ &=\frac{2\sigma}{\pi}\int_{-\infty}^{+\infty}\frac{3(u-t_0)^2-\sigma^2}{(\sigma^2+(u-t_0)^2)^3}\Big(g(t_0)+g'(t_0)(u-t_0)+\mu(u)(u-t_0)\Big)du\end{aligned}$$

$$= g'(t_0)\frac{2\sigma}{\pi}\int_{-\infty}^{+\infty}\frac{(3(u-t_0)^2-\sigma^2)(u-t_0)}{(\sigma^2+(u-t_0)^2)^3}du$$
$$+\frac{2\sigma}{\pi}\int_{-\infty}^{+\infty}\frac{(3(u-t_0)^2-\sigma^2)(u-t_0)}{(\sigma^2+(u-t_0)^2)^3}\mu(u)du =: I'_\sigma + I''_\sigma.$$

Moreover,

$$\frac{2\sigma}{\pi}\int_{-\infty}^{+\infty}\frac{(3(u-t_0)^2-\sigma^2)(u-t_0)}{(\sigma^2+(u-t_0)^2)^3}du = \frac{2\sigma}{\pi}\int_{-\infty}^{+\infty}\frac{(3v^2-1)v}{(1+v^2)^3}du = 0.$$

Thus $I'_\sigma = 0$. Now we estimate I''_σ:

$$I''_\sigma = \frac{2\sigma}{\pi}\Big(\int_{|u-t_0|\le\delta} + \int_{|u-t_0|>\delta}\Big)\frac{(3(u-t_0)^2-\sigma^2)(u-t_0)}{(\sigma^2+(u-t_0)^2)^3}\mu(u)du =: J'_\sigma + J''_\sigma.$$

As above, we get $J''_\sigma = O(\sigma)$ when $\sigma \longrightarrow 0+$. It remains to estimate J'_σ. We have

$$\sigma|J'_\sigma| \le \varepsilon\frac{2\sigma^2}{\pi}\int_{|u-t_0|\le\delta}\frac{|3(u-t_0)^2-\sigma^2||u-t_0|}{(\sigma^2+(u-t_0)^2)^3}du$$
$$\le \varepsilon\frac{2}{\pi}\int_{-\infty}^{+\infty}\frac{|3v^2-1||v|}{(1+v^2)^3}dv =: \mathrm{const}\,\varepsilon.$$

Finally,

$$\sigma\left|\frac{\partial^2 G}{\partial t^2}(\sigma+it_0)\right| \le \mathrm{const}_1\,\sigma^2 + \mathrm{const}_2\,\varepsilon,$$

which completes the proof. □

Lemma 8.2.14. *If a finite derivative $g'(t_0)$ exists, then $\varphi_1(\sigma+it_0) = o(1/\sigma)$ when $\sigma \longrightarrow 0+$.*

Proof. We have

$$\frac{\partial G}{\partial t}(\sigma+it_0) = -\sum_{n=0}^{\infty} e^{-\sigma b^n}\sin(b^n t_0) =: \Lambda(\sigma),$$
$$\frac{\partial^2 G}{\partial t^2}(\sigma+it_0) = -\sum_{n=0}^{\infty} b^n e^{-\sigma b^n}\cos(b^n t_0), \quad \sigma>0. \tag{8.2.5}$$

Observe that

$$\Lambda'(\sigma) = \sum_{n=0}^{\infty} b^n e^{-\sigma b^n}\sin(b^n t_0), \quad \Lambda''(\sigma) = -\sum_{n=0}^{\infty} b^{2n} e^{-\sigma b^n}\sin(b^n t_0), \quad \sigma>0.$$

Consequently, by Lemma 8.2.9,

$$|\Lambda'(\sigma)| \le \varphi_1(\sigma) = O(1/\sigma), \quad |\Lambda''(\sigma)| \le \varphi_2(\sigma) = O(1/\sigma^2) \text{ when } \sigma \longrightarrow 0+.$$

Now we apply Lemma 8.2.10 to the function $\Phi := \Lambda - g'(t_0)$. In view of Lemma 8.2.13, we get $\Lambda'(\sigma) = o(1/\sigma)$ when $\sigma \longrightarrow 0+$. Finally, in view of Lemma 8.2.13 and (8.2.5), we conclude that

$$\varphi_1(\sigma + it_0) = \sum_{n=0}^{\infty} b^n e^{-\sigma b^n} \cos(b^n t_0) - i \sum_{n=0}^{\infty} b^n e^{-\sigma b^n} \sin(b^n t_0)$$
$$= o(1/\sigma) \text{ when } \sigma \longrightarrow 0+. \qquad \square$$

Proof of Theorem 8.2.12. Suppose that a finite derivative $\boldsymbol{C}'_{1/b,b}(\frac{t_0}{2\pi})$ exists for some $t_0 \in \mathbb{R}$. We take in Lemma 8.2.11 $\varrho := 1$ and $\Psi(\sigma) := \varphi_1(\sigma + it_0)$. By Lemma 8.2.14, $\varphi_1(\sigma + it_0) = o(1/\sigma)$ when $\sigma \longrightarrow 0+$. Lemma 8.2.9 implies that $\Psi^{(p)}(\sigma) = O(1/\sigma^{\varrho+p})$ when $\sigma \longrightarrow 0+$ ($p \in \mathbb{N}$). Thus, by Lemma 8.2.11, $\Psi^{(p)}(\sigma) = o(1/\sigma^{\varrho+p})$ when $\sigma \longrightarrow 0+$, i.e., $\varphi_{1+p}(\sigma + it_0) = o(1/\sigma^{1+p})$ when $\sigma \longrightarrow 0+$ ($p \in \mathbb{N}$). Let $p \in \mathbb{N}$ be such that $1 + p \geq \varrho_0$, where ϱ_0 is as in Lemma 8.2.8. Then

$$\Big(\frac{1+p}{b^m}\Big)^{1+p} \Big|\varphi_1\Big(\frac{1+p}{b^m} + it_0\Big)\Big| \geq \frac{1}{2}, \quad m \in \mathbb{N};$$

a contradiction. $\square$

Remark 8.2.15. Notice that for the case $b \in \mathbb{N}$, G.H. Hardy gives in [Har16] simpler proofs of Theorems 8.2.1 and 8.2.12 based on the Poisson integral formula.

8.3 Baouche–Dubuc Method

Theorem 8.3.1 (cf. [BD92]; see also Theorem 3.11.1)**.** *If $ab > 1$ and $\alpha := -\frac{\log a}{\log b} \in (0, 1)$, then $\boldsymbol{W}_{1,a,b,\boldsymbol{\theta}}$ is α-anti-Hölder continuous uniformly with respect to $x \in \mathbb{R}$ and $\boldsymbol{\theta}$ (cf. Remark 3.2.1*(h)*). In particular, $\boldsymbol{W}_{1,a,b,\boldsymbol{\theta}} \in \mathfrak{M}(\mathbb{R}) \subset \mathfrak{ND}_{\pm}(\mathbb{R})$ (cf. Remark 2.5.4*(a)*).*

Remark 8.3.2. In fact, A. Baouche and S. Dubuc in [BD92] considered only the case $\boldsymbol{\theta} = 0$ and proved that $\boldsymbol{W}_{1,a,b,0}$ is weakly α-anti-Hölder continuous uniformly with respect to $x \in \mathbb{R}$.

Proof of Theorem 8.3.1. By Remark 3.2.1(h), we have only to check that there exist $\varepsilon > 0$ and $\delta_0 \in (0, 1]$ such that for all $\boldsymbol{\theta}$ and $\delta \in (0, \delta_0)$, there exists a $t \in (0, \delta]$ such that $|\boldsymbol{W}_{1,a,b,\boldsymbol{\theta}}(t) - \boldsymbol{W}_{1,a,b,\boldsymbol{\theta}}(0)| > \varepsilon\delta^\alpha$.

Put $f := \boldsymbol{W}_{1,a,b,\boldsymbol{\theta}}$. Let $L, m \in \mathbb{N}$, $N \in 2\mathbb{N}$, be such that $b^L < \frac{N}{2}$ and $L < m$. Let $h := \frac{N}{2b^m}$. To simplify notation, put $A_n = 2\pi b^n$. Note that $A_m h = \pi N$. Take an $n \in \mathbb{N}_0$. First observe that

$$\int_0^h \cos(A_m t + \theta_m) dt = \frac{1}{A_m} \sin(A_m t + \theta_m)\Big|_0^h$$
$$= \frac{1}{A_m}\Big(\sin(A_m h + \theta_m) - \sin\theta_m\Big) = 0.$$

Let

$$I := \frac{2}{h} \int_0^h f(t) \cos(2\pi b^m t + \theta_m) dt = \frac{2}{h} \int_0^h (f(t) - f(0)) \cos(A_m t + \theta_m) dt.$$

We have

$$\frac{2}{h} \int_0^h \cos^2(A_m t + \theta_m) dt = \frac{1}{h}\Big(t + \frac{1}{2A_m} \sin(2A_m t + 2\theta_m)\Big)\Big|_0^h$$
$$= 1 + \frac{1}{2A_m h}\Big(\sin(2A_m h + 2\theta_m) - \sin(2\theta_m)\Big) = 1.$$

Moreover, for $n \neq m$, we get

$$\begin{aligned}
&\frac{2}{h}\int_0^h \cos(A_n t+\theta_n)\cos(A_m t+\theta_m)dt \\
&=\frac{1}{h}\int_0^h \Big(\cos((A_n+A_m)t+\theta_n+\theta_m)+\cos((A_n-A_m)t+\theta_n-\theta_m)\Big)dt \\
&=\frac{1}{h}\Big(\frac{\sin((A_n+A_m)t+\theta_n+\theta_m)}{A_n+A_m}+\frac{\sin((A_n-A_m)t+\theta_m-\theta_m)}{A_n-A_m}\Big)\Big|_0^h \\
&=\frac{1}{h}\frac{1}{A_n^2-A_m^2}\Big(A_n\sin((A_n+A_m)h+\theta_n+\theta_m)-A_n\sin(\theta_n+\theta_m) \\
&\qquad -A_m\sin((A_n+A_m)h+\theta_n+\theta_m)+A_m\sin(\theta_n+\theta_m) \\
&\qquad +A_n\sin((A_n-A_m)h+\theta_n-\theta_m)-A_n\sin(\theta_n-\theta_m) \\
&\qquad +A_m\sin((A_n-A_m)h+\theta_n-\theta_m)-A_m\sin(\theta_n-\theta_m)\Big) \\
&=\frac{1}{h}\frac{2}{A_n^2-A_m^2}\Big(A_n\sin(A_nh+\theta_n)\cos\theta_m-A_n\sin\theta_n\cos\theta_m \\
&\qquad -A_m\cos(A_nh+\theta_n)\sin\theta_m+A_m\cos\theta_n\sin\theta_m\Big) \\
&=\frac{1}{h}\frac{4}{A_n^2-A_m^2}\Big(A_n\cos\theta_m\cos(\theta_n+\tfrac{1}{2}A_nh)\sin(\tfrac{1}{2}A_nh) \\
&\qquad +A_m\sin\theta_m\sin(\theta_n+\tfrac{1}{2}A_nh)\sin(\tfrac{1}{2}A_nh)\Big).
\end{aligned}$$

Hence

$$\Big|\frac{2}{h}\int_0^h \cos(A_n t+\theta_n)\cos(A_m t+\theta_m)dt\Big| \leq \frac{1}{h}\frac{4}{|A_n-A_m|}|\sin(\tfrac{1}{2}A_nh)|.$$

Thus

$$\begin{aligned}
|I-a^m| &\leq \sum_{n=0}^{m-L-1}\frac{2a^nb^n}{b^m-b^n}+\sum_{n=m-L,\,n\neq m}^{\infty}\frac{2a^n}{\pi|b^n-b^m|h} \\
&\leq \sum_{n=0}^{m-L-1}\frac{2a^nb^n}{b^m-b^{m-1}}+\sum_{n=m-L}^{\infty}\frac{2a^n}{\pi(b^m-b^{m-1})h} \\
&\leq \frac{2a^{m-L}b^{m-L}}{(ab-1)(b^m-b^{m-1})}+\frac{2a^{m-L}}{\pi(1-a)(b^m-b^{m-1})h} \\
&\leq \frac{2a^{m-L}b^{-L}}{(ab-1)(1-\frac{1}{b})}+\frac{4a^{m-L}}{N\pi(1-a)(1-\frac{1}{b})}<sa^m,
\end{aligned}$$

where

$$s:=\frac{2}{(ab)^L(1-\frac{1}{b})}\Big(\frac{1}{ab-1}+\frac{1}{\pi(1-a)}\Big).$$

Fix $L,N\in\mathbb{N}$, $N\equiv 0 \pmod 2$, such that $s<1$ and $b^L<\frac{N}{2}$. Put $c:=\frac{1-s}{2}$. Then $I>2ca^m$ for $m>L$. Consequently, for every $m>L$, there exists a $t_m\in(0,h]$ such that $|f(t_m)-f(0)|>ca^m$. Now take a $\delta\in(0,\frac{N}{2b^L})$ and let $m\in\mathbb{N}$, $m>L$, be such that $h=\frac{N}{2b^m}\leq\delta<\frac{N}{2b^{m-1}}$. Let $\varepsilon:=ca(\frac{2}{N})^\alpha$. Then

$$|f(t_m) - f(0)| > ca^m = \frac{c}{b^{m\alpha}} = \frac{c2^\alpha}{N^\alpha b^\alpha}\Big(\frac{N}{2b^{m-1}}\Big)^\alpha > \varepsilon\delta^\alpha. \qquad \square$$

8.4 Kairies–Girgensohn Method

The main aim of this section is to discuss the case $ab = 1$, at least when $b \in \mathbb{N}_2$. The general case remains open.

Theorem 8.4.1 (cf. [Gir94]). *If $a \in (0,1)$, $b \in \mathbb{N}_2$ with $ab \geq 1$, and $\theta \in \mathbb{R}$, then $\boldsymbol{W}_{1,a,b,\theta} \in \mathcal{ND}_\pm(\mathbb{R})$.*

The proof will be based on studying systems of functional equations (see Sect. 4.2 in the discussion of the Takagi function) and the Schauder coefficients of solutions of such a system.

To adjust the function $\boldsymbol{W}_{1,a,b,\theta}$ in order to obtain simpler formulas, we put

$$\widetilde{\boldsymbol{W}}_{a,b,\theta}(x) := \sum_{n=0}^{\infty} a^n \sin(b\pi b^n x + \theta), \quad x \in \mathbb{R}.$$

Note that $\boldsymbol{W}_{1,a,b,\theta}(x) = \widetilde{\boldsymbol{W}}_{a,b,\theta+\frac{\pi}{2}}(\frac{2x}{b})$, $x \in \mathbb{R}$. Hence it suffices to verify that $\widetilde{\boldsymbol{W}}_{a,b,\theta} \in \mathcal{ND}_\pm(I)$.

Fix a, b, θ as above. Then $\widetilde{\boldsymbol{W}}_{a,b,\theta}$ satisfies the following system of functional equations on $\mathbb{I}$ (Exercise):

$$g\Big(\frac{x+j}{b}\Big) = (-1)^{jb} a g(x) + (-1)^j \sin(\pi x + \theta), \quad x \in \mathbb{I},\ j = 0, \dots, b-1. \tag{8.4.1}$$

8.4.1 A System of Functional Equations

In this part, we will discuss a system of functional equations that generalizes (8.4.1) from above, namely we study solutions of the following system:

$$g\Big(\frac{x+j}{b}\Big) = a_j g(x) + g_j(x), \quad x \in \mathbb{I},\ j = 0, \dots, b-1, \tag{8.4.2}$$

where $a_j \in \mathbb{R}$ with $|a_j| < 1$ and $g_j : \mathbb{I} \longrightarrow \mathbb{R}$, $j = 0, \dots, b-1$.

Note that if $a_j = (-1)^{jb} a$, $a \in (0,1)$, and $g_j(x) := (-1)^j \sin(\pi x + \theta)$, $x \in \mathbb{I}$, $j = 0, \dots, b-1$, then (8.4.2) is exactly (8.4.1) from above.

If $f : \mathbb{I} \longrightarrow \mathbb{R}$ is a solution of (8.4.2), then

$$g_0(0) = (1 - a_0) f(0), \qquad g_{b-1}(1) = (1 - a_{b-1}) f(1),$$
$$f\Big(\frac{j}{b}\Big) = a_{j-1} f(1) + g_{j-1}(1), \qquad f\Big(\frac{j}{b}\Big) = a_j f(0) + g_j(0),$$

where $j = 1, \dots, b-1$. In particular,

$$a_{j-1} \frac{g_{b-1}(1)}{1 - a_{b-1}} + g_{j-1}(1) = a_j \frac{g_0(0)}{1 - a_0} + g_j(0), \quad j = 1, \dots, b-1. \tag{8.4.3}$$

Theorem 8.4.2. *If* (8.4.3) *is true and all the* g_j *are continuous, then there exists exactly one* $f \in \mathcal{C}(\mathbb{I})$ *solving* (8.4.2).

Remark 8.4.3. Applying this result to (8.4.1), it follows that $\widetilde{\boldsymbol{W}}_{a,b,\theta}$ is the only solution of (8.4.1). Moreover, one may study (8.4.2) without knowing the explicit form of its solution.

Proof of Theorem 8.4.2. We introduce the metric space

$$X := \Big\{u \in \mathcal{C}(\mathbb{I}) : u(0) = \frac{g_0(0)}{1-a_0},\ u(1) = \frac{g_{b-1}(1)}{1-a_{b-1}}\Big\}$$

with $d(u', u'') := \|u' - u''\|_I$. Then the mapping $T : X \longrightarrow X$ given by

$$Tu(x) := a_j u(bx - j) + g_j(bx - j), \qquad \text{if } x \in \Big[\frac{j}{b}, \frac{j+1}{b}\Big],\ j = 0, \ldots, b-1,$$

is well defined (use (8.4.3)). Note that every continuous solution of (8.4.2) is a fixed point of T. Moreover,

$$d(Tu', Tu'') \le \max\{|a_j| : j = 0, \ldots, b-1\} \cdot d(u', u''),$$

i.e., T is a contraction. Therefore, by virtue of the Banach fixed-point theorem, there exists exactly one function $u_0 \in X$ with $Tu_0 = u_0$. Hence u_0 is the only continuous function satisfying (8.4.2). □

8.4.2 The Faber–Schauder Basis of $\mathcal{C}(\mathbb{I})$

Let $b \in \mathbb{N}_2$. We introduce the following functions $\sigma_{0,0}$, $\sigma_{1,0}$, and $\sigma_{i,j,n}$ ($i = 0, \ldots, b^{n-1}-1$, $j = 1, \ldots, b-1, n \in \mathbb{N}$) on $\mathbb{I}$:

$$\begin{aligned}
&\sigma_{0,0}(x) := 1 - x, \quad \sigma_{1,0}(x) := x,\\
&\sigma_{i,j,n} := \text{ the polygonal line with nodes at}\\
&(0,0),\ \Big(\frac{ib+j-1}{b^n}, 0\Big),\ \Big(\frac{ib+j}{b^n}, 1\Big),\ \Big(\frac{ib+j+1}{b^n}, 0\Big),\ (1,0).
\end{aligned}$$

Note that all these functions depend on b; nevertheless, for simplicity, we omit the extra index b, here and in our further discussions.

Theorem 8.4.4. *If* $f \in \mathcal{C}(\mathbb{I})$, *then* f *has the following unique expansion:*

$$f = \gamma_{0,0}(f)\sigma_{0,0} + \gamma_{1,0}(f)\sigma_{1,0} + \sum_{n=1}^{\infty} \sum_{i=0}^{b^{n-1}-1} \sum_{j=1}^{b-1} \gamma_{i,j,n}(f)\sigma_{i,j,n}$$

on $\mathbb{I}$, *where the series is uniformly convergent.*

The Schauder coefficients $\gamma_{0,0}(f)$, $\gamma_{1,0}(f)$, $\gamma_{i,j,n}(f)$ *satisfy the following relations:*

$$\begin{aligned}
&\gamma_{0,0}(f) = f(0), \quad \gamma_{1,0}(f) = f(1),\\
&\gamma_{i,j,n}(f) = f\Big(\frac{ib+j}{b^n}\Big) - \frac{b-j}{b} f\Big(\frac{i}{b^{n-1}}\Big) - \frac{j}{b} f\Big(\frac{i+1}{b^{n-1}}\Big).
\end{aligned}$$

Moreover, if

$$f_0 := \gamma_{0,0}(f)\sigma_{0,0} + \gamma_{1,0}(f)\sigma_{1,0}, \quad f_n := \sum_{i=0}^{b^{n-1}-1} \sum_{j=1}^{b-1} \gamma_{i,j,n}(f)\sigma_{i,j,n}, \quad n \in \mathbb{N},$$

then f_n is a polygonal line with nodes at most at the points $\frac{m}{b^n}$, and there we have $f_n(\frac{m}{b^n}) = f(\frac{m}{b^n})$, $m = 0, \ldots, b^n$.

Remark 8.4.5. In the case $b = 2$, this basis (given by the $\sigma_{0,0}, \sigma_{1,0}, \sigma_{i,j,n}$) may be found in papers of Faber (see [Fab08, Fab10]).

Proof of Theorem 8.4.4. Step 1°. We first prove the uniform convergence under the assumption that f_n is linear on the intervals $[\frac{m}{b^n}, \frac{m+1}{b^n}]$, $m = 0, \ldots, b^n - 1$, and $f_n(\frac{m}{b^n}) = f(\frac{m}{b^n})$, $m = 0, \ldots, b^n$ $(n \in \mathbb{N})$. Let $\varepsilon > 0$ and choose $n_0 \in \mathbb{N}$ such that

$$|f(x') - f(x'')| \leq \frac{\varepsilon}{2} \quad \text{for all } x', x'' \in \mathbb{I},\ |x' - x''| < \frac{1}{b^{n_0}}.$$

Let $n > n_0$. Take an $m \in \{0, 1, \ldots, b^n\}$ with $x_1 := \frac{m}{b^n} \leq x \leq \frac{m+1}{b^n} =: x_2$. Then $|x - x_1| \leq x_2 - x_1 = \frac{1}{b^n} < \frac{1}{b^{n_0}}$. Thus, $|f(x) - f(x_1)| < \frac{\varepsilon}{2}$. In view of our assumption, we have that $f_n|_{[x_1,x_2]}$ is linear and $f_n(x_1) = f(x_1)$, $f(x_2) = f_n(x_2)$. Hence

$$\begin{aligned} |f(x) - f_n(x)| &\leq |f(x) - f(x_1)| + |f(x_1) - f_n(x)| < \frac{\varepsilon}{2} + |f_n(x_1) - f_n(x)| \\ &\leq \frac{\varepsilon}{2} + |f_n(x_2) - f_n(x_1)| = \frac{\varepsilon}{2} + |f(x_2) - f(x_1)| < \varepsilon. \end{aligned}$$

Therefore, $\|f - f_n\|_{\mathbb{I}} < \varepsilon$ whenever $n > n_0$.

Step 2°. Note that by definition, f_n is piecewise linear with nodes at most at the points $\frac{m}{b^n}$, $m = 0, \ldots, b^n$. Moreover, $\sigma_{i,j,k}(0) = \sigma_{i,j,k}(1) = 0$ for all admissible indices with i, j for $k \in \mathbb{N}$. Hence, $f(0) = f_0(0) = \gamma_{0,0}(0)$ and $f(1) = f_{0,0}(1) = \gamma_{1,0}(f)$.

Assume now that the formula for the Schauder coefficients for $n-1$, $n \in \mathbb{N}$, and $f(\frac{m}{b^{n-1}}) = f_{n-1}(\frac{m}{b^{n-1}})$, $m = 0, \ldots, b^{n-1}$, are verified. If $\mathbb{N} \ni k > n$, then

$$\sigma_{i,j,k}\Big(\frac{ib+j}{b^n}\Big) = \sigma_{i,j,k}\Big(\frac{ib^{k-n}b + jb^{k-n}}{b^k}\Big) - 0.$$

Moreover, if $(s,t) \neq (i,j)$ are admissible indices for n, then $\sigma_{s,t,n}\Big(\frac{ib+j}{b^n}\Big) = 0$. Therefore, $f = f_{n-1} + \gamma_{i,j,n}(f)\sigma_{i,j,n}$; in particular,

$$f\Big(\frac{ib+j}{b^n}\Big) = f_{n-1}\Big(\frac{ib+j}{b^n}\Big) + \gamma_{i,j,n}(f) = f_n\Big(\frac{ib+j}{b^n}\Big).$$

Recall now that f_{n-1} is linear over the interval $\Big[\frac{i}{b^{n-1}}, \frac{i+1}{b^{n-1}}\Big]$, which together with the induction assumption implies that

$$\begin{aligned} f_{n-1}\Big(\frac{ib+j}{b^n}\Big) &= \frac{b-j}{b} f_{n-1}\Big(\frac{i}{b^{n-1}}\Big) + \frac{j}{b} f_{n-1}\Big(\frac{i+1}{b^{n-1}}\Big) \\ &= \frac{b-j}{b} f\Big(\frac{i}{b^{n-1}}\Big) + \frac{j}{b} f\Big(\frac{i+1}{b^{n-1}}\Big). \end{aligned}$$

Finally, merging the last two equations gives the claimed formula for $\gamma_{i,j,n}(f)$. □

Applying the former result on the form of the Schauder coefficients, we get the following lemma, which will be used later in the discussion of nowhere differentiability properties of the Weierstrass function.

Lemma 8.4.6. *Let $f \in \mathcal{C}(\mathbb{I})$, $n \in \mathbb{N}$, $i \in \{0, \ldots, b^{n-1}-1\}$. Then:*

(a) *if f is concave on $[\frac{i}{b^{n-1}}, \frac{i+1}{b^{n-1}}]$, then $\gamma_{i,j,n}(f) \geq 0$ for all $j \in \{1, \ldots, b-1\}$;*

(b) *if f is differentiable and f' is convex on $[\frac{i}{b^{n-1}}, \frac{i+2}{b^{n-1}}]$ with $i < b^{n-1}-1$, then $\gamma_{i,j,n}(f) \geq \gamma_{i+1,j,n}(f)$ for all $j \in \{1, \ldots, b-1\}$, i.e., the numbers $\gamma_{i,j,n}(f)$ are decreasing with respect to i;*

(c) *if f is differentiable and f' is convex on $[\frac{i}{b^{n-1}}, \frac{i+1}{b^{n-1}}]$, then*

$$\gamma_{i,j,n}(f) \geq \gamma_{i,b-j,n}(f), \quad j = 1, \ldots, \left\lfloor \frac{b}{2} \right\rfloor;$$

(d) *if $f(x) = \pm f(1-x)$, $x \in \mathbb{I}$, then $\gamma_{i,j,n}(f) = \pm \gamma_{b^{n-1}-1-i,b-j,n}(f)$ for all $j = 1, \ldots, b-1$.*

Proof. (a) Let $\frac{i}{b^{n-1}} \leq x' < y < x'' \leq \frac{i+1}{b^{n-1}}$. Then concavity implies that $\boldsymbol{\Delta} f(x', y) \geq \boldsymbol{\Delta} f(y, x'')$. In particular, we obtain

$$\gamma_{i,j,n}(f) = \frac{j(b-j)}{b^{n+1}} \Big(\boldsymbol{\Delta} f\Big(\frac{ib}{b^n}, \frac{ib+j}{b^n}\Big) - \boldsymbol{\Delta} f\Big(\frac{ib+j}{b^n}, \frac{(i+1)b}{b^n}\Big)\Big) \geq 0.$$

(b) For $x \in [\frac{i}{b^{n-1}}, \frac{i+1}{b^{n-1}}]$ put $\widetilde{f}(x) := f(x+h)$, where $h := \frac{1}{b^{n-1}}$. Applying that f' is assumed to be a convex function implies

$$\begin{aligned} f'(x_2) &= f'\Big(\frac{h}{x_2+h-x_1} x_1 + \frac{x_2-x_1}{x_2+h-x_1}(x_2+h)\Big) \\ &\leq \frac{h}{x_2+h-x_1} f'(x_1) + \frac{x_2-x_1}{x_2+h-x_1} f'(x_2+h) \\ f'(x_1+h) &= f'\Big(\frac{x_2-x_1}{x_2+h-x_1} x_1 + \frac{h}{x_2+h-x_1}(x_2+h)\Big) \\ &\leq \frac{x_2-x_1}{x_2+h-x_1} f'(x_1) + \frac{h}{x_2+h-x_1} f'(x_2+h), \end{aligned}$$

where $\frac{i}{b^{n-1}} \leq x_1 < x_2 \leq \frac{i+1}{b^{n-1}}$. Adding these two inequalities gives

$$f'(x_2) + f'(x_1+h) \leq f'(x_1) + f'(x_2+h),$$

implying that $\widetilde{f}'(x_2) \leq \widetilde{f}'(x_1)$. Thus, $\widetilde{f}$ is concave on the interval $[\frac{i}{b^{n-1}}, \frac{i+1}{b^{n-1}}]$. Applying (a) leads to

$$\begin{aligned} 0 \leq \gamma_{i,j,n}(\widetilde{f}) &= \Big(f\Big(\frac{ib+j}{b^n}\Big) - f\Big(\frac{ib+j}{b^n} + \frac{1}{b^{n-1}}\Big)\Big) \\ &\quad - \frac{b-j}{b}\Big(f\Big(\frac{i}{b^{n-1}}\Big) - f\Big(\frac{i}{b^{n-1}} + \frac{1}{b^{n-1}}\Big)\Big) \\ &\quad - \frac{j}{b}\Big(f\Big(\frac{i+1}{b^{n-1}}\Big) - f\Big(\frac{i+1}{b^{n-1}} + \frac{1}{b^{n-1}}\Big)\Big) \\ &= \gamma_{i,j,n}(f) - \gamma_{i+1,j,n}(f). \end{aligned}$$

(c) Put $h := \frac{2i+1}{b^{n-1}}$ and $\widetilde{f}(x) := f(x) - f(h-x)$, $x \in [\frac{i}{b^{n-1}}, \frac{i+1}{b^{n-1}}]$. Let $\frac{i}{b^{n-1}} \leq x_1 < x \leq \frac{2i+1}{2b^{n-1}}$. Since f' is assumed to be convex, we get

$$\begin{aligned} f'(x_2) &= f'\Big(\frac{h-x_1-x_2}{h-2x_1}x_1 + \frac{x_2-x_1}{h-2x_1}(h-x_1)\Big) \\ &\le \frac{h-x_1-x_2}{h-2x_1}f'(x_1) + \frac{x_2-x_1}{h-2x_1}f'(h-x_1), \\ f'(h-x_2) &= f'\Big(\frac{x_2-x_1}{h-2x_1}x_1 + \frac{h-x_1-x_2}{h-2x_1}(h-x_1)\Big) \\ &\le \frac{x_2-x_1}{h-2x_1}f'(x_1) + \frac{h-x_1-x_2}{h-2x_1}f'(h-x_1). \end{aligned}$$

Adding these two inequalities yields

$$f'(x_2) + f'(h-x_1) \le f'(x_1) + f'(h-x_1),$$

implying that $\widetilde{f}'(x_2) \le \widetilde{f}'(x_1)$. Hence, the function $\widetilde{f}$ is concave on $[\frac{i}{b^{n-1}}, \frac{2i+1}{2b^{n-1}}]$. Therefore, one obtains that

$$\boldsymbol{\Delta}\widetilde{f}(x_1,x_2) \ge \boldsymbol{\Delta}\widetilde{f}(x_1,x_3), \quad \frac{i}{b^{n-1}} \le x_1 < x_2 < x_3 \le \frac{2i+1}{2b^{n-1}}.$$

Note that $\widetilde{f}(\frac{2i+1}{2b^{n-1}} + x) = -\widetilde{f}(\frac{2i+1}{2b^{n-1}} - x)$. So we have $\boldsymbol{\Delta}\widetilde{f}(\frac{i}{b^{n-1}}, \frac{i+1}{b^{n-1}}) = \boldsymbol{\Delta}\widetilde{f}(\frac{i}{b^{n-1}}, \frac{2i+1}{2b^{n-1}})$. Using the formula for the Schauder coefficients yields

$$\begin{aligned} \gamma_{i,j,n}(\widetilde{f}) &= \frac{j}{b^n}\Big(\boldsymbol{\Delta}\widetilde{f}\Big(\frac{ib}{b^n}, \frac{ib+j}{b^n}\Big) - \boldsymbol{\Delta}\widetilde{f}\Big(\frac{i}{b^{n-1}}, \frac{i+1}{b^{n-1}}\Big)\Big) \\ &= \frac{j}{b^n}\Big(\boldsymbol{\Delta}\widetilde{f}\Big(\frac{ib}{b^n}, \frac{ib+j}{b^n}\Big) - \boldsymbol{\Delta}\widetilde{f}\Big(\frac{i}{b^{n-1}}, \frac{2i+1}{2b^{n-1}}\Big)\Big) \ge 0 \end{aligned}$$

or

$$\begin{aligned} 0 \le \gamma_{i,j,n}(\widetilde{f}) &= \Big(f\Big(\frac{ib+j}{b^n}\Big) - f\Big(\frac{2i+j}{b^{n-1}} - \frac{ib+j}{b^n}\Big)\Big) \\ &\quad - \frac{b-j}{b}\Big(f\Big(\frac{i}{b^{n-1}}\Big) - f\Big(\frac{2i+1}{b^{n-1}} - \frac{i}{b^{n-1}}\Big)\Big) \\ &\quad - \frac{j}{b}\Big(f\Big(\frac{i+1}{b^{n-1}}\Big) - f\Big(\frac{2i+1}{b^{n-1}} - \frac{i+1}{b^{n-1}}\Big)\Big) \\ &= \gamma_{i,j,n}(f) - \gamma_{i,b-j,n}(f), \end{aligned}$$

which completes the proof.

(d) The proof is a direct application of the form of the Schauder coefficients, and therefore, it is left as an Exercise.

□

8.4.3 Nowhere Differentiability and the Schauder Coefficients

Generalizing a result of Faber (see [Fab08, Fab10]), it is possible to formulate necessary conditions for the Schauder coefficients of a function $f \in \mathcal{C}(\mathbb{I})$ with $f'_+(x_0) \in \mathbb{R}$ for some $x_0 \in [0,1)$. Fix $b \in \mathbb{N}_2$. Let $f \in \mathcal{C}(\mathbb{I})$, $q, n \in \mathbb{N}$ with $q \le b^{n-1}$, and $j \in \{1, \ldots, b-1\}$. Put

$$\widetilde{\gamma}^{(q)}_{j,n}(f) := \min\{\max\{\gamma_{s,j,n}(f) : i \le s \le i+q-1\} : \ 0 \le i \le b^{n-1} - q\},$$

$$\widetilde{\delta}_{j,n}^{(q)}(f) := b^n \widetilde{\gamma}_{j,n}^{(q)}(f), \quad \delta_{i,j,n}(f) := b^n \gamma_{i,j,n}(f).$$

Theorem 8.4.7. *If $f \in \mathcal{C}(\mathbb{I})$ and $x_0 \in [0,1)$ are such that $f'_+(x_0) \in \mathbb{R}$, then*

$$\lim_{n\to\infty} \widetilde{\delta}_{j,n}^{(q)}(f) = 0, \quad q \in \mathbb{N},\ j = 1, \dots, b-1.$$

Remark 8.4.8. This result makes it possible, at least in principle, to exclude the possibility that a function f has a finite right-sided derivative.

Proof of Theorem 8.4.7. Let

$$x_0 = \sum_{k=1}^{\infty} \frac{\xi_k}{b^k}, \quad 0 \le \xi_k < b,$$

be the b-adic representation of x_0 with the condition that there are infinitely many j's with $\xi_j \neq 9$. Fix $s \in \{1, \dots, q\}$. For $n \in \mathbb{N}_2$, put

$$u_n := \sum_{k=1}^{n-1} \frac{\xi_k}{b^k} + \frac{s}{b^{n-1}},$$

$$v'_n := \sum_{k=1}^{n-1} \frac{\xi_k}{b^k} + \frac{s+1}{b^{n-1}},$$

$$v''_n := \sum_{k=1}^{n-1} \frac{\xi_k}{b^k} + \frac{s}{b^{n-1}} + \frac{j}{b^n}.$$

Note that if $n \ge n_0$ for some $n_0 \in \mathbb{N}_2$, then $u_n, v'_n, v''_n \in \mathbb{I}$. Moreover, we have

$$0 < v'_n - x_0 \le \frac{q+1}{b^{n-1}} \underset{n\to\infty}{\longrightarrow} 0, \quad 0 < v''_n - x_0 \le \frac{qb+j}{b^n} \underset{n\to\infty}{\longrightarrow} 0,$$
$$u_n - x_0 \le q(v'_n - u_n), \quad u_n - x_0 \le \frac{qb}{j}(v''_n - u_n).$$

Applying Remark 2.1.4(a), we get $\lim_{n\to\infty}(\boldsymbol{\Delta} f(u_n, v'_n) - \boldsymbol{\Delta} f(u_n, v''_n)) = 0$. To evaluate this difference, fix an $n \ge n_0$ and recall that f_k, $k < n$, is linear on the segments $[\frac{m}{b^{n-1}}, \frac{m+1}{b^{n-1}}]$. Therefore, $\boldsymbol{\Delta} f_k(u_n, v'_n) = \boldsymbol{\Delta} f_k(u_n, v''_n)$. Moreover, define $r_n := \sum_{k=n+1}^{\infty} f_k$ and note that $r_n(u_n) = r_n(v'_n) = r_n(v''_n) = 0$. Hence, we have

$$0 = \lim_{n\to\infty} (\boldsymbol{\Delta} f(u_n, v'_n) - \boldsymbol{\Delta} f(u_n, v''_n)) = \lim_{n\to\infty} (\boldsymbol{\Delta} f_n(u_n, v'_n) - \boldsymbol{\Delta} f_n(u_n, v''_n)).$$

Since $f_n(u_n) = f_n(v'_n) = 0$, we end up with $0 = -\lim_{n\to\infty} \boldsymbol{\Delta} f_n(u_n, v''_n)$. Put $i_n := \xi_1 b^{n-2} + \dots + \xi_{n-1}$. Then $\boldsymbol{\Delta} f_n(u_n, v''_n) = \frac{b^n}{j} \gamma_{i_n+s,j,n}(f)$. Hence, the sequences $\left(b^n \gamma_{i_n+s,j,n}(f)\right)_{n\in\mathbb{N}_{n_0}}$ converge to 0, $s = 1, \dots, q$. In particular, if n is large, then

$$\widetilde{\delta}_{j,n}^{(q)}(f) \le \max\{|\delta_{i_n+s,j,n}(f)| : s = 1, \dots, q\} \underset{n\to\infty}{\longrightarrow} 0,$$

which completes the proof. □

In order to apply the former theorem, one needs an effective way to calculate the Schauder coefficients.

8.4.4 Schauder Coefficients of Solutions of a System of Functional Equations

Given $b \in \mathbb{N}_2$, we now study the continuous solution $f \in \mathcal{C}(\mathbb{I})$ of the generalized system of functional equations (8.4.2) satisfying (8.4.3) with continuous functions g_j.

Theorem 8.4.9. *Let f be as before. Then:*

(a) $\gamma_{0,0}(f) = \frac{g_0(0)}{1-a_0}, \quad \gamma_{1,0}(f) = \frac{g_{b-1}(1)}{1-a_{b-1}}$;

(b) $\gamma_{0,j,1}(f) = \frac{b-j}{b}\Big((a_j - 1)\gamma_{0,0}(f) + g_j(0)\Big) + \frac{j}{b}\Big((a_{j-1} - 1)\gamma_{1,0}(f) + g_{j-1}(1)\Big),$
$j = 1, \dots, b-1$;

(c) $\gamma_{i+\nu b^{n-1},j,n+1}(f) = a_\nu \gamma_{i,j,n}(f) + \gamma_{i,j,n}(g_\nu),$
$n \in \mathbb{N},\ i = 0, \dots, b^{n-1} - 1,\ j = 1, \dots, b-1,\ \nu = 0, \dots, b-1.$

Proof. (a) Use (8.4.3) and Theorem 8.4.4.
(b) We know that $\gamma_{0,j,1}(f) = f(\frac{j}{b}) - \frac{b-j}{b} f(0) - \frac{j}{b} f(1)$. Using (8.4.3) leads to

$$f\Big(\frac{j}{b}\Big) = \frac{b-j}{b}\Big(a_j f(0) + g_j(0)\Big) + \frac{j}{b}\Big(a_{j-1} f(1) + g_{j-1}(1)\Big).$$

Together with (a), it gives (b).
(c) Recall the representation of $\gamma_{i+\nu b^{n-1},j,n}(f)$ in Theorem 8.4.4. Applying to the three terms on the right-hand side of that representation the functional equations in (8.4.2) gives

$$\begin{aligned} f\Big(\frac{(i+\nu b^{n-1})b + j}{b^{n+1}}\Big) &= f\Big(\frac{(ib+j)b^{-n} + \nu}{b}\Big) = a_\nu f\Big(\frac{ib+j}{b^n}\Big) + g_\nu\Big(\frac{ib+j}{b^n}\Big), \\ f\Big(\frac{i+\nu b^{n-1}}{b^n}\Big) &= f\Big(\frac{ib^{-n+1} + \nu}{b}\Big) = a_\nu f\Big(\frac{i}{b^{n-1}}\Big) + g_\nu\Big(\frac{i}{b^{n-1}}\Big), \\ f\Big(\frac{i+\nu b^{n-1}+1}{b^n}\Big) &= f\Big(\frac{(i+1)b^{-n+1} + \nu}{b}\Big) = a_\nu f\Big(\frac{i+1}{b^{n-1}}\Big) + g_\nu\Big(\frac{i+1}{b^{n-1}}\Big), \end{aligned}$$

which verifies (c).

□

In order to show that f has nowhere a finite or infinite right-sided derivative on $[0,1)$, it suffices to find j and q such that the sequence $(\widetilde{\delta}^{(q)}_{j,n}(f))_n$ does not converge to 0. To do so, Theorem 8.4.9 will be useful.

Put

$$\begin{aligned} &\alpha := \min\{|a_j| : j = 0, \dots, b-1\}, \\ &\widetilde{\delta}_{j,n} = \widetilde{\delta}_{j,n}(g_0, \dots, g_{b-1}) := \max\{\delta_{i,j,n}(g_\nu) : \ i = 0, \dots, b^{n-1} - 1, \\ &\qquad\qquad\qquad\qquad \nu = 0, \dots, b-1\}, \quad n \in \mathbb{N},\ j = 1, \dots, b-1. \end{aligned}$$

Lemma 8.4.10. *Let $n_0 \in \mathbb{N}$, $i_0 \in \{0, \dots, b^{n-1} - 1\}$, and $j_0 \in \{1, \dots, b-1\}$. Assume that $\alpha > 0$. Then*

$$|\delta_{i_0+rb^{n_0-1},j_0,n_0+k}(f)| \geq (\alpha b)^k \Big(|\delta_{i_0,j_0,n_0}(f)| - b\sum_{s=0}^{k-1}\frac{1}{(\alpha b)^{s+1}}\widetilde{\delta}_{j_0,n_0+s}\Big),$$

where $k, r \in \mathbb{N}_0$ with $i_0+rb^{n_0-1} \leq b^{n_0+k-1}-1$ (note that the empty sum is by definition equal to 0).

Proof. We will use induction over $k \in \mathbb{N}_0$.

Step 1°. Let $k = 0$. Then $r = 0$, and the claimed inequality is obviously true.

Step 2°. Assume that the lemma is true for $k \in \mathbb{N}_0$. We have to verify that the claim is true for all $\delta_{i_0+rb^{n_0-1}+\nu b^{n_0+k-1},j_0,n_0+k+1}(f)$, $\nu = 0, \ldots, b-1$, if it is true for $\delta_{i_0+rb^{n_0-1},j_0,n_0+k}(f)$. Using Theorem 8.4.9 and the induction hypothesis gives

$$\begin{aligned}
&|\delta_{i_0+rb^{n_0-1}+\nu b^{n_0+k-1},j_0,n_0+k+1}(f)| \\
&= |ba_\nu \delta_{i_0+rb^{n_0-1},j_0,n_0+k}(f) + b\delta_{i_0+rb^{n_0-1},j_0,n_0+k}(f)| \\
&\geq (\alpha b)|\delta_{i_0+rb^{n_0-1},j_0,n_0+k}(f)| - b|\delta_{i_0+rb^{n_0-1},j_0,n_0+k}(f)| \\
&\geq (\alpha b)^{k+1}\Big(|\delta_{i_0,j_0,n_0}(f)| - b\sum_{s=0}^{k-1}\frac{1}{(\alpha b)^{s+1}}\widetilde{\delta}_{j_0,n_0+s}\Big) - b\widetilde{\delta}_{j_0,n_0+k} \\
&\geq (\alpha b)^{k+1}\Big(|\delta_{i_0,j_0,n_0}(f)| - b\sum_{s=0}^{k}\frac{1}{(\alpha b)^{s+1}}\widetilde{\delta}_{j_0,n_0+s}\Big).
\end{aligned}$$

□

Put

$$M_{j,n}^{(\alpha)}(f) := \sum_{s=0}^{\infty}\frac{1}{(\alpha b)^{s+1}}\widetilde{\delta}_{j,n+s} \in [0,\infty].$$

Corollary 8.4.11. *Let $n_0 \in \mathbb{N}$, $i_0 \in \{0, \ldots, b^{n-1}-1\}$, $j_0 \in \{1, \ldots, b-1\}$, and $k, r \in \mathbb{N}_0$. Assume that $\alpha > 0$. Then*

$$|\delta_{i_0+rb^{n-1},j_0,n_0+k}(f)| \geq (\alpha b)^k\Big(|\delta_{i_0,j_0,n_0}(f)| - bM_{j_0,n_0}^{(\alpha)}(f)\Big).$$

Finally, we end up with a criterion that may be helpful in proving nowhere differentiability properties of the solution f of (8.4.2).

Theorem 8.4.12. *Assume that $\alpha b \geq 1$. If there are suitable indices i_0, j_0, and n_0 with $|\delta_{i_0,j_0,n_0}(f)| > bM_{j_0,n_0}^{(\alpha)}(f)$, then f has nowhere on $[0,1)$ a right-sided (finite) derivative.*

Proof. An immediate consequence of Corollary 8.4.11 and the assumption is that

$$|\delta_{i_0+rb^{n-1},j_0,n_0+k}(f)| \geq (\alpha b)^k C,$$

where $C := |\delta_{i_0,j_0,n_0}(f)| - bM_{j_0,n_0}^{(\alpha)}(f) > 0$. Hence, the left-hand terms cannot converge to 0 if k tends to ∞. Put

$$q := b^{n-1} \text{ and } t_k := \min\{|\delta_{i_0+rb^{n-1},j,n+k}(f)| : r \in \mathbb{N}\}.$$

Then $(t_k)_k$ does not converge to zero. Finally, recall that $\widetilde{\delta}_{j_0,n_0+k}^{(q)}(f) \geq t_k$. Thus, $(\widetilde{\delta}_{j_0,n_0+k}^{(q)})_{k\in\mathbb{N}}$ also does not tend to zero. Hence, Theorem 8.4.7 implies the nowhere differentiability of f. □

To be able to apply this kind of result, we obviously need that $M_{j_0,n_0}^{(\alpha)}(f) < \infty$. Conditions on (8.4.2) under which this holds will be discussed in the next lemma.

Lemma 8.4.13. *Let* $\alpha b \geq 1$.

(a) *If all the functions* g_ν *in* (8.4.2) *are differentiable on* $\mathbb{I}$ *with* $\|g'_\nu\|_{\mathbb{I}} \leq L_1$ *and* $\alpha b > 1$, *then*

$$M_{j,n}^{(\alpha)}(f) \leq 2L_1 \frac{j(b-j)}{b(\alpha b - 1)}, \quad n \in \mathbb{N},\ j = 1, \dots, b-1.$$

(b) *If all the functions* g_ν *in* (8.4.2) *are twice differentiable on* $\mathbb{I}$ *with* $\|g''_\nu\|_{\mathbb{I}} \leq L_2$, *then*

$$M_{j,n}^{(\alpha)}(f) \geq \frac{L_2 j(b-j)}{2b^{n-1}(\alpha b^2 - 1)}, \quad n \in \mathbb{N},\ j = 1, \dots, b-1.$$

Proof. (a) By assumption, we have $|g_\nu(x') - g_\nu(x'')| \leq L_1|x' - x''|$ for $x', x'' \in \mathbb{I}$ for all ν. Now fix an $i \in \{0, \dots, b^{n-1} - 1\}$ and a $\nu \in \{0, \dots, b-1\}$. Then

$$\begin{aligned} |\gamma_{i,j,n}(g_\nu)| &\leq \frac{b-j}{b}\left|g_\nu\left(\frac{ib+j}{b^n}\right) - g_\nu\left(\frac{i}{b^{n-1}}\right)\right| + \frac{j}{b}\left|g_\nu\left(\frac{i+1}{b^{n-1}}\right) - g_\nu\left(\frac{ib+j}{b^n}\right)\right| \\ &\leq \frac{b-j}{b} L_1 \frac{j}{b^n} + \frac{j}{b} L_1 \frac{b-j}{b^n} = 2L_1 \frac{j(b-j)}{b^{n+1}}. \end{aligned}$$

Thus, $|\delta_{i,j,n}(g_\nu)| \leq 2L_1 \frac{j(b-j)}{b}$. Since i and ν were arbitrary, we get $\widetilde{\delta}_{j,n} \leq 2L_1 \frac{j(b-j)}{b}$. Plugging this estimate into the definition of $M_{j,n}^{(\alpha)}(f)$ leads finally to $M_{j,n}^{(\alpha)}(f) \leq \frac{1}{\alpha b - 1}$.

(b) Fix a $\nu \in \{0, \dots, b-1\}$ and let $x', x'' \in \mathbb{I}$. Then

$$g_\nu(x') - g_\nu(x'') = g'_\nu(x'')(x' - x'') + \frac{g''_\nu(\xi)}{2}(x' - x'')^2,$$

where ξ lies between x' and x''. Using Theorem 8.4.4 implies

$$\begin{aligned} \gamma_{i,j,n}(g_\nu) &= -\frac{b-j}{b}\left(g_\nu\left(\frac{i}{b^{n-1}}\right) - g_\nu\left(\frac{ib+j}{b^n}\right)\right) - \frac{j}{b}\left(f\left(\frac{i+1}{b^{n-1}}\right) - g_\nu\left(\frac{ib+j}{b^n}\right)\right) \\ &= -\frac{b-j}{b}\left(g'_\nu\left(\frac{ib+j}{b^n}\right)\frac{-j}{b^n} + g''_\nu(\xi_1)\frac{j^2}{2b^{2n}}\right) \\ &\quad - \frac{j}{b}\left(g'_\nu\left(\frac{ib+j}{b^n}\right)\frac{b-j}{b^n} + g''_\nu(\xi_2)\frac{(b-j)^2}{2b^{2n}}\right) \\ &= -\frac{g''_\nu(\xi_1)(b-j)j^2}{2bb^{2n}} - g''_\nu(\xi_2)\frac{j(b-j)^2}{2bb^{2n}}, \end{aligned}$$

where ξ_1 (resp. ξ_2) lies between $\frac{i}{b^{n-1}}$ and $\frac{ib+j}{b^n}$ (resp. $\frac{ib+j}{b^n}$ and $\frac{i+1}{b^{n-1}}$). By virtue of the assumptions in (b), it follows that

$$|\delta_{i,j,n}(g_\nu)| \leq \frac{L_2}{2}\left|\frac{(b-j)j^2}{bb^n} + \frac{j(b-j)^2}{bb^n}\right| = \frac{L_2 j(b-j)}{2b^n}.$$

If one puts this estimate into the definition of $M_{j,n}^{(\alpha)}(f)$, then the claim in (b) is an easy consequence.

□

8.4.5 Nowhere Differentiability of $W_{1,a,b,\theta}$ for $ab \geq 1$, $b \in \mathbb{N}_2$

After this long journey along Schauder bases and systems of functional equations, we are now in a position to prove Theorem 8.4.1.

Proof of Theorem 8.4.1. As mentioned at the beginning of this section, it suffices to show that the function $f := \widetilde{W}_{a,b,\theta}$ with $ab \geq 1$ has nowhere on $[0,1)$ a right-sided derivative. By virtue of Theorem 8.4.12, it suffices to find suitable indices i,j,n such that $|\delta_{i,j,n}(f)| > bM_{j,n}^{(1/b)}(f) \geq bM_{j,n}^{(\alpha)}(f)$. Put $M_{j,n}(f) := M_{j,n}^{(1/b)}(f)$.

Step 1°. Let us first discuss the case $b=2$. Thus $f = \widetilde{W}_{a,b,\theta}$ and the data in the associated functional system are given by $a_0 = a_1 = a$ (i.e., $\alpha = a$), $g_0(x) := \sin(\pi x + \theta)$, and $g_1(x) = -\sin(\pi x + \theta)$, $x \in \mathbb{I}$. Then Lemma 8.4.13(b) leads to $M_{1,n}(f) \leq \frac{\pi^2}{2^n}$, $n \in \mathbb{N}$. Fix $n=3$. Then we have the following list of $\delta_{i,j,n}(f)$ (use Theorem 8.4.9):

$$\gamma_{0,0}(f) = \tfrac{\sin\theta}{1-a}, \quad \gamma_{1,0}(f) = \tfrac{-\sin(\pi+\theta)}{1-a} = \tfrac{\sin\theta}{1-a};$$

$$\delta_{0,1,1}(f) = \left((a-1)\tfrac{\sin\theta}{1-a} - \sin\theta\right) + \left((a-1)\tfrac{\sin\theta}{1-a} - \sin\theta\right) = -4\sin\theta,$$

$$\delta_{0,1,2}(f) = 2a\delta_{0,1,1}(f) + 2\delta_{0,1,1}(g_0) = -8a\sin\theta + 4\cos\theta,$$

$$\delta_{1,1,2}(f) = 2a\delta_{0,1,1}(f) + 2\delta_{0,1,1}(g_1) = -8a\sin\theta - 4\cos\theta,$$

$$\delta_{0,1,3}(f) = 2a\delta_{0,1,2}(f) + 2\delta_{0,1,2}(g_0) = -16a^2\sin\theta + 8a\cos\theta + 16\sin(\theta + \tfrac{\pi}{4})\sin^2\tfrac{\pi}{8},$$

$$\delta_{1,1,3}(f) = 2a\delta_{1,1,2}(f) + 2\delta_{1,1,2}(g_0) = -16a^2\sin\theta - 8a\cos\theta + 16\sin(\theta + \tfrac{3\pi}{4})\sin^2\tfrac{\pi}{8},$$

$$\delta_{2,1,3}(f) = 2a\delta_{0,1,2}(f) + 2\delta_{0,1,2}(g_1) = -16a^2\sin\theta + 8a\cos\theta - 16\sin(\theta + \tfrac{\pi}{4})\sin^2\tfrac{\pi}{8},$$

$$\delta_{3,1,3}(f) = 2a\delta_{1,1,2}(f) + 2\delta_{1,1,2}(g_1) = -16a^2\sin\theta - 8a\cos\theta - 16\sin(\theta + \tfrac{3\pi}{4})\sin^2\tfrac{\pi}{8}.$$

By assumption, we have $1 > a \geq \frac{1}{2}$. Then:

if $\theta \in [0, \frac{\pi}{4}]$, then $|\delta_{3,1,3}(f)| = 16a^2\sin\theta + 8a\cos\theta + 16\sin(\theta + \frac{3\pi}{4})\sin^2\frac{\pi}{8}$
$$\geq 4\sin\theta + 4\cos\theta \geq 4 > \tfrac{\pi^2}{4} \geq 2M_{1,3}(f);$$

if $\theta \in [\frac{\pi}{4}, \frac{\pi}{2}]$, then $|\delta_{1,1,3}(f)| = 16a^2\sin\theta + 8a\cos\theta - 16\sin(\theta + \frac{3\pi}{4})\sin^2\frac{\pi}{8}$
$$\geq 4\sin\theta + 4\cos\theta \geq 4 > \tfrac{\pi^2}{4} \geq 2M_{1,3}(f);$$

if $\theta \in [\frac{\pi}{2}, \frac{3\pi}{4}]$, then $|\delta_{2,1,3}(f)| = 16a^2\sin\theta - 8a\cos\theta - 16\sin(\theta + \frac{3\pi}{4})\sin^2\frac{\pi}{8}$
$$\geq 4\sin\theta - 4\cos\theta \geq 4 > \tfrac{\pi^2}{4} \geq 2M_{1,3}(f);$$

if $\theta \in [\frac{3\pi}{4}, \pi]$, then $|\delta_{0,1,3}(f)| = 16a^2\sin\theta - 8a\cos\theta - 16\sin(\theta + \frac{3\pi}{4})\sin^2\frac{\pi}{8}$
$$\geq 4\sin\theta - 4\cos\theta \geq 4 > \tfrac{\pi^2}{4} \geq 2M_{1,3}(f).$$

Hence the condition from above has been verified, which implies that f has no right-sided derivative on $[0,1)$.

Step 2°. Now we assume that $b \in \mathbb{N}_3$, and here we will discuss three different cases.

Case 1°. Let $f(x) := \widetilde{W}_{a,b,0}(x) = \sum_{n=1}^{\infty} a^n \sin(b\pi b^n x)$. Then the data in the corresponding (8.4.2) are given by $a_j = (-1)^{jb}$ and $g_j = (-1)^j \sin(\pi x)$, $x \in I$. Note that $\|g_j''\|_I \leq \pi^2$. Therefore, Lemma 8.4.13 leads to $M_{j,n}(f) \leq \frac{\pi^2}{2b^{n-1}}$. Then

$$\begin{aligned}\delta_{1,1,2}(f) &= (-1)^b ab\delta_{0,1,1}(f) + b\delta_{0,1,1}(g_1) \\ &= 0 - b^2 \sin^2 \frac{\pi}{b} \le -2b \le -6 < -\frac{\pi^2}{2} < -bM_{1,2}(f).\end{aligned}$$

Hence Theorem 8.4.12 applies.

Case 2°. Let $f(x) := \widetilde{\boldsymbol{W}}_{a,b,\pi/2}(x) = \sum_{n=0}^{\infty} a^n \cos(b\pi b^n x)$, $x \in \mathbb{I}$. Then the data in the corresponding (8.4.2) are given by $a_j = (-1)^{jb}$ and $g_j(x) = (-1)^j \cos(\pi x)$, $x \in \mathbb{I}$. Note that $|\gamma_{i,j,n}(g_j)| = |\gamma_{i,j,n}(g)|$, where $g(x) := \cos(\pi x)$, $x \in \mathbb{I}$ (see Theorem 8.4.4).

Then:

$$\begin{aligned}M_{1,n}(f) = M_{1,n}^{(1/b)}(f) &= \sum_{k=0}^{\infty} \widetilde{\delta}_{1,n+k} \\ &= \sum_{k=n}^{\infty} \max\{|\delta_{i,1,k}(g_\nu)| : 0 \le i \le b^{n-1}-1,\ 0 \le \nu \le b-1\} \\ &= \sum_{k=n}^{\infty} b^k \max\{|\gamma_{i,1,k}(g_\nu)| : 0 \le i \le b^{n-1}-1,\ 0 \le \nu \le b-1\} \\ &= \sum_{k=n}^{\infty} b^k \max\{|\gamma_{i,1,k}(g)| : 0 \le i \le b^{n-1}-1\} \\ &\overset{(1)}{=} \sum_{k=n}^{\infty} b^k \max\{|\gamma_{0,1,k}(g)|, |\gamma_{b^{n-1}-1,1,k}(g)|\} \\ &\overset{(2)}{=} \sum_{k=n}^{\infty} b^k \max\{\gamma_{0,1,k}(g), -\gamma_{b^{n-1}-1,1,k}(g)\} \\ &\overset{(3)}{=} \sum_{k=n}^{\infty} b^k \max\{\gamma_{0,1,k}(g), \gamma_{0,b-1,k}(g)\} \\ &\overset{(4)}{=} \sum_{k=n}^{\infty} \gamma_{0,1,k}(g) \\ &= \lim_{N\to\infty} \sum_{k=n}^{N} \Big(\cos\frac{1}{b^k} - \frac{b-1}{b} \quad \frac{1}{b}\cos\frac{\pi}{b^{k-1}}\Big) \\ &= \lim_{N\to\infty} \Big(\sum_{k=n}^{N} b^k\Big(\cos\frac{1}{b^N} - \cos 0\Big) - \sum_{k=n}^{N} b^{k-1}\Big(1 - \cos\frac{\pi}{b^{k-1}}\Big)\Big) \\ &= \lim_{N\to\infty} \Big(b^N\Big(\cos\frac{1}{b^N} - \cos 0\Big) - b^{n-1}\Big(\cos\frac{\pi}{b^{n-1}} - 1\Big)\Big) \\ &= \cos' 0 + b^{n-1}\Big(1 - \cos\frac{\pi}{b^{n-1}}\Big) = b^{n-1}\Big(1 - \cos\frac{\pi}{b^{n-1}}\Big).\end{aligned}$$

The equality (1), (2), (3), resp. (4) from above is a consequence of Lemma 8.4.6(b), (a), (d), resp. (c).

In particular, we get $bM_{1,2}(f) = b^2(1 - \cos\frac{\pi}{b}) \le \frac{3b}{2}$ (use that $\cos(\pi x) \ge 1 - \frac{3}{2}x$ for $x \in [0, \frac{1}{3}]$).

Assume first that b is even. We will calculate $\delta_{1,1,2}(f)$.

Using the recurrence formula (see Theorem 8.4.9), we have $\gamma_{0,0}(f) = \frac{1}{1-a}$ and $\gamma_{1,0}(f) = \frac{1}{1-a}$. Thus the next step leads to

$$\delta_{0,1,1}(f) = b\Big(\frac{b-1}{b}\Big(\frac{a-1}{1-a} + g_1(0)\Big) + \frac{1}{b}\Big(\frac{a-1}{1-a} + g_0(1)\Big)\Big) = -2b.$$

Therefore, using that $ab \geq 1$, we obtain

$$\begin{aligned}\delta_{1,1,2}(f) &= ab\delta_{0,1,1}(f) + b\delta_{0,1,1}(g_1)\\ &= -2ab^2 - b^2\Big(\cos\frac{\pi}{b} - \frac{b-1}{b}\cos 0 - \frac{1}{b}\cos\pi\Big)\\ &= -2b^2\Big(a+\frac{1}{b}\Big) + b^2\Big(1-\cos\frac{\pi}{b}\Big) = -2b^2\Big(a+\frac{1}{b}\Big) + bM_{1,2}(f)\\ &\leq -4b + \frac{3b}{2} = -\frac{5b}{2} \leq -bM_{1,2}(f).\end{aligned}$$

Assuming now that b is odd, a similar procedure as before leads to

$$\delta_{1,1,2}(f) = 2ab\frac{b-1}{1-a} - 2b + b^2\Big(1-\cos\frac{\pi}{b}\Big) = 2b^2\frac{a-\frac{1}{b}}{1-a} + bM_{1,2}(f) \geq bM_{1,2}(f).$$

Therefore, in both cases we have the inequality $|\delta_{1,1,2}(f)| \geq bM_{1,2}(f)$. If the strict inequality is true, then Theorem 8.4.12 applies directly. If $|\delta_{1,1,2}(f)| \geq bM_{1,2}(f)$ holds, we get

$$\begin{aligned}bM_{1,3}^{(a)}(f) &= ab^2M_{1,2}^{(a)}(f) - b\widetilde{\delta}_{1,2}\\ &< ab|\delta_{1,1,2}(f)| - b|\delta_{1,1,2}(g_0)|\\ &\leq |ab\delta_{1,1,2}(f) + b\delta_{1,1,2}(g_0)| = |\delta_{1,1,3}(f)|.\end{aligned}$$

So we end up with $|\delta_{1,1,3}(f)| > bM_{1,3}(f)$, which allows us to apply Theorem 8.4.12. Hence, f has nowhere on $[0,1)$ a right-sided derivative.

Case 3°. Let $f := \widetilde{\boldsymbol{W}}_{a,b,\theta}$. Then $f = \cos\theta\cdot f_1 + \sin\theta\cdot f_2$, where $f_1 := \widetilde{\boldsymbol{W}}_{a,b,0}$ and $f_2 := \widetilde{\boldsymbol{W}}_{a,b,\pi/2}$. We may assume that $\cos\theta\cdot\sin\theta \neq 0$. Recall (cf. (8.4.2)) that f_j solves the following system:

$$h\Big(\frac{x+\nu}{b}\Big) = a_\nu^{(k)}h(x) + g_\nu^{(k)}(x), \quad x \in \mathbb{I},\ \nu = 0,\ldots,b-1, \tag{8.4.4}$$

with the associated data $a_\nu^{(1)} = a_\nu^{(2)} = (-1)^{\nu b}a$, $g_\nu^{(1)}(x) = (-1)^\nu\sin(\pi x)$, and $g_\nu^{(2)}(x) = (-1)^\nu\cos(\pi x)$, $x \in \mathbb{I}$, $\nu = 0,\ldots,b-1$. Then f is the uniquely determined solution of the following system:

$$h\Big(\frac{x+\nu}{b}\Big) = a_\nu h(x) + g_\nu(x), \quad x \in \mathbb{I},\ \nu = 0,\ldots,b-1, \tag{8.4.5}$$

where $a_\nu = (-1)^{\nu b}a$ and $g_\nu(x) = \cos\theta\cdot g_\nu^{(1)} + \sin\theta\cdot g_\nu^{(2)}$, $\nu = 0,\ldots,b-1$.

By virtue of Theorem 8.4.12, it suffices to find suitable indices i,j,n such that $\delta_{i,j,n}(f) > bM_{j,n}^{(a)}(f)$, where the data $\delta_{i,j,n}(f)$ and $M_{j,n}^{(a)}(f)$ are taken with respect to the system (8.4.5).

By virtue of Theorem 8.4.4, it is clear that

$$\delta_{i,j,n}(f) = \cos\theta\cdot\delta_{i,j,n}(f_1) + \sin\theta\cdot\delta_{i,j,n}(f_2),$$

where $\delta_{i,j,n}(f_k)$ is understood with respect to (8.4.4). Moreover,

$$M_{j,n}^{(a)}(f) \le |\cos\theta| \cdot M_{j,n}^{(a)}(f_1) + |\sin\theta| \cdot M_{j,n}^{(a)}(f_2),$$

where $M_{,j,n}^{(a)}(f_k)$ is understood with respect to (8.4.4).

Now fix the indices $i_0 = 0$, $j_0 = b-1$, and $n_0 = 2$. Then Theorem 8.4.9 leads to

$$\begin{aligned}\delta_{0,b-1,2}(f_1) &= ab\delta_{0,b-1,1}(f_1) + b\delta_{0,b-1,1}(g_0^{(1)}) \\ &= b\delta_{0,b-1,1}(g_0^{(1)}) = -b\delta_{0,1,1}(g_1^{(1)}) \\ &= -\delta_{1,1,2}(f_1).\end{aligned}$$

Here we have used that $\delta_{0,b-1,1}(f_1) = 0$, $g_0^{(1)} = -g_1^{(1)}$, and Lemma 8.4.6(d).

A similar calculation gives

$$\begin{aligned}\delta_{0,b-1,2}(f_2) &= ab\delta_{0,b-1,1}(f_2) + b\delta_{0,b-1,1}(g_0^{(2)}), \\ &= \begin{cases} ab\delta_{0,1,1}(f_2) + b\delta_{0,1,1}(g_1^{(2)}), & \text{if } b \text{ is even} \\ -ab\delta_{0,1,1}(f_2) + b\delta_{0,1,1}(g_1^{(2)}), & \text{if } b \text{ is odd} \end{cases} \\ &= \delta_{1,1,2}(f_2).\end{aligned}$$

Recall that $\delta_{1,1,2}(f_j)$ is negative, $k = 1, 2$. If now $\cos\theta$ and $\sin\theta$ have the same sign, then

$$\begin{aligned}bM_{1,2}(f) &\le |\cos\theta| \cdot M_{1,2}(f_1) + |\sin\theta| \cdot M_{1,2}(f_2) \\ &< |\cos\theta| \cdot |\delta_{1,1,2}(f_1)| + |\sin\theta| \cdot |\delta_{1,1,2}(f_2)| \\ &= |\cos\theta \cdot \delta_{1,1,2}(f_1) + \sin\theta \cdot \delta_{1,1,2}(f_2)| \\ &= |\delta_{1,1,2}(f)|.\end{aligned}$$

In the remaining case, in which $\cos\theta$ and $\sin\theta$ have opposite signs, one is led to

$$\begin{aligned}bM_{b-1,2}(f) &\le |\cos\theta| \cdot M_{b-1,2}(f_1) + |\sin\theta| \cdot M_{b-1}(f_2) \\ &= |\cos\theta| \cdot M_{1,2}(f_1) + |\sin\theta| \cdot M_{1,2}(f_2) \\ &< |\cos\theta| \cdot |\delta_{1,1,2}(f_1)| + |\sin\theta| \cdot |\delta_{1,1,2}(f_2)| \\ &= |\cos\theta| \cdot |-\delta_{0,b-1,2}(f_1)| + |\sin\theta| \cdot |\delta_{0,b-1,2}(f_2)| \\ &= |\cos\theta \cdot \delta_{0,b-1,2}(f_1) + \sin\theta \cdot \delta_{0,b-1,2}(f_2)| \\ &= |\delta_{0,b-1,2}(f)|,\end{aligned}$$

which completes the proof. □

8.5 Weierstrass-Type Functions from a General Point of View

The Weierstrass-type function $\boldsymbol{W}_{p,a,b,\boldsymbol{\theta}}$ from Chap. 3 may be considered a special case of the following more general family of functions:

$$\boldsymbol{F}_{\Phi,p,\boldsymbol{a},\boldsymbol{b},\boldsymbol{\theta}}(x) := \sum_{n=0}^{\infty} a_n \Phi^p(b_n x + \theta_n), \quad x \in \mathbb{R},$$

where:

- $\varPhi : \mathbb{R} \longrightarrow \mathbb{I}$ is such that $\varPhi(-x) = \varPhi(x)$, $\varPhi(x+1) = \varPhi(x)$, and $|\varPhi(x) - \varPhi(y)| \leq |x - y|$, $x, y \in \mathbb{R}$, (8.5.1)
- $p \in \mathbb{N}$,
- $\boldsymbol{a} := (a_n)_{n=0}^{\infty} \subset \mathbb{C}_*$ and $\sum_{n=0}^{\infty} |a_n| < +\infty$,
- $\boldsymbol{b} := (b_n)_{n=0}^{\infty} \subset \mathbb{R}_{>0}$, $b_{n+1} > b_n$, $n \in \mathbb{N}_0$, and $\sum_{n=0}^{\infty} |a_n| b_n = +\infty$,
- $\boldsymbol{\theta} := (\theta_n)_{n=0}^{\infty} \subset \mathbb{R}$. (8.5.2)

Remark 8.5.1. (a) Let $\nu(x) := \cos(2\pi x)$, $x \in \mathbb{R}$. Then

$$\boldsymbol{F}_{\nu,p,(a^n)_{n=0}^{\infty},(b^n)_{n=0}^{\infty},\frac{1}{2\pi}\boldsymbol{\theta}} = \boldsymbol{W}_{p,a,b,\boldsymbol{\theta}}.$$

(b) Let $\psi(x) := \operatorname{dist}(x, \mathbb{Z})$, $x \in \mathbb{R}$. Then $\boldsymbol{F}_{\psi,1,(a^n)_{n=0}^{\infty},(b^n)_{n=0}^{\infty},\boldsymbol{\theta}} = \boldsymbol{T}_{a,b,\boldsymbol{\theta}}$, which is a Takagi–van der Waerden-type function (cf. § 4.1), and $\boldsymbol{F}_{\psi,1,\boldsymbol{a},\boldsymbol{b},\boldsymbol{\theta}} = \boldsymbol{T}_{\boldsymbol{a},\boldsymbol{b},\boldsymbol{\theta}}$, which is a generalized Takagi–van der Waerden function (cf. Theorem 4.3.1).

We fix a function $\varPhi$ with (8.5.1). To simplify notation, we will use the following conventions (similarly as in § 3.1):

- if $\varPhi$ is fixed, then $\boldsymbol{F}_{p,\boldsymbol{a},\boldsymbol{b},\boldsymbol{\theta}} = \boldsymbol{F}_{\varPhi,p,\boldsymbol{a},\boldsymbol{b},\boldsymbol{\theta}}$;
- if $\varPhi, p, \boldsymbol{a}, \boldsymbol{b}$ are fixed, then $\boldsymbol{F}_{\boldsymbol{\theta}} := \boldsymbol{F}_{\varPhi,p,\boldsymbol{a},\boldsymbol{b},\boldsymbol{\theta}}$;
- if $\varPhi, p, \boldsymbol{a}, \boldsymbol{b}, \boldsymbol{\theta}$ are fixed, then $\boldsymbol{F} := \boldsymbol{F}_{\varPhi,p,\boldsymbol{a},\boldsymbol{b},\boldsymbol{\theta}}$.

Functions of the type $\boldsymbol{F}_{\varPhi,p,\boldsymbol{a},\boldsymbol{b},\boldsymbol{\theta}}$ have many common properties that are listed below.

Remark 8.5.2. (a) $\sum_{n=0}^{\infty} \sup_{x\in\mathbb{R}} |a_n \varPhi^p(b_n x + \theta_n)| \leq \sum_{n=0}^{\infty} |a_n| =: A$. Consequently, $\boldsymbol{F} \in \mathcal{C}(\mathbb{R}, \mathbb{C})$ and $|F(x)| \leq A$, $x \in \mathbb{R}$.

(b) $\boldsymbol{F}_{\boldsymbol{\theta}}(x + x_0) = \boldsymbol{F}_{x_0\boldsymbol{b}+\boldsymbol{\theta}}(x)$, $\boldsymbol{F}_{\boldsymbol{\theta}}(-x) = \boldsymbol{F}_{-\boldsymbol{\theta}}(x)$, $x, x_0 \in \mathbb{R}$.

(c) The function $\boldsymbol{F}_{p,\boldsymbol{a},\boldsymbol{b},\boldsymbol{\theta}}$ may be formally defined also when $B := \sum_{n=0}^{\infty} |a_n| b_n < +\infty$. However, in this case, $\boldsymbol{F}_{\boldsymbol{\theta}}$ is Lipschitz continuous, uniformly with respect to $\boldsymbol{\theta}$. In particular, $\boldsymbol{F}_{\boldsymbol{\theta}}$ is almost everywhere differentiable. Such functions are of course irrelevant from our point of view.

Indeed,

$$\begin{aligned}
|\boldsymbol{F}_{\boldsymbol{\theta}}(x+h) - \boldsymbol{F}_{\boldsymbol{\theta}}(x)| &\leq \sum_{n=0}^{\infty} |a_n| |\varPhi^p(b_n(x+h) + \theta_n) - \varPhi^p(b_n x + \theta_n)| \\
&\overset{\substack{\text{mean value}\\\text{theorem}}}{\leq} \sum_{n=0}^{\infty} |a_n| p |\varPhi(b_n(x+h) + \theta_n) - \varPhi(b_n x + \theta_n)| \\
&\overset{(8.5.1)}{\leq} p \sum_{n=0}^{\infty} |a_n| b_n |h| = pB|h|, \quad x, h \in \mathbb{R}.
\end{aligned}$$

Observe that if moreover, $\varPhi \in \mathcal{C}^1(\mathbb{R})$ and $C := \sup_{x\in\mathbb{R}} |\varPhi'(x)| < +\infty$ (e.g., $\varPhi = \nu$), then $\boldsymbol{F} \in \mathcal{C}^1(\mathbb{R}, \mathbb{C})$. Indeed,

$$\sum_{n=0}^{\infty} \sup_{x\in\mathbb{R}} |a_n (\varPhi^p(b_n x + \theta_n))'| \leq \sum_{n=0}^{\infty} |a_n| p b_n C = pBC.$$

(d) For every $p, \boldsymbol{a}, \boldsymbol{b}$, and every $\beta \in (0, 1]$, the following conditions are equivalent (cf. Remark 3.2.1(e)):

(i) $\boldsymbol{F_\theta}$ is β-Hölder continuous uniformly with respect to $\boldsymbol{\theta}$, i.e.,

$$\exists_{c>0}\ \forall_{\boldsymbol{\theta}} : |\boldsymbol{F_\theta}(x+h) - \boldsymbol{F_\theta}(x)| \le c|h|^\beta, \quad x, h \in \mathbb{R};$$

(ii) $\boldsymbol{F_\theta}$ is *right-sided β-Hölder continuous at* 0 *uniformly with respect to* $\boldsymbol{\theta}$, i.e.,

$$\exists_{c,\ \delta_0>0}\ \forall_{\boldsymbol{\theta}} : |\boldsymbol{F_\theta}(h) - \boldsymbol{F_\theta}(0)| \le c|h|^\beta, \quad h \in (0, \delta_0).$$

(e) If $|a_n| \le a^n$, $b_n \le b^n$, $n \in \mathbb{N}_0$, where $0 < a < 1$, $ab > 1$, and $\alpha := -\frac{\log a}{\log b}$, then $\boldsymbol{F}_{p,\boldsymbol{a},\boldsymbol{b},\boldsymbol{\theta}}$ is α-Hölder continuous uniformly with respect to $\boldsymbol{\theta}$ and

$$|\boldsymbol{F}_{p,\boldsymbol{a},\boldsymbol{b},\boldsymbol{\theta}}(x+h) - \boldsymbol{F}_{p,\boldsymbol{a},\boldsymbol{b},\boldsymbol{\theta}}(x)| \le p\,\mathrm{const}(a,b)|h|^\alpha, \quad x, h \in \mathbb{R}.$$

Indeed, let $0 < h \le 1$ and let $N = N(h) \in \mathbb{N}_0$ be such that $b^N h \le 1 < b^{N+1}h$. Then

$$|\boldsymbol{F_\theta}(h) - \boldsymbol{F_\theta}(0)| \le p\sum_{n=0}^{\infty} |a_n||\Phi(b_n(x+h)+\theta_n) - \Phi(b_n x + \theta_n)|$$
$$p\Big(\sum_{n=0}^{N-1} |a_n| b_n h + \sum_{n=N}^{\infty} |a_n| 2\Big) \le p\Big(\sum_{n=0}^{N-1} (ab)^n h + 2\sum_{n=N}^{\infty} a^n\Big)$$
$$= p\Big(\frac{(ab)^N - 1}{ab-1} h + 2\frac{a^N}{1-a}\Big) < p\Big(\frac{1}{ab-1} + \frac{2}{1-a}\Big)a^N \le cph^\alpha,$$

where c depends only on a and b. Using (d), we get the result.

(f) For every $p, \boldsymbol{a}, \boldsymbol{b}$, and $\beta \in (0,1]$, the following conditions are equivalent (cf. Remark 3.2.1(h)):

(i) $\boldsymbol{F_\theta}$ is *β-anti-Hölder continuous uniformly with respect to $x \in \mathbb{R}$ and $\boldsymbol{\theta}$*, i.e.,

$$\exists_{\varepsilon>0}\ \forall_{\boldsymbol{\theta},\ x\in\mathbb{R},\ \delta\in(0,1)}\ \exists_{h_\pm\in(0,\delta]} : |\boldsymbol{F_\theta}(x \pm h_\pm) - \boldsymbol{F_\theta}(x)| > \varepsilon\delta^\beta;$$

(ii) $\exists_{\varepsilon,\ \delta_0>0}\ \forall_{\boldsymbol{\theta},\ \delta\in(0,\delta_0)}\ \exists_{h_+\in(0,\delta]} : |\boldsymbol{F_\theta}(h_+) - \boldsymbol{F_\theta}(0)| > \varepsilon\delta^\beta$.

(g) The following conditions are equivalent (use (b)):

(i) $\boldsymbol{F_\theta} \in \mathfrak{ND}(\mathbb{R})$ (resp. $\boldsymbol{F_\theta} \in \mathfrak{ND}^\infty(\mathbb{R})$) for every $\boldsymbol{\theta}$;
(ii) for every $\boldsymbol{\theta}$, a finite (resp. finite or infinite) derivative $\boldsymbol{F}'_{\boldsymbol{\theta}}(0)$ does not exist.

(h) The following conditions are equivalent (use (b)):

(i) $\boldsymbol{F_\theta} \in \mathfrak{ND}_\pm(\mathbb{R})$ (resp. $\boldsymbol{F_\theta} \in \mathfrak{ND}^\infty_\pm(\mathbb{R})$) for every $\boldsymbol{\theta}$;
(ii) for every $\boldsymbol{\theta}$, a finite (resp. finite or infinite) right-sided derivative $(\boldsymbol{F_\theta})'_+(0)$ does not exist.

Our main aim is to discuss nowhere differentiability of the *generalized Weierstrass-type function*

$$\boldsymbol{W}_{p,\boldsymbol{a},\boldsymbol{b},\boldsymbol{\theta}}(x) := \sum_{n=0}^{\infty} a_n \cos^p(2\pi b_n x + \theta_n), \quad x \in \mathbb{R},$$

where p, $\boldsymbol{a}$, $\boldsymbol{b}$, $\boldsymbol{\theta}$ are as in (8.5.2). Functions of the above general type were studied parallel to the classical Weierstrass functions; cf. § 3.5.1.

We will see that the nowhere differentiability of the function $\boldsymbol{W}_{p,\boldsymbol{a},\boldsymbol{b},\boldsymbol{\theta}}$ is strictly related to the nowhere differentiability of the exponential function

$$\boldsymbol{E}_{\boldsymbol{a},\boldsymbol{b}}(x) := \sum_{n=0}^{\infty} a_n e^{2\pi i b_n x}, \quad x \in \mathbb{R},$$

which will be studied in § 8.6.

While studying the functions $\boldsymbol{W}_{p,\boldsymbol{a},\boldsymbol{b},\boldsymbol{\theta}}$ *and* $\boldsymbol{E}_{\boldsymbol{a},\boldsymbol{b}}$*, we always assume that* $p, \boldsymbol{a}, \boldsymbol{b}, \boldsymbol{\theta}$ *satisfy* (8.5.2). A special role is played by the cases $p = 1$ or/and ($\boldsymbol{\theta} = 0$ or $\boldsymbol{\theta} = -\frac{\pi}{2}$), i.e., by the functions

$$\boldsymbol{C}_{\boldsymbol{a},\boldsymbol{b}}(x) := \boldsymbol{W}_{1,\boldsymbol{a},\boldsymbol{b},0}(x) = \sum_{n=0}^{\infty} a_n \cos(2\pi b_n x),$$

$$\boldsymbol{S}_{\boldsymbol{a},\boldsymbol{b}}(x) := \boldsymbol{W}_{1,\boldsymbol{a},\boldsymbol{b},-\frac{\pi}{2}}(x) = \sum_{n=0}^{\infty} a_n \sin(2\pi b_n x), \quad x \in \mathbb{R}.$$

We begin with an extension of Remarks 3.2.1 and 8.5.2 for the function $\boldsymbol{E}_{\boldsymbol{a},\boldsymbol{b}}$.

Remark 8.5.3 (Details Are Left to the Reader). (a) $\boldsymbol{E}_{\boldsymbol{a},\boldsymbol{b}} \in \mathcal{C}(\mathbb{R}, \mathbb{C})$ and $|\boldsymbol{E}_{\boldsymbol{a},\boldsymbol{b}}(x)| \le A := \sum_{n=0}^{\infty} |a_n|$, $x \in \mathbb{R}$.

(b) The function $\boldsymbol{E}_{\boldsymbol{a},\boldsymbol{b}}$ may be formally defined also in the case that $\sum_{n=0}^{\infty} |a_n| b_n < +\infty$. However, in this case, $\boldsymbol{E}_{\boldsymbol{a},\boldsymbol{b}} \in \mathcal{C}^1(\mathbb{R}, \mathbb{C})$.

(c) $\boldsymbol{E}_{\boldsymbol{a},\boldsymbol{b}}(x + x_0) = \boldsymbol{E}_{(e^{2\pi i b_n x_0} a_n)_{n=0}^{\infty}, \boldsymbol{b}}(x)$.

(d) For every $\boldsymbol{a}, \boldsymbol{b}$, the following conditions are equivalent:

(i) $\boldsymbol{E}_{\boldsymbol{a},\boldsymbol{b}}$ is β-Hölder continuous uniformly with respect to $(\arg a_n)_{n=0}^{\infty}$, i.e.,

$$\exists_{c>0}\ \forall_{\boldsymbol{a}'=(a'_n)_{n=0}^{\infty}:\ |a'_n|=|a_n|,\ n\in\mathbb{N}_0} : |\boldsymbol{E}_{\boldsymbol{a}',\boldsymbol{b}}(x+h) - \boldsymbol{E}_{\boldsymbol{a}',\boldsymbol{b}}(x)| \le c|h|^{\beta}, \quad x, h \in \mathbb{R};$$

(ii) $\boldsymbol{E}_{\boldsymbol{\theta}}$ is β-Hölder continuous at 0 uniformly with respect to $(\arg a_n)_{n=0}^{\infty}$, i.e.,

$$\exists_{c,\ \delta_0>0}\ \forall_{\boldsymbol{a}'=(a'_n)_{n=0}^{\infty}:\ |a'_n|=|a_n|,\ n\in\mathbb{N}_0} : |\boldsymbol{E}_{\boldsymbol{a}',\boldsymbol{b}}(h) - \boldsymbol{E}_{\boldsymbol{a}',\boldsymbol{b}}(0)| \le ch^{\beta}, \quad |h| < \delta_0.$$

(e) If $|a_n| \le a^n$, $b_n \le b^n$, $n \in \mathbb{N}_0$, where $0 < a < 1$, $ab > 1$, and $\alpha := -\frac{\log a}{\log b}$, then $\boldsymbol{E}_{\boldsymbol{a},\boldsymbol{b}}$ is α-Hölder continuous and

$$|\boldsymbol{E}_{\boldsymbol{a},\boldsymbol{b}}(x+h) - \boldsymbol{E}_{\boldsymbol{a},\boldsymbol{b}}(x)| \le \mathrm{const}(a, b)|h|^{\alpha}, \quad x, h \in \mathbb{R}.$$

(f) The following conditions are equivalent:

(i) $\boldsymbol{E}_{\boldsymbol{a}',\boldsymbol{b}} \in \mathcal{ND}(\mathbb{R})$ (resp. $\boldsymbol{E}_{\boldsymbol{a}',\boldsymbol{b}} \in \mathcal{ND}^{\infty}(\mathbb{R})$) for any $\boldsymbol{a}' = (a'_n)_{n=0}^{\infty}$ with $|a'_n| = |a_n|$, $n \in \mathbb{N}_0$;

(ii) for every $\boldsymbol{a}' = (a'_n)_{n=0}^{\infty}$ with $|a'_n| = |a_n|$, $n \in \mathbb{N}_0$, a finite (resp. finite or infinite) derivative $\boldsymbol{E}'_{\boldsymbol{a}',\boldsymbol{b}}(0)$ does not exist.

(g) Assume additionally that $(b_n)_{n=0}^{\infty} \subset \mathbb{N}$. Then $\boldsymbol{E}_{\boldsymbol{a},\boldsymbol{b}}(x) = f(e^{2\pi i x})$, $x \in \mathbb{R}$, where f is given by the power series

$$f(z) := \sum_{n=0}^{\infty} a_n z^{b_n}, \quad z \in \overline{\mathbb{D}}. \tag{8.5.3}$$

Obviously, f is holomorphic in $\mathbb{D}$ and continuous in $\overline{\mathbb{D}}$.
- If $\boldsymbol{E}_{\boldsymbol{a},\boldsymbol{b}} \in \mathcal{ND}(\mathbb{R})$, then $\mathbb{D}$ must be the domain of convergence of (8.5.3); cf. Proposition 3.5.8.
- If $\frac{b_{n+1}}{b_n} \geq \lambda > 1$, $n \in \mathbb{N}_0$, then (8.5.3) is a Hadamard lacunary power series, and its domain of convergence coincides with $\mathbb{D}$ (cf. [Boa10, RS02]).

Such an approach has been used, e.g., in [Bel71, Bel73, Bel75].

8.6 Johnsen's Method

Roughly speaking, the aim of this section, based on [Joh10], is to apply Fourier transform methods (cf. § A.3) to study the nowhere differentiability of Weierstrass-type functions.

Remark 8.6.1. Suppose that we have a function $\widetilde{\chi} \in \mathcal{C}_0^\infty(\mathbb{R}, \mathbb{I})$ such that $\operatorname{supp} \widetilde{\chi} \subset \mathbb{R}_{>0}$. Define $\chi := \mathfrak{F}^{-1}(\widetilde{\chi})$, i.e.,

$$\chi(t) := \int_{\mathbb{R}} \widetilde{\chi}(\tau) e^{2\pi i t\tau} d\tau, \quad t \in \mathbb{R} \quad (\text{cf. § A.3}).$$

Obviously, by Proposition A.3.3, we have $\widetilde{\chi} = \widehat{\chi} = \mathfrak{F}(\chi)$, i.e.,

$$\widetilde{\chi}(\tau) = \widehat{\chi}(\tau) = \int_{\mathbb{R}} \chi(t) e^{-2\pi i t\tau} dt, \quad \tau \in \mathbb{R}.$$

We know that $\chi \in \mathcal{C}^\infty(\mathbb{R}) \cap L^\infty(\mathbb{R})$ and $t^k\chi \in L^1(\mathbb{R})$ for every $k \in \mathbb{N}$ (cf. Remark A.3.2). Moreover, $\widehat{\chi}^{(k)} = (-2\pi i)^k \widehat{t^k \chi}$ (cf. Remark A.3.2). In particular,

$$\int_{\mathbb{R}} \chi(t) t^k e^{-2\pi i \tau} dt = \frac{\widehat{\chi}^{(k)}(\tau)}{(-2\pi i)^k} = 0, \quad \tau \leq 0,\ k \in \mathbb{N}_0.$$

Remark 8.6.2. Let $\varphi : \mathbb{R} \longrightarrow \mathbb{C}$ be a bounded continuous function and let $t_0 \in \mathbb{R}$.

(a) If a finite $\varphi'(t_0)$ exists, then we put $\boldsymbol{\Delta}\varphi(t_0, t_0) := \varphi'(t_0)$. Observe that $\boldsymbol{\Delta}\varphi(t_0, \cdot) \in \mathcal{C}(\mathbb{R}, \mathbb{C}) \cap L^\infty(\mathbb{R})$.

(b) If $\varphi \in \mathcal{H}^\beta(\mathbb{R}; t_0)$, then we define

$$H_\beta\varphi(t_0, t) := \frac{|\varphi(t) - \varphi(t_0)|}{|t - t_0|^\beta}, \quad t \in \mathbb{R} \setminus \{t_0\}.$$

Observe that $H_\beta\varphi(t_0, \cdot) \in \mathcal{C}(\mathbb{R} \setminus \{t_0\}, \mathbb{C}) \cap L^\infty(\mathbb{R} \setminus \{t_0\})$.

The following result is a generalization of Theorem 2.1 in [Joh10].

Theorem 8.6.3. *Let*

$$J(t) := \sum_{n,s=0}^{\infty} a_{n,s} e^{2\pi i Q_s b_n t}, \quad t \in \mathbb{R},$$

where:
- $(a_{n,s})_{(n,s)\in\mathbb{N}_0\times\mathbb{N}_0} \subset \mathbb{C}$ *and* $\sum_{n,s=0}^{\infty} |a_{n,s}| < +\infty$,
- $(Q_s)_{s=1}^{\infty} \subset \mathbb{R}_*$, $Q_0 > Q_1 > \cdots > Q_r > 0 > Q_i$, $i \geq r+1$ *(for some* $r \in \mathbb{N}_0$*)*,
- $(b_n)_{n=0}^{\infty} \subset \mathbb{R}_{>0}$ *and* $\frac{b_{n+1}}{b_n} \geq \lambda > \frac{Q_0}{Q_r}$, $n \in \mathbb{N}_0$,
- $a_{n,q} b_n \not\to 0$ *for a* $q \in \{0, \ldots, r\}$.

Then $J \in \mathbf{ND}(\mathbb{R})$. *Moreover, if* $\sup_{n\in\mathbb{N}_0} |a_{n,q}|b_n^\beta = +\infty$ *for some* $q \in \{0,\dots,r\}$ *and* $\beta \in (0,1]$, *then* $J \in \mathbf{NH}^\beta(\mathbb{R})$.

Remark 8.6.4. Theorem 2.1 in [Joh10] is the case $q = r = 0$ and $a_{n,s} = 0$ for $s > 0$.

Proof of Theorem 8.6.3. It is clear that $J \in \mathcal{C}(\mathbb{R},\mathbb{C}) \cap L^\infty(\mathbb{R})$. Suppose that $J'(t_0)$ exists (for some $t_0 \in \mathbb{R}$). Observe that $Q_s \in (\frac{Q_0}{\lambda}, \lambda Q_r)$, $s = 0,\dots,r$. Let $\widetilde{\chi} \in \mathcal{C}_0^\infty(\mathbb{R},\mathbb{I})$ be such that $\widetilde{\chi}(Q_q) = 1$, $\widetilde{\chi}(Q_s) = 0$, $s \neq q$, and $\operatorname{supp}\widetilde{\chi} \subset [\frac{Q_0}{\lambda}, \lambda Q_r]$. Moreover, let χ be as in Remark 8.6.1. Take an arbitrary $k \in \mathbb{N}_0$ and calculate

$$\begin{aligned}
b_k \int_{\mathbb{R}} \chi(b_k t) J(t_0 - t)dt &= b_k \int_{\mathbb{R}} \chi(b_k t)\Big(\sum_{n,s=0}^{\infty} a_{n,s} e^{2\pi i Q_s b_n (t_0 - t)}\Big)dt \\
&= \sum_{n,s=0}^{\infty} a_{n,s} e^{2\pi i Q_s b_n t_0} \int_{\mathbb{R}} b_k \chi(b_k t) e^{-2\pi i Q_s b_n t} dt \\
&= \sum_{n,s=0}^{\infty} a_{n,s} e^{2\pi i Q_s b_n t_0} \int_{\mathbb{R}} \chi(u) e^{-2\pi i Q_s \frac{b_n}{b_k} u} du \\
&= \sum_{n,s=0}^{\infty} a_{n,s} e^{2\pi i Q_s b_n t_0} \widehat{\chi}\Big(Q_s \frac{b_n}{b_k}\Big) \overset{(*)}{=} a_{k,q} e^{2\pi i b_k Q_q t_0},
\end{aligned}$$

where (*) is a consequence of the following facts:

- if $s > r$, then $Q_s \frac{b_n}{b_k} < 0$, and therefore $\widehat{\chi}\big(Q_s \frac{b_n}{b_k}\big) = 0$;
- if $s \in \{0,\dots,r\}$ and $n > k$, then $Q_s \frac{b_n}{b_k} \geq Q_s \lambda \geq Q_r \lambda$, and hence $\widehat{\chi}\big(Q_s \frac{b_n}{b_k}\big) = 0$;
- if $s \in \{0,\dots,r\}$ and $n < k$, then $Q_s \frac{b_n}{b_k} \leq \frac{Q_s}{\lambda} \leq \frac{Q_0}{\lambda}$, and hence $\widehat{\chi}\big(Q_s \frac{b_n}{b_k}\big) = 0$;
- if $s \in \{0,\dots,r\}$ and $n = k$, then $\widehat{\chi}(Q_s) \neq 0$ iff $s = q$.

Recall that $b_k \int_{\mathbb{R}} \chi(b_k t)dt = \int_{\mathbb{R}} \chi(u)du = \widehat{\chi}(0) = 0$ (Remark 8.6.1). Hence we get

$$\begin{aligned}
a_{k,q} e^{2\pi i b_k Q_q t_0} &= b_k \int_{\mathbb{R}} \chi(b_k t) J(t_0 - t)dt = b_k \int_{\mathbb{R}} \chi(b_k t)(J(t_0 - t) - J(t_0))dt \\
&= \int_{\mathbb{R}} \chi(u)(J(t_0 - u/b_k) - J(t_0))du = \int_{\mathbb{R}} \chi(u)(-u/b_k)\mathbf{\Delta} J(t_0, t_0 - u/b_k)du,
\end{aligned}$$

where $\mathbf{\Delta} J(t_0,\cdot)$ is as in Remark 8.6.2(a). Consequently, by the Lebesgue theorem, we have

$$\begin{aligned}
-a_{k,q} b_k e^{2\pi i b_k Q_q t_0} &= \int_{\mathbb{R}} \chi(u) u \mathbf{\Delta} J(t_0, t_0 - u/b_k)du \\
&\underset{k\to+\infty}{\longrightarrow} J'(t_0) \int_{\mathbb{R}} \chi(u) u du = J'(t_0) \frac{1}{(-2\pi i)} \widehat{\chi}'(0) = 0.
\end{aligned}$$

Hence $a_{k,q} b_k \longrightarrow 0$; a contradiction.

Suppose that $J \in \mathbf{H}^\beta(\mathbb{R}, t_0)$ (for some $t_0 \in \mathbb{R}$) and let $H_\beta J(t_0,\cdot) \leq C$, where $H_\beta J(t_0,\cdot)$ is as in Remark 8.6.2(b). Then we get

$$|a_{k,q}| b_k^\beta \leq \int_{\mathbb{R}} |\chi(u)||u|^\beta H_\beta J(t_0, t_0 - u/b_k)du \leq C \int_{\mathbb{R}} |\chi(u)||u|^\beta du < +\infty.$$

Hence $\sup_{k\in\mathbb{N}_0}|a_{k,q}|b_k^\beta < +\infty$; a contradiction. □

In the case $q = r = 0$, we immediately get the following result.

Theorem 8.6.5. *Assume that* $\boldsymbol{a}, \boldsymbol{b}$ *satisfy* (8.5.2), $\frac{b_{n+1}}{b_n} \geq \lambda > 1$, $n \in \mathbb{N}_0$, *and* $a_n b_n \not\to 0$. *Then* $\boldsymbol{E}_{\boldsymbol{a},\boldsymbol{b}} \in \mathfrak{N}\mathfrak{D}(\mathbb{R})$. *If, moreover,* $\sup_{n\in\mathbb{N}}|a_n|b_n^\beta = +\infty$ *for some* $\beta \in (0,1]$, *then* $\boldsymbol{E}_{\boldsymbol{a},\boldsymbol{b}} \in \mathfrak{N}\mathfrak{H}^\beta(\mathbb{R})$.

Theorem 8.6.6. *Assume that* $p, \boldsymbol{a}, \boldsymbol{b}, \boldsymbol{\theta}$ *satisfy* (8.5.2),

$$\frac{b_{n+1}}{b_n} \geq \lambda > \begin{cases} p, & \text{if } p \equiv 1 \pmod 2 \\ \frac{p}{2}, & \text{if } p \equiv 0 \pmod 2 \end{cases}, \qquad n \in \mathbb{N}_0,$$

and $a_n b_n \not\to 0$. *Then* $\boldsymbol{W}_{p,\boldsymbol{a},\boldsymbol{b},\boldsymbol{\theta}} \in \mathfrak{N}\mathfrak{D}(\mathbb{R})$. *If, moreover,* $\sup_{n\in\mathbb{N}}|a_n|b_n^\beta = +\infty$ *for some* $\beta \in (0,1]$, *then* $\boldsymbol{W}_{p,\boldsymbol{a},\boldsymbol{b},\boldsymbol{\theta}} \in \mathfrak{N}\mathfrak{H}^\beta(\mathbb{R})$.

Proof. We have

$$\boldsymbol{W}_{p,\boldsymbol{a},\boldsymbol{b},\boldsymbol{\theta}}(t) = \sum_{n=0}^\infty a_n\Big(\frac{1}{2}(e^{i(2\pi b_n t+\theta_n)} + e^{-i(2\pi b_n t+\theta_n)})\Big)^p$$
$$= \sum_{n=0}^\infty a_n \frac{1}{2^p}\sum_{s=0}^p \binom{p}{s} e^{i(p-2s)\theta_n} e^{2\pi i b_n(p-2s)t}.$$

For $s > p$, put $a_{n,s} := 0$. For $s \in \{0,\dots,p\}$, let $a_{n,s} := a_n\frac{1}{2^p}\binom{p}{s}e^{i(p-2s)\theta_n}$, $Q_s := p - 2s$, $q = r := \begin{cases} \frac{p-1}{2}, & \text{if } p \equiv 1 \pmod 2 \\ \frac{p}{2} - 1, & \text{if } p \equiv 0 \pmod 2 \end{cases}$. Then

$$\frac{Q_0}{Q_r} = \frac{p}{p-2r} = \begin{cases} p, & \text{if } p \equiv 1 \pmod 2 \\ \frac{p}{2}, & \text{if } p \equiv 0 \pmod 2 \end{cases} < \lambda,$$

and therefore Theorem 8.6.3 applies. □

Theorem 8.6.7. *Assume that* $p, a, b, \boldsymbol{\theta}$ *satisfy* (3.1.2) *and*

$$b > \begin{cases} p, & \text{if } p \equiv 1 \pmod 2 \\ \frac{p}{2}, & \text{if } p \equiv 0 \pmod 2 \end{cases}.$$

Then $\boldsymbol{W}_{p,a,b,\boldsymbol{\theta}} \in \mathfrak{N}\mathfrak{D}(\mathbb{R})$. *If, moreover,* $ab^\beta > 1$ *for some* $\beta \in (0,1]$, *then* $\boldsymbol{W}_{p,a,b,\boldsymbol{\theta}} \in \mathfrak{N}\mathfrak{H}^\beta(\mathbb{R})$.

In particular, the above theorem extends Hardy's results (cf. Theorems 8.2.1 and 8.2.12).

Corollary 8.6.8 (Darboux-Type Functions). *If*

$$\boldsymbol{a} := (1/n!)_{n=0}^\infty, \quad \boldsymbol{b} := ((n+1)!)_{n=0}^\infty,$$

then $\boldsymbol{W}_{p,\boldsymbol{a},\boldsymbol{b},\boldsymbol{\theta}} \in \mathfrak{N}\mathfrak{H}^1(\mathbb{R}) \subset \mathfrak{N}\mathfrak{D}(\mathbb{R})$ *(for arbitrary* $\boldsymbol{\theta}$*)*.

Remark 8.6.9. The classical *Darboux function* is the case $p = 1$ and $\boldsymbol{\theta} = -\frac{\pi}{2}$ (cf. [Dar79]):

$$\boldsymbol{W}_{1,\boldsymbol{a},\boldsymbol{b},-\frac{\pi}{2}}(t) = \sum_{n=0}^\infty \frac{1}{n!}\sin(2\pi(n+1)!t), \quad t \in \mathbb{R}.$$

Proof of Corollary 8.6.8. We have $\frac{b_{n+1}}{b_n} = n+2 \geq p+1 > p$ for $n \geq p-1$. Moreover, $a_n b_n = n+1$. Thus we may apply Theorem 8.6.6. □

Corollary 8.6.10. *Let*

$$f(t) := \sum_{n=1}^{\infty} \frac{a^n}{(2n-1)!!} \cos^p(2\pi(2n-1)!!t + \theta_n), \quad t \in \mathbb{R},$$

where $(2n-1)!! := 1 \cdot 3 \cdot 5 \cdots (2n-1)$, $a \geq 1$, $p \in \mathbb{N}$, $\boldsymbol{\theta} = (\theta_n)_{n=1}^{\infty} \subset \mathbb{R}$. *Then* $f \in \mathfrak{ND}(\mathbb{R})$. *Moreover,* $f \in \mathfrak{NH}^1(\mathbb{R})$ *if* $a > 1$.

Proof. Put $a_n := \frac{a^n}{(2n-1)!!}$, $b_n := (2n-1)!!$, $n \in \mathbb{N}$. Then $\frac{a_{n+1}}{a_n} = \frac{a}{2n+1} \longrightarrow 0$, which implies that $\sum_{n=1}^{\infty} a_n < +\infty$. Moreover, $\frac{b_{n+1}}{b_n} = 2n+1 > p$ for $n \gg 1$ and $a_n b_n = a^n$. Thus Theorem 8.6.6 applies. □

Corollary 8.6.11. *Let*

$$f(t) := \sum_{n=1}^{\infty} \frac{1}{2^{n!}} \cos^p(2\pi 2^{(2n)!} t + \theta_n), \quad t \in \mathbb{R} \quad (p \in \mathbb{N}).$$

Then $f \in \mathfrak{NH}^1(\mathbb{R})$. *In particular,* $f \in \mathfrak{ND}(\mathbb{R})$.

The case $p = 1$, $\boldsymbol{\theta} = 0$ has been studied in [Cat83].

Corollary 8.6.12. *Let*

$$f(t) = \sum_{n=0}^{\infty} a_n \cos^p(2\pi b_n t + \theta_n) \sin^{p'}(2\pi b_n t + \theta'_n), \quad t \in \mathbb{R},$$

where $\boldsymbol{a}, \boldsymbol{b}$ *satisfy* (8.5.2), $p, p' \in \mathbb{N}$, $(\theta_n)_{n=0}^{\infty}, (\theta'_n)_{n=0}^{\infty} \subset \mathbb{R}$,

$$\frac{b_{n+1}}{b_n} \geq \lambda > \begin{cases} p+p', & \text{if } p+p' \equiv 1 \pmod 2 \\ \frac{p+p'}{2}, & \text{if } p+p' \equiv 0 \pmod 2 \end{cases}, \quad n \in \mathbb{N}_0,$$

and $a_n b_n \not\to 0$. *Then* $f \in \mathfrak{ND}(\mathbb{R})$, *and if* $\sup_{n \in \mathbb{N}} |a_n| b_n^{\beta} = +\infty$ *for some* $\beta \in (0,1]$, *then* $f \in \mathfrak{NH}^{\beta}(\mathbb{R})$.

Some special cases of the above result were proved already in [Muk34].

Proof. We have

$$\begin{aligned} f(t) &= \sum_{n=0}^{\infty} a_n \Big(\frac{1}{2}(e^{i(2\pi b_n t + \theta_n)} + e^{-i(2\pi b_n t + \theta_n)})\Big)^p \times \\ &\quad \times \Big(\frac{1}{2i}(e^{i(2\pi b_n t + \theta'_n)} - e^{-i(2\pi b_n t + \theta'_n)})\Big)^{p'} \\ &= \sum_{n=0}^{\infty} a_n \frac{1}{2^p} \sum_{s=0}^{p} \binom{p}{s} e^{i(p-2s)\theta_n} e^{2\pi i b_n (p-2s)t} \times \\ &\quad \times \frac{1}{(2i)^{p'}} \sum_{s'=0}^{p'} \binom{p'}{s'} (-1)^{s'} e^{i(p'-2s')\theta'_n} e^{2\pi i b_n (p-2s')t} \end{aligned}$$

$$= \sum_{n=0}^{\infty} a_n \frac{1}{2^p} \frac{1}{(2i)^{p'}} \sum_{s=0}^{p} \sum_{s'=0}^{p'} \binom{p}{s} \binom{p'}{s'} \times$$
$$\times (-1)^{s'} e^{i((p-2s)\theta_n + (p'-2s')\theta'_n)} e^{2\pi i b_n (p-2s+p'-2s')t}$$
$$= \sum_{n=0}^{\infty} a_n \sum_{\varrho=0}^{p+p'} a_{n,\varrho} e^{2\pi i Q_\varrho b_n t},$$

where

$$a_{n,\varrho} := a_n \frac{1}{2^p} \frac{1}{(2i)^{p'}} \sum_{\substack{s\in\{0,\dots,p\}\\ s'\in\{0,\dots,p'\}\\ s+s'=\varrho}} \binom{p}{s} \binom{s'}{p'} (-1)^{s'} e^{i((p-2s)\theta_n + (p'-2s')\theta'_n)},$$
$$Q_\varrho := p + p' - 2\varrho.$$

Observe that

$$\sum_{n=0}^{\infty} a_n \sum_{\varrho=0}^{p+p'} |a_{n,\varrho}| \le \sum_{n=0}^{\infty} |a_n| \frac{1}{2^p} \frac{1}{2^{p'}} \sum_{\substack{s\in\{0,\dots,p\}\\ s'\in\{0,\dots,p'\}\\ s+s'=\varrho}} \binom{p}{s} \binom{s'}{p'} \le \sum_{n=0}^{\infty} |a_n| < +\infty.$$

Now we argue as in the proof of Theorem 8.6.6. □

Remark 8.6.13. Observe that Theorem 8.6.6 gives a very effective method of finding nowhere differentiable functions. Nevertheless, the requirement $\frac{b_{n+1}}{b_n} \ge \lambda > 1$, $n \in \mathbb{N}_0$, is very restrictive. We would like to study, for instance, functions of the form

$$t \longmapsto \sum_{n=1}^{\infty} \frac{1}{n^p} e^{2\pi i n^q t} \quad (p > 1,\ q > 0),$$

where Theorem 8.6.6 does not work. We need a more subtle tool.

Theorem 8.6.14. *Assume that* $\boldsymbol{a}$, $\boldsymbol{b}$, $\boldsymbol{\theta}$ *satisfy* (8.5.2) *and* $a_n \Delta b_n \not\to 0$, *where*

$$\Delta b_n := \min\{b_n - b_{n-1}, b_{n+1} - b_n\} \quad (b_{-1} := 0).$$

Then $\boldsymbol{W}_{1,\boldsymbol{a},\boldsymbol{b},\boldsymbol{\theta}}, \boldsymbol{E}_{\boldsymbol{a},\boldsymbol{b}} \in \mathcal{ND}(\mathbb{R})$. *If, moreover,* $\sup_{n\in\mathbb{N}_0} |a_n|(\Delta b_n)^\beta = +\infty$ *for some* $\beta \in (0,1]$, *then* $\boldsymbol{W}_{1,\boldsymbol{a},\boldsymbol{b},\boldsymbol{\theta}}, \boldsymbol{E}_{\boldsymbol{a},\boldsymbol{b}} \in \mathcal{NH}^\beta(\mathbb{R})$.

Proof. Let $f := \boldsymbol{W}_{1,\boldsymbol{a},\boldsymbol{b},\boldsymbol{\theta}}$. It is clear that $f \in \mathcal{C}(\mathbb{R},\mathbb{C}) \cap L^\infty(\mathbb{R})$. Suppose that $f'(t_0)$ exists. Let $\widetilde{\chi} \in \mathcal{C}_0^\infty(\mathbb{R},\mathbb{I})$ be such that $\widetilde{\chi}(0) = 1$ and $\operatorname{supp}\widetilde{\chi} \subset (-\frac{1}{2},\frac{1}{2})$. Define χ as in Remark 8.6.1. Let $\widetilde{\chi}_k(\tau) := \widetilde{\chi}(\frac{\tau - b_k}{\Delta b_k})$. Observe that

$$\operatorname{supp}\widetilde{\chi}_k \subset \Big(b_k - \frac{1}{2}(b_k - b_{k-1}), b_k + \frac{1}{2}(b_{k+1} - b_k)\Big).$$

Define $\chi_k(t) := (\mathfrak{F}^{-1}\widetilde{\chi}_k)(t) = \int_{\mathbb{R}} \widetilde{\chi}_k(\tau) e^{2\pi i t\tau} d\tau$. Note that

$$\chi_k(t) = \int_{\mathbb{R}} \widetilde{\chi}\Big(\frac{\tau - b_k}{\Delta b_k}\Big) e^{2\pi i t\tau} d\tau = \int_{\mathbb{R}} \widetilde{\chi}(u) e^{2\pi i t(b_k + \Delta b_k u)} \Delta b_k du$$
$$= \Delta b_k e^{2\pi i t b_k} \chi(\Delta b_k t).$$

We have

$$\int_{\mathbb{R}} \chi_k(t) f(t_0 - t) dt = \int_{\mathbb{R}} \chi_k(t) \Big(\sum_{n=0}^{\infty} a_n \cos(2\pi b_n(t_0 - t) + \theta_n) \Big) dt$$
$$= \sum_{n=0}^{\infty} a_n \int_{\mathbb{R}} \chi_k(t) \cos(2\pi b_n(t_0 - t) + \theta_n) dt$$
$$= \sum_{n=0}^{\infty} a_n \int_{\mathbb{R}} \chi_k(t) \frac{1}{2} \Big(e^{i(2\pi b_n(t_0 - t) + \theta_n)} + e^{-i(2\pi b_n(t_0 - t) + \theta_n)} \Big) dt$$
$$= \sum_{n=0}^{\infty} a_n \frac{1}{2} \Big(e^{i(2\pi b_n t_0 + \theta_n)} \int_{\mathbb{R}} \Delta b_k e^{2\pi i t b_k} \chi(\Delta b_k t) e^{-2\pi i b_n t} dt$$
$$+ e^{-i(2\pi i b_n t_0 + \theta_n)} \int_{\mathbb{R}} \Delta b_k e^{2\pi i t b_k} \chi(\Delta b_k t) e^{2\pi i b_n t} dt \Big)$$
$$= \sum_{n=0}^{\infty} a_n \frac{1}{2} \Big(e^{i(2\pi b_n t_0 + \theta_n)} \int_{\mathbb{R}} \chi(u) e^{-2\pi i \frac{b_n - b_k}{\Delta b_k} u} du$$
$$+ e^{-i(2\pi b_n t_0 + \theta_n)} \int_{\mathbb{R}} \chi(u) e^{2\pi i \frac{b_n + b_k}{\Delta b_k} u} du \Big)$$
$$= \sum_{n=0}^{\infty} a_n \frac{1}{2} \Big(e^{i(2\pi b_n t_0 + \theta_n)} \widehat{\chi}\Big(\frac{b_n - b_k}{\Delta b_k} \Big)$$
$$+ e^{-i(2\pi i b_n t_0 + \theta_n)} \widehat{\chi}\Big(- \frac{b_n + b_k}{\Delta b_k} \Big) \Big) = \frac{a_k}{2} e^{i(2\pi b_k t_0 + \theta_k)}.$$

Note that

$$\int_{\mathbb{R}} \chi_k(t) dt = \int_{\mathbb{R}} \Delta b_k e^{2\pi i t b_k} \chi(\Delta b_k t) dt = \int_{\mathbb{R}} e^{2\pi i u \frac{b_k}{\Delta b_k}} \chi(u) du$$
$$= \widehat{\chi}\Big(- \frac{b_k}{\Delta b_k} \Big) = 0.$$

Hence

$$\frac{a_k}{2} e^{i(2\pi b_k t_0 + \theta_k)} = \int_{\mathbb{R}} \chi_k(t)(f(t_0 - t) - f(t_0)) dt$$
$$= \int_{\mathbb{R}} \Delta b_k e^{2\pi i t b_k} \chi(\Delta b_k t)(f(t_0 - t) - f(t_0)) dt$$
$$= \int_{\mathbb{R}} e^{2\pi i u \frac{b_k}{\Delta b_k}} \chi(u)(f(t_0 - u/\Delta b_k) - f(t_0)) du$$
$$= \int_{\mathbb{R}} e^{2\pi i u \frac{b_k}{\Delta b_k}} \chi(u)(-u/\Delta b_k) \boldsymbol{\Delta} f(t_0, t_0 - u/\Delta b_k) du.$$

Moreover,

$$\int_{\mathbb{R}} e^{2\pi i u \frac{b_k}{\Delta b_k}} \chi(u) u du = \frac{1}{-2\pi i} \widehat{\chi}'\Big(-\frac{b_k}{\Delta b_k}\Big) = 0.$$

Then, by the Lebesgue theorem,

$$-\frac{a_k \Delta b_k}{2} e^{i(2\pi b_k t_0 + \theta_k)} = \int_{\mathbb{R}} e^{2\pi i u \frac{b_k}{\Delta b_k}} \chi(u) u \boldsymbol{\Delta} f(t_0, t_0 - u/\Delta b_k) du$$
$$= \int_{\mathbb{R}} e^{2\pi i u \frac{b_k}{\Delta b_k}} \chi(u) u (\boldsymbol{\Delta} f(t_0, t_0 - u/\Delta b_k) - f'(t_0)) du \underset{k\to+\infty}{\longrightarrow} 0,$$

where $\boldsymbol{\Delta} f(t_0, \cdot)$ is as in Remark 8.6.2(a). Hence $a_k \Delta b_k \longrightarrow 0$; a contradiction.

Suppose that $f \in \mathcal{H}^\beta(\mathbb{R}; t_0)$ and let $H_\beta f(t_0, \cdot) \le C$. Then we get

$$\tfrac{1}{2}|a_k(\Delta b_k)^\alpha| \le \int_{\mathbb{R}} |\chi(u)||u|^\alpha H_\beta f(t_0, t_0 - u/b_k) du \le C \int_{\mathbb{R}} |\chi(u)||u|^\beta du,$$

where $H_\beta f(t_0, \cdot)$ is as in Remark 8.6.2(b). Hence $\sup_{k\in\mathbb{N}_0} |a_k|(\Delta b_k)^\beta < +\infty$; a contradiction.

The proof for $\boldsymbol{E}_{\boldsymbol{a},\boldsymbol{b}}$ is analogous—EXERCISE. □

Theorem 8.6.15 (Riemann-Type Functions). *For $p > 1$, $q > 0$, let $\boldsymbol{F} \in \{\boldsymbol{E}, \boldsymbol{C}, \boldsymbol{S}\}$ with*

$$\boldsymbol{E}(t) := \sum_{n=1}^{\infty} \frac{1}{n^p} e^{2\pi i n^q t}, \quad C_{p,q}(t) = \boldsymbol{C}(t) := \operatorname{Re} \boldsymbol{E}(t) = \sum_{n=1}^{\infty} \frac{1}{n^p} \cos(2\pi n^q t),$$
$$S_{p,q}(t) = \boldsymbol{S}(t) := \operatorname{Im} \boldsymbol{E}(t) = \sum_{n=1}^{\infty} \frac{1}{n^p} \sin(2\pi n^q t), \quad t \in \mathbb{R}.$$

Then:

(a) *if $0 < q < p - 1$, then $\boldsymbol{F} \in \mathcal{C}^1(\mathbb{R})$;*
(b) *if $q \ge p + 1$, then $\boldsymbol{F} \in \mathcal{ND}(\mathbb{R})$;*
(c) *if $q > p + 1$ then $\boldsymbol{F} \in \mathcal{NH}^1(\mathbb{R})$;*
(d) *if $q > 1$ and $\alpha \in (0, 1]$ are such that $\alpha > \frac{p}{q-1}$, then $\boldsymbol{F} \in \mathcal{NH}^\alpha(\mathbb{R})$;*
(e) *if $q > p - 1$ and $0 < \alpha \le \frac{p-1}{q}$, then $\boldsymbol{F} \in \mathcal{H}^\alpha(\mathbb{R})$.*

Remark 8.6.16. (a) The classical *Riemann function* is the case $p = q = 2$:

$$\boldsymbol{R}(t) := \sum_{n=1}^{\infty} \frac{1}{n^2} \sin(\pi n^2 t), \quad t \in \mathbb{R} \quad \text{(cf. Chap. 13 and Fig. 8.1).}$$

(b) The functions $C_{p,q}$ and $S_{p,q}$ were also studied for other configurations of parameters $p > 1, q > 0$. For example:

- If $1 < p < \frac{5}{2}$, then finite derivatives $C'_{p,2}(t)$, $S'_{p,2}(t)$ do not exist at any $t \in \mathbb{R} \setminus \mathbb{Q}$ (cf. [Har16], Theorem 4.31).
- A finite derivative $S'_{p,p+1/2}(t)$ does not exist at any $t \in \mathbb{Q}$ (cf. [Lut86], § 6). In particular, $S_{3/2,2} \in \mathcal{ND}(\mathbb{R})$.

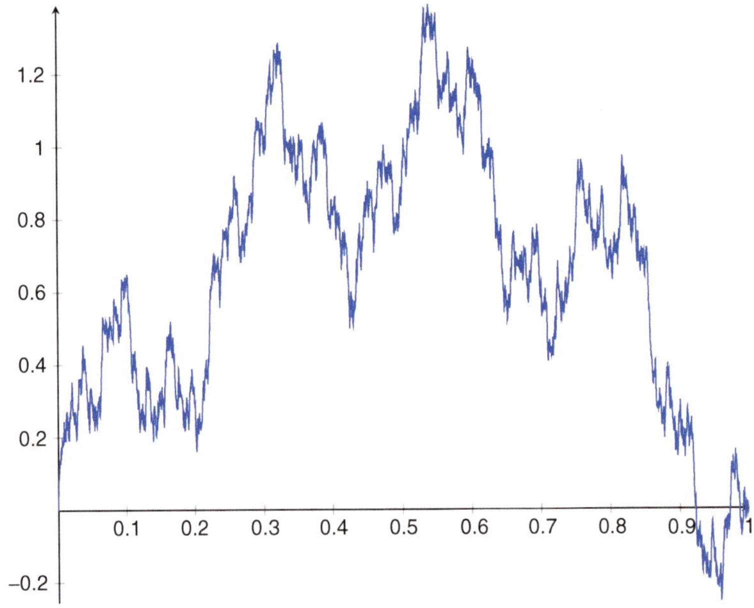

Fig. 8.1 Riemann-type function $\mathbb{I} \ni x \longmapsto \sum\limits_{n=1}^{\infty} \frac{\sin(\pi n^3 x)}{n^2}$

Proof of Theorem 8.6.15. Put $a_n := \frac{1}{n^p}$, $b_n := n^q$.

(a) is obvious.

(b) and (c) follow from Theorem 8.6.14. Indeed, using the mean value theorem, we get

$$\begin{aligned} a_n \Delta b_n \geq \frac{n^q - (n-1)^q}{n^p} &= \frac{1}{n^{p-q}}\left(1 - \left(1 - \frac{1}{n}\right)^q\right) \\ &= \frac{1}{n^{p-q}} q \xi_n^{q-1} \frac{1}{n} \geq \frac{q}{n^{p-q+1}}\left(1 - \frac{1}{n}\right)^{q-1}, \end{aligned}$$

and the right-hand side tends to q when $q = p+1$, and to $+\infty$ when $q > p+1$.

(d) We have

$$\begin{aligned} a_n (\Delta b_n)^\alpha \geq \frac{(n^q - (n-1)^q)^\alpha}{n^p} &= \frac{1}{n^{p-\alpha q}}\left(1 - \left(1 - \frac{1}{n}\right)^q\right)^\alpha \\ &= \frac{1}{n^{p-\alpha q}}\left(q \xi_n^{q-1} \frac{1}{n}\right)^\alpha \geq \frac{q^\alpha}{n^{p-\alpha q+\alpha}}\left(1 - \frac{1}{n}\right)^{\alpha(q-1)}. \end{aligned}$$

(e) For $t \in \mathbb{R}$ and $0 < |h| < 1$, let $N = N(h)$ be defined by the relation $N \leq |h|^{-1/q} < N+1$. Then we have

$$\begin{aligned}|\boldsymbol{C}(t+h)-\boldsymbol{C}(t)| &= \Big|\sum_{n=1}^{\infty}\frac{1}{n^p}(\cos(2\pi n^q(t+h))-\cos(2\pi n^q t))\Big|\\ &\le 2\sum_{n=1}^{\infty}\frac{1}{n^p}|\sin(\pi n^q h)| \le 2\pi\sum_{n=1}^{N}\frac{1}{n^{p-q}}|h| + 2\sum_{n=N+1}^{\infty}\frac{1}{n^p}\\ &\overset{(*)}{\le} 2\pi N^{q-p+1}|h| + \frac{2N^{1-p}}{p-1} \le \operatorname{const}|h|^{\frac{p-1}{q}},\end{aligned}$$

where (*) follows from the estimate

$$\begin{aligned}\sum_{n=N+1}^{\infty}\frac{1}{n^p} \le \int_N^{\infty}\frac{dx}{x^p} = \frac{x^{1-p}}{1-p}\Big|_N^{\infty} = \frac{N^{1-p}}{p-1} \le \frac{(\frac{N+1}{2})^{1-p}}{p-1}\\ \le \frac{1}{p-1}\Big(\frac{|h|^{-1/q}}{2}\Big)^{1-p} = \frac{2^{p-1}}{p-1}|h|^{\frac{p-1}{q}}.\end{aligned}$$

The proof for $\boldsymbol{S}$ is analogous—EXERCISE.

□

Theorem 8.6.14 permits us to study functions generated by very slowly increasing sequences $(b_n)_{n=1}^{\infty}$. The following corollaries will illustrate this phenomenon.

Corollary 8.6.17. *For $b \ge a > 1$, let $\boldsymbol{F} \in \{\boldsymbol{E}, \boldsymbol{C}, \boldsymbol{S}\}$ with*

$$\begin{aligned}\boldsymbol{E}(t) &:= \sum_{n=2}^{\infty}\frac{1}{n\log^a n}e^{2\pi i n^2 t\log^b n},\\ \boldsymbol{C}(t) &:= \operatorname{Re}\boldsymbol{E}(t) = \sum_{n=2}^{\infty}\frac{1}{n\log^a n}\cos(2\pi n^2 t\log^b n),\\ \boldsymbol{S}(t) &:= \operatorname{Im}\boldsymbol{E}(t) = \sum_{n=2}^{\infty}\frac{1}{n\log^a n}\sin(2\pi n^2 t\log^b n), \quad t\in\mathbb{R}.\end{aligned}$$

Then $\boldsymbol{F} \in \mathcal{ND}(\mathbb{R})$. Moreover, if $b > a > 1$, then $\boldsymbol{F} \in \mathcal{NH}^1(\mathbb{R})$.

Proof. We use Theorem 8.6.14 with $a_n := \frac{1}{n\log^a n}$, $b_n := n^2\log^b n$. Observe that the function $(1,+\infty) \ni x \overset{B}{\longmapsto} x^2\log^b x$ is convex. In particular, $b_n - b_{n-1} \le b_{n+1} - b_n$. Thus, $\Delta b_n = b_n - b_{n-1} \ge 2(n-1)\log^b(n-1)$. Consequently, we get

$$a_n\Delta b_n \ge \frac{2(n-1)\log^b(n-1)}{n\log^a n} = 2\frac{n-1}{n}\Big(\frac{\log(n-1)}{\log n}\Big)^b\log^{b-a} n. \qquad □$$

Corollary 8.6.18. *Let $\boldsymbol{F} \in \{\boldsymbol{E}, \boldsymbol{C}, \boldsymbol{S}\}$ with*

$$\boldsymbol{E}(t) := \sum_{n=n_0}^{\infty}a_n e^{2\pi i\frac{n}{|a_n|}t}, \quad \boldsymbol{C}(t) := \sum_{n=n_0}^{\infty}a_n\cos\Big(2\pi\frac{n}{|a_n|}t\Big),$$

$$\boldsymbol{S}(t) := \sum_{n=n_0}^{\infty}a_n\sin\Big(2\pi\frac{n}{|a_n|}t\Big), \quad t\in\mathbb{R},$$

where $(a_n)_{n=n_0}^{\infty} \subset \mathbb{C}_*$, $\sum_{n=n_0}^{\infty} |a_n| < +\infty$, $|a_{n+1}| \leq |a_n|$, $n \in \mathbb{N}_{n_0}$, *and there exists a convex function* $B : [n_0, +\infty) \longrightarrow \mathbb{R}$ *such that* $B(n) = \frac{n}{|a_n|}$, $n \in \mathbb{N}_{n_0}$. *Then* $\boldsymbol{F} \in \mathfrak{ND}(\mathbb{R})$.

Proof. We use Theorem 8.6.14 with $b_n := \frac{n}{|a_n|}$. Then for $n \geq \mathbb{N}_{n_0+1}$, we have

$$|a_n|\Delta b_n = |a_n|(b_n - b_{n-1}) = |a_n|\Big(\frac{n}{|a_n|} - \frac{n-1}{|a_{n-1}|}\Big) \geq 1. \qquad \square$$

For $p \in \mathbb{N}_2$, define $\operatorname{Exp}_p := \underbrace{\exp \circ \cdots \circ \exp}_{p\times}$, $L_p := \underbrace{\log \circ \cdots \circ \log}_{p\times}$; note that L_p is defined on the interval $(\operatorname{Exp}_{p-2}(1), +\infty)$ (with $\operatorname{Exp}_0(1) := 1$). Observe that

$$L_p'(x) = \frac{1}{xL_1(x) \cdots L_{p-1}(x)}, \qquad x > \operatorname{Exp}_{p-1}(1).$$

Corollary 8.6.19. *For* $p \in \mathbb{N}_2$ *and* $a > 1$, *let* $\boldsymbol{F} \in \{\boldsymbol{E}, \boldsymbol{C}, \boldsymbol{S}\}$ *with*

$$\boldsymbol{E}(t) := \sum_{n=n_0}^{\infty} \frac{1}{nL_1(n) \cdots L_{p-1}(n)L_p^a(n)} e^{2\pi i n^2 t L_1(n) \cdots L_{p-1}(n) L_p^a(n)},$$
$$\boldsymbol{C}(t) := \operatorname{Re} \boldsymbol{E}(t), \quad \boldsymbol{S}(t) := \operatorname{Im} \boldsymbol{E}(t), \quad t \in \mathbb{R},$$

where $n_0 > \operatorname{Exp}_{p-1}(1)$. *Then* $\boldsymbol{F} \in \mathfrak{ND}(\mathbb{R})$.

Proof. We apply Corollary 8.6.18 with $B(x) := x^2 L_1(x) \cdots L_{p-1}(x) L_p^a(x)$.
Indeed, let

$$A(x) := \frac{1}{xL_1(x) \cdots L_{p-1}(x)L_p^a(x)},$$
$$\psi(x) := \frac{1}{1-a} L_p^{1-a}(x), \quad x > \operatorname{Exp}_{p-1}(1).$$

Then $\psi' = A$. Thus for $x_0 > \operatorname{Exp}_{p-1}(1)$, we have

$$\int_{x_0}^{\infty} A(x)x dx = \int_{x_0}^{\infty} \psi'(x) dx = \psi(x)\Big|_{x_0}^{\infty} = -\psi(x_0) < +\infty.$$

Consequently, $\sum_{n=n_0}^{\infty} A(n) < +\infty$. Moreover,

$$B'(x) = 2xL_1(x) \cdots L_{p-1}(x)L_p^a(x)$$
$$+ x\Big(\sum_{s=1}^{p-2} L_{s+1}(x) \cdots L_{p-1}(x) + 1\Big)L_p^a(x) + xaL_p^{a-1}(x), \quad x > \operatorname{Exp}_{p-1}(1).$$

Thus B' is increasing, which implies that B is convex. $\square$

8.7 Hata's Method

The aim of the section, based on [Hat88b, Hat88a, Hat94], is to discuss some subtler nowhere differentiability properties of the Weierstrass-type functions $\boldsymbol{W}_{1,a,b,\theta}$ with $ab > 1$ and $\boldsymbol{\theta} = \theta =$ const. In particular, we will prove that:

(a) There exists a zero-measure set $\Xi \subset \mathbb{R}$ such that every $x \in \mathbb{R} \setminus \Xi$ is a knot point of $\boldsymbol{W}_{1,a,b,\theta}$ for arbitrary $\theta \in \mathbb{R}$—Theorem 8.7.4.
(b) Let $\psi^* \in (0, \frac{\pi}{2})$ be such that $\tan \psi^* = \pi + \psi^*$ ($\psi^* \approx 1.3518$). If $ab \geq 1 + \frac{1}{\cos \psi^*} \approx 5.6034$, then $\boldsymbol{W}_{1,a,b,\theta} \in \boldsymbol{\mathcal{ND}}^{\infty}(\mathbb{R})$—Theorem 8.7.6.
(c) Notice that M. Hata in [Hat88b] proved a weaker result stating that $\boldsymbol{W}_{1,a,b,\theta} \in \boldsymbol{\mathcal{ND}}^{\infty}(\mathbb{R})$, provided that $ab \geq 1 + \pi^2 \approx 10.8696$ (in fact, looking at Hata's proof gives $ab \geq 10.7425$).

8.7.1 Nowhere Differentiability of the Weierstrass-Type Functions: Finite One-Sided Derivatives

Let $f(x) := \boldsymbol{W}_{1,a,b,\theta}(\frac{x}{2})$ with $0 < a < 1$, $ab > 1$, $\theta \in \mathbb{R}$ (cf. § 3.1). *We fix a and b* and put $\alpha := -\frac{\log a}{\log b} \in (0, 1)$.

For $x \in \mathbb{R}$, $\varepsilon, \eta > 0$, define

$$\begin{aligned}
E_x^+(\varepsilon) &:= \{s \in \mathbb{R} : f(x+s) - f(x) \geq \varepsilon |s|^\alpha\}, \\
E_x^-(\varepsilon) &:= \{s \in \mathbb{R} : f(x+s) - f(x) \leq -\varepsilon |s|^\alpha\}, \\
E_x(\varepsilon) &:= \{s \in \mathbb{R} : |f(x+s) - f(x)| \geq \varepsilon |s|^\alpha\} = E_x^+(\varepsilon) \cup E_x^-(\varepsilon), \\
E_x^\pm(\varepsilon, \eta) &:= E_x^\pm(\varepsilon) \cap [0, \eta], \quad E_x^\pm(\varepsilon, -\eta) := E_x^\pm(\varepsilon) \cap [-\eta, 0], \\
E_x(\varepsilon, \eta) &:= E_x(\varepsilon) \cap [0, \eta], \quad E_x(\varepsilon, -\eta) := E_x(\varepsilon) \cap [-\eta, 0].
\end{aligned}$$

Remark 8.7.1. (a) All the above sets are Borel measurable.
(b) If $0 < \varepsilon' < \varepsilon''$, then $E_x^\pm(x, \varepsilon'') \subset E_x^\pm(x, \varepsilon')$.
(c) $E_x^+(\varepsilon) \cap E_x^-(\varepsilon) = \{0\}$.

Recall (cf. Remark 3.5.6) that a point $x_0 \in \mathbb{R}$ is a *knot point of f* if

$$D^+ f(x_0) = D^- f(x_0) = +\infty, \quad D_+ f(x_0) = D_- f(x_0) = -\infty.$$

Remark 8.7.2. Let $x \in \mathbb{R}$, $\varepsilon > 0$, and let $(\delta_m)_{m=1}^\infty \subset \mathbb{R}_{>0}$ be such that $\delta_m \longrightarrow 0$. Then (Exercise):

(a) If $E_x^+(\varepsilon, \delta_m) \neq \{0\}$, $m \in \mathbb{N}$, then $D^+ f(x) = +\infty$.
(b) If $E_x^-(\varepsilon, \delta_m) \neq \{0\}$, $m \in \mathbb{N}$, then $D_+ f(x) = -\infty$.
(c) If $E_x^-(\varepsilon, -\delta_m) \neq \{0\}$, $m \in \mathbb{N}$, then $D^- f(x) = +\infty$.
(d) If $E_x^+(\varepsilon, -\delta_m) \neq \{0\}$, $m \in \mathbb{N}$, then $D_- f(x) = -\infty$.
(e) If $E_x(\varepsilon, \delta_m) \neq \{0\}$, $m \in \mathbb{N}$, then $D^+ f(x) = +\infty$ or $D_+ f(x) = -\infty$.
(f) If $E_x(\varepsilon, -\delta_m) \neq \{0\}$, $m \in \mathbb{N}$, then $D^- f(x) = +\infty$ or $D_- f(x) = -\infty$.
(g) If $E_x^+(\varepsilon, \pm\delta_m) \neq \{0\}$ and $E_x^-(\varepsilon, \pm\delta_m) \neq \{0\}$, $m \in \mathbb{N}$, then x is a knot point of f.

Recall (cf. Remark 3.2.1(g)) that there exists a $K_0 = K_0(a, b) > 0$ such that

$$|f(x+h) - f(x)| \leq K_0 |h|^\alpha, \quad x, h \in \mathbb{R}. \tag{8.7.1}$$

Notice that in the theorems below, all the constants $K_i, \varepsilon_i, \eta_i, C_i > 0$ will depend only on a, b (and will be independent of θ). Moreover, once they have been defined, the constants $K_i, \varepsilon_i, \eta_i, C_i$ remain the same for the entire section.

Theorem 8.7.3. *There exist $\varepsilon_0 = \varepsilon_0(a, b)$, $\eta_0 = \eta_0(a, b)$, and $C_0 = C_0(a, b) > 0$ such that*

$$\mathcal{L}(E_x(\varepsilon_0, \pm\eta)) \geq C_0\eta, \quad x \in \mathbb{R},\ \eta \in (0, \eta_0]. \tag{8.7.2}$$

In particular, $f \in \mathbf{ND}_\pm(\mathbb{R})$ (cf. Remark 8.7.2(e),(f)).

Proof. Step 1°. There exist $\eta_1 = \eta_1(a, b)$ and $C_1 = C_1(a, b) > 0$ such that

$$\int_0^\eta |f(x \pm s) - f(x)|ds \geq C_1\eta^{1+\alpha}, \quad x \in \mathbb{R},\ \eta \in (0, \eta_1]. \tag{8.7.3}$$

Indeed, define $r : \mathbb{R} \longrightarrow \mathbb{R}$,

$$r(x) := \sum_{n=1}^{\infty} a^{-n}(\cos(\pi b^{-n}x + \theta) - \cos\theta), \quad x \in \mathbb{R}.$$

Observe that

$$\sum_{n=1}^{\infty} a^{-n}|\cos(\pi b^{-n}x + \theta) - \cos\theta| \leq 2\sum_{n=1}^{\infty} a^{-n}|\sin(\pi b^{-n}x/2)|$$
$$\leq \pi\sum_{n=1}^{\infty}(ab)^{-n}|x| = \frac{\pi}{ab - 1}|x| =: K_1|x| < +\infty, \quad x \in \mathbb{R}.$$

In particular, $r \in \mathcal{C}(\mathbb{R})$. Using the same method (EXERCISE), we get

$$|r(x + h) - r(x)| \leq K_1|h|, \quad x, h \in \mathbb{R}. \tag{8.7.4}$$

Let $g := f + r - \frac{\cos\theta}{1-a}$. Observe (EXERCISE) that

$$g(x) = ag(bx), \quad x \in \mathbb{R}. \tag{8.7.5}$$

Define

$$I_{n,\ell,T}(u) := \frac{b^\ell}{2n}\int_{T-n/b^\ell}^{T+n/b^\ell} u(s)e^{-i\pi b^\ell s}ds, \quad u \in \mathcal{C}(\mathbb{R}, \mathbb{C}),\ n, \ell \in \mathbb{N},\ T \in \mathbb{R}.$$

Note that $I_{n,\ell,T}(1) = 0$. Put $T_{n,\ell,j} := T + (2j - n)/b^\ell$. We have

$$|I_{n,\ell,T}(r)| = \left|\frac{b^\ell}{2n}\sum_{j=0}^{n-1}\int_{T_{n,\ell,j}}^{T_{n,\ell,j+1}} r(s)e^{-i\pi b^\ell s}ds\right|$$
$$= \left|\frac{b^\ell}{2n}\sum_{j=0}^{n-1}\int_0^{2/b^\ell} r(T_{n,\ell,j} + s)e^{-i\pi b^\ell(T_{n,\ell,j}+s)}ds\right|$$
$$= \left|\frac{b^\ell}{2n}\sum_{j=0}^{n-1}\int_0^{2/b^\ell} r(T_{n,\ell,j} + s)e^{-i\pi b^\ell s}ds\right|$$

$$= \left| \frac{b^\ell}{2n} \sum_{j=0}^{n-1} \int_0^{2/b^\ell} \Big(r(T_{n,\ell,j} + s) - r(T_{n,\ell,j}) \Big) e^{-i\pi b^\ell s} ds \right|$$

$$\overset{(8.7.4)}{\leq} \frac{b^\ell}{2n} \sum_{j=0}^{n-1} \int_0^{2/b^\ell} K_1 s ds \leq \frac{b^\ell}{2} K_1 \frac{1}{2} \left(\frac{2}{b^\ell}\right)^2 = \frac{K_1}{b^\ell};$$

$$\begin{aligned} I_{n,\ell,T}(f) &= \frac{b^\ell}{2n} \sum_{k=0}^{\infty} a^k \int_{T-n/b^\ell}^{T+n/b^\ell} \cos(\pi b^k (x+s) + \theta) e^{-i\pi b^\ell s} ds \\ &= \frac{b^\ell}{4n} \sum_{k=0}^{\infty} a^k \int_{T-n/b^\ell}^{T+n/b^\ell} (e^{i(\pi b^k (x+s) + \theta - \pi b^\ell s)} + e^{-i(\pi b^k (x+s) + \theta + \pi b^\ell s)}) ds \\ &= \frac{1}{2} a^\ell e^{i(\pi b^\ell x + \theta)} + \frac{b^\ell}{4n} \sum_{\substack{k=0 \\ k \neq \ell}}^{\infty} a^k \int_{T-n/b^\ell}^{T+n/b^\ell} e^{i(\pi b^k (x+s) + \theta - \pi b^\ell s)} ds \\ &\qquad + \frac{b^\ell}{4n} \sum_{k=0}^{\infty} a^k \int_{T-n/b^\ell}^{T+n/b^\ell} e^{-i(\pi b^k (x+s) + \theta + \pi b^\ell s)} ds \\ &=: \frac{1}{2} a^\ell e^{i(\pi b^\ell x + \theta)} + R_{n,\ell,T}. \end{aligned}$$

Observe that

$$\begin{aligned} |R_{n,\ell,T}| &\leq \frac{b^\ell}{4n} \sum_{\substack{k=0 \\ k \neq \ell}}^{\infty} a^k \frac{2}{\pi |b^k - b^\ell|} + \frac{b^\ell}{4n} \sum_{k=0}^{\infty} a^k \frac{2}{\pi (b^k + b^\ell)} \\ &= \frac{1}{2n\pi} \Bigg(\sum_{\substack{k=0 \\ k \neq \ell}}^{\infty} a^k \frac{1}{|b^{k-\ell} - 1|} + \sum_{k=0}^{\infty} a^k \frac{1}{b^{k-\ell} + 1} \Bigg) =: \frac{1}{n} A_\ell. \end{aligned}$$

Hence

$$|I_{n,\ell,T}(g)| \geq |I_{n,\ell,T}(f)| - |I_{n,\ell,T}(r)| \geq \frac{1}{2} a^\ell - \frac{1}{n} A_\ell - \frac{K_1}{b^\ell} = a^\ell \left(\frac{1}{2} - \frac{K_1}{(ab)^\ell} \right) - \frac{1}{n} A_\ell.$$

Take an $L = L(a,b)$ such that $\frac{K_1}{(ab)^L} < \frac{1}{6}$. Consequently, $|I_{n,L,T}(g)| \geq \frac{1}{3} a^L - \frac{1}{n} A_L$. Now take an $N = N(a,b)$ so big that $c_0 := \frac{1}{3} a^L - \frac{1}{N} A_L > 0$. Define $h_0 := 2N/b^L$, $h_m := h_0/b^m$. Then

$$\begin{aligned} &\int_0^{h_m} |g(x \pm s) - g(x)| ds \overset{(8.7.5)}{=} \int_0^{h_0/b^m} a^m |g(b^m (x \pm s)) - g(b^m x)| ds \\ &= \left(\frac{a}{b}\right)^m \int_0^{h_0} |g(b^m x \pm s) - g(b^m x)| ds \\ &\geq \left(\frac{a}{b}\right)^m \left| \int_0^{h_0} (g(b^m x \pm s) - g(b^m x)) e^{-i\pi b^L s} ds \right| \\ &= \left(\frac{a}{b}\right)^m \left| \int_0^{h_0} g(b^m x \pm s) e^{-i\pi b^L s} ds \right| \\ &= \left(\frac{a}{b}\right)^m \left| \int_{b^m x}^{b^m x \pm h_0} g(s) e^{\pm i\pi b^L (b^m x - s)} ds \right| \end{aligned}$$

$$= \Big(\frac{a}{b}\Big)^m \Big| \int_{b^m x}^{b^m x \pm 2N/b^L} g(s) e^{-i\pi b^L s} ds \Big| = \Big(\frac{a}{b}\Big)^m h_0 |I_{N,L,b^m x \pm N/b^\ell}(g)|$$
$$\geq b^{-m(1+\alpha)} h_0 c_0 = c_0 h_0^{-\alpha} h_m^{1+\alpha} := c_1 h_m^{1+\alpha}.$$

On the other hand,

$$\int_0^{h_m} |r(x \pm s) - r(x)| ds \leq \int_0^{h_m} K_1 s ds = K_1 \frac{1}{2} h_m^2 =: c_2 h_m^2.$$

Thus

$$\int_0^{h_m} |f(x \pm s) - f(x)| ds \geq \int_0^{h_m} |g(x \pm s) - g(x)| ds - \int_0^{h_m} |r(x \pm s) - r(x)| ds$$
$$\geq (c_1 - c_2 h_m^{1-\alpha}) h_m^{1+\alpha} \geq \frac{c_1}{2} h_m^{1+\alpha}, \quad m \geq m_0 = m_0(a,b).$$

Put $\eta_1 := h_{m_0}$. Take an $\eta \in (0, \eta_1]$ and let $M = M(\eta) \in \mathbb{N}$ be such that $h_M \leq \eta < h_{M-1}$. Then

$$\int_0^{\eta} |f(x \pm s) - f(x)| ds \geq \frac{c_1}{2} h_M^{1+\alpha} = \frac{c_1}{2} \Big(\frac{h_{M-1}}{b}\Big)^{1+\alpha} = \frac{a}{2b} c_1 h_{M-1}^{1+\alpha}$$
$$> \frac{a}{2b} c_1 \eta^{1+\alpha} =: C_1 \eta^{1+\alpha}.$$

Step 2°. Using Step 1° and (8.7.1), for $x \in \mathbb{R}$, $\varepsilon > 0$, and $\eta \in (0, \eta_1]$, we get

$$C_1 \eta^{1+\alpha} \leq \int_0^{\eta} |f(x \pm s) - f(x)| ds$$
$$\leq \varepsilon \eta^\alpha (\eta - \mathcal{L}(E_x(\varepsilon, \pm\eta))) + K_0 \int_{E_x(\varepsilon, \pm\eta)} s^\alpha ds$$
$$\leq \varepsilon \eta^\alpha (\eta - \mathcal{L}(E_x(\varepsilon, \pm\eta))) + K_0 \eta^\alpha \mathcal{L}(E_x(\varepsilon, \pm\eta))$$
$$\leq \eta^{1+\alpha} \Big(\varepsilon + (K_0 - \varepsilon) \frac{\mathcal{L}(E_x(\varepsilon, \pm\eta))}{\eta}\Big).$$

Consequently, if $\varepsilon_0 := \frac{1}{2} \min\{C_1, K_0\}$, then

$$\frac{\mathcal{L}(E_x(\varepsilon_0, \pm\eta))}{\eta} \geq \frac{C_1 - \varepsilon_0}{K_0 - \varepsilon_0} =: C_0 > 0, \quad \eta \in (0, \eta_1]. \qquad \square$$

8.7.2 Knot Points of Weierstrass-Type Functions

Theorem 8.7.4. *There exist $\varepsilon_2 = \varepsilon_2(a,b), \eta_2 = \eta_2(a,b), C_2 = C_2(a,b) > 0$, and a zero-measure set $\Xi_2 = \Xi_2(a,b) \subset \mathbb{R}$ such that*

$$\limsup_{\eta \to 0+} \min\Big\{\frac{\mathcal{L}(E_x^+(\varepsilon_2, \pm\eta))}{\eta}, \frac{\mathcal{L}(E_x^-(\varepsilon_2, \pm\eta))}{\eta}\Big\} \geq C_2, \quad x \in \mathbb{R} \setminus \Xi_2.$$

In particular, each point $x \in \mathbb{R} \setminus \Xi_2$ is a knot point of f (cf. Remark 8.7.2).

Proof. Define

$$I_n(x\pm) := \int_{1/b^{n+1}}^{1/b^n} \frac{f(x\pm s)-f(x)}{s^{1+\alpha}}ds, \quad M(x\pm) := \liminf_{n\to+\infty}\Big|\frac{1}{n}\sum_{j=0}^{n-1} I_j(x\pm)\Big|,$$
$$d^\pm(\varepsilon, x\mu) := \limsup_{\eta\to 0+} \frac{\mathcal{L}(E_x^\pm(\varepsilon,\mu\eta))}{\eta}, \quad \mu\in\{-,+\}.$$

Step 1°. Let

$$\Xi_2^\mu := \Big\{x\in\mathbb{R} : M(x\mu) \le \frac{\log b}{9b}\varepsilon_0 C_0^2\Big\}, \quad \mu\in\{-,+\}. \tag{8.7.6}$$

Then there exists an $\varepsilon_3 = \varepsilon_3(a,b) > 0$ such that

$$d^\pm(\varepsilon_3, x\mu) \ge \frac{\varepsilon_0 C_0^2}{4(\varepsilon_0 C_0 + 2bK_0)} =: C_3, \quad \mu\in\{-,+\},\ x\in\Xi_2^\mu. \tag{8.7.7}$$

Indeed, we consider the case $\mu=+$ (the case $\mu=-$ is left for the reader as an EXERCISE). Suppose that there exist sequences $(\delta_m)_{m=1}^\infty \subset (0,\varepsilon_0)$, $\delta_m \longrightarrow 0$, $(x_m)_{m=1}^\infty \subset \Xi_2^+$ such that $\min\{d^+(\delta_m, x_m+), d^-(\delta_m, x_m+)\} < C_3$, $m\in\mathbb{N}$. We may assume that $d^-(\delta_m, x_m+) < C_3$, $m\in\mathbb{N}$ (the other case is left for the reader as an EXERCISE). Thus there exists a sequence $(\tau_m)_{m=1}^\infty \subset (0,\eta_0)$ such that

$$\mathcal{L}(E_{x_m}^-(\delta_m,\eta)) < C_3\eta, \quad \eta\in(0,\tau_m),\ m\in\mathbb{N}. \tag{8.7.8}$$

In view of (8.7.2), we get

$$\frac{\mathcal{L}(E_{x_m}^+(\varepsilon_0,\eta))}{\eta} \ge \frac{\mathcal{L}(E_{x_m}^+(\delta_m,\eta))}{\eta} \ge C_0 - \frac{\mathcal{L}(E_{x_m}^-(\delta_m,\eta))}{\eta} \overset{(8.7.8)}{\ge} C_0 - C_3,$$
$$\eta\in(0,\tau_m),\ m\in\mathbb{N}. \tag{8.7.9}$$

Take an $m\in\mathbb{N}$ and let $L = L(m), N\in\mathbb{N}$ be such that

$$\frac{1}{b^L} \le \tau_m < \frac{1}{b^{L-1}}, \quad \frac{1}{b^N} \le \frac{C_0}{2} < \frac{1}{b^{N-1}}. \tag{8.7.10}$$

Then for $\ell \ge L$, we get

$$\int_{1/b^{\ell+N}}^{1/b^\ell} \frac{f(x_m+s)-f(x_m)}{s^{1+\alpha}}ds = \int_{E_A(\delta_m)} + \int_{E_B(\delta_m)} + \int_{E_C(\delta_m)} =: I_A + I_B + I_C,$$

where

$$E_A(\varepsilon) := E_x^+(\varepsilon)\cap[1/b^{\ell+N}, 1/b^\ell], \quad E_B(\varepsilon) := E_x^-(\varepsilon)\cap[1/b^{\ell+N}, 1/b^\ell],$$
$$E_C(\varepsilon) = [1/b^{\ell+N}, 1/b^\ell]\setminus(E_A(\varepsilon)\cup E_B(\varepsilon)).$$

Since $\delta_m \le \varepsilon_0$, we have

$$\begin{aligned} I_A &\ge \int_{E_A(\varepsilon_0)} \frac{f(x_m+s)-f(x_m)}{s^{1+\alpha}}ds \ge \int_{E_A(\varepsilon_0)} \frac{\varepsilon_0}{s}ds \ge \varepsilon_0 b^\ell \mathcal{L}(E_A(\varepsilon_0)) \\ &\ge \varepsilon_0 b^\ell (\mathcal{L}(E^+_{x_m}(\varepsilon_0, 1/b^\ell)) - 1/b^{\ell+N}) \overset{(8.7.9)}{\ge} \varepsilon_0(C_0 - C_3 - 1/b^N) \\ &\overset{(8.7.10)}{\ge} \varepsilon_0(C_0/2 - C_3). \end{aligned}$$

Moreover,

$$\begin{aligned} |I_B| &\le \int_{E_B(\delta_m)} \frac{K_0}{s}ds \le K_0 b^{\ell+N}\mathcal{L}(E_B(\delta_m)) \le K_0 b^N \frac{\mathcal{L}(E^-_{x_m}(\delta_m, 1/b^\ell))}{1/b^\ell} \\ &\overset{(8.7.8)}{<} K_0 b^N C_3 \overset{(8.7.10)}{\le} K_0 b \frac{2}{C_0} C_3, \end{aligned}$$

$$|I_C| \le \int_{E_C(\delta_m)} \frac{\delta_m}{s}ds \le \delta_m \int_{1/b^{\ell+N}}^{1/b^\ell} \frac{ds}{s} = \delta_m \log(b^N).$$

Thus

$$\begin{aligned} \sum_{j=\ell}^{\ell+N-1} I_j(x_m+) &= \int_{1/b^{\ell+N}}^{1/b^\ell} \frac{f(x_m+s)-f(x_m)}{s^{1+\alpha}}ds \ge I_A - |I_B| - |I_C| \\ &\ge \frac{\varepsilon_0 C_0}{4} - \delta_m N \log b. \end{aligned}$$

Finally,

$$\begin{aligned} M(x_m+) &\overset{(*)}{=} \liminf_{k\to+\infty} \left| \frac{1}{kN} \sum_{\ell=L}^{k} \sum_{j=\ell}^{\ell+N-1} I_j(x_m+) \right| \\ &\ge \frac{1}{N}\left(\frac{\varepsilon_0 C_0}{4} - \delta_m N \log b\right) \ge \frac{\log b}{8b}\varepsilon_0 C_0^2 - \delta_m \log b, \end{aligned}$$

where (*) follows from the lemma below.

Lemma 8.7.5. *Let $(a_n)_{n=0}^\infty \subset \mathbb{C}$ be a bounded sequence. Then for arbitrary $L, N \in \mathbb{N}$, we have*

$$\liminf_{n\longrightarrow+\infty} \left| \frac{1}{n} \sum_{j=0}^{n-1} a_j \right| = \liminf_{k\to+\infty} \left| \frac{1}{kN} \sum_{\ell=L}^{k} \sum_{j=\ell}^{\ell+N-1} a_j \right|,$$

$$\limsup_{n\longrightarrow+\infty} \left| \frac{1}{n} \sum_{j=0}^{n-1} a_j \right| = \limsup_{k\to+\infty} \left| \frac{1}{kN} \sum_{\ell=L}^{k} \sum_{j=\ell}^{\ell+N-1} a_j \right|.$$

Proof. Assume that $|a_n| \le B$, $n \in \mathbb{N}_0$. We have only to observe (EXERCISE) that for $k > L$, we have

$$\frac{1}{kN}\sum_{\ell=L}^{k}\sum_{j=\ell}^{\ell+N-1}a_j=\frac{1}{kN}\Big(R_k+N\sum_{j=0}^{k+N-1}a_j\Big)=\frac{R_k}{kN}+\frac{k+N}{k}\Big(\frac{1}{k+N}\sum_{j=0}^{k+N-1}a_j\Big)$$

and $|R_k|\le(N^2+NL)B$. □

Consequently, there exists an m such that $M(x_m+) > \frac{\log b}{9b}\varepsilon_0 C_0^2$; a contradiction (cf. (8.7.6)).

Now in view of Step 1°, we have only to prove that $\Xi_2:=\mathbb{R}\setminus\Xi_2^\pm$ is of zero measure.

Step 2°. We have

$$\begin{aligned}\sum_{j=0}^{n-1}I_j(x\pm)&=\int_{1/b^n}^{1}\frac{f(x\pm s)-f(x)}{s^{1+\alpha}}ds\\&=\sum_{k=0}^{\infty}a^k\int_{1/b^n}^{1}\frac{\cos(\pi b^k(x\pm s)+\theta)-\cos(\pi b^k x+\theta)}{s^{1+\alpha}}ds\\&=\sum_{k=0}^{\infty}a^k\int_{b^k/b^n}^{b^k}\frac{\cos(\pi b^k x+\theta\pm\pi s)-\cos(\pi b^k x+\theta)}{s^{1+\alpha}}b^{k\alpha}ds\\&=\mathrm{Re}\Big(\sum_{k=0}^{\infty}e^{i(\pi b^k x+\theta)}\int_{b^{k-n}}^{b^k}\frac{e^{\pm i\pi s}-1}{s^{1+\alpha}}ds\Big).\end{aligned}$$

For $m\in\mathbb{Z}$, $n\in\mathbb{N}$, put

$$\begin{aligned}G_m^\pm&:=\int_{b^{m-1}}^{b^m}\frac{e^{\pm i\pi s}-1}{s^{1+\alpha}}ds,\\S_n(t)&:=e^{it}+e^{ibt}+\cdots+e^{ib^{n-1}t},\\S_{n,\theta}(t)&:=S_n(t)e^{i\theta}=e^{i(t+\theta)}+e^{i(bt+\theta)}+\cdots+e^{i(b^{n-1}t+\theta)},\\A_{m,n}(x)&:=\begin{cases}S_{n,\theta}(\pi b^m x), & \text{if } m\ge 0\\ S_{n+m,\theta}(\pi x), & \text{if } -n+1\le m\le -1\\ 0, & \text{if } m\le -n\end{cases}.\end{aligned}$$

Then

$$\begin{aligned}\sum_{k=0}^{\infty}e^{i(\pi x^k+\theta)}\int_{b^{k-n}}^{b^k}\frac{e^{\pm i\pi s}-1}{s^{1+\alpha}}ds&=\sum_{k=0}^{\infty}e^{i(\pi b^k x+\theta)}\sum_{m=k-n+1}^{k}G_m^\pm\\&=\sum_{m=-\infty}^{\infty}A_{m,n}(x)G_m^\pm.\end{aligned}$$

It is known (cf. [KSZ48]) that there exists a zero-measure set $\Xi_0\subset\mathbb{R}$ such that

$$\lim_{n\to+\infty}\frac{1}{n}S_n(\pi x)=0,\quad x\in\mathbb{R}\setminus\Xi_0.$$

Let $x\in\mathbb{R}\setminus\Xi_0$. Since

$$|S_{n+m,\theta}(t)-S_{n,\theta}(t)|=|S_{n+m}(t)-S_n(t)|\le|m|,\quad m\ge -n+1,$$

we get $\frac{1}{n}S_{n+m,\theta}(\pi x) \longrightarrow 0$ uniformly with respect to θ (for $m \geq -n+1$). Since

$$|S_{n,\theta}(b^m t) - S_{n+m,\theta}(t)| = |S_n(b^m t) - S_{n+m}(t)| \leq m, \quad m \geq 0,$$

we get $\frac{1}{n}S_{n,\theta}(\pi b^m x) \longrightarrow 0$ uniformly with respect to θ (for $m \geq 0$). Hence for $x \in \mathbb{R} \setminus \Xi_0$, we have

$$\begin{aligned}\lim_{n\to+\infty} \sum_{j=0}^{n-1} I_j(x\pm) &= \operatorname{Re}\Big(\lim_{n\to+\infty} \frac{1}{n} \sum_{m=-\infty}^{\infty} A_{m,n}(x) G_m^{\pm} \Big) \\ &\stackrel{(\dagger)}{=} \operatorname{Re}\Big(\sum_{m=-\infty}^{\infty} \lim_{n\to+\infty} \frac{1}{n} A_{m,n}(x) G_m^{\pm} \Big) = 0,\end{aligned}$$

where (†) follows from the fact that

$$\sum_{m=-\infty}^{\infty} \frac{1}{n} |A_{m,n}(x) G_m^{\pm}| \leq \sum_{m=-\infty}^{\infty} |G_m^{\pm}| \leq \int_0^{\infty} \frac{2ds}{s^{1+\alpha}} < +\infty.$$

In particular, $M(x\pm) = 0$, $x \in \mathbb{R} \setminus \Xi_0$. Thus $\mathbb{R} \setminus \Xi_0 \subset \Xi_2^{\pm}$ (cf. (8.7.6)). Using (8.7.7) completes the proof. □

8.7.3 Nowhere Differentiability of Weierstrass-Type Functions: Infinite Derivatives

Theorem 8.7.6. *Let $\psi^* \in (0, \frac{\pi}{2})$ be such that $\tan \psi^* = \pi + \psi^*$ ($\psi^* \approx 1.3518$). If $ab \geq 1 + \frac{1}{\cos \psi^*} \approx 5.6034$, then there exist $\varepsilon_4 = \varepsilon_4(a,b), \eta_4 = \eta_4(a,b), C_4 = C_4(a,b) > 0$ such that for every $x \in \mathbb{R}$, we have*

$$\begin{aligned}\limsup_{\eta\to 0+} \min \Big\{ & \frac{\mathcal{L}(E_x^+(\varepsilon_4, \eta))}{\eta} + \frac{\mathcal{L}(E_x^-(\varepsilon_4, -\eta))}{\eta}, \\ & \frac{\mathcal{L}(E_x^+(\varepsilon_4, -\eta))}{\eta} + \frac{\mathcal{L}(E_x^-(\varepsilon_4, \eta))}{\eta} \Big\} \geq C_4. \end{aligned} \tag{8.7.11}$$

In particular, $f \in \mathfrak{M}(\mathbb{R}) \cap \mathfrak{ND}^{\infty}(\mathbb{R}) \subset \mathfrak{ND}_{\pm}(\mathbb{R}) \cap \mathfrak{ND}^{\infty}(\mathbb{R})$ (cf. Theorem 8.3.1 and Remark 8.7.2).

Proof. Put

$$\begin{aligned} d^*(\varepsilon, x) &:= \limsup_{\eta\to 0+} \Big(\frac{\mathcal{L}(E_x^+(\varepsilon, \eta))}{\eta} + \frac{\mathcal{L}(E_x^-(\varepsilon, -\eta))}{\eta} \Big), \\ d_*(\varepsilon, x) &:= \limsup_{\eta\to 0+} \Big(\frac{\mathcal{L}(E_x^+(\varepsilon, -\eta))}{\eta} + \frac{\mathcal{L}(E_x^-(\varepsilon, \eta))}{\eta} \Big). \end{aligned}$$

Let $\mathbb{S} := \{p \in L^{\infty}(0,1) : \|p\|_{\infty} = 1\}$. For $p \in L^{\infty}(0,1)$, put $\widetilde{p}(x) := p(x - \lfloor x \rfloor)$, $x \in \mathbb{R}$. Define

$$F_x(p) := \limsup_{n\to+\infty} \left| \frac{1}{n} \int_{1/b^n}^{1} \frac{f(x+s)-f(x-s)}{s^{1+\alpha}} \widetilde{p}\Big(\frac{\log s}{\log b}\Big) ds \right|,$$
$$x \in \mathbb{R},\ p \in L^\infty(0,1).$$

Observe that

$$\left| \frac{1}{n} \int_{1/b^n}^{1} \frac{f(x+s)-f(x-s)}{s^{1+\alpha}} \widetilde{p}\Big(\frac{\log s}{\log b}\Big) ds \right|$$
$$\le \frac{1}{n} \int_{1/b^n}^{1} \frac{2^\alpha K_0}{s} \|p\|_\infty ds = (2^\alpha K_0 \log b)\|p\|_\infty.$$

From now on, fix an $x \in \mathbb{R}$.

Step 1°. Assume that $\gamma(x) := \sup_{p\in\mathbb{S}} F_x(p) - F_x(1) > 0$. Let $\varepsilon_5 := \frac{\gamma(x)}{8b}$. Then

$$c_3 := \min\{d^*(\varepsilon_5, x), d_*(\varepsilon_5, x)\} \ge \frac{\gamma(x)}{8bK_0} =: C_5. \tag{8.7.12}$$

Indeed, suppose that $c_3 := d_*(\varepsilon_5, x)$ (the case $c_3 := d^*(\varepsilon_5, x)$ is left for the reader as an EXERCISE). Take a $\delta > 0$. Then there exists an $L = L(\delta) \in \mathbb{N}$ such that

$$\frac{\mathcal{L}(E_x^+(\varepsilon, -\eta))}{\eta} + \frac{\mathcal{L}(E_x^-(\varepsilon, \eta))}{\eta} < c_3 + \delta, \quad \eta \in \Big(0, \frac{1}{b^L}\Big].$$

Let $E_*(x) := (E_x^-(\varepsilon_5) \cup (-E_x^+(\varepsilon_5))) \cap \mathbb{R}_+$,

$$\psi_x : \mathbb{R}_+ \longrightarrow \mathbb{R}_+, \quad \psi_x(s) := \begin{cases} 2K_0 s^\alpha, & \text{if } s \in E_*(x) \\ 2\varepsilon_5 s^\alpha, & \text{if } s \notin E_*(x) \end{cases}.$$

Then

$$f(x+s) - f(x-s) = (f(x+s) - f(x)) - (f(x-s) - f(x))$$
$$\ge \begin{cases} -2K_0 s^\alpha, & \text{if } s \in E_*(x) \\ -\varepsilon_5 s^\alpha - \varepsilon_5 |-s|^\alpha, & \text{if } s \notin E_*(x) \end{cases} = -\psi_x(s), \quad s \in \mathbb{R}_+. \tag{8.7.13}$$

Take $\ell \ge L$ and $p \in \mathbb{S}$. Then

$$\left| \int_{1/b^{\ell+1}}^{1/b^\ell} \frac{\psi_x(s)}{s^{1+\alpha}} \widetilde{p}\Big(\frac{\log s}{\log b}\Big) ds \right| \le \int_{1/b^{\ell+1}}^{1/b^\ell} \frac{2\varepsilon_5}{s} ds + \int_{E_*(x)\cap[1/b^{\ell+1}, 1/b^\ell]} \frac{2K_0}{s} ds$$
$$\le 2\varepsilon_5 \log b + 2K_0 b^{\ell+1} \mathcal{L}(E_*(x) \cap [0, 1/b^\ell])$$
$$\le 2\varepsilon_5 b + 2K_0 b^{\ell+1} (\mathcal{L}(E_x^-(\varepsilon_5, 1/b^\ell)) + \mathcal{L}(E_x^+(\varepsilon_5, -1/b^\ell)))$$
$$\le \frac{\gamma(x)}{4} + 2bK_0(c_3 + \delta).$$

Consequently,

$$\begin{aligned}
&\limsup_{n\to+\infty}\Big|\frac{1}{n}\int_{1/b^n}^{1}\frac{\psi_x(s)}{s^{1+\alpha}}\,\widetilde{p}\Big(\frac{\log s}{\log b}\Big)ds\Big| \\
&=\limsup_{n\to+\infty}\Big|\frac{1}{n}\int_{1/b^L}^{1}\frac{\psi_x(s)}{s^{1+\alpha}}\,\widetilde{p}\Big(\frac{\log s}{\log b}\Big)ds+\frac{1}{n}\int_{1/b^n}^{1/b^L}\frac{\psi_x(s)}{s^{1+\alpha}}\,\widetilde{p}\Big(\frac{\log s}{\log b}\Big)ds\Big| \\
&=\limsup_{n\to+\infty}\Big|\frac{1}{n}\int_{1/b^n}^{1/b^L}\frac{\psi_x(s)}{s^{1+\alpha}}\,\widetilde{p}\Big(\frac{\log s}{\log b}\Big)ds\Big| \\
&\le\limsup_{n\to+\infty}\frac{n-L}{n}\Big(\frac{\gamma(x)}{4}+2bK_0(c_3+\delta)\Big)=\frac{\gamma(x)}{4}+2bK_0(c_3+\delta).
\end{aligned}$$

Since $\delta>0$ was arbitrary, we get

$$\limsup_{n\to+\infty}\Big|\frac{1}{n}\int_{1/b^n}^{1}\frac{\psi_x(s)}{s^{1+\alpha}}\,\widetilde{p}\Big(\frac{\log s}{\log b}\Big)ds\Big|\le\frac{\gamma(x)}{4}+2bK_0c_3,\quad p\in\mathbb{S}.$$

Thus,

$$\begin{aligned}
F_x(p)&\le\frac{\gamma(x)}{4}+2bK_0c_3 \\
&\quad+\limsup_{n\to+\infty}\Big|\frac{1}{n}\int_{1/b^n}^{1}\frac{f(x+s)-f(x-s)+\psi_x(s)}{s^{1+\alpha}}\,\widetilde{p}\Big(\frac{\log s}{\log b}\Big)ds\Big| \\
&\le\frac{\gamma(x)}{4}+2bK_0c_3+\limsup_{n\to+\infty}\frac{1}{n}\int_{1/b^n}^{1}\Big|\frac{f(x+s)-f(x-s)+\psi_x(s)}{s^{1+\alpha}}\Big|ds \\
&\overset{(8.7.13)}{=}\frac{\gamma(x)}{4}+2bK_0c_3+\limsup_{n\to+\infty}\frac{1}{n}\int_{1/b^n}^{1}\frac{f(x+s)-f(x-s)+\psi_x(s)}{s^{1+\alpha}}ds \\
&\le\frac{\gamma(x)}{2}+4bK_0c_3+\limsup_{n\to+\infty}\Big|\frac{1}{n}\int_{1/b^n}^{1}\frac{f(x+s)-f(x-s)}{s^{1+\alpha}}ds\Big| \\
&=\frac{\gamma(x)}{2}+4bK_0c_3+F_x(1).
\end{aligned}$$

It follows that $0<\gamma(x)\le\frac{\gamma(x)}{2}+4bK_0c_3$. Finally, $c_3\ge\frac{\gamma(x)}{8bK_0}=C_5$ (cf. (8.7.12)).

Step 2°. (Cf. Step 1° of the proof of Theorem 8.7.4.) Let

$$\varepsilon_6:=\frac{\varepsilon_0C_0^2}{8b}<\varepsilon_0,\quad C_6:=\frac{1}{4\varepsilon_0C_0+8bK_0}.$$

Then

$$\min\{d^*(\varepsilon_6,x),d_*(\varepsilon_6,x)\}\ge C_6\Big(\varepsilon_0C_0^2-\frac{4b}{\log b}F_x(1)\Big). \tag{8.7.14}$$

Indeed, we may assume that $c_4(x):=\varepsilon_0C_0^2-\frac{4b}{\log b}F_x(1)>0$. Suppose that $d_*(\varepsilon_6,x)<C_6c_4(x)$ (the case $d^*(\varepsilon_6,x)<C_6c_4(x)$ is left for the reader as an Exercise). There exists an $L=L(x)\in\mathbb{N}$ such that

$$\frac{\mathcal{L}(E_x^+(\varepsilon_6,-\eta))}{\eta}+\frac{\mathcal{L}(E_x^-(\varepsilon_6,\eta))}{\eta}<C_6c_4(x),\quad \eta\in\Big(0,\frac{1}{b^L}\Big)\subset(0,\eta_0).$$

Using (8.7.2), we get

$$\begin{aligned}\frac{\mathcal{L}(E_x^+(\varepsilon_0,\eta))}{\eta}&\geq C_0-\frac{\mathcal{L}(E_x^-(\varepsilon_0,-\eta))}{\eta}\geq C_0-\frac{\mathcal{L}(E_x^-(\varepsilon_6,-\eta))}{\eta}\\&\geq C_0-C_6c_4(x),\quad \eta\in\Big(0,\frac{1}{b^L}\Big).\end{aligned}$$

Analogously,

$$\frac{\mathcal{L}(E_x^-(\varepsilon_0,-\eta))}{\eta}\geq C_0-C_6c_4(x),\quad \eta\in\Big(0,\frac{1}{b^L}\Big).$$

Let $N\in\mathbb{N}$ be such that $\frac{1}{b^N}\leq\frac{C_0}{2}<\frac{1}{b^{N-1}}$. Take an $\ell\geq L$. Then

$$\int_{1/b^{\ell+N}}^{1/b^\ell}\frac{f(x+s)-f(x)}{s^{1+\alpha}}ds=\int_{E_A(\varepsilon_6)}+\int_{E_B(\varepsilon_6)}+\int_{E_C(\varepsilon_6)}=:I_A+I_B+I_C,$$

where

$$\begin{aligned}E_A(\varepsilon)&:=E_x^+(\varepsilon)\cap[1/b^{\ell+N},1/b^\ell],\quad E_B(\varepsilon):=E_x^-(\varepsilon)\cap[1/b^{\ell+N},1/b^\ell],\\E_C(\varepsilon)&=[1/b^{\ell+N},1/b^\ell]\setminus(E_A(\varepsilon)\cup E_B(\varepsilon)).\end{aligned}$$

Since $\varepsilon_6\leq\varepsilon_0$, we have

$$\begin{aligned}I_A&\geq\int_{E_A(\varepsilon_0)}\frac{f(x+s)-f(x)}{s^{1+\alpha}}ds\geq\int_{E_A(\varepsilon_0)}\frac{\varepsilon_0}{s}ds\geq\varepsilon_0b^\ell\mathcal{L}(E_A(\varepsilon_0))\\&\geq\varepsilon_0b^\ell(\mathcal{L}(E_x^+(\varepsilon_0,1/b^\ell))-1/b^{\ell+N})\geq\varepsilon_0(C_0-C_6c_4(x)-1/b^N)\\&\geq\varepsilon_0(C_0/2-C_6c_4(x)).\end{aligned}$$

Moreover,

$$\begin{aligned}|I_B|&\leq\int_{E_B(\varepsilon_6)}\frac{K_0}{s}ds\leq K_0b^{\ell+N}\mathcal{L}(E_B(\varepsilon_6))\leq K_0b^N\frac{\mathcal{L}(E_x^-(\varepsilon_6,1/b^\ell))}{1/b^\ell}\\&<K_0b^NC_6c_4(x)\leq K_0b\frac{2}{C_0}C_6c_4(x),\end{aligned}$$

$$\begin{aligned}|I_C|&\leq\int_{E_C(\varepsilon_6)}\frac{\varepsilon_6}{s}ds\leq\varepsilon_6\int_{1/b^{\ell+N}}^{1/b^\ell}\frac{ds}{s}=\varepsilon_6\log(b^N)\leq\frac{\varepsilon_0C_0^2}{8b}b^N\\&<\frac{\varepsilon_0C_0^2}{8b}b\frac{2}{C_0}=\frac{\varepsilon_0C_0}{4}.\end{aligned}$$

Thus

$$\int_{1/b^{\ell+N}}^{1/b^\ell}\frac{f(x+s)-f(x)}{s^{1+\alpha}}ds\geq I_A-|I_B|-|I_C|\geq\frac{1}{4}\Big(\varepsilon_0C_0-\frac{c_4(x)}{C_0}\Big).$$

Using the same methods, one can prove (EXERCISE) that

$$\int_{1/b^{\ell+N}}^{1/b^{\ell}} \frac{f(x)-f(x-s)}{s^{1+\alpha}}ds \geq \frac{1}{4}\Big(\varepsilon_0 C_0 - \frac{c_4(x)}{C_0}\Big).$$

Consequently,

$$\sum_{j=\ell}^{\ell+N-1}(I_j(x+)-I_j(x-)) = \int_{1/b^{\ell+N}}^{1/b^{\ell}} \frac{f(x+s)-f(x-s)}{s^{1+\alpha}}ds$$
$$\geq \frac{1}{2}\Big(\varepsilon_0 C_0 - \frac{c_4(x)}{C_0}\Big) > 0,$$

and therefore, using Lemma 8.7.5, we get

$$F_x(1) = \limsup_{k\to+\infty}\Big|\frac{1}{kN}\sum_{\ell=L}^{k}\sum_{j=\ell}^{\ell+N-1}(I_j(x+)-I_j(x-))\Big|$$
$$\geq \frac{1}{2N}\Big(\varepsilon_0 C_0 - \frac{c_4(x)}{C_0}\Big) > \frac{\log b}{4b}(\varepsilon_0 C_0^2 - c_4(x)) = F_x(1);$$

a contradiction.

Step 3°. (Cf. Step 2° of the proof of Theorem 8.7.4.) Take a $p \in \mathbb{S}$. Then

$$\int_{1/b^n}^{1} \frac{f(x+s)-f(x-s)}{s^{1+\alpha}}\,\widetilde{p}\Big(\frac{\log s}{\log b}\Big)ds$$
$$= -2\sum_{k=0}^{\infty} a^k \sin(\pi b^k x+\theta)\int_{1/b^n}^{1}\frac{\sin(\pi b^k s)}{s^{1+\alpha}}\,\widetilde{p}\Big(\frac{\log s}{\log b}\Big)ds$$
$$= -2\sum_{k=0}^{\infty} a^k \sin(\pi b^k x+\theta)\int_{b^k/b^n}^{b^k}\frac{\sin(\pi u)}{u^{1+\alpha}}b^{k\alpha}\,\widetilde{p}\Big(\frac{\log u - k\log b}{\log b}\Big)du$$
$$= -2\sum_{k=0}^{\infty} \sin(\pi b^k x+\theta)\int_{b^{k-n}}^{b^k}\frac{\sin(\pi s)}{s^{1+\alpha}}\,\widetilde{p}\Big(\frac{\log s}{\log b}\Big)ds.$$

For $m \in \mathbb{Z}$, $n \in \mathbb{N}$, put

$$G_m(p) := \int_{b^{m-1}}^{b^m}\frac{\sin(\pi s)}{s^{1+\alpha}}\,\widetilde{p}\Big(\frac{\log s}{\log b}\Big)ds,$$
$$S_n(t) := e^{it} + e^{ibt} + \cdots + e^{ib^{n-1}t},$$
$$s_{n,\theta}(t) := \operatorname{Im}(S_n(t)e^{i\theta}) = \sin(t+\theta) + \sin(bt+\theta) + \cdots + \sin(b^{n-1}t+\theta),$$
$$B_{m,n}(x) := \begin{cases} s_{n,\theta}(\pi b^m x), & \text{if } m \geq 0 \\ s_{n+m,\theta}(\pi x), & \text{if } -n+1 \leq m \leq -1\ . \\ 0, & \text{if } m \leq -n \end{cases}$$

Then

$$\sum_{k=0}^{\infty}\sin(\pi b^k x+\theta)\int_{b^{k-n}}^{b^k}\frac{\sin(\pi s)}{s^{1+\alpha}}ds=\sum_{k=0}^{\infty}\sin(\pi b^k x+\theta)\sum_{m=k-n+1}^{k}G_m(p)$$
$$=\sum_{m=-\infty}^{\infty}B_{m,n}(x)G_m(p).$$

Let $(n_j)_{j=1}^{\infty}$ be such that

$$F_x(1)=\lim_{j\to+\infty}\Big|\frac{1}{n_j}\int_{1/b^{n_j}}^{1}\frac{f(x+s)-f(x-s)}{s^{1+\alpha}}ds\Big|.$$

Since $\frac{1}{n}|s_{n,\theta}(x)|\le 1$, we may assume that $\frac{1}{n_j}s_{n_j,\theta}(\pi x)\longrightarrow s_\theta^*(x)$ when $j\longrightarrow+\infty$. Similarly as in Step 2° of the proof of Theorem 8.7.4, we get $\frac{1}{n_j}s_{n_j+m,\theta}(\pi x)\longrightarrow s_\theta^*(x)$ uniformly with respect to θ (for every $m\ge -n+1$) and $\frac{1}{n_j}s_{n_j,\theta}(\pi b^m x)\longrightarrow s_\theta^*(x)$ uniformly with respect to θ (for every $m\ge 0$). It follows that

$$\lim_{j\to+\infty}\Big|\frac{1}{n_j}\int_{1/b^{n_j}}^{1}\frac{f(x+s)-f(x-s)}{s^{1+\alpha}}\,\widetilde{p}\Big(\frac{\log s}{\log b}\Big)ds\Big|$$
$$=2\lim_{j\to+\infty}\Big|\sum_{m=-\infty}^{\infty}\frac{1}{n_j}B_{m,n_j}(x)G_m(p)\Big|$$
$$\overset{(**)}{=}2\Big|\sum_{m=-\infty}^{\infty}\lim_{j\to+\infty}\frac{1}{n_j}B_{m,n_j}(x)G_m(p)\Big|$$
$$=2\Big|\sum_{m=-\infty}^{\infty}s_\theta^*(x)G_m(p)\Big|=2|s_\theta^*(x)|\Big|\int_0^{\infty}\frac{\sin(\pi s)}{s^{1+\alpha}}\,\widetilde{p}\Big(\frac{\log s}{\log b}\Big)ds\Big|,$$

where (**) follows from the fact that

$$\sum_{m=-\infty}^{\infty}\frac{1}{n_j}|B_{m,n_j}(x)G_m(p)|\le\sum_{m=-\infty}^{\infty}|G_m(p)|\le\int_0^{\infty}\frac{ds}{s^{1+\alpha}}<+\infty.$$

Consequently,

$$\gamma(x)\ge F_x(p)-F_x(1)\ge\lim_{j\to+\infty}\Big|\frac{1}{n_j}\int_{1/b^{n_j}}^{1}\frac{f(x+s)-f(x-s)}{s^{1+\alpha}}\,\widetilde{p}\Big(\frac{\log s}{\log b}\Big)ds\Big|$$
$$-\lim_{j\to+\infty}\Big|\frac{1}{n_j}\int_{1/b^{n_j}}^{1}\frac{f(x+s)-f(x-s)}{s^{1+\alpha}}ds\Big|$$
$$=2|s_\theta^*(x)|\Big(\Big|\int_0^{\infty}\frac{\sin(\pi s)}{s^{1+\alpha}}\,\widetilde{p}\Big(\frac{\log s}{\log b}\Big)ds\Big|-\int_0^{\infty}\frac{\sin(\pi s)}{s^{1+\alpha}}ds\Big)$$
$$\ge 2|s_\theta^*(x)|\int_0^{\infty}\frac{\sin(\pi s)}{s^{1+\alpha}}\Big(\widetilde{p}\Big(\frac{\log s}{\log b}\Big)-1\Big)ds=:2|s_\theta^*(x)|U(p). \tag{8.7.15}$$

Note that $U(p)$ is independent of x and θ.

Step 4°. If $U^*:=\sup_{p\in\mathbb{S}}U(p)>0$, then (8.7.11) is fulfilled.

Indeed, let

$$c_5 := \frac{\frac{\log b}{16b}\varepsilon_0 C_0^2}{\int_0^\infty \frac{\sin(\pi s)}{s^{1+\alpha}} ds}.$$

If $|s_\theta^*(x)| \leq c_5$, then

$$F_x(1) \leq 2c_5 \int_0^\infty \frac{\sin(\pi s)}{s^{1+\alpha}} ds = \frac{\log b}{8b}\varepsilon_0 C_0^2,$$

and therefore, $c_4(x) = \varepsilon_0 C_0^2 - \frac{4b}{\log b}F_x(1) \geq \frac{\varepsilon_0 C_0^2}{2}$. Thus, using (8.7.14), we get (8.7.11) with $\varepsilon_4 := \varepsilon_6$ and $C_4 := \frac{C_6\varepsilon_0 C_0^2}{2}$.

If $|s_\theta^*(x)| > c_5$, then $\gamma(x) > 2c_5 U^* > 0$. Thus, using (8.7.12), we get (8.7.11) with $\varepsilon_4 := \varepsilon_5$ and $C_4 := \frac{c_5 U^*}{4bK_0}$.

Consequently, it remains to prove that $U^* > 0$, provided that $ab \geq 1 + \frac{1}{\cos\psi^*} =: c^*$.

Step 5°. Define

$$G(t) := \sum_{n=-\infty}^{\infty} a^n \sin(\pi b^n t), \quad t \in \mathbb{R}.$$

Note that $G \in \mathcal{C}(\mathbb{R})$. Let $M := \min_{1\leq t\leq b} G(t)$. First, we will show that $U^* > 0$, provided that $M < 0$.

Indeed, if $M < 0$, then there exist $1 < \mu < \nu < b$ such that

$$0 > \int_\mu^\nu \frac{G(s)}{s^{1+\alpha}} ds = \sum_{n=-\infty}^{\infty} a^n \int_\mu^\nu \frac{\sin(\pi b^n s)}{s^{1+\alpha}} ds = \sum_{n=-\infty}^{\infty} \int_{\mu b^n}^{\nu b^n} \frac{\sin(\pi s)}{s^{1+\alpha}} ds.$$

Observe that $\nu b^n \leq \mu b^{n+1}$. Define $p : \mathbb{I} \longrightarrow \mathbb{R}$,

$$p(t) := \begin{cases} 0, & \text{if } \frac{\log\mu}{\log b} < t < \frac{\log\nu}{\log b} \\ 1, & \text{otherwise} \end{cases}.$$

Then, using (8.7.15), we get

$$0 > \sum_{n=-\infty}^{\infty} \int_{\mu b^n}^{\nu b^n} \frac{\sin(\pi s)}{s^{1+\alpha}} ds = -\int_0^\infty \frac{\sin(\pi s)}{s^{1+\alpha}} \Big(\widetilde{p}\Big(\frac{\log s}{\log b}\Big) - 1\Big) ds = -U(p).$$

Hence $U^* > 0$.

Step 6°. Now in view of Step 4°, we have only to prove that $M < 0$, provided that $ab \geq c^*$. First observe that $\psi^* \approx 1.3518 > 1.2566 \approx \frac{2}{5}\pi$. Consider the function

$$\Big(\frac{2\pi}{5}, \frac{\pi}{2}\Big) \ni \psi \overset{g}{\longmapsto} 1 + \frac{\pi + \psi}{\sin\psi}.$$

One may easily check (Exercise) that $g(\psi) \geq g(\psi^*) = 1 + \frac{1}{\cos\psi^*} > 5$. We will prove that for every $\psi \in (\frac{2\pi}{5}, \frac{\pi}{2})$, if $ab \geq 1 + g(\psi)$, then there exists a point $x^* \in (1, 2)$ such that $G(x^*) < 0$. The point $\psi = \psi^*$ gives the best estimate. Thus, fix a $\psi \in (\frac{2\pi}{5}, \frac{\pi}{2})$ and assume that $ab \geq 1 + g(\psi)$. In particular, $b > 5$. Set $q_1 := 1$. We define inductively a sequence of odd natural numbers $q_1 < q_2 < \dots$. Let $\gamma_n := b(q_n + \frac{\psi}{\pi})$. Obviously, there exists a $p_n \in \mathbb{Z}$ such that $|p_n - \frac{\gamma_n - 1}{2}| \leq \frac{1}{2}$. We set $q_{n+1} := 2p_n + 1$. We have $|q_{n+1} - \gamma_n| \leq 1$. Note that

$$q_{n+1} \geq \gamma_n - 1 = b\Big(q_n + \frac{\psi}{\pi}\Big) - 1 > 5q_n - 1 \geq 4q_n.$$

Let

$$I_n := \Big[\frac{q_n}{b^{n-1}}, \frac{q_n + 1}{b^{n-1}}\Big].$$

Observe that $I_{n+1} \subset I_n$. In fact,

$$\frac{q_{n+1}}{b} \geq \frac{\gamma_n - 1}{b} = q_n + \frac{\psi}{\pi} - \frac{1}{b} > q_n + \frac{\psi}{\pi} - \frac{1}{5} > q_n,$$

and therefore, $\frac{q_{n+1}}{b} \geq q_n$. Similarly,

$$\frac{q_{n+1} + 1}{b} \leq \frac{\gamma_n + 2}{b} = q_n + \frac{\psi}{\pi} + \frac{2}{b} < q_n + \frac{\psi}{\pi} + \frac{2}{5} < q_n + 1.$$

Thus $\bigcap_{n=1}^{\infty} I_n = \{x^*\} \subset (1, 2)$. We have

$$b^{n-1}x^* - q_n \geq \frac{q_{n+1}}{b} - q_n \geq \frac{\psi}{\pi} - \frac{1}{b} > \frac{\psi}{\pi} - \frac{2}{b} > \frac{\psi}{\pi} - \frac{2}{5} > 0,$$
$$b^{n-1}x^* - q_n \leq \frac{q_{n+1} + 1}{b} - q_n \leq \frac{\psi}{\pi} + \frac{2}{b} < 1, \quad n \in \mathbb{N}.$$

Hence

$$\sin(\pi b^{n-1}x^*) \leq -\sin\left(\psi - \frac{2\pi}{b}\right), \quad n \in \mathbb{N}.$$

Consequently,

$$\begin{aligned}
G(x^*) &= \sum_{n=0}^{\infty} a^n \sin(\pi b^n x^*) + \sum_{n=1}^{\infty} a^{-n} \sin(\pi b^{-n} x^*) \\
&< -\sum_{n=0}^{\infty} a^n \sin\left(\psi - \frac{2\pi}{b}\right) + \sum_{n=1}^{\infty} a^{-n} \pi b^{-n} x^* \\
&= -\frac{1}{1-a} \sin\left(\psi - \frac{2\pi}{b}\right) + \frac{\pi x^*}{ab - 1} \\
&\leq -\frac{1}{1-a} \sin\left(\psi - \frac{2\pi}{b}\right) + \frac{1}{ab-1}\left(\pi + \psi + \frac{2\pi}{b}\right).
\end{aligned}$$

To get $G(x^*) < 0$, it suffices to have

$$-\frac{1}{1-a} \sin\left(\psi - \frac{2\pi}{b}\right) + \frac{1}{ab-1}\left(\pi + \psi + \frac{2\pi}{b}\right) \leq 0,$$

or equivalently,

$$ab \geq \frac{\pi + \psi + \frac{2\pi}{b} + \sin(\psi - \frac{2\pi}{b})}{\frac{1}{b}(\pi + \psi + \frac{2\pi}{b}) + \sin(\psi - \frac{2\pi}{b})} = F\Big(\frac{2\pi}{b}\Big),$$

where

$$F(t) := \frac{\pi + \psi + t + \sin(\psi - t)}{\frac{t}{2\pi}(\pi + \psi + t) + \sin(\psi - t)}, \quad 0 \leq t \leq \psi.$$

One can check that F is decreasing (Exercise). Thus, $F(t) \leq F(0) = g(\psi)$, $0 \leq t \leq \psi$. □

8.8 Summary

In a concentrated tabular form, the best known results may be summarized as follows (recall that $\mathcal{ND}^\infty \subset \mathcal{ND} \supset \mathcal{ND}_\pm \supset \mathcal{M} \supset \mathcal{M} \cap \mathcal{ND}^\infty$):

$\mathcal{ND}^\infty$	$\mathcal{ND}$	$\mathcal{ND}_\pm$	$\mathcal{M}$	$\mathcal{M} \cap \mathcal{ND}^\infty$
$p=1$, $\boldsymbol{\theta}=0$, b odd $ab>1+\frac{3}{2}\pi(1-a)$ Theorem 3.8.1		$p=1$, $\boldsymbol{\theta}$ arb., b even $b\geq 14$, $a=1/b$ Theorem 3.6.1		p,b odd, $\boldsymbol{\theta}=0$ $ab>1+\frac{3}{2}p\pi$ Theorem 3.5.1
$p=1$, $\boldsymbol{\theta}=0$, $b\in 2\mathbb{N}\setminus(3\mathbb{N})$ $ab>1+\frac{16\pi}{9}(1-a)$ Theorem 3.9.5		p arb., $\boldsymbol{\theta}$ arb. $(a<a_1(p),\ b>\Psi_1(a))$ or $(a<a_2(p),\ b>\Psi_2(a))$ Theorem 3.7.1		
$p=1$, $\boldsymbol{\theta}=0$, $b>3$ $ab>1+\frac{(3b-1)\pi(1-a)}{2(b-1)\cos(\frac{\pi}{b-1})}$ Theorem 3.9.9	$\boldsymbol{\theta}$ arb., $ab\geq 1$ (p odd, $b>p$) or (p even, $b>p/2$) Theorem 8.6.7	$p=1$, $\boldsymbol{\theta}=$const $b\in\mathbb{N}_2$, $ab\geq 1$ Theorem 8.4.1	$p=1$, $\boldsymbol{\theta}$ arb. $ab>1$ Theorem 8.3.1	$p=1$, $\boldsymbol{\theta}=$const $ab>5.6034$ Theorem 8.7.6

? It is seen that many cases remain undecided ?

Chapter 9
Takagi–van der Waerden-Type Functions II

Summary. In this chapter, using more developed tools, we extend results stated in Chap. 4.

9.1 Introduction

Recall (cf. § 4.1) that

$$\boldsymbol{T}_{a,b,\boldsymbol{\theta}}(x) := \sum_{n=0}^{\infty} a^n \boldsymbol{\psi}(b^n x + \theta_n), \quad x \in \mathbb{R},$$

where $0 < a < 1$, $ab \geq 1$, $\boldsymbol{\theta} = (\theta_n)_{n=0}^{\infty} \subset \mathbb{R}$, and $\boldsymbol{\psi}(x) := \operatorname{dist}(x, \mathbb{Z})$, $x \in \mathbb{R}$.

Let us summarize some results proved so far:

(1) $\boldsymbol{T} = \boldsymbol{T}_{1/2,2,0} \in \mathcal{ND}_{\pm}(\mathbb{I})$ (Theorem 4.2.1).
(2) $\boldsymbol{T}_{1/b,b,\boldsymbol{\theta}} \in \mathcal{ND}_{\pm}(\mathbb{R})$, provided that $b \geq 10$ (Theorem 4.3.1).
(3) $\boldsymbol{T}_{a,b,0} \in \mathcal{ND}_{\pm}(\mathbb{R})$, provided that $ab \geq 1$ and $b \in \mathbb{N}_2$ (Theorem 4.3.2); in particular, $\boldsymbol{T}_{1/b,b,0} \in \mathcal{ND}_{\pm}(\mathbb{R})$, provided that $b \in \mathbb{N}_2$.

Now we like to go further and obtain the following results:

(4) If $ab > 1$, then $\boldsymbol{T}_{a,b,\boldsymbol{\theta}} \in \mathcal{M}(\mathbb{R}) \subset \mathcal{ND}_{\pm}(\mathbb{R})$ (Theorem 9.2.1).
(5) A characterization of the set of all $x \in \mathbb{R}$ such that $\boldsymbol{T}'_{\pm}(x) \in \{-\infty, +\infty\}$ (Theorems 9.3.1 and 9.3.4).

[?] The question whether $\boldsymbol{T}_{1/b,b,\boldsymbol{\theta}} \in \mathcal{ND}_{\pm}(\mathbb{R})$ for $1 < b < 10$ remains open [?]

9.2 The Case $ab > 1$

Theorem 9.2.1 (cf. [BD94]). *If $ab > 1$, then $\boldsymbol{T}_{a,b,\boldsymbol{\theta}}$ is α-anti-Hölder continuous uniformly with respect to $x \in \mathbb{R}$ and $\boldsymbol{\theta}$ ($\alpha := -\frac{\log a}{\log b}$). Consequently, $\boldsymbol{T}_{a,b,\boldsymbol{\theta}} \in \mathcal{M}(\mathbb{R}) \subset \mathcal{ND}_{\pm}(\mathbb{R})$ for every $\boldsymbol{\theta}$.*

Proof. Assume that $ab > 1$ and let $\boldsymbol{T}_{\boldsymbol{\theta}} := \boldsymbol{T}_{a,b,\boldsymbol{\theta}}$. Remark 8.5.2(f) reduces the proof to the following condition:

M. Jarnicki, P. Pflug, *Continuous Nowhere Differentiable Functions*, Springer Monographs in Mathematics, DOI 10.1007/978-3-319-12670-8_9

$$\exists_{\varepsilon,\, \delta_0>0}\ \forall_{\boldsymbol{\theta},\, \delta\in(0,\delta_0)}\ \exists_{h_+\in(0,\delta]} : |\boldsymbol{T}_{\boldsymbol{\theta}}(h_+) - \boldsymbol{T}_{\boldsymbol{\theta}}(0)| > \varepsilon\delta^\alpha.$$

We are going to prove that there exist $c > 0$ and $N \in \mathbb{N}$ such that for an arbitrary $\boldsymbol{\theta}$, we have

$$\frac{b^m}{N}\int_0^{\frac{N}{b^m}} \boldsymbol{T}_{\boldsymbol{\theta}}(t)\cos 2\pi(b^m t + \theta_m)dt < -ca^m, \quad m \in \mathbb{N}_0. \tag{9.2.1}$$

Suppose for a moment that (9.2.1) has been proven. Since

$$\int_0^{\frac{N}{b^m}} \cos 2\pi(b^m t + \theta_m)dt = 0, \quad m \in \mathbb{N}_0,$$

we get (for arbitrary $\boldsymbol{\theta}$)

$$\frac{b^m}{N}\int_0^{\frac{N}{b^m}} (\boldsymbol{T}_{\boldsymbol{\theta}}(t) - \boldsymbol{T}_{\boldsymbol{\theta}}(0))\cos 2\pi(b^m t + \theta_m)dt < -ca^m, \quad m \in \mathbb{N}_0. \tag{9.2.2}$$

Put $\delta_0 := 1$. Fix a $\delta \in (0,1)$ and let $m \in \mathbb{N}_0$ be such that $\frac{N}{b^m} < \delta \le b\frac{N}{b^m}$. Then (9.2.2) implies that there exists an $h_+ \in (0, \frac{N}{b^m}) \subset (0,\delta)$ such that

$$|\boldsymbol{T}_{\boldsymbol{\theta}}(h_+) - \boldsymbol{T}_{\boldsymbol{\theta}}(0)| > ca^m = \frac{c}{b^{m\alpha}} = \frac{c}{(Nb)^\alpha}\Big(b\frac{N}{b^m}\Big)^\alpha \ge \varepsilon\delta^\alpha$$

with $\varepsilon := \frac{ca}{N^\alpha}$, which completes the proof.

We move to the proof of (9.2.1). Define

$$J_k(x,\theta,N) := \frac{a^k}{Nb^k}\int_x^{x+Nb^k} \psi(t)\cos 2\pi(b^{-k}t + \theta)dt, \quad k \in \mathbb{Z},\ x,\theta \in \mathbb{R},\ N \in \mathbb{N}.$$

Direct calculations give

$$\begin{aligned}
&\frac{b^m}{N}\int_0^{\frac{N}{b^m}} \boldsymbol{T}_{\boldsymbol{\theta}}(t)\cos 2\pi(b^m t + \theta_m)dt \\
&= \sum_{n=0}^{\infty} \frac{a^n b^m}{N}\int_0^{\frac{N}{b^m}} \psi(b^n t + \theta_n)\cos 2\pi(b^m t + \theta_m)dt \\
&\overset{n=k+m}{=} \sum_{k=-m}^{\infty} \frac{a^{k+m}b^m}{N}\int_0^{\frac{N}{b^m}} \psi(b^{k+m}t + \theta_{k+m})\cos 2\pi(b^m t + \theta_m)dt \\
&\overset{u=b^{k+m}t+\theta_{k+m}}{=} \sum_{k=-m}^{\infty} \frac{a^{k+m}b^m}{N}\int_{\theta_{k+m}}^{b^{k+m}\frac{N}{b^m}+\theta_{k+m}} \psi(u)\cos 2\pi\Big(b^m\frac{u-\theta_{m+k}}{b^{k+m}} + \theta_m\Big)\frac{du}{b^{k+m}} \\
&= a^m \sum_{k=-m}^{\infty} J_k(x_k, \theta'_k, N) \text{ with } x_k := \theta_{m+k},\ \theta'_k = \theta_m - b^{-k}\theta_{m+k}.
\end{aligned} \tag{9.2.3}$$

The remaining part of the proof of (9.2.1) will be divided into the following ten steps.

Step 1°. $|J_k(x,\theta,N)| \le \begin{cases} \frac{a^k}{2}, & \text{if } k \ge 0 \\ \frac{(ab)^k}{\pi}, & \text{if } k < 0 \end{cases}, \quad x,\theta \in \mathbb{R},\ N \in \mathbb{N}.$

Indeed, the case $k \geq 0$ is obvious. If $k < 0$, then integration by parts gives

$$\begin{aligned}
&|J_k(x,\theta,N)| \\
&= \frac{a^k}{2\pi N}\Big|\psi(t)\sin 2\pi(b^{-k}t+\theta)\Big|_x^{x+Nb^k} - \int_x^{x+Nb^k}\psi'(t)\sin 2\pi(b^{-k}t+\theta)dt\Big| \\
&\leq \frac{a^k}{2\pi N}(Nb^k + Nb^k) = \frac{(ab)^k}{\pi}.
\end{aligned}$$

$$\text{Step } 2^o. \quad \Big|\sum_{n=0}^{p}\cos(x+ny)\Big| \leq \frac{1}{|\sin(y/2)|}, \qquad x,y\in\mathbb{R},\ p\in\mathbb{N}_0.$$

Indeed,

$$\Big|\sum_{n=0}^{p}\cos(x+ny)\Big| \leq \Big|\sum_{n=0}^{p}e^{i(x+ny)}\Big| = \Big|\frac{1-e^{i(p+1)y}}{1-e^{iy}}\Big| \leq \frac{1}{|\sin(y/2)|}.$$

Step 3°. If $m := b^{-k} \in \mathbb{Z}$, then

$$J_k(x,\theta,N) \underset{N\to+\infty}{\longrightarrow} -\frac{a^k(1-(-1)^m)\cos 2\pi\theta}{2\pi^2 m^2}$$

uniformly with respect to $x,\theta\in\mathbb{R}$.

Indeed, let $p(N) := \lfloor \frac{N}{m} \rfloor$. Then

$$\begin{aligned}
J_k(x,\theta,N) &= \frac{ma^k}{N}\int_x^{x+\frac{N}{m}}\psi(t)\cos 2\pi(mt+\theta)dt \\
&= \frac{ma^k}{N}\Big(p(N)\int_0^1\psi(t)\cos 2\pi(mt+\theta)dt + \int_{x+p(N)}^{x+\frac{N}{m}}\psi(t)\cos 2\pi(mt+\theta)dt\Big) \\
&= a^k\Big(\frac{p(N)}{\frac{N}{m}}\int_0^1\psi(t)\cos 2\pi(mt+\theta)dt + \frac{m}{N}\int_{x+p(N)}^{x+\frac{N}{m}}\psi(t)\cos 2\pi(mt+\theta)dt\Big).
\end{aligned}$$

Observe that

$$\frac{m}{N}\Big|\int_{x+p(N)}^{x+\frac{N}{m}}\psi(t)\cos 2\pi(mt+\theta)dt\Big| \leq \frac{1}{2}\Big(1-\frac{p(N)}{\frac{N}{m}}\Big) \underset{N\to+\infty}{\longrightarrow} 0.$$

Thus

$$\begin{aligned}
J_k(x,\theta,N) &\underset{N\to+\infty}{\longrightarrow} a^k\int_0^1\psi(t)\cos 2\pi(mt+\theta)dt \\
&= \frac{a^k}{2\pi m}\Big(\psi(t)\sin 2\pi(mt+\theta)\Big|_0^1 - \int_0^1\psi'(t)\sin 2\pi(mt+\theta)dt\Big) \\
&= \frac{a^k}{(2\pi m)^2}\Big(\cos 2\pi(mt+\theta)\Big|_0^{1/2} - \cos 2\pi(mt+\theta)\Big|_{1/2}^1\Big) \\
&= -\frac{a^k(1-(-1)^m)\cos 2\pi\theta}{2\pi^2 m^2} \quad \text{uniformly with respect to } x,\theta\in\mathbb{R}.
\end{aligned}$$

Step 4°. If $b^{-k} \notin \mathbb{Z}$, then

$$J_k(x,\theta,N) \underset{N\to+\infty}{\longrightarrow} 0 \text{ uniformly with respect to } x,\theta\in\mathbb{R}.$$

Indeed, let $p(N) := \lfloor Nb^k \rfloor$. Then

$$J_k(x,\theta,N) = \frac{a^k}{Nb^k}\Big(\int_x^{x+1} \psi(t) \sum_{r=0}^{p(N)-1} \cos 2\pi(b^{-k}(t+r)+\theta)dt + \int_{x+p(N)}^{x+Nb^k} \psi(t)\cos 2\pi(b^{-k}t+\theta)dt\Big).$$

Obviously,

$$\frac{a^k}{Nb^k}\Big|\int_{x+p(N)}^{x+Nb^k} \psi(t)\cos 2\pi(b^{-k}t+\theta)dt\Big| \le \frac{a^k}{2}\Big(1-\frac{p(N)}{Nb^k}\Big) \underset{N\to+\infty}{\longrightarrow} 0$$

uniformly with respect to $x,\theta\in\mathbb{R}$.

On the other hand, using Step 2°, we get

$$\frac{a^k}{Nb^k}\Big|\int_x^{x+1} \psi(t) \sum_{r=0}^{p(N)-1} \cos 2\pi(b^{-k}(t+r)+\theta)dt\Big| \le \frac{a^k}{2Nb^k|\sin \pi b^{-k}|} \underset{N\to+\infty}{\longrightarrow} 0$$

uniformly with respect to $x,\theta\in\mathbb{R}$.

Step 5°. If $d\in\mathbb{Q}_{>0}$ and $d^h\in\mathbb{N}$ for some $h\in\mathbb{N}$, then $d\in\mathbb{N}$.

Indeed, let $d=\frac{p}{q}$, where $p,q\in\mathbb{N}$ are relatively prime. Then $p^h=d^hq^h$, which implies that $q=1$.

Step 6°. Let $S:=\{k\in\mathbb{N} : b^k\in\mathbb{N}\}$. If $S\neq\varnothing$, then $S=r\mathbb{N}$, where $r:=\min S$.

Indeed, suppose that $k\in S\setminus r\mathbb{N}$, $k=rq+h$, $q,h\in\mathbb{N}$, $0<h<r$. Then $b^h=\frac{b^k}{(b^r)^q}\in\mathbb{Q}$ and $(b^h)^r=(b^r)^h\in\mathbb{N}$. Consequently, by Step 5°, $b^h\in\mathbb{N}$; a contradiction.

Step 7°. $J_0(x,0,N)=-\frac{1}{\pi^2}$.

Indeed, using the proof of Step 3°, we get

$$J_0(x,0,N)=\frac{1}{N}\int_0^N \psi(t)\cos 2\pi t dt = \int_0^1 \psi(t)\cos 2\pi t dt = -\frac{1}{\pi^2}.$$

Fix an $\eta>0$ such that $3\eta<\frac{1}{\pi^2}$. Moreover, if $S\neq\varnothing$ and $S=r\mathbb{N}$ (cf. Step 6°), then we require that η be so small that $3\eta+\frac{1}{\pi^2 ab(b^r-1)}<\frac{1}{\pi^2}$.

Step 8°. There exists an $M\in\mathbb{N}$ such that

$$\sum_{|k|>M} |J_k(x,\theta,N)|<\eta, \quad x,\theta\in\mathbb{R},\ N\in\mathbb{N}.$$

Indeed, we only need to use Step 1°.

Step 9°. If $S=\varnothing$, then (9.2.1) is satisfied.

Indeed, by Step 4°, there exists an $N \in \mathbb{N}$ such that $|J_k(x,\theta,N)| < \frac{\eta}{M}$ for all $x,\theta \in \mathbb{R}$ and $0 < |k| \le M$. Consequently, $\sum_{k\neq 0} |J_k(x,\theta,N)| < 3\eta$ for all $x,\theta \in \mathbb{R}$. Thus, using (9.2.3) and Step 7°, we conclude that

$$\frac{b^m}{N}\int_x^{x+\frac{N}{b^m}} \boldsymbol{T}_{\boldsymbol{\theta}}(t)\cos 2\pi(b^m t+\theta_m)dt < \Big(3\eta - \frac{1}{\pi^2}\Big)a^m =: -\varepsilon a^m,$$
$$x \in \mathbb{R},\ m \in \mathbb{N}_0.$$

Step 10°. If $S \neq \varnothing$, then (9.2.1) is satisfied.

Indeed, by Step 6° we have $S = r\mathbb{N}$ for some $r \in \mathbb{N}_2$. Analogously as in Step 9°, we find an N_1 such that

$$|J_k(x,\theta,N)| < \frac{\eta}{M},\quad 0<|k|\le M,\ -k\notin S,\ N\ge N_1,\ x,\theta\in\mathbb{R}$$

(in particular, the above inequality holds for all $1 \le k \le M$). Using Step 3°, we find an $N \ge N_1$ such that

$$|J_k(x,\theta,N)| < \frac{a^k b^{2k}}{\pi^2} + \frac{\eta}{M} \le \frac{b^k}{\pi^2 ab} + \frac{\eta}{M},\quad -M\le k\le -1,\ -k\in S,\ x,\theta\in\mathbb{R}.$$

Thus

$$\sum_{k\neq 0}|J_k(x,\theta,N)| < 3\eta + \sum_{p=1}^{\infty}\frac{1}{\pi^2 abb^{pr}} = 3\eta + \frac{1}{\pi^2 ab(b^r-1)} < \frac{1}{\pi^2},\quad x,\theta\in\mathbb{R},$$

and we finish the proof as in Step 9°. □

9.3 Infinite Unilateral Derivatives of $\boldsymbol{T}_{1/2,2,0}$

Recall (cf. Theorems 4.2.1 and 4.3.2) that $\boldsymbol{T} \in \mathcal{ND}_{\pm}(\mathbb{R})$. The aim of this section is to characterize the points $x \in \mathbb{R}$ for which infinite one-sided derivatives exist. In fact, since $\boldsymbol{T}(x+1) = \boldsymbol{T}(x)$ and $\boldsymbol{T}(-x) = \boldsymbol{T}(x)$, $x \in \mathbb{R}$, we get $\boldsymbol{T}'_{\pm}(x) = \boldsymbol{T}'_{\pm}(x+1)$ and $\boldsymbol{T}'_{\pm}(x) = -\boldsymbol{T}'_{\mp}(-x)$. In particular, $\boldsymbol{T}'_{\pm}(x) = -\boldsymbol{T}'_{\mp}(1-x)$. Hence, it suffices to consider only $\boldsymbol{T}'_{+}(x)$ for $x \in [0,1)$. We will discuss the following two cases:

(a) (Cf. Theorem 9.3.1) $x \in \mathbb{I}$ is a dyadic rational, i.e., $x = \frac{k}{2^m}$ with $m \in \mathbb{N}$, $k \in \mathbb{N}_0$, $k \le 2^m$.
(b) (Cf. Theorem 9.3.4) $x \in (0,1)$ is not a dyadic rational.

We begin with the simpler case of x a dyadic rational.

Theorem 9.3.1 (cf. [BA36]). *If $x \in \mathbb{I}$ is a dyadic rational, then $\boldsymbol{T}'_{\pm}(x) = \pm\infty$.*

Proof. We have only to show that $\boldsymbol{T}'_{+}(x) = +\infty$ for $x = \frac{k}{2^m}$ with $m \in \mathbb{N}$, $k \in \mathbb{N}_0$, $k < 2^m$. Define

$$S_k(t) := \sum_{n=0}^{k}\frac{1}{2^n}\varphi(2^n t),\quad t\in\mathbb{R},\ k\in\mathbb{N}_0.$$

Observe that

- Since $|\psi(t)-\psi(u)|\leq|t-u|$ for all $t,u\in\mathbb{R}$, we get $|\boldsymbol{\Delta}S_{m-1}(x,x+h)|\leq m$.
- If $n\geq m$, then $\psi(2^n x)=\psi(k2^{n-m})=0$.
- If $m\leq n\leq m+p-1$ and $0<h<\frac{1}{2^{m+p}}$, then $2^n h<\frac{1}{2^{m+p-n}}\leq\frac{1}{2}$, and therefore, $\frac{1}{2^n}\frac{\psi(2^n(x+h))}{h}=\frac{\psi(2^n h)}{2^n h}=1$.

Consequently,

$$\boldsymbol{\Delta T}(x,x+h)\geq p-m \text{ for } 0<h<\frac{1}{2^{m+p}},\ p\in\mathbb{N},$$

which immediately implies that $\boldsymbol{T}'_+(x)=+\infty$. □

From now on, we assume that $x\in(0,1)$ is not a dyadic rational. We will use the following two representations of x:

$$x=\sum_{n=1}^{\infty}\frac{1}{2^{a_n}}=\sum_{k=1}^{\infty}\frac{\varepsilon_k}{2^k},$$

where $(a_n)_{n=1}^{\infty}\subset\mathbb{N}$, $a_n<a_{n+1}$, $n\in\mathbb{N}$, $(\varepsilon_k)_{k=1}^{\infty}\subset\{0,1\}$.

Remark 9.3.2. (a) $\varepsilon_k=1\iff k\in\{a_1,a_2,\dots\}$.

(b) It is excluded that:
- there exists a $p\in\mathbb{N}$ such that $a_{p+k}=a_p+k$ for all $k\in\mathbb{N}$, or
- there exists a $p\in\mathbb{N}$ such that $\varepsilon_k=1$ for all $k\geq p$.

Equivalently: $\sup\{n\in\mathbb{N}:a_n+1<a_{n+1}\}=\sup\{k\in\mathbb{N}:\varepsilon_k=0\}=+\infty$.

(c) $1-x$ is not dyadic.

We will use the following notation:
$I_n:=\varepsilon_1+\dots+\varepsilon_n$, $O_n:=n-I_n$, $D_n:=O_n-I_n=2O_n-n=n-2I_n$.
Let $\Xi_n(x):=(a_{n+1}-a_n)-\log_2(a_{n+1}-a_n)-(a_n-2n)$.

Remark 9.3.3. (a) If $a_n\leq k<a_{n+1}$, then $I_k=n$, and hence $D_k=k-2I_k=k-2n\geq a_n-2n=D_{a_n}$. In particular,

$$D_k\longrightarrow+\infty\iff D_{a_n}\longrightarrow+\infty. \qquad (9.3.1)$$

(b) Since $\Xi_n(x)\geq-(a_n-2n)$, we conclude that

$$\Xi_n(x)\longrightarrow-\infty\implies a_n-2n\longrightarrow+\infty. \qquad (9.3.2)$$

Theorem 9.3.4 (cf. [Krü07, AK10])**.** *If $x\in(0,1)$ is not dyadic, $x=\sum_{n=1}^{\infty}\frac{1}{2^{a_n}}$, $1-x=\sum_{n=1}^{\infty}\frac{1}{2^{b_n}}$, where $(a_n)_{n=1}^{\infty}\subset\mathbb{N}$, $(b_n)_{n=1}^{\infty}\subset\mathbb{N}$, $a_n<a_{n+1}$, $b_n<b_{n+1}$, $n\in\mathbb{N}$ (cf. Remark 9.3.2), then:*

(1) $\boldsymbol{T}'_+(x)=+\infty\iff a_n-2n\longrightarrow+\infty$;
(2) $\boldsymbol{T}'_-(x)=+\infty\iff \Xi_n(x)\longrightarrow-\infty$;
(3) $\boldsymbol{T}'_+(x)=-\infty\iff \Xi_n(1-x)\longrightarrow-\infty$;
(4) $\boldsymbol{T}'_-(x)=-\infty\iff b_n-2n\longrightarrow+\infty$;
(5) $\boldsymbol{T}'(x)=+\infty\iff \Xi_n(x)\longrightarrow-\infty$;
(6) $\boldsymbol{T}'(x)=-\infty\iff \Xi_n(1-x)\longrightarrow-\infty$.

Remark 9.3.5. Observe that:

- Statement (1) is equivalent to (4).
- Statement (3) is equivalent to (2).

- (5) is a direct consequence of (1), (2), and (9.3.2).
- Analogously, (6) is a consequence of (3) and (4).

Thus, we need to prove only (1) and (3).

The proofs will be given in Sect. 9.4. First, we present some examples and auxiliary results.

Example 9.3.6 (cf. [AK10]). (a) (Cf. [Krü07], Proposition 5.3) Assume that the number of consecutive 0's in the expansion of x is bounded, i.e., $a_{n+1} - a_n \leq M$, $n \in \mathbb{N}$. Then $\Xi_n(x) \leq M - (a_n - 2n)$. Thus, using (5), we get the implication $a_n - 2n \longrightarrow +\infty \Longrightarrow \boldsymbol{T}'(x) = +\infty$.

(b) Let $\lambda_n := \frac{a_{n+1}}{a_n}$ and assume that $\limsup_{n\to+\infty} \lambda_n > 2$. Then $\Xi_n(x) \not\to -\infty$.
Indeed, if $\lambda_n > 2$, then

$$\begin{aligned}\Xi_n(x) &= (\lambda_n - 2)a_n + 2n - \log_2((\lambda_n - 1)a_n)\\ &\geq 2n + \log_2((\lambda_n - 2)a_n) - \log_2((\lambda_n - 1)a_n) \overset{\substack{\text{mean value}\\ \text{theorem}}}{\geq} 2n - \frac{1}{(\lambda_n - 2)\log 2}.\end{aligned}$$

(c) If for some $0 < \varepsilon \leq 1$ we have $\limsup_{n\to+\infty} \lambda_n = 2 - \varepsilon$ and $\liminf_{n\to+\infty} \frac{a_n}{n} > \frac{2}{\varepsilon}$, then $\Xi_n(x) \longrightarrow -\infty$.
Indeed, take an $\varepsilon' \in (0, \varepsilon)$ such that $\liminf_{n\to+\infty} \frac{a_n}{n} > \frac{2}{\varepsilon'}$ and let $N \in \mathbb{N}$ be such that $\lambda_n \leq 2 - \varepsilon'$ and $\frac{a_n}{n} \geq \frac{2}{\varepsilon'}$ for $n \geq N$. Then for $n \geq N$, we get

$$\begin{aligned}\Xi_n(x) &= \Big(\lambda_n - 2 + \frac{2n}{a_n}\Big)a_n - \log_2((\lambda_n - 1)a_n)\\ &\leq (-\varepsilon' + \varepsilon')a_n - \log_2((\lambda_n - 1)a_n) = -\log_2((\lambda_n - 1)a_n).\end{aligned}$$

(d) The criterion from (c) applies, for example, to the following particular cases (Exercise):

- $a_n := 3n$ (with $\varepsilon := 1$);
- $a_n := p(n)$, where $p(x) = a_s x^s + \cdots + a_0$, $s \geq 2$, $a_s > 0$, and $p(n) < p(n+1)$, $n \in \mathbb{N}$ (with $\varepsilon := 1$);
- $a_n := \lfloor \alpha^n \rfloor$, where $1 < \alpha < 2$ (with $\varepsilon := 2 - \alpha$);
- $a_n :=$ the nth prime number (with $\varepsilon := 1$; hint: $\lim_{n\to+\infty} \frac{a_n}{n\log n} = 1$).

(e) If $a_n = 2^n$, then $\Xi_n(x) = 2^{n+1} - 2 \cdot 2^n + 2n - \log_2(2^{n+1} - 2^n) = n \longrightarrow +\infty$.

(f) If $a_n = 2^n + n$, then

$$\begin{aligned}\Xi_n(x) &= 2^{n+1} + n + 1 - 2(2^n + n) + 2n - \log_2(2^{n+1} + n + 1 - (2^n + n))\\ &= n + 1 - \log_2(2^n + 1) = 1 - \log_2(1 + 2^{-n}) \longrightarrow 1.\end{aligned}$$

(g) If $a_n = 2^n + (1 + \varepsilon)n$ ($\varepsilon > 0$), then

$$\begin{aligned}\Xi_n(x) &= 2^{n+1} + (1 + \varepsilon)(n + 1) - 2(2^n + (1 + \varepsilon)n) + 2n\\ &\quad - \log_2(2^{n+1} + (1 + \varepsilon)(n + 1) - (2^n + (1 + \varepsilon)n))\\ &= 1 + \varepsilon - \varepsilon n - \log_2(1 + 2^{-n}(1 + \varepsilon)) \longrightarrow -\infty.\end{aligned}$$

(h) Define

$$d_1(x) := \lim_{n\to+\infty} \frac{1}{n}\sum_{k=1}^{n} \varepsilon_k = \lim_{n\to+\infty} \frac{I_n}{n}, \quad d_0(x) := 1 - d_1(x) = \lim_{n\to+\infty} \frac{O_n}{n},$$

provided that the limit exists. Theorem 9.3.4(5)(6) implies the following corollaries:

(i) If $0 < d_1(x) < \frac{1}{2}$, then $\boldsymbol{T}'(x) = +\infty$.
(ii) If $\frac{1}{2} < d_1(x) < 1$, then $\boldsymbol{T}'(x) = -\infty$.
(iii) If $d_1(x) = 0$ and $\limsup_{n\to+\infty} \frac{a_{n+1}}{a_n} < 2$, then $\boldsymbol{T}'(x) = +\infty$.
(iv) If $d_1(x) = 1$ and $\limsup_{n\to+\infty} \frac{b_{n+1}}{b_n} < 2$, then $\boldsymbol{T}'(x) = -\infty$.

Indeed, observe that $d_1(x) = \lim_{n\to+\infty} \frac{n}{a_n}$. Moreover, if $0 < d_1(x) < 1$, then $\lim_{n\to+\infty} \frac{a_{n+1}}{a_n} = 1$. Consequently, (i) and (iii) follow from (c) and (5). Implications (ii) and (iv) are left to the reader as an EXERCISE.

Some other examples will be given in Sect. 9.5.

Let $p \in \mathbb{N}$ and $h \in (\frac{1}{2^{p+1}}, \frac{1}{2^p}]$ be such that $x + h < 1$. (9.3.3)
Write $x+h=\sum_{k=1}^{\infty} \frac{\varepsilon'_k}{2^k}$, where $\varepsilon'_k \in \{0,1\}$, $k \in \mathbb{N}$. We assume that $\sup\{k \in \mathbb{N} : \varepsilon'_k=0\}=+\infty$.
Let $X_n(x) := 1 - 2\varepsilon_n = (-1)^{\varepsilon_n}$, $X_n(x+h) := 1 - 2\varepsilon'_n = (-1)^{\varepsilon'_n}$, $n \in \mathbb{N}$.

Lemma 9.3.7. *Let h be as in* (9.3.3). *Then*

$$\begin{aligned} &\psi(2^n(x+h)) - \psi(2^n x) \\ &= 2^{n-1} \sum_{k=n+1}^{\infty} \frac{1}{2^k}(X_{n+1}(x)X_k(x) - X_{n+1}(x+h)X_k(x+h)), \quad n \in \mathbb{N}_0. \end{aligned}$$

Proof. We have (cf. the proof of Proposition 4.1.4)

$$\begin{aligned} \psi(2^n x) &= 2^n \Big(\varepsilon_{n+1} \sum_{k=n+1}^{\infty} \frac{1}{2^k} + (-1)^{\varepsilon_{n+1}} \sum_{k=n+1}^{\infty} \frac{\varepsilon_k}{2^k}\Big) \\ &= 2^n \sum_{k=n+1}^{\infty} \frac{1}{2^k}(\varepsilon_{n+1} + (1 - 2\varepsilon_{n+1})\varepsilon_k) \\ &= 2^n \sum_{k=n+1}^{\infty} \frac{1}{2^k}\Big(-\frac{1}{2}(1 - 2\varepsilon_{n+1})(1 - 2\varepsilon_k) + \frac{1}{2}\Big) \\ &= -2^{n-1} \sum_{k=n+1}^{\infty} \frac{1}{2^k} X_{n+1}(x)X_k(x) + \frac{1}{2}. \end{aligned}$$

□

Define

$$k_0 = k_0(h) := \begin{cases} \max\{k \in \mathbb{N} : \varepsilon_1 = \varepsilon'_1, \ldots, \varepsilon_k = \varepsilon'_k\}, & \text{if } \varepsilon_1 = \varepsilon'_1 \\ 0, & \text{if } \varepsilon_1 \neq \varepsilon'_1 \end{cases}.$$

Remark 9.3.8. (a) $k_0 \leq p$.
Indeed, suppose that $k_0 \geq p+1$. Then

$$h = \sum_{k=p+2}^{\infty} \frac{\varepsilon'_k - \varepsilon_k}{2^k} \leq \sum_{k=p+2}^{\infty} \frac{1}{2^k} = \frac{1}{2^{p+1}};$$

a contradiction.

(b) $\varepsilon_{k_0+1} = 0$ and $\varepsilon'_{k_0+1} = 1$.
Indeed, suppose that $\varepsilon_{k_0+1} = 1$ and $\varepsilon'_{k_0+1} = 0$. Then

$$\begin{aligned} h &= -\frac{1}{2^{k_0+1}} + \sum_{k=k_0+2}^{\infty} \frac{\varepsilon'_k - \varepsilon_k}{2^k} \\ &\leq -\frac{1}{2^{k_0+1}} + \sum_{k=k_0+2}^{\infty} \frac{1}{2^k} = -\frac{1}{2^{k_0+1}} + \frac{1}{2^{k_0+1}} = 0; \end{aligned}$$

a contradiction.
Consequently,

- $X_{k_0+1}(x) = 0$ and $X_{k_0+1}(x+h) = 1$;
- if $k_0 = p-1$, then $O_p = O_{k_0} + 1$.

(c) If $k_0 \leq p-2$, then $\varepsilon_k = 1$ and $\varepsilon'_k = 0$ for $k = k_0+2, \dots, p$.
Indeed, let $h = \sum_{k=1}^{\infty} \frac{\varepsilon''_k}{2^k}$. Observe that $\varepsilon''_1 = \dots = \varepsilon''_{p-1} = 0$. Suppose that $\varepsilon_k = \varepsilon'_k$ for some $k \in \{k_0+2, \dots, p\}$. Then, since $\varepsilon''_{k-1} = 0$, we conclude that $\varepsilon_{k-1} = \varepsilon'_{k-1}$. After a finite number of steps, we get $\varepsilon_{k_0+1} = \varepsilon'_{k_0+1}$, which contradicts (b). Thus $\varepsilon_k \neq \varepsilon'_k$, $k = k_0+1, \dots, p$. Suppose that $\varepsilon_\ell = 0$ and $\varepsilon'_\ell = 1$ for some $\ell \in \{k_0+2, \dots, p\}$. Using (b), we get

$$\begin{aligned} h &= \frac{1}{2^{k_0+1}} + \sum_{k=k_0+2}^{p} \frac{\varepsilon'_k - \varepsilon_k}{2^k} + \sum_{k=p+1}^{\infty} \frac{\varepsilon'_k - \varepsilon_k}{2^k} \\ &\geq \frac{1}{2^{k_0+1}} + \sum_{k-k_0+2,\ k\neq\ell}^{p} \frac{\varepsilon'_k - \varepsilon_k}{2^k} + \frac{1}{2^\ell} - \frac{1}{2^p} \\ &\geq \frac{1}{2^{k_0+1}} - \sum_{k=k_0+2}^{p} \frac{1}{2^k} + \frac{2}{2^\ell} - \frac{1}{2^p} \\ &\geq \frac{1}{2^{k_0+1}} - \Big(\frac{1}{2^{k_0+1}} - \frac{1}{2^p}\Big) + \frac{1}{2^{p-1}} - \frac{1}{2^p} = \frac{1}{2^{p-1}} > \frac{1}{2^p}; \end{aligned}$$

a contradiction.
Consequently,

- if $k_0 + 2 \leq p$, then $X_k(x) = 1$ and $X_k(x+h) = 0$ for $k = k_0+2, \dots, p$;
- if $k_0 \leq p-2$, then $O_p = O_{k_0} + 1$.

(d) Assume that $1 \leq q < p$ and $\varepsilon_q = 0$, $\varepsilon_{q+1} = \dots = \varepsilon_p = 1$. Then $k_0 = q-1$ (use (b) and (c)).

(e) If $h = \frac{1}{2^p}$, then $k_0 \leq p-1$ and $\varepsilon_k = \varepsilon'_k$ for all $k > p$.
Indeed, we have $h = \sum_{k=1}^{\infty} \frac{\delta_{p,k}}{2^k}$. If we add $x+h$, then it is clear that $\varepsilon_k = \varepsilon'_k$ for all $k > p$. Moreover, $\varepsilon'_p = 1 - \varepsilon_p$, so $k_0 \leq p-1$.

(f) If $h = \frac{1}{2^p}$ and $\varepsilon_p = 0$, then $k_0 = p-1$ (use (c) and (e)).

(g) $\lim_{h\to 0+} k_0(h) = +\infty$.
Indeed, for every $k \in \mathbb{N}$, there exists a $q > k$ such that $\varepsilon_q = 0$. Then for every $p \geq q$ and $\frac{1}{2^{p+1}} < h \leq \frac{1}{2^p}$, using (c), we get $k_0(h) \geq q - 1 \geq k$.

Lemma 9.3.9. *Let h be as in (9.3.3). If $k_0 = k_0(h) \geq 1$, then*

$$\begin{aligned}\boldsymbol{T}(x+h) - \boldsymbol{T}(x) &= \Big(hD_{k_0}\Big) + \Big(-(p-k_0-2)\sum_{k=p+1}^{\infty}\frac{1}{2^k}(1-\varepsilon_k-\varepsilon_k')\Big)\\ &\quad + \Big(\frac{1}{2}\sum_{n=p}^{\infty}\sum_{k=n+1}^{\infty}\frac{1}{2^k}(X_{n+1}(x)X_k(x) - X_{n+1}(x+h)X_k(x+h))\Big)\\ &=: A(h) + B(h) + C(h).\end{aligned}$$

Remark 9.3.10. Observe that

$$\begin{aligned}|B(h)| &\leq |p-k_0-2|\sum_{k=p+1}^{\infty}\frac{3}{2^k} = |p-k_0-2|\frac{3}{2^p} \overset{(9.3.3)}{\leq} 6|p-k_0-2|h,\\ |C(h)| &\leq \sum_{n=p}^{\infty}\sum_{k=n+1}^{\infty}\frac{1}{2^k} = \frac{1}{2^{p-1}} \overset{(9.3.3)}{<} 4h.\end{aligned} \tag{9.3.4}$$

In particular,

$$\lim_{h\to 0+} \boldsymbol{\Delta T}(x, x+h) = \pm\infty \iff \lim_{h\to 0+}\Big(D_{k_0(h)} + \frac{B(h)}{h}\Big) = \pm\infty.$$

Proof of Lemma 9.3.9. Using Lemma 9.3.7 and Remark 9.3.8, we get

$$\begin{aligned}&\boldsymbol{T}(x+h) - \boldsymbol{T}(x) - C(h)\\ &= \frac{1}{2}\sum_{n=0}^{k_0-1}\sum_{k=n+1}^{\infty}\frac{1}{2^k}(X_{n+1}(x)X_k(x) - X_{n+1}(x+h)X_k(x+h))\\ &\quad + \frac{1}{2}\sum_{n=k_0}^{p-1}\sum_{k=n+1}^{\infty}\frac{1}{2^k}(X_{n+1}(x)X_k(x) - X_{n+1}(x+h)X_k(x+h))\\ &= \sum_{n=0}^{k_0-1} X_{n+1}(x)\frac{1}{2}\sum_{k=n+1}^{\infty}\frac{1}{2^k}(X_k(x) - X_k(x+h))\\ &\quad + \frac{1}{2}\sum_{k=k_0+1}^{\infty}\frac{1}{2^k}(X_k(x) + X_k(x+h))\\ &\quad + \frac{1}{2}\sum_{n=k_0+1}^{p-1}\sum_{k=n+1}^{\infty}\frac{1}{2^k}(-X_k(x) - X_k(x+h))\\ &= \sum_{n=1}^{k_0}(1-2\varepsilon_n)\sum_{k=n+1}^{\infty}\frac{1}{2^k}(\varepsilon_k' - \varepsilon_k)\\ &\quad + \frac{1}{2}\sum_{k=p+1}^{\infty}\frac{1}{2^k}(1-2\varepsilon_k + 1 - 2\varepsilon_k')\end{aligned}$$

$$- \frac{1}{2}(p-1-k_0) \sum_{k=p+1}^{\infty} \frac{1}{2^k}(1-2\varepsilon_k+1-2\varepsilon'_k)$$

$$= (k_0 - 2I_{k_0}) \sum_{k=1}^{\infty} \frac{1}{2^k}(\varepsilon'_k - \varepsilon_k) - (p-k_0-2) \sum_{k=p+1}^{\infty} \frac{1}{2^k}(1-\varepsilon_k-\varepsilon'_k)$$

$$= D_{k_0}(x+h-x) - (p-k_0-2) \sum_{k=p+1}^{\infty} \frac{1}{2^k}(1 \quad \varepsilon_k - \varepsilon'_k)$$

$$= A(h) + B(h).$$

□

Lemma 9.3.11. *Let h be as in (9.3.3). Assume that $k_0 = k_0(h) \le p-1$. Then*

$$\sum_{k=p+1}^{\infty} \frac{1}{2^k}(1-\varepsilon_k-\varepsilon'_k) \le h. \tag{9.3.5}$$

Moreover, if $\varepsilon_{p+m+1} = 0$ for some $m \in \mathbb{N}_0$, then

$$\sum_{k=p+1}^{\infty} \frac{1}{2^k}(1-\varepsilon_k-\varepsilon'_k) \ge -h\Big(1-\frac{1}{2^m}\Big). \tag{9.3.6}$$

Proof. If $k_0 = p-1$, then (using Remark 9.3.8(b)) we get

$$\sum_{k=p+1}^{\infty} \frac{\varepsilon_k}{2^k} + h = \frac{1}{2^p} + \sum_{k=p+1}^{\infty} \frac{\varepsilon'_k}{2^k}.$$

If $k_0 \le p-2$, then (using Remark 9.3.8(b)(c)) we get

$$\sum_{k=k_0+2}^{p} \frac{1}{2^k} + \sum_{k=p+1}^{\infty} \frac{\varepsilon_k}{2^k} + h = \frac{1}{2^{k_0+1}} + \sum_{k=p+1}^{\infty} \frac{\varepsilon'_k}{2^k}.$$

In both cases, we have

$$\sum_{k=p+1}^{\infty} \frac{\varepsilon_k}{2^k} + h = \frac{1}{2^p} + \sum_{k=p+1}^{\infty} \frac{\varepsilon'_k}{2^k}.$$

Hence,

$$h - \sum_{k=p+1}^{\infty} \frac{1}{2^k}(1-\varepsilon_k-\varepsilon'_k) = \frac{1}{2^p} + \sum_{k=p+1}^{\infty} \frac{1}{2^k}(2\varepsilon'_k - 1) = \sum_{k=p+1}^{\infty} \frac{2\varepsilon'_k}{2^k} \ge 0.$$

Moreover,

$$h\Big(1-\frac{1}{2^m}\Big) + \sum_{k=p+1}^{\infty} \frac{1}{2^k}(1-\varepsilon_k-\varepsilon'_k) \ge h - \frac{1}{2^{m+p}} + \sum_{k=p+1}^{\infty} \frac{1}{2^k}(1-\varepsilon_k-\varepsilon'_k)$$

$$= \frac{1}{2^p} + \sum_{k=p+1}^{\infty} \frac{\varepsilon'_k}{2^k} - \sum_{k=p+1}^{\infty} \frac{\varepsilon_k}{2^k} - \frac{1}{2^{m+p}} + \sum_{k=p+1}^{\infty} \frac{1}{2^k}(1-\varepsilon_k-\varepsilon'_k)$$

$$= \frac{1}{2^p} - \frac{1}{2^{m+p}} + \sum_{k=p+1}^{\infty} \frac{1}{2^k}(1 - 2\varepsilon_k)$$
$$\geq \frac{1}{2^p} - \frac{1}{2^{m+p}} + \sum_{k=p+1}^{\infty} \frac{1}{2^k}(-1) + \frac{2}{2^{p+m+1}} = 0. \qquad \square$$

Lemma 9.3.12. *For $c \geq 1$, let*

$$f(m) = f_c(m) := \Big(1 - \frac{1}{2^m}\Big)(c - m), \quad m \in \mathbb{N}_0,$$

and let $m^* = m^*_c := \max\{m \in \mathbb{N} : f(m) = \max f(\mathbb{N}_0)\}$. Then

$$\log_2 c - 2 < m^* \leq \log_2 c + 1. \tag{9.3.7}$$

Proof. First observe that $f(0) = 0$. If $c = 1$, then $f(1) = 0$ and $f(m) < 0$ for $m \geq 2$. Thus $m^* = 1$. If $c > 1$, then $f(1) = \frac{1}{2}(c-1) > 0$, and hence $m^* \geq 1$.

Note that $f(m+1) - f(m) = \frac{1}{2^{m+1}}(c + 1 - m) - 1$. Thus

$$f(m+1) \geq f(m) \Longleftrightarrow 2^{m+1} + m \leq c + 1.$$

Consequently,

$$2^{m^*} + m^* - 1 \leq c + 1 \Longrightarrow 2^{m^*} \leq c + 2 - m^* \leq c + 1 \leq 2c$$
$$\Longrightarrow m^* \leq \log_2 c + 1,$$
$$2^{m^*+1} + m^* > c + 1 \Longrightarrow 2^{m^*+2} > 2^{m^*+1} + m^* > c + 1 > c$$
$$\Longrightarrow m^* > \log_2 c - 2. \qquad \square$$

9.4 Proof of Theorem 9.3.4

We are going to prove statements (1) and (3) of Theorem 9.3.4.

Proof of (1)($\Longleftarrow$). We assume that $a_n - 2n \longrightarrow +\infty$. Then $D_n \longrightarrow +\infty$ (cf. (9.3.1)). Let $\frac{1}{2^{p+1}} < h \leq \frac{1}{2^p}$ and $k_0 := k_0(h)$.

- If $k_0 \leq p - 2$, then

$$B(h) = -\Big(\sum_{k=p+1}^{\infty} \frac{1}{2^k}(1 - \varepsilon_k - \varepsilon'_k)\Big)(p - k_0 - 2) \overset{(9.3.5)}{\geq} -h(p - k_0 - 2).$$

By Remark 9.3.8(b)(c), we have $D_{k_0} = 2O_{k_0} - k_0 = 2O_p - 2 - k_0$. Consequently,

$$D_{k_0} + \frac{B(h)}{h} \geq 2O_p - 2 - k_0 - p + k_0 + 2 = 2O_p - p = D_p.$$

- If $k_0 = p - 1$, then by (9.3.10), we have

$$D_{k_0} + \frac{B(h)}{h} \geq 2O_p - 2 - k_0 - 6 = 2O_p - (p-1) - 8 = D_p - 7.$$

- If $k_0 = p$, then by (9.3.10), we have

$$D_{k_0} + \frac{B(h)}{h} \geq 2O_{k_0} - k_0 - 12 = 2O_p - p - 12 = D_p - 12.$$

Thus, $\boldsymbol{T}'_+(x) = +\infty$. □

Proof of (1)($\Longrightarrow$). Recall that $a_n - 2n = D_{a_n}$ and $D_k \longrightarrow +\infty \iff D_{a_n} \longrightarrow +\infty$ (cf. (9.3.1)). Suppose that $D_k \not\longrightarrow +\infty$. Then there exists a subsequence $(n_k)_{k=1}^{\infty}$ such that $\lim_{k\to+\infty} D_{n_k} = g < +\infty$. Let $p_k := \min\{n \geq n_k : \varepsilon_n = 0\}$, $h_k := \frac{1}{2^{p_k}}$. Thus $k_0(h_k) = p_k - 1$ (cf. Remark 9.3.8(f)). Consequently,

$$\begin{aligned} D_{k_0(h_k)} + \frac{B(h_k)}{h_k} &\leq 2O_{k_0(h_k)} - k_0(h_k) + 6 = 2(O_{p_k} - 1) - (p_k - 1) + 6 \\ &= 2(O_{n_k} - 2) - p_k + 7 \leq 2O_{n_k} - n_k + 3 = D_{n_k} + 3; \end{aligned}$$

a contradiction. □

Proof of (3)($\Longleftarrow$). We assume that $\Xi_n(1-x) \longrightarrow -\infty$. Hence $b_n - 2n \longrightarrow +\infty$ (cf. (9.3.2)), and therefore $D_{a_n} \longrightarrow -\infty$. Consequently, $D_k \longrightarrow -\infty$ (cf. (9.3.1)).

Let $\frac{1}{2^{p+1}} < h \leq \frac{1}{2^p}$, and let $n \in \mathbb{N}$ be such that $b_n \leq p < b_{n+1}$. Let $m := b_{n+1} - p - 1 \in \mathbb{N}_0$. Then $\varepsilon_{p+m+1} = 0$.

- If $k_0 = p$, then $D_{k_0} + \frac{B(h)}{h} \leq D_{k_0} + 12$.
- If $k_0 = p - 1$, then $D_{k_0} + \frac{B(h)}{h} \leq D_{k_0} + 6$.
- If $k_0 \leq p - 2$, then $k_0 = b_n - 1$ (cf. Remark 9.3.8(b)(c)(d)). Since $O_{b_n} = n$, we get $D_{k_0} = 2O_{k_0} - k_0 \leq 2O_{b_n} - k_0 = 2n - k_0$. Using (9.3.6), we get

$$\frac{B(h)}{h} \leq \Big(1 - \frac{1}{2^m}\Big)(p - k_0 - 2) \leq \Big(1 - \frac{1}{2^m}\Big)(p - k_0).$$

Thus

$$\begin{aligned} D_{k_0} + \frac{B(h)}{h} &\leq 2n - k_0 + \Big(1 - \frac{1}{2^m}\Big)(p - k_0) \\ &\overset{\substack{k_0 = b_n - 1 \\ p = b_{n+1} - m - 1}}{=} 2n - b_n + \Big(1 - \frac{1}{2^m}\Big)(b_{n+1} - b_n - m) + 1. \end{aligned}$$

Put $m_n := m^*_{b_{n+1} - b_n}$ (Lemma 9.3.12). Then

$$\begin{aligned} &2n - b_n + f_{b_{n+1}-b_n}(m) \leq 2n - b_n + f_{b_{n+1}-b_n}(m_n) \\ &\leq 2n - b_n + b_{n+1} - b_n - m_n \leq b_{n+1} - 2b_n + 2n - m_n \\ &\overset{(9.3.7)}{\leq} b_{n+1} - 2b_n + 2n - \log_2(b_{n+1} - b_n) + 2 = \Xi_n(1-x) + 2. \end{aligned}$$

Finally, $\boldsymbol{T}'_+(x) = -\infty$. □

Proof of (3)($\Longrightarrow$). Suppose that there exist a subsequence $(b_{n_s})_{s=1}^{\infty}$ and $M \in \mathbb{R}$ such that $\Xi_{n_s}(1-x) > M$, $s \in \mathbb{N}$. We consider the following two cases.

- $D_n \longrightarrow -\infty$, or equivalently, $b_n - 2n \longrightarrow +\infty$. Fix any $s \in \mathbb{N}$, and let $m_s := m^*_{b_{n_s+1} - b_{n_s}}$ (Lemma 9.3.12). Note that $m_s < b_{n_s+1}$. Define $p_s := b_{n_s+1} - m_s$, $h_s := \frac{1}{2^{p_s}}$. Then

$$b_{n_s+1} - m_s - b_{n_s} \geq b_{n_s+1} - b_{n_s} - \log_2(b_{n_s+1} - b_{n_s}) - 1$$
$$> M + (b_{n_s} - 2n_s) - 1 \underset{s\to+\infty}{\longrightarrow} +\infty.$$

Hence $b_{n_s} < p_s < b_{n_s+1}$ for $s \geq s_0 \gg 1$. Consequently, $k_0(h_s) = b_{n_s} - 1$ (cf. Remark 9.3.8(d)) and

$$D_{k_0(s)} = 2O_{k_0(h_s)} - k_0(h_s) = 2n_s - b_{n_s} - 1, \quad s \geq s_0.$$

Moreover, $p_s - k_0(h_s) - 2 = b_{n_s+1} - m_s - b_{n_s} + 1 - 2 = b_{n_s+1} - b_{n_s} - m_s - 1$. Observe that $\varepsilon_k = \varepsilon'_k = 1$ for $k = p_s + 1, \dots, b_{n_s+1} - 1$. Hence

$$\sum_{k=p_s+1}^{\infty} \frac{1}{2^k}(1 - \varepsilon_k - \varepsilon'_k) \leq -\sum_{k=p_s+1}^{b_{n_s+1}-1} \frac{1}{2^k} + \sum_{k=b_{n_s+1}}^{\infty} \frac{1}{2^k}$$
$$= -\frac{1}{2^{p_s}} + \frac{1}{2^{b_{n_s+1}-1}} + \frac{1}{2^{b_{n_s+1}-1}}$$
$$= -\frac{1}{2^{p_s}}\Big(1 - \frac{1}{2^{b_{n_s+1}-2-p_s}}\Big) = -h_s\Big(1 - \frac{1}{2^{m_s-2}}\Big).$$

Consequently, for $s \geq s_0$, we get $\frac{B(h_s)}{h_s} \geq \Big(1 - \frac{1}{2^{m_s-2}}\Big)(b_{n_s+1} - b_{n_s} - m_s - 1)$. Finally,

$$D_{k_0(h_s)} + \frac{B(h_s)}{h_s} \geq 2n_s - b_{n_s} - 1 + \Big(1 - \frac{1}{2^{m_s-2}}\Big)(b_{n_s+1} - b_{n_s} - m_s - 1)$$
$$\geq b_{n_s+1} - 2b_{n_s} + 2n_s - m_s - \frac{1}{2^{m_s-2}}(b_{n_s+1} - b_{n_s}) - 2$$
$$\overset{(9.3.7)}{\geq} b_{n_s+1} - 2b_{n_s} + 2n_s - \log_2(b_{n_s+1} - b_{n_s}) - \frac{1}{2^{m_s-2}}(b_{n_s+1} - b_{n_s}) - 3$$
$$\geq M - \frac{1}{2^{m_s-2}}(b_{n_s+1} - b_{n_s}) - 3 \overset{(9.3.7)}{\geq} M - 19;$$

a contradiction.

- $\limsup_{n\to+\infty} D_n > -\infty$. Take a subsequence $(n_k)_{k=1}^{\infty}$ such that $\lim_{k\to+\infty} D_{n_k} = g > -\infty$. Let $p_k := \max\{n \leq n_k : \varepsilon_n = 0\}$, $h_k := \frac{1}{2^{p_k}}$. Then $D_{p_k} \geq D_{n_k}$ and $k_0(h_k) = p_k - 1$ (cf. Remark 9.3.8(b)(c)). Consequently, $D_{k_0(h_k)} = D_{p_k} - 1 \geq D_{n_k} - 1$, and therefore,

$$D_{k_0(h_k)} + \frac{B(h_k)}{h_k} \geq D_{k_0(h_k)} - 6 \geq D_{n_k} - 7;$$

a contradiction. □

9.5 The Case of Normal Numbers

By Proposition A.9.3 and Theorems 9.3.4 and 4.3.2, we get the following theorem.

Theorem 9.5.1. *The set*

$$\{x \in (0,1) : \boldsymbol{T}'_+(x),\ \boldsymbol{T}'_-(x) \text{ do not exist (in the finite or infinite sense)}\}$$

is of full measure.

Example 9.5.2. (a) If $a_n := 2n + \lfloor\sqrt{n}\rfloor$, $x := \sum_{n=1}^{\infty} \frac{1}{2^{a_n}}$, then x is a normal number (cf. § A.9) with $\Xi_n(x) \longrightarrow -\infty$ (and hence, by Theorem 9.3.4(5), $\boldsymbol{T}'(x) = +\infty$).
Indeed, it is clear that $\lim_{n\to+\infty} \frac{n}{a_n} = \frac{1}{2}$. Hence, $d_1(x) = \frac{1}{2}$. Moreover,

$$\begin{aligned}\Xi_n(x) &= 2(n+1) + \lfloor\sqrt{n+1}\rfloor - 2n - 2\lfloor\sqrt{n}\rfloor + 2n \\ &\quad - \log_2(2(n+1) + \lfloor\sqrt{n+1}\rfloor - 2n - \lfloor\sqrt{n}\rfloor) \\ &= -\lfloor\sqrt{n}\rfloor + (\lfloor\sqrt{n+1}\rfloor - \lfloor\sqrt{n}\rfloor + 2) - \log_2(\lfloor\sqrt{n+1}\rfloor - \lfloor\sqrt{n}\rfloor + 2) \\ &\qquad \longrightarrow -\infty.\end{aligned}$$

Similarly for $a_n := 2n + \lfloor\log n\rfloor$ (Exercise).

(b) There exists a normal number x such that $a_n - 2n \longrightarrow +\infty$ (and hence, by Theorem 9.3.4(1), $\boldsymbol{T}'_+(x) = +\infty$), but $\Xi_n(x) \not\longrightarrow -\infty$ (and hence, by Theorems 9.3.4(2) and 4.3.2, $\boldsymbol{T}'_-(x)$ does not exist).
Indeed, let $a_1 := 3$ and

$$a_{n+1} := \begin{cases} 2n + 3\lfloor\sqrt{n}\rfloor, & \text{if } a_n \le 2n + \lfloor\sqrt{n}\rfloor \\ a_n + 1, & \text{otherwise} \end{cases}, \quad n \in \mathbb{N}.$$

Directly from the definition we get $2n + \lfloor\sqrt{n}\rfloor - 1 \le a_n \le 2n + 3\lfloor\sqrt{n}\rfloor$.
In fact, for $n = 1$, the inequalities are trivial. Suppose that they hold for some n.
In the case $a_n \le 2n + \lfloor\sqrt{n}\rfloor$, we have

$$\begin{aligned} a_{n+1} &= 2n + 3\lfloor\sqrt{n}\rfloor \le 2(n+1) + 3\lfloor\sqrt{n+1}\rfloor, \\ a_{n+1} &= 2n + 3\lfloor\sqrt{n}\rfloor \ge 2(n+1) + \lfloor\sqrt{n+1}\rfloor - 1. \end{aligned}$$

In the case $a_n \ge 2n + \lfloor\sqrt{n}\rfloor + 1$, we have

$$\begin{aligned} a_{n+1} &= a_n + 1 \le 2n + 3\lfloor\sqrt{n}\rfloor + 1 \le 2(n+1) + 3\lfloor\sqrt{n}\rfloor, \\ a_{n+1} &= a_n + 1 \ge 2n + \lfloor\sqrt{n}\rfloor + 2 \ge 2(n+1) + \lfloor\sqrt{n+1}\rfloor - 1. \end{aligned}$$

Hence $a_n - 2n \ge \lfloor\sqrt{n}\rfloor - 1 \longrightarrow +\infty$ and $\frac{a_n}{n} \longrightarrow \frac{1}{2}$ (thus $d_1(x) = \frac{1}{2}$). If n is such that $a_n \le 2n + \lfloor\sqrt{n}\rfloor$, then

$$\begin{aligned}\Xi_n(x) &= 2n + 3\lfloor\sqrt{n}\rfloor - 2a_n + 2n - \log_2(2n + 3\lfloor\sqrt{n}\rfloor - a_n) \\ &\ge 4n + 3\lfloor\sqrt{n}\rfloor - 2(2n + \lfloor\sqrt{n}\rfloor) - \log_2(2n + 3\lfloor\sqrt{n}\rfloor - (2n + \lfloor\sqrt{n}\rfloor - 1)) \\ &\ge \lfloor\sqrt{n}\rfloor - \log_2(2\lfloor\sqrt{n}\rfloor + 1) \ge \lfloor\sqrt{n}\rfloor - \log_2(\lfloor\sqrt{n}\rfloor) - 2.\end{aligned}$$

It remains to observe that it is impossible that for some $n \in \mathbb{N}$, we have $a_n + k > 2(n+k) + \lfloor\sqrt{n+k}\rfloor$, $k \in \mathbb{N}_0$.

Chapter 10
Bolzano-Type Functions II

Summary. In this chapter, using more advanced tools, we extend results stated in Chap. 5.

10.1 Bolzano-Type Functions

If we apply the general construction from § 5.1 to $N = 4$, $\varphi_1 = \frac{3}{8}$, $\varphi_2 = \frac{1}{2}$, $\varphi_3 = \frac{7}{8}$, $\Phi_1 = \frac{5}{8}$, $\Phi_2 = \frac{1}{2}$, $\Phi_3 = \frac{9}{8}$, then we get the *classical Bolzano function* $\boldsymbol{B} : \mathbb{I} \longrightarrow \mathbb{R}$. Recall that we already know that $\boldsymbol{B} \in \mathcal{ND}(\mathbb{I})$ (cf. Theorem 5.1.2). We have (cf. § 5.1):

$$\Big(\frac{1}{8}\Big)^n \le |\Delta(S_{n,i})| \le \Big(\frac{5}{8}\Big)^n, \quad \Big(\frac{1}{8}\Big)^n \le \delta(J_{n,i}) \le \Big(\frac{3}{8}\Big)^n,$$
$$1 \le |\varkappa(S_{n,i})| \le \Big(\frac{5}{3}\Big)^n, \quad i = 1, \dots, 4^n,$$
$$\max_{x \in \mathbb{I}} |L_{n+1}(x) - L_n(x)| \le \frac{1}{4} \max\Big\{|\Delta(S_{n,i})| : i = 1, \dots, 4^n\Big\} \le \frac{1}{4}\Big(\frac{5}{8}\Big)^n.$$

We point out that the name *Bolzano function* is sometimes assigned to a different function; cf., e.g., [Brž49, Kow23, Sin28].

Let $\mathcal{M}$ be the set of all local extrema of the function $\boldsymbol{B}$. Moreover, let $\mathcal{P}$ denote the set of all points $x_0 \in (0,1)$ such that the determining sequence $(S_n)_{n=1}^{\infty}$ for x_0 (cf. Remark 5.1.1(d)) satisfies the following condition:

$$\forall_{s \in \mathbb{N}}\ \exists_{n \ge s} :$$
$$\operatorname{type}(J_{n+1}) = 3,\ \operatorname{type}(J_{n+2}) = 2,\ \operatorname{type}(J_{n+3}) = 1,\ \operatorname{type}(J_{n+4}) = 4.$$

One can prove (Exercise) that the set $\mathcal{P}$ is uncountable (cf. [Jar81]).

Theorem 10.1.1 (cf. [Jar22, Jar81]; see also [Brž49, Kow23]). $\boldsymbol{B} \in \mathcal{ND}^{\infty}((0,1)) \cap \mathcal{ND}_{\pm}(\mathbb{I})$. *Moreover:*

- *if $x \in \mathcal{M} \cap (0,1)$ is a local maximum, then $\boldsymbol{B}'_-(x) = +\infty$ and $\boldsymbol{B}'_+(x) = -\infty$;*
- *if $x \in \mathcal{M} \cap (0,1)$ is a local minimum, then $\boldsymbol{B}'_-(x) = -\infty$ and $\boldsymbol{B}'_+(x) = +\infty$;*
- *if $x = \frac{3}{7}$, then $D_-f(x), D^-f(x), D_+f(x), D^+f(x) \in \mathbb{R}$;*
- *if $x \in \mathcal{P}$, then $D_-f(x) = D_+f(x) = -\infty, D^-f(x) = D^+f(x) = +\infty$, i.e., x is a knot point (cf. Remark 3.5.6).*

M. Jarnicki, P. Pflug, *Continuous Nowhere Differentiable Functions*, Springer Monographs in Mathematics, DOI 10.1007/978-3-319-12670-8_10

? It is an open problem whether other Bolzano-type functions from Sect. 5.3 have similar properties ?

Proof of Theorem 10.1.1. Put $f := \boldsymbol{B}$. If $(S_n)_{n=1}^{\infty}$, $J_n := J(S_n)$ is a determining sequence, then to simplify notation, we will write $\delta_n := \Delta(J_n)$, $\Delta_n := \Delta(S_n)$, $\varkappa_n := \varkappa(S_n)$. Recall that $\mathfrak{N}$ denotes the set of nodes (§ 5.1). The proof, based entirely on [Jar81], will be divided into 11 steps.

Step 1°. *If $x_0 \notin \mathfrak{N}$, then a finite derivative $f'(x_0)$ does not exist.*
Recall (Remark 5.1.3(b)) that $M_\Delta = \frac{5}{8}$, $\Sigma = \{1,3\}$, and condition (5.1.1) is satisfied. Thus the result follows directly from Theorem 5.1.2.

Step 2°. *If $x_0 \in \mathfrak{N}_p \setminus \{1\}$, then a finite right-sided derivative $f'_+(x_0)$ does not exist.*
Suppose that $f'_+(x_0) \in \mathbb{R}$ exists. Let $(S_n)_{n=1}^{\infty}$ be the determining sequence of type (L) for x_0. Then for $n \geq p$, we get

$$\varkappa_n = \boldsymbol{\Delta} L_n(a_n, b_n) = \boldsymbol{\Delta} f(x_0, b_n) \underset{n \to +\infty}{\longrightarrow} f'_+(x_0).$$

Observe that if $\operatorname{type}(J_{n_0}) \in \{2,4\}$ for some $n_0 \geq p$, then $\operatorname{type}(J_n) = 1$ for all $n > n_0$. Thus, we are always in case (A) (cf. the proof of Theorem 5.1.2); a contradiction.

Step 3°. *If $x_0 \in \mathfrak{N}_p \setminus \{0\}$, then a finite or infinite left-sided derivative $f'_-(x_0)$ does not exist.*
Suppose that $f'_-(x_0) \in \overline{\mathbb{R}}$ exists. Let $(S_n)_{n=1}^{\infty}$ be the determining sequence of type (R) for x_0. Then for $n \geq p$, we get

$$\varkappa_n = \boldsymbol{\Delta} L_n(a_n, b_n) = \boldsymbol{\Delta} f(x_0, a_n) \underset{n \to +\infty}{\longrightarrow} f'_-(x_0).$$

Observe that if $\operatorname{type}(J_{n_0}) \in \{1,3\}$ for some $n_0 \geq p$, then $\operatorname{type}(J_n) = 4$ for all $n > n_0$. Thus, we are always in case (B); a contradiction.

Step 4°. $(\mathfrak{N} \setminus \{0\}) \cap \mathfrak{M} = \varnothing$.
Suppose that $x_0 \in (\mathfrak{N} \setminus \{0\}) \cap \mathfrak{M}$ and let $S_n, J_n = [a_n, b_n]$ be as in Step 3°. Then we are in case (B), and hence $\boldsymbol{\Delta} f(x_0, a_{n_s - 1}) = -\boldsymbol{\Delta} f(x_0, a_{n_s})$ for all $s \in \mathbb{N}$. This means that the differential quotient oscillates in every neighborhood of x_0; a contradiction.

Step 5°. *Let $S \in \mathfrak{S}_p$, $J := J(S) = [a,b]$, $x_0 := a + \frac{4}{5}\delta(J)$. Then $f(x_0) = f(a) + \frac{4}{3}\Delta(S)$. If $\varkappa(S) > 0$, then the minimum of f in J is realized at $x = a$ and the maximum at $x = x_0$. If $\varkappa(S) < 0$, then the maximum of f in J is realized at $x = a$ and the minimum at $x = x_0$.*
Consequently, by Step 4°,

$$\mathfrak{M} \setminus \{0\} = \Big\{a + \frac{4}{5}\delta(J) : J := J(S) = [a,b],\ S \in \mathfrak{S}_p,\ p \in \mathbb{N}\Big\}.$$

In particular, $\mathfrak{M}$ *is countable and dense in* $\mathbb{I}$.
Indeed, suppose that $\varkappa(S) > 0$ (the case $\varkappa(S) < 0$ is left to the reader as an Exercise). Let $x_*, x^* \in J$ be such that $\min_J f = f(x_*)$, $\max_J f = f(x^*)$. We wish to prove that $x_* = a$ and $x^* = x_0$. It is clear (Exercise) that $x_* = a$.
Let $(S_n)_{n=1}^{\infty}$ be a determining sequence for x^* with $S_p = S$, $J_p = J$, $J_{p+1} = [a + \frac{1}{2}\delta(J), b - \frac{1}{8}\delta(J)]$. Obviously, x^* also realizes the maximum in the interval J_{p+1} and $\varkappa_{p+1} > 0$. Thus

$$a_{p+2} = a_{p+1} + \frac{1}{2}\delta_{p+1} = a + \frac{1}{2}\delta(J) + \frac{1}{2} \cdot \frac{3}{8}\delta(J),$$

$$b_{p+2} = b_{p+1} - \frac{1}{8}\delta_{p+1} = b - \frac{1}{8}\delta(J) - \frac{1}{8} \cdot \frac{3}{8}\delta(J).$$

After a finite number of steps, we get

$$a_{p+k} = a + \frac{1}{2}\delta(J)\sum_{j=0}^{k-1}\left(\frac{3}{8}\right)^j \underset{k\to+\infty}{\longrightarrow} a + \frac{4}{5}\delta(J) = x_0,$$

$$b_{p+k} = b - \frac{1}{8}\delta(J)\sum_{j=0}^{k-1}\left(\frac{3}{8}\right)^j \underset{k\to+\infty}{\longrightarrow} b - \frac{1}{5}\delta(J) = a + \frac{4}{5}\delta(J) = x_0.$$

Moreover,

$$f(a_{p+k}) = f(a) + \frac{1}{2}\Delta(S)\sum_{j=0}^{k-1}\left(\frac{5}{8}\right)^j \underset{k\to+\infty}{\longrightarrow} f(a) + \frac{4}{3}\Delta(S) = f(x_0).$$

Step 6°. *If $x_0 \notin \mathbf{N}$, then an infinite derivative $f'(x_0)$ does not exist.*

Suppose that $f'(x_0) \in \{-\infty, +\infty\}$ exists. Let $(S_n)_{n=1}^{\infty}$ be the determining sequence for x_0. If (B) is satisfied, then we are done. Thus we may assume that there exists an $n_0 \in \mathbb{N}$ such that type$(J_n) \in \{1, 3\}$ for $n \geq n_0$. Recall that $\varkappa_n = (\frac{5}{3})^{n-n_0}\varkappa_{n_0}$ for $n \geq n_0$. Thus

$$\frac{1}{\varkappa_{n_0}}f'(x_0) = \lim_{n\to+\infty}\frac{1}{\varkappa_{n_0}}\boldsymbol{\Delta}f(a_n, b_n) = \lim_{n\to+\infty}\left(\frac{5}{3}\right)^{n-n_0} = +\infty.$$

We may exclude the case that type$(J_n) = 1$ for $n \gg 1$, because in such a case, we must have $x_0 \in \mathbf{N}$. We consider the following three possibilities:

(C) There exists a sequence $(n_s)_{s=1}^{\infty}$, $n_1 \geq n_0$, such that type$(J_{n_s+1}) = $ type$(J_{n_s+2}) = 3$ for all $s \in \mathbb{N}$.

Fix an $s \in \mathbb{N}$ and let $m := n_s$. Let J be the next interval from the right to J_m and let $S := S(J)$. Observe that $\delta(J) = \frac{1}{3}\delta_m$, $\varkappa(S) = -\frac{3}{5}\varkappa_m$, and $\Delta(S) = -\frac{1}{5}\Delta_m$.

Suppose that $\varkappa_{n_0} > 0$ (the case $\varkappa_{n_0} < 0$ is left to the reader as an EXERCISE). Since type$(J_{m+1}) = $ type$(J_{m+2}) = 3$, we conclude that

$$f(x_0) \geq f(a_m) + \frac{1}{2}\Delta_m + \frac{1}{2} \cdot \frac{5}{3} \cdot \frac{3}{8}\Delta_m = f(a_m) + \frac{13}{16}\Delta_m.$$

Define

$$x_m := b_m + \frac{4}{5}\delta(J) = b_m + \frac{4}{15}\delta_m.$$

Note that $0 < x_m - x_0 < x_m - a_m = \frac{19}{15}\delta_m$. Using Step 5°, we have

$$f(x_m) = f(b_m) + \frac{4}{3}\Delta(S) = f(a_m) + \Delta_m - \frac{4}{3} \cdot \frac{1}{5}\Delta_m = f(a_m) + \frac{11}{15}\Delta_m.$$

Hence

$$\boldsymbol{\Delta}f(x_0, x_m) = \frac{f(x_m) - f(x_0)}{x_m - x_0} \leq \left(\frac{11}{15} - \frac{13}{16}\right)\varkappa_m\frac{15}{19}.$$

Thus

$$\frac{1}{\varkappa_{n_0}}\boldsymbol{\Delta} f(x_0, x_m) < -\frac{19}{15\cdot 16}\Big(\frac{5}{3}\Big)^{m-n_0}\frac{15}{19}.$$

Consequently,

$$\frac{1}{\varkappa_{n_0}} f'(x_0) = \lim_{s\to+\infty}\frac{1}{\varkappa_{n_0}}\boldsymbol{\Delta} f(x_0, x_{n_s}) = -\infty;$$

a contradiction.

(D) There exists a sequence $(n_s)_{s=1}^{\infty}$, $n_1 \geq n_0$, such that $\operatorname{type}(J_{n_s+1}) = \operatorname{type}(J_{n_s+2}) = 1$ for all $s \in \mathbb{N}$.
This case is left to the reader as an EXERCISE.

(E) There exists an $n_1 \in \mathbb{N}$ such that $\operatorname{type}(J_{n_1+2n}) = 3$ and $\operatorname{type}(J_{n_1+2n+1}) = 1$ for all $n \in \mathbb{N}_0$.

Fix an $n \in \mathbb{N}_0$ and let $m := n_1 + 2n$. Then we have

$$a_{m+1} = a_m, \quad a_{m+2} = a_{m+1} + \frac{1}{2}\delta_{m+1} = a_m + \frac{1}{2}\cdot\frac{3}{8}\delta_m.$$

Hence

$$a_{m+2k} = a_m + \delta_m \frac{1}{2}\sum_{j=1}^{k}\Big(\frac{3}{8}\Big)^{2j-1} \underset{k\to+\infty}{\longrightarrow} a_m + \frac{12}{55}\delta_m,$$

$$f(a_{m+2k}) = f(a_m) + \Delta_m \frac{1}{2}\sum_{j=1}^{k}\Big(\frac{5}{8}\Big)^{2j-1} \underset{k\to+\infty}{\longrightarrow} f(a_m) + \frac{20}{39}\Delta_m.$$

Thus

$$x_0 = a_m + \frac{12}{55}\delta_m, \quad f(x_0) = f(a_m) + \frac{20}{39}\Delta_m.$$

Let $x_m := a_m + \frac{1}{2}\delta_m$. Then $f(x_m) = f(a_m) + \frac{1}{2}\Delta_m$. Hence

$$\boldsymbol{\Delta} f(x_0, x_m) = \frac{f(a_m) + \frac{1}{2}\Delta_m - (f(a_m) + \frac{20}{39}\Delta_m)}{a_m + \frac{1}{2}\delta_m - (a_m + \frac{12}{55}\delta_m)} = \frac{\frac{1}{2} - \frac{20}{39}}{\frac{1}{2} - \frac{12}{55}}\varkappa_m$$
$$= -\frac{1}{31}\cdot\frac{50}{39}\varkappa_m,$$

and we get a contradiction as in case (C).

Step 7°. *Let $x_0 \in \mathbf{M} \cap (0, 1)$. If x_0 is a local maximum, then $f'_-(x_0) = +\infty$ and $f'_+(x_0) = -\infty$. If x_0 is a local minimum, then $f'_-(x_0) = -\infty$ and $f'_+(x_0) = +\infty$.*

Let $x_0 = a + \frac{4}{5}\delta(J)$ be as in Step 5° and let $(S_n)_{n=1}^{\infty}$ be a determining sequence for x_0. Observe that $a_p = a$, $b_p = b$, $a_{n+1} = a_n + \frac{1}{2}\delta_n$, $b_{n+1} = b_n - \frac{1}{8}\delta_n$, and $\varkappa_n = (\frac{5}{3})^{n-p}\varkappa_p$ for all $n \geq p$. Assume that $\varkappa_p > 0$ (the case $\varkappa_p < 0$ is left for the reader as an EXERCISE).

If $x' \in [a_n, a_{n+1}]$, then $f(x') \leq f(x_0) - \frac{1}{2}\Delta_n$. If $x'' \in [b_{n+1}, b_n]$, then $f(x'') \leq f(a_n) + \frac{9}{8}\Delta_n$. Note that $0 < x_0 - x' < \delta_n$ and $0 < x'' - x_0 < \delta_n$. Hence

$$\boldsymbol{\Delta} f(x_0, x') = \frac{f(x_0) - f(x')}{x_0 - x'} > \frac{1}{2}\varkappa_n = \frac{1}{2}\Big(\frac{5}{3}\Big)^{n-p}\varkappa_p.$$

Consequently, $f'_-(x_0) = +\infty$. On the other hand,

$$\begin{aligned}\boldsymbol{\Delta} f(x_0, x'') &= \frac{f(x'') - f(x_0)}{x'' - x_0} \\ &\leq \frac{f(a_n) + \frac{9}{8}\Delta_n - (f(a_n) + \frac{4}{3}\Delta_n)}{\delta_n} < -\Big(\frac{4}{3} - \frac{9}{8}\Big)\Big(\frac{5}{3}\Big)^{n-p} \varkappa_p.\end{aligned}$$

Hence $f'_+(x_0) = -\infty$.

Step 8°. *Let x_0 be such that for its determining sequence $(J_n)_{n=1}^\infty$, we have* $\operatorname{type}(J_n) = 2$ *for every $n \in \mathbb{N}$. Then $x_0 = \frac{3}{7}$ and $D_-f(x_0)$, $D^-f(x_0)$, $D_+f(x_0)$, $D^+f(x_0)$ are finite.*

We have $a_{n+1} = a_n + \frac{3}{8}\delta_n$ and $\delta_n = (\frac{1}{8})^n$, $n \in \mathbb{N}$. Consequently, $x_0 = \frac{3}{8}\sum_{n=0}^\infty (\frac{1}{8})^n = \frac{3}{7}$. Moreover, $\varkappa_n = (-1)^n$ and

$$f(a_{n+1}) = f(a_n) + (-1)^n \frac{5}{8}\delta_n, \quad n \in \mathbb{N}.$$

Hence, $f(x_0) = \frac{5}{8}\sum_{n=0}^\infty (-\frac{1}{8})^n = \frac{5}{9}$. Observe that $a_{n+k} = a_n + \frac{3}{8}\sum_{s=n}^{n+k-1} (\frac{1}{8})^s$, so $x_0 = a_n + \frac{3}{7}\delta_n$. Similarly, $f(x_0) = f(a_n) + (-1)^n \frac{5}{9}\delta_n$, $n \in \mathbb{N}$. Fix an $n \in \mathbb{N}$ and take an arbitrary $x \in [a_{2n}, a_{2n+1}]$. Then $(\frac{3}{7} - \frac{3}{8})\delta_{2n} \leq x_0 - x \leq \delta_{2n}$ and $f(a_{2n}) \leq f(x) \leq f(a_{2n}) + \frac{5}{6}\Delta_{2n}$. Thus $|\boldsymbol{\Delta} f(x_0, x)| \leq C = \text{const}$, where C is independent of n and x. In a similar way (EXERCISE), one gets $|\boldsymbol{\Delta} f(x_0, x)| \leq \text{const}$ for $x \in [a_{2n+1}, a_{2n+2}]$. Consequently, $D_-f(x_0)$ and $D^-f(x_0)$ are finite. An analogous argument (EXERCISE) shows that $D_+f(x_0)$ and $D^+f(x_0)$ are finite.

Step 9°. *If $x_0 \in \mathfrak{P}$, then $D_-f(x_0) = D_+f(x_0) = -\infty, D^-f(x_0) = D^+f(x_0) = +\infty$.*

Let $S = [(a, A), (b, B)]$ be an arbitrary segment, $J := [a, b]$, $L(x) = A + \varkappa(S)(x - a)$, $x \in J$. Then for every point $x_0 \in J$, there exists a point $x_0^* \in J$ such that $|x_0 - x_0^*| = \frac{1}{2}\delta(J)$ and $\boldsymbol{\Delta} L(x_0, x_0^*) = \varkappa(S)$. We say that x_0^* is *conjugate to x_0 with respect to S*.

Take an $x_0 \in \mathfrak{P}$ and let $(S_n)_{n=1}^\infty$ be a determining sequence for x_0. Let x_n^* be conjugate to x_0 with respect to S_n, $n \in \mathbb{N}$.

Take an $s \in \mathbb{N}$ and let $n = n(s) \geq s$ be as in the definition of the set $\mathfrak{P}$. We have $\varkappa_{n+1} = \frac{5}{3}\varkappa_n$, $\varkappa_{n+2} = -\frac{5}{3}\varkappa_n$, $\varkappa_{n+3} = -(\frac{5}{3})^2\varkappa_n$. Observe that $x_n^* < x_0$, $x_{n+1}^* > x_0$, $x_{n+2}^* > x_0$, $x_{n+3}^* < x_0$. Moreover, $\boldsymbol{\Delta} f(x_0, x_n^*) = \varkappa_n$, $\boldsymbol{\Delta} f(x_0, x_{n+1}^*) = \frac{5}{3}\varkappa_n$, $\boldsymbol{\Delta} f(x_0, x_{n+2}^*) = -\frac{5}{3}\varkappa_n$, $\boldsymbol{\Delta} f(x_0, x_{n+3}^*) = -(\frac{5}{3})^2\varkappa_n$. Observe that if $s \longrightarrow +\infty$, then $n(s) \longrightarrow +\infty$, $|\varkappa_{n(s)}| \longrightarrow +\infty$, $x_{n(s)}^* \longrightarrow x_0-$, $x_{n(s)+3}^* \longrightarrow x_0-$, $x_{n(s)+1}^* \longrightarrow x_0+$, and $x_{n(s)+2}^* \longrightarrow x_0+$, which directly implies the required result.

Step 10°. *For every $x_0 \in (0, 1]$, a finite left-sided derivative $f'_-(x_0)$ does not exist.*

By Step 3°, we may assume that $x_0 \notin \mathbf{N}$. Let $(S_n)_{n=1}^\infty$ be a determining sequence for x_0.

Suppose that there exists a sequence $(n_s)_{s=1}^\infty$ such that $\operatorname{type}(J_{n_s}) = 2$, $s \in \mathbb{N}$. Assume that $n_s = m + 1$ for some m. Assume that $\varkappa_m > 0$ (the case $\varkappa_m < 0$ is left to the reader as an EXERCISE). Then $f(a_m) + \frac{1}{2}\Delta_m - \frac{1}{3} \cdot \frac{1}{8}\Delta_m \leq f(x_0) \leq f(a_m) + \frac{5}{8}\Delta_m$, and hence $\boldsymbol{\Delta} f(x_0, a_m) > (\frac{1}{2} - \frac{1}{24})\varkappa_m$. Let $x_m := a_m + \frac{4}{5} \cdot \frac{3}{8}\delta_m < x_0$. Then $f(x_m) = f(a_m) + \frac{4}{5} \cdot \frac{5}{8}\Delta_m$. Hence $\boldsymbol{\Delta} f(x_0, x_m) < -(\frac{5}{6} - \frac{5}{8})\varkappa_m$. Thus a finite left-sided derivative $f'_-(x_0)$ does not exist.

A similar argument works if $\operatorname{type}(J_{n_s}) = 4$, $s \in \mathbb{N}$ (EXERCISE).

Thus we may assume that $\operatorname{type}(J_n) \in \{1, 3\}$ for $n \gg 1$, but it is excluded that $\operatorname{type}(J_n) = 1$ for $n \gg 1$. Then there exists a sequence $(n_s)_{s=1}^\infty$ such that $\operatorname{type}(J_{n_s+1}) = 3$, $s \in \mathbb{N}$. Let $x_{n_s}^*$ be conjugate to x_0 with respect to S_{n_s}. Then $x_{n_s}^* < x_0$ and $\boldsymbol{\Delta} f(x_0, x_{n_s}^*) = \varkappa_{n_s}$. Thus a finite $f'_-(x_0)$ does not exist.

Step 11°. *For every $x_0 \in [0,1)$, a finite right-sided derivative $f'_+(x_0)$ does not exist.*

Suppose that there exists a sequence $(n_s)_{s=1}^{\infty}$ such that $\text{type}(J_{n_s+1}) = 1$, $s \in \mathbb{N}$. Let $x^*_{n_s}$ be conjugate to x_0 with respect to S_{n_s}. Then $x_0 < x^*_{n_s}$ and $\boldsymbol{\Delta} f(x_0, x^*_{n_s}) = \varkappa_{n_s}$. Thus a finite $f'_+(x_0)$ does not exist.

Hence we may assume that $\text{type}(J_n) \in \{2,3,4\}$ for $n \gg 1$, but it is excluded that $\text{type}(J_n) = 4$ for $n \gg 1$. Thus, suppose that $\text{type}(J_{n+1}) = 2$ or $\text{type}(J_{n+1}) = 3$ for infinitely many n's. Fix such an n. Then $(x_0, f(x_0))$ lies in the rectangle determined by the lines $x = a_n + \frac{3}{8}\delta_n, a_n$, $x = a_n + \frac{7}{8}\delta_n$, $y = f(a_n) + \frac{11}{24}\Delta_n$, and $y = f(a_n) + \frac{4}{3}\Delta_n$. Let

$$
\begin{aligned}
P_1 &:= (b_n, f(b_n)),\\
P_2 &:= (b_n - \tfrac{1}{16}\delta_n, f(b_n) + \tfrac{1}{16}\Delta_n),\\
P_3 &:= (b_n - \tfrac{1}{64}\delta_n, f(b_n) - \tfrac{1}{64}\Delta_n).
\end{aligned}
$$

For $(\xi, \eta) \in R := [\frac{3}{8}, \frac{7}{8}] \times [\frac{11}{24}, \frac{4}{3}]$, let $P(\xi, \eta) := (b_n + \xi\delta_n, f(b_n) + \eta\Delta_n)$, and let $s_i(\xi, \eta)$ be the slope of the segment $[P(\xi, \eta), P_i]$. We have

$$
s_1(\xi, \eta) = \frac{\eta - 1}{\xi - 1}\varkappa_n, \quad s_2(\xi, \eta) = \frac{\eta - 1 - \frac{1}{16}}{\xi - 1 + \frac{1}{16}}\varkappa_n, \quad s_3(\xi, \eta) = \frac{\eta - 1 + \frac{1}{64}}{\xi - 1 + \frac{1}{64}}\varkappa_n.
$$

Put $\varphi := \frac{1}{|\varkappa_n|}(\max\{s_1, s_2, s_3\} - \min\{s_1, s_2, s_3\})$. It clear that $\varphi : R \longrightarrow \mathbb{R}_{>0}$ is continuous and independent of n. Hence there exists a point $(\xi_0, \eta_0) \in R$ such that

$$
0 < C = \varphi(\xi_0, \eta_0) = \min \varphi(R),
$$

where C is independent of n. Let $\mu, \nu \in \{1,2,3\}$ be such that $s_\mu(\xi_0, \eta_0) - s_\nu(\xi_0, \eta_0) = C|\varkappa_n|$. Put $x'_n := P_\mu(x_0, \eta_0)$, $x''_n := P_\nu(\xi_0, \eta_0)$. Then $\boldsymbol{\Delta} f(x_0, x'_n) - \boldsymbol{\Delta} f(x_0, x''_n) \geq C|\varkappa_n|$, which easily implies that a finite right-sided derivative $f'_+(x_0)$ does not exist. □

Chapter 11
Besicovitch Functions

Summary. In Chap. 7, it was shown that $\mathcal{B}(\mathbb{I})$ is of first category in $\mathcal{C}(\mathbb{I})$, i.e., most functions in $\mathcal{C}(\mathbb{I})$ have somewhere on $\mathbb{I}$ an infinite one-sided derivative. In the first part of this chapter, the construction of concrete functions belonging to $\mathcal{B}(\mathbb{I})$, resp. $\mathcal{BM}(\mathbb{I})$, is discussed. The remaining part deals with a categorial argument proving that the set $\mathcal{BM}(\mathbb{I})$ is in some sense even a large set.

11.1 Morse's Besicovitch Function

Recall first that according to a result of S. Saks (see Theorem 7.5.1), the set $\mathcal{B}(\mathbb{I})$ is of first category in $\mathcal{C}(\mathbb{I})$. Therefore, most of the functions in $\mathcal{C}(\mathbb{I})$ have somewhere on $\mathbb{I}$ an infinite one-sided derivative. Nevertheless, in 1924, A.S. Besicovitch found the first effective geometric construction of a function belonging to $\mathcal{B}(\mathbb{I})$. Later, in 1928, E.D. Pepper treated the same function, trying to clarify certain details (see [Pep28]). Nevertheless, details of their proofs remained unclear. In this section, we present a 1938 construction given by A.P. Morse (see [Mor38]) of a *Besicovitch–Morse function* $\boldsymbol{M}$ on $\mathbb{I}$, i.e., a function $\boldsymbol{M} \in \mathcal{BM}(\mathbb{I})$ (see Sect. 11.1.4).

11.1.1 Preparation

Put

$$\lambda_n := \frac{1}{2} + \frac{n}{2(|n|+3)}, \qquad n \in \mathbb{Z}.$$

M. Jarnicki, P. Pflug, *Continuous Nowhere Differentiable Functions*, Springer Monographs in Mathematics, DOI 10.1007/978-3-319-12670-8_11

Then (EXERCISE):

- $0 < \lambda_{n+1} - \lambda_n < \lambda_n^2 < \lambda_n, \quad \lambda_n + \lambda_{-n} = 1, \quad n \in \mathbb{Z}$.
  Indeed, if $n \in \mathbb{Z}$, then

$$\begin{aligned} 0 < \lambda_{n+1} - \lambda_n &= \frac{1}{2}\frac{3}{(|n|+3)(|n+1|+3)} = \Big(\frac{3}{2(|n|+3)}\Big)^2 \frac{2(|n|+3)}{3(|n+1|+3)} \\ &\le \Big(\frac{3+n+|n|}{2(|n|+3)}\Big)^2 \frac{2(|n+1|+3+1)}{3(|n+1|+3)} \\ &= \lambda_n^2 \frac{2}{3}\Big(1 + \frac{1}{|n+1|+3}\Big) \le \frac{8}{9}\lambda_n^2 < \lambda_n^2. \end{aligned}$$

- $\lambda_n \underset{n\to+\infty}{\longrightarrow} 1, \quad \lambda_n \underset{n\to-\infty}{\longrightarrow} 0.$
- $\lambda_{2n+1} < \sqrt{\lambda_{2n}}, \quad n \in \mathbb{Z}$.
  Indeed, if $n \in \mathbb{N}_0$, then

$$\begin{aligned} \lambda_{2n+1}^2 &= \Big(\frac{1}{2} + \frac{1}{2}\frac{2n+1}{2n+1+3}\Big)^2 = \frac{1}{4}\Big(\frac{4n+5}{2n+4}\Big)^2 \\ &\le \frac{1}{4}\frac{(4n+5)^2}{(2n+3)(2n+5)} < \frac{1}{2}\frac{4n+3}{2n+3} = \lambda_{2n}. \end{aligned}$$

  If $n = -m$, $m \in \mathbb{N}$, then

$$\begin{aligned} \lambda_{2n+1}^2 &= \Big(\frac{1}{2} + \frac{-2m+1}{2(|-2m+1|+3)}\Big)^2 = \frac{1}{4}\Big(\frac{3}{2m+2}\Big)^2 \\ &\le \frac{1}{4}\frac{9}{(2m+3)(2m+1)} < \frac{1}{2}\frac{3}{2m+3} = \lambda_{2n}. \end{aligned}$$

Moreover, we have the following result.

Lemma 11.1.1. *There exists a closed, nowhere dense subset $E \subset \mathbb{I}$ with*

$$\mathcal{L}(E \cap [\lambda_n, \lambda_n + h]) > 0, \quad \mathcal{L}(E \cap [\lambda_n - h, \lambda_n]) > 0, \quad n \in \mathbb{Z},\ h > 0.$$

Proof. Step 1°. Fix a sequence $(\eta_n)_{n=1}^\infty$ such that

$$1 > 2\eta_1 > 4\eta_2 > \dots > 2^k\eta_k > \dots$$

(for example, take $\eta_k := \frac{1}{2^k}\frac{ks+1}{k+1}$, where $s \in (0,1)$). Then we construct a Cantor-like set along the following standard lines:

From $\mathbb{I}$, we remove the open concentric interval $I_{1,1}$ of length $1-2\eta_1$. What remains are two closed intervals $J_{1,1}$ and $J_{1,2}$, each of length η_1. Now we remove a concentric open interval $I_{2,j}$ of length $\eta_1 - 2\eta_2$ of the intervals $J_{1,j}$, $j = 1, 2$. Then four closed intervals $J_{2,j}$, $j = 1, 2, 3, 4$, remain, each of length η_2. Now we continue this construction. At the nth step, we end up with 2^n intervals $J_{n,k}$, $k = 1, \dots, 2^n$, each of length η_n. Put $C := \bigcap_{n=1}^\infty \Big(\bigcup_{k=1}^{2^n} J_{n,k}\Big)$. Obviously, C is a closed subset of $\mathbb{I}$ without inner points. Moreover, we get $\mathcal{L}(C) = \lim_{n\to\infty} 2^n\eta_n$ (in the special case from above we have $\mathcal{L}(C) = s$). Moreover, $\mathcal{L}([0,\eta_n] \cap C) = \lim_{k\to\infty} 2^k\eta_{n+k} = \mathcal{L}(C)/2^n$. Because of the symmetry with respect to $1/2$, we have also the following identity:

$$\mathcal{L}(C \cap [1-\eta_n, 1]) = \mathcal{L}(C)/2^n.$$

Step 2°. Fix a Cantor-like set C as above with $2^k\eta_k = \frac{ks+1}{k+1}$. Then $\mathcal{L}(C) = s > 0$. So it remains to put $E := \bigcup_{n\in\mathbb{Z}}\big(\lambda_n + (\lambda_{n+1}-\lambda_n)C\big)$. □

Definition 11.1.2. Let $\theta : [0,2] \longrightarrow \mathbb{R}$ be defined as

$$\theta(x) := \begin{cases} \gamma_n + (\gamma_{n+1}-\gamma_n)\frac{\mathcal{L}(E\cap[\lambda_n,x])}{\mathcal{L}(E\cap[\lambda_n,\lambda_{n+1}])}, & \text{if } x\in[\lambda_n,\lambda_{n+1}],\ n\in\mathbb{Z} \\ 1, & \text{if } x = 1 \\ 0, & \text{if } x = 0 \\ \theta(2-x), & \text{if } 1\le x\le 2 \end{cases},$$

where E is a set satisfying the properties of Lemma 11.1.1 and

$$\gamma_n := \begin{cases} \lambda_n, & \text{if } n\in 2\mathbb{Z}+1 \\ \sqrt{\lambda_n}, & \text{if } n\in 2\mathbb{Z} \end{cases}.$$

Remark 11.1.3. We collect some simple properties of the function θ.

(a) θ is well defined and continuous on $(0,1)$.
(b) If $\lambda_{2n-1}\le x\le\lambda_{2n}$, then $\lambda_{2n-1}\le\theta(x)\le\sqrt{\lambda_{2n}}$ (use that $\lambda_{2n-1}<\sqrt{\lambda_{2n}}$); if $\lambda_{2n}\le x\le\lambda_{2n+1}$, then $\lambda_{2n+1}\le\theta(x)\le\sqrt{\lambda_{2n}}$ (use that $\lambda_{2n+1}<\sqrt{\lambda_{2n}}$), $n\in\mathbb{Z}$; in particular, θ is continuous on $\mathbb{I}$ and therefore on $[0,2]$.
(c) If $K(\theta) := \operatorname{int}\{x\in[0,2] : \theta \text{ is differentiable at } x \text{ with } \theta'(x)=0\}$, then $K(\theta)$ is an open dense subset of $(0,2)$ (use that the set E is nowhere dense).
(d) If $x\in(0,1)$, then $\frac{x}{2}\le\theta(x)\le\frac{x+3}{4}$. Indeed, if $\lambda_{2n-1}\le x\le\lambda_{2n}$, then

$$\begin{aligned}\frac{x}{2} &\le \frac{\lambda_{2n}}{2}\le\lambda_{2n-1}=\gamma_{2n-1}\le\theta(x)\le\sqrt{\lambda_{2n}}\\ &= \sqrt{1-\lambda_{-2n}}\le\sqrt{1-\frac{\lambda_{-2n+1}}{2}}=\sqrt{\Big(\frac{1+\lambda_{2n-1}}{2}\Big)\cdot 1}\\ &\le \frac{\frac{1+\lambda_{2n-1}}{2}+1}{2}\le\frac{3+x}{4},\end{aligned}$$

where in the penultimate inequality, the standard relation between the geometric and algebraic means is used.
If $\lambda_{2n}\le x\le\lambda_{2n+1}$, then $x\le\lambda_{2n+1}\le\theta(x)\le\sqrt{\lambda_{2n}}\le\sqrt{x}\le\frac{3+x}{4}$.

Lemma 11.1.4. (a) *If* $x_0\in(0,2)$, *then* $\limsup_{\xi\to x_0}|\boldsymbol{\Delta}\theta(x_0,\xi)|<+\infty$.
(b) *If* $n\in\mathbb{Z}$, *then* $\lambda_n\notin K(\theta)$. *In particular,* $1\notin K(\theta)$.
(c) *Let* $0\le\alpha<\beta\le 2$ *be such that* $(\alpha,\beta)\subset K(\theta)$. *Then* $\theta(m)\ge(\beta-\alpha)^{1/2}$, *where* $m := \frac{\alpha+\beta}{2}$.

Proof. (a) Step 1°. Let $x_0 = 1$. If $0<\xi<1$, then one finds an $n\in\mathbb{Z}$ such that $\lambda_n\le\xi\le\lambda_{n+1}$. Using Remark 11.1.3(b), it follows that:
if $n = 2k$ for some $k\in\mathbb{Z}$, then

$$|\theta(\xi)-\theta(1)| = 1-\theta(\xi)\le 1-\lambda_{2k+1} = 1-\lambda_n\le 1-\xi,$$

and if $n = 2k-1$ for some $k\in\mathbb{Z}$, then

$$|\theta(\xi)-\theta(1)| = 1-\theta(\xi)\le 1-\lambda_n=\lambda_{-n}\le 2\lambda_{-n-1} = 2(1-\lambda_{n+1})\le 2(1-\xi).$$

If $1 < \xi < 2$, then $0 < 2 - \xi < 1$, and thus

$$|\theta(\xi) - \theta(1)| = |\theta(2-\xi) - \theta(1)| \le 2(1 - (2-\xi)) = 2(\xi - 1).$$

Then both inequalities give $\limsup_{\xi \to 1} |\boldsymbol{\Delta}\theta(1, \xi)| \le 2$.

Step 2°. Let $x_0 \in (0,1)$ be such that there exists an $n \in \mathbb{Z}$ such that $\lambda_n < x_0 < \lambda_{n+1}$. Take a $\xi \in (\lambda_n, \lambda_{n+1})$, $\xi \ne x_0$. Then

$$|\theta(\xi) - \theta(x_0)| \le |\gamma_{n+1} - \gamma_n| \frac{|\xi - x_0|}{\mathcal{L}(E \cap [\lambda_n, \lambda_{n+1}])} \le C|\xi - x_0|$$

(use that $|\gamma_{n+1} - \gamma_n| \le 1$).

Now let $x_0 = \lambda_n$ for a suitable n. If $\xi \in (\lambda_n, \lambda_{n+1})$, then

$$|\theta(\xi) - \theta(x_0)| \le |\gamma_{n+1} - \gamma_n| \frac{|\xi - x_0|}{\mathcal{L}(E \cap [\lambda_n, \lambda_{n+1}])} \le C|\xi - x_0|$$

for a possibly different constant C. An analogous estimate is true if $\xi \in (\lambda_{n-1}, \lambda_n)$ with a new constant C. Hence $\limsup_{\xi \to x_0} |\boldsymbol{\Delta}\theta(x_0, \xi)| < +\infty$.

Step 3°. The case $x_0 \in (1,2)$ is left for the reader as an EXERCISE.

(b) Assume that there is an $n \in \mathbb{Z}$ with $\lambda_n \in K(\theta)$. Then there is an interval $J := (\alpha', \beta') \subset K(\theta)$ with $\lambda_n \in J$. Thus the function θ has to be identically equal to γ_n on J, contradicting the fact that $\mathcal{L}([\lambda_n, \cdot))$ is not identically 0 on $J \cap [\lambda_n, \lambda_{n+1}]$.

In particular, $1 \notin K(\theta)$.

(c) By (b), we may assume for the interval (α, β) that $(\alpha, \beta) \subset (0,1)$ and that $(\alpha, \beta) \subset (\lambda_N, \lambda_{N+1})$ for a suitable N. Recall that $\lambda_{n+1} - \lambda_n \le \lambda_n^2$, $n \in \mathbb{N}$. If N is odd, then

$$\theta(m) \ge \lambda_N > (\lambda_{N+1} - \lambda_N)^{1/2} \ge (\beta - \alpha)^{1/2}.$$

If N is even, then $\theta(m) \ge \lambda_{N+1} > \lambda_N > (\beta - \alpha)^{1/2}$.

□

For future use, put $H(\theta) := [0,2] \setminus K(\theta)$. Note that 1 and all the λ_n belong to $H(\theta)$.

11.1.2 A Class of Continuous Functions and Its Properties

Definition 11.1.5. By $\mathfrak{A}$ we denote the set of all functions $f \in \mathcal{C}((-1,1))$ satisfying the following properties:

(a) $0 \le f(x) = f(-x), \quad x \in (-1,1)$;
(b) if $K(f) := \operatorname{int}\{x \in (-1,1) : f'(x) \text{ exists and } f'(x) = 0\}$, then $K(f)$ is dense in $(-1,1)$;
(c) if $P(f) := \{x \in (-1,1) : f(x) > 0\}$, then $P(f)$ is dense in $(-1,1)$;
(d) if $(\alpha, \beta) \subset P(f)$, then there exists an $h \ge \sqrt{\frac{\beta - \alpha}{2}}$ such that $f(x) = h\theta(\frac{2x - 2\alpha}{\beta - \alpha})$, $x \in (\alpha, \beta)$.

In particular, $\widetilde{\theta} \in \mathfrak{A}$, where $\widetilde{\theta}(x) := \theta(1 + x)$, $x \in [-1,1]$.

Lemma 11.1.6. *Let $f \in \mathfrak{A}$. Then:*

(a) $K(f) \subset P(f)$;
(b) *if* $(\alpha, \beta) \subset K(f)$ *with* $-1 < \alpha < \beta < 1$, *then* $f(m) \geq (\beta - \alpha)^{1/2}$ *for* $m := \frac{\alpha+\beta}{2}$;

Proof. (a) Use the fact that $K(f)$ and $P(f)$ are dense in $(-1, 1)$.
(b) By assumption, there exist numbers α', β' with $-1 < \alpha' < \alpha < \beta < \beta' < 1$ such that $(\alpha', \beta') \subset P(f)$ (use the continuity of f). Put

$$m' := \frac{\alpha' + \beta'}{2}, \quad \alpha'' := \frac{2(\alpha - \alpha')}{\beta' - \alpha'}, \quad \beta'' := \frac{2(\beta - \alpha')}{\beta' - \alpha'}, \quad m'' := \frac{\alpha'' + \beta''}{2}.$$

By virtue of Definition 11.1.5(c) we obtain an $h \geq (\frac{\beta' - \alpha'}{2})^{1/2}$ such that

$$f(x) = h\theta\Big(\frac{2x - 2\alpha'}{\beta' - \alpha'}\Big), \quad x \in (\alpha', \beta');$$

in particular, one has $h = f(m')$.

Recall that $(\alpha, \beta) \subset K(f)$, i.e., f is identically constant on (α, β). Therefore, θ itself is identically constant on the interval $(\frac{2\alpha - 2\alpha'}{\beta' - \alpha'}, \frac{2\beta - 2\alpha'}{\beta' - \alpha'}) \ni m''$. Therefore,

$$\begin{aligned} f(m) &= f(m')\theta\Big(\frac{2m - 2\alpha'}{\beta' - \alpha'}\Big) = f(m')\theta(m'') \geq f(m')(\beta'' - \alpha'')^{1/2} \\ &\geq \Big(\frac{\beta' - \alpha'}{2}\Big)^{1/2} (\beta'' - \alpha'')^{1/2} = (\beta - \alpha)^{1/2}. \end{aligned}$$

□

Remark 11.1.7. Let $f \in \mathfrak{A}$ and $J = (\alpha, \beta) \subset K(f)$. Then $-1 < \alpha$ and $\beta < 1$. Otherwise, let, for example, $\alpha = -1$. Then $f|_J \equiv c > 0$. Simultaneously, $f(x) = h\theta(\frac{2x - 2\alpha}{\beta' - \alpha})$, $x \in J$, where $\beta' \geq \beta$ and $h > 0$. Therefore,

$$c = \lim_{x \to -1+} f(x) = h \lim_{x \to -1+} \theta\Big(\frac{2x - 2\alpha}{\beta' - \alpha}\Big) = 0;$$

a contradiction. A similar argument works for the remaining case.

11.1.3 A New Function $\overline{f}$ for Every $f \in \mathfrak{A}$

Let us first fix the following convention to simplify formulas: if $J = (\alpha, \beta)$ is any bounded open interval, then we denote by m_J its midpoint. Moreover, if $A \subset [-1, 1]$ is some open subset and if $s > 0$, then

$$A^{(s)} := \{-1, +1\} \cup \{x \in (-1, 1) : \exists_{a<b} : x \in (a, b) \subset A, \ b - a > s\}.$$

Now choose a function $f \in \mathfrak{A}$. To such a function we associate a new function, denoted by $\overline{f}$, via the following definition.

Definition 11.1.8. For $x \in (-1, 1)$, put

$$\overline{f}(x) := \begin{cases} 0, & \text{if } x \in H(f) := (-1, 1) \setminus K(f) \\ h\theta(\frac{2x - 2\alpha}{\beta - \alpha}), & \text{if } x \in J = (\alpha, \beta) \subset K(f), \ J \text{ is maximal} \end{cases},$$

where

$$h := \min\Big\{\frac{f(m_J)}{\sqrt{2}}, \Big(\inf K(f)^{(\beta-\alpha)} \cap [m_J, 1] - \sup K(f)^{(\beta-\alpha)} \cap [-1, m_J]\Big)^{1/2}\Big\}.$$

In the following, we will discuss various properties of this new function.

Lemma 11.1.9. *Let $f \in \mathfrak{A}$, and let $J = (\alpha, \beta) \subset K(f)$ be maximal. Then:*

(a) $\overline{f}(m_J) \geq \sqrt{\frac{\beta-\alpha}{2}}$;
(b) *if $x \in J$, then*

$$0 < \overline{f}(-x) = \overline{f}(x) = \overline{f}(m_J)\theta\Big(\frac{2x - 2\alpha}{\beta - \alpha}\Big) \leq \overline{f}(m_J) \leq \frac{f(x)}{\sqrt{2}}.$$

Proof. Put $\alpha_0 := \sup K(f)^{(\beta-\alpha)} \cap [-1, m_J]$ and $\beta_0 := \inf K(f)^{(\beta-\alpha)} \cap [m_J, 1]$.
(a) Then

$$\overline{f}(m_J) = h\theta\Big(\frac{2m_J - 2\alpha}{\beta - \alpha}\Big) = h\theta(1) = h$$
$$= \min\Big\{(\beta_0 - \alpha_0)^{1/2}, \frac{f(m_J)}{\sqrt{2}}\Big\} \leq \frac{f(x)}{\sqrt{2}}$$

for all $x \in (\alpha, \beta)$ (recall that f is identically constant on (α, β)). Moreover, using Lemma 11.1.6(b),

$$\overline{f}(m_J) = h = \min\Big\{(\beta_0 - \alpha_0)^{1/2}, \frac{f(m_J)}{\sqrt{2}}\Big\} \geq \sqrt{\frac{\beta - \alpha}{2}} > 0.$$

In particular, $\overline{f}(x) = \overline{f}(m_J)\theta(\frac{2x-2\alpha}{\beta-\alpha}) > 0$ for all $x \in J$.

(b) It remains to verify that $\overline{f}(x) = \overline{f}(-x)$, $x \in J$. Fix an $x \in J$. Then $-J \subset K(f)$ is a maximal subinterval with $-x \in -J$. Therefore,

$$\overline{f}(-x) = \widetilde{h}\theta\Big(\frac{2x + 2\beta}{-\alpha - (-\beta)}\Big),$$

where $\widetilde{h} := \min\{f(m_J)/\sqrt{2}, (\widetilde{\beta}_0 - \widetilde{\alpha}_0)^{1/2}\}$ with (using the symmetry of f)

$$\widetilde{\beta}_0 := \inf K(f)^{(\beta-\alpha)} \cap [-m_J, 1] = -\sup K(f)^{(\beta-\alpha)} \cap [-1, m_j] = -\alpha_0,$$
$$\widetilde{\alpha}_0 := \sup K(f)^{(\beta-\alpha)} \cap [-1, -m_J] = -\inf K(f)^{(\beta-\alpha)} \cap [m_J, 1] = -\beta_0.$$

Finally, applying $\theta(1 + \xi) = \theta(1 - \xi)$, $\xi \in (0, 1)$, leads to

$$\overline{f}(-x) = \overline{f}(m_J)\theta\Big(\frac{-2x + 2\beta}{\beta - \alpha}\Big) = \overline{f}(m_J)\theta\Big(\frac{2x - 2\alpha}{\beta - \alpha}\Big) = \overline{f}(x). \qquad \square$$

Corollary 11.1.10. *Let $f \in \mathfrak{A}$. Then:*

(a) $0 \le \overline{f}(x) = \overline{f}(-x) \le f(x)/\sqrt{2}$, $x \in (-1,1)$;
(b) $K(f) = P(\overline{f})$ *and* $H(f) = Z(\overline{f}) := \{x \in (-1,1) : \overline{f}(x) = 0\}$.

Proof. (a) Using the former lemma it suffices to verify the claim for all $x \in H(f)$. Because of the symmetry, we have $x \in H(f)$ if and only if $-x \in H(f)$. Thus, $\overline{f}(x) = \overline{f}(-x) = 0$.
(b) By virtue of Lemma 11.1.9(b), we have $K(f) \subset P(\overline{f})$. Directly from the definition it follows that $H(f) \subset Z(\overline{f})$. To get the equalities, use the fact that $P(\overline{f}) \cup Z(\overline{f}) = (-1,1) = K(f) \cup H(f)$ and that in each case, both sets are disjoint.

□

Theorem 11.1.11. *Let $f \in \mathfrak{A}$. Then:*

(a) *if the interval $J = (\alpha, \beta) \subset K(f)$ is maximal, then $\alpha, m_J \in H(\overline{f})$ and*

$$\overline{f}(\alpha) = \lim_{\xi \to \alpha+} \overline{f}(\xi) = 0 \le \liminf_{H(\overline{f}) \ni \xi \to \alpha+} \boldsymbol{\Delta}\overline{f}(\alpha, \xi) < \limsup_{H(\overline{f}) \ni \xi \to \alpha+} \boldsymbol{\Delta}\overline{f}(\alpha, \xi) = +\infty;$$

(b) *if $x \in H(f) \cap P(f)$, then*

$$\overline{f}(x) = \lim_{\xi \to x+} \overline{f}(\xi) = 0 \le \liminf_{H(\overline{f}) \ni \xi \to x+} \boldsymbol{\Delta}\overline{f}(x, \xi) < \limsup_{H(\overline{f}) \ni \xi \to x+} \boldsymbol{\Delta}\overline{f}(x, \xi) = +\infty.$$

Proof. (a) We know that by definition, $\overline{f}(x) = \overline{f}(m_J)\theta(\frac{2x-2\alpha}{\beta-\alpha})$, $x \in J$, and that $\lambda_n \in H(\theta)$ (see Lemma 11.1.4(b)). Therefore, $\alpha \in H(\overline{f})$. The same argument, using the fact that $1 \in H(\theta)$, leads to $m_j \in H(\overline{f})$.
Knowing that $\overline{f}(\alpha) = 0$ leads to

$$\lim_{\xi \to \alpha+} \overline{f}(m_J)\theta\Big(\frac{2\xi - 2\alpha}{\beta - \alpha}\Big) = \overline{f}(m_J)\theta(0) = 0 = \overline{f}(\alpha).$$

Moreover, we know that $\overline{f}(\xi) - \overline{f}(\alpha) \ge 0$, which implies the first inequality.
Finally, put $\xi_n := \alpha + \lambda_n(\beta - \alpha)/2$, $n \in \mathbb{Z}$. Recall that $\lambda_n \in H(\theta)$. Thus $\xi_n \in H(\overline{f})$. Then

$$\frac{\overline{f}(\xi_n) - \overline{f}(\alpha)}{\xi_n - \alpha} = \frac{h\theta(\lambda_n)}{\lambda_n}\frac{2}{\beta - \alpha} = \begin{cases} \frac{2h}{\beta-\alpha}, & \text{if } n \in 2\mathbb{Z}+1 \\ \frac{\sqrt{\lambda_n}}{\lambda_n}\frac{2h}{\beta-\alpha}, & \text{if } n \in 2\mathbb{Z} \end{cases}.$$

Letting $n \longrightarrow -\infty$ implies the final claim in (a).

(b) Applying (a), we may assume, without loss of generality, that $x \in P(f) \cap H(f) \cap H(f) \cap (x, 1]$.

Step 1°. Put

$$M(r) := \sup_{t \in (-1,1)} \Big(\inf K(f)^{(r)} \cap [t,1] - \sup K(f)^{(r)} \cap [-1,t]\Big).$$

Recall that $K(f)$ is open. Therefore, $K(f)^{(r)}$ is an increasing sequence of sets with $K(f)^{(r)} \underset{r \to 0+}{\nearrow} K(f)$. Moreover, $M(r) \underset{r \to 0+}{\longrightarrow} 0$. In fact, let ε be a given positive number.

Then choose points $-1 = a_0 < a_1 < \cdots < a_N = 1$ such that $a_j - a_{j-1} < \varepsilon/2$, $j = 1, \ldots, N$. Since all these intervals intersect $K(f)$, we obtain a positive δ such that $K(f)^{(\delta)} \cap [a_{j-1}, a_j] \neq \varnothing$, $j = 1, \ldots, N$. Hence, $M(r) < \varepsilon$ for all positive $r < \delta$.

Let $(\xi_n)_{n=1}^{\infty}$ be a sequence in $K(f)$ with $\xi_n \searrow x$. Choose maximal intervals $J_n = (\alpha_n, \beta_n) \subset K(f)$ containing ξ_n. Then

$$0 = \overline{f}(x) \leq \overline{f}(\xi_n) \leq \overline{f}(m_{J_n}) \leq \big(M(\beta_n - \alpha_n)\big)^{1/2} \underset{n\to+\infty}{\longrightarrow} 0,$$

where the first inequality follows from Corollary 11.1.10 and the last one from Definition 11.1.8.

Step 2°. Since $x \in P(f)$, i.e., $f(x) > 0$, there exists a positive ε such that $f \geq f(x)/2$ on $(x - \varepsilon, x + \varepsilon) \subset (-1, 1)$. Then we find an $r_0 > 0$ such that $\sqrt{2M(r)} < f(x)/2$, $r \in (0, r_0)$. Finally, recall that x is an accumulation point of $H(f)$ from the right. Therefore, we can choose a sequence $(x_j)_{j=1}^{\infty} \subset H(f) \cap (x, x + \varepsilon)$ such that $x_j \underset{j\to\infty}{\searrow} x$.

Let $(\alpha_1, \beta_1) \subset K(f) \cap (x_2, x_1)$ be a maximal subinterval with length δ_1. Then there are only a finite number of maximal intervals of $K(f) \cap (x, \beta_1)$ with length greater or equal to δ_1. Denote the one of these intervals with maximal length d_1 by (a_1, b_1). Then $[x, m_1] \cap K(f)^{(b_1 - a_1)} = \varnothing$, where $m_1 := (a_1 + b_1)/2$. Fix a j_1 with $x_{j_1} < a_1$ and $x_{j_1} - x_{j_1+1} < d_1/2$ and then take a maximal interval $(\alpha_2, \beta_2) \subset K(f) \cap (x_{j_1+1}, x_{j_1})$. Repeating the above construction leads to an interval (a_2, b_2) with $b_2 \leq \beta_2$ of maximal length d_2 with respect to $\beta_2 - \alpha_2$. Thus, $[x, m_2] \cap K(f)^{(b_2 - a_2)} = \varnothing$, where $m_2 := (a_2 + b_2)/2$.

Continuing the construction leads to a sequence of maximal intervals $(a_n, b_n) \subset K(f)$ and midpoints $m_n := (a_n + b_n)/2$ satisfying the following properties:

- $m_n \longrightarrow x+$;
- $[x, m_n] \cap K(f)^{(b_n - a_n)} = \varnothing$;
- $2M(b_n - a_n) \leq f(m_n)$.

Therefore, $\overline{f}(m_n) \leq \big(M(\beta_n - \alpha_n)\big)^{1/2} < f(m_n)/\sqrt{2}$, $n \in \mathbb{N}$. Thus Definition 11.1.8 implies that

$$\begin{aligned}\overline{f}(m_n) &= \Big(\inf K(f)^{(\beta_n - \alpha_n)} \cap [m_n, 1] - \sup K(f)^{(\beta_n - \alpha_n)} \cap [-1, m_n]\Big)^{1/2} \\ &\geq (m_n - \sup K(f)^{(\beta_n - \alpha_n)} \cap [-1, x])^{1/2} \geq (m_n - x)^{1/2},\end{aligned}$$

which gives

$$\frac{\overline{f}(m_n) - \overline{f}(x)}{m_n - x} \geq \frac{(m_n - x)^{1/2}}{m_n - x} \underset{n\to+\infty}{\longrightarrow} +\infty,$$

while $0 = \frac{\overline{f}(\alpha_n) - \overline{f}(x)}{\alpha_n - x} \underset{n\to+\infty}{\longrightarrow} 0$. Recalling that $\alpha_n, m_n \in H(\overline{f})$ completes the proof of the last two inequalities in (b). □

Applying all the previous results finally gives the following theorem.

Theorem 11.1.12. *If $f \in \mathfrak{A}$, then $\overline{f} \in \mathfrak{A}$.*

Proof. Take an $f \in \mathfrak{A}$.

Step 1°. So far, we know that $\overline{f}$ is right continuous at every point from $H(f) \cap P(f)$ (see Theorem 11.1.11(b)), at every point of $H(f) \cap Z(f)$ (use Corollary 11.1.10(a) and the continuity of f), and at all points of $K(f)$ (use Lemma 11.1.9 and the continuity of θ). The symmetry of $\overline{f}$ is a consequence of Corollary 11.1.10(a), and it gives the continuity of $\overline{f}$. Thus $\overline{f}$ fulfills the condition in Definition 11.1.5(a).

Step 2°. By definition, we know that $K(\theta)$ is dense in $(0,2)$. Therefore, using Lemma 11.1.9(b), we have that $K(\overline{f})$ is dense in $K(f)$, which itself is dense in $(-1,1)$. Hence, $K(\overline{f})$ is dense in $(-1,1)$.

Moreover, $P(\overline{f}) = K(f)$ (see Corollary 11.1.10(b)) and therefore, $P(\overline{f})$ is dense in $(-1,1)$.

Step 3°. To see that $\overline{f}$ fulfills also the last condition in Definition 11.1.5, apply Corollary 11.1.10(b) and Lemma 11.1.9(a). □

11.1.4 A Besicovitch–Morse Function

Let $F_0(x) := \theta(x+1)$, $x \in (-1,1)$, and if $n \in \mathbb{N}_0$, then $F_{n+1} := \overline{F_n}$. Recall that $F_0 \in \mathfrak{A}$ and therefore, using Theorem 11.1.12, $F_n \in \mathfrak{A}$, $n \in \mathbb{N}$. In particular, the functions F_n are symmetric, nonnegative, and continuous on $(-1,1)$. Moreover, Corollary 11.1.10(a) implies that

$$F_{n+1}(x) \le F_n(x)/\sqrt{2} \le \cdots \le F_0(x)/(\sqrt{2})^{n+1} \le 1/(\sqrt{2})^{n+1}, \quad n \in \mathbb{N}.$$

Hence, the series

$$\boldsymbol{M}(x) := \sum_{\nu=0}^{\infty} (-1)^{\nu} F_{\nu}(x), \quad x \in (-1,1),$$

is uniformly convergent, i.e., $\boldsymbol{M}$ is a *symmetric continuous function* on the interval $(-1,1)$. We will see that $\boldsymbol{M}$ is an example of a Besicovitch–Morse function.

Theorem 11.1.13. *The function $\boldsymbol{M}$ is symmetric and continuous on $(-1,1)$, and it satisfies*

$$\liminf_{\xi\to x+} |\boldsymbol{\Delta M}(x,\xi)| < \limsup_{\xi\to x+} |\boldsymbol{\Delta M}(x,\xi)| = +\infty, \quad x \in (-1,1).$$

If $x \in \bigcap_{n=0}^{\infty} P(F_n)$, then

$$\liminf_{\xi\to x+} |\boldsymbol{\Delta M}(x,\xi)| = 0.$$

Remark 11.1.14. (a) Since $\boldsymbol{M}$ is symmetric, the corresponding statements for the left-sided limits are also true.

(b) Recall that the sets $P(F_n)$ are open dense subsets of the interval $(-1,1)$. Hence their intersection is a residual subset of $(-1,1)$.

(c) It remains to observe that if $f(x) := \boldsymbol{M}(-\frac{1}{2}+x)$, $x \in \mathbb{I}$, then $f \in \boldsymbol{\mathfrak{BM}}(\mathbb{I})$.

Proof of Theorem 11.1.13. Step 1°. Let $x_0 \in \bigcup_{n=0}^{\infty} Z(F_n)$. Because of Corollary 11.1.10(b), we know that $Z(F_k) \subset Z(F_{k+1})$, $k \in \mathbb{N}_0$. Therefore, there exists an $N \in \mathbb{N}$ such that $x_0 \in Z(F_N) \setminus Z(F_{N-1})$. Thus, $x_0 \in P(F_{N-1}) \subset P(F_{N-2}) \subset \cdots \subset P(F_0)$. Now applying the property (c) in Definition 11.1.5 and Lemma 11.1.4(a), it follows that

$$\limsup_{\xi\to x_0+} |\boldsymbol{\Delta} F_k(x_0,\xi)| < +\infty, \quad k = 0,\ldots,N-1.$$

Hence

$$\limsup_{\xi\to x_0+} |\boldsymbol{\Delta} S_{N-1}(x_0,\xi)| < +\infty,$$

where $S_M := \sum_{\mu=0}^{M} (-1)^{\mu} F_{\mu}$ and $R_M := \boldsymbol{M} - S_M$, $M \in \mathbb{N}_0$.

Moreover, if $\xi \in Z(F_{N+1})$, then $\boldsymbol{\Delta} R_N(x_0,\xi) = 0$. Applying that $H(F_k) = Z(F_{k+1})$, $k \in \mathbb{N}_0$ (see Corollary 11.1.10(b)), and Theorem 11.1.11(b), we obtain

$$\begin{aligned}
0 \le \liminf_{Z(F_{N+1})\ni\xi\to x_0+} \boldsymbol{\Delta} F_N(x_0,\xi) &= \liminf_{H(F_N)\ni\xi\to x_0+} \boldsymbol{\Delta} F_N(x_0,\xi) \\
< \limsup_{H(F_N)\ni\xi\to x_0+} \boldsymbol{\Delta} F_N(x_0,\xi) &= \limsup_{Z(F_{N+1})\ni\xi\to x_0+} \boldsymbol{\Delta} F_N(x_0,\xi) = +\infty.
\end{aligned}$$

Thus,

$$\begin{aligned}
&\liminf_{\xi\to x_0+} |\boldsymbol{\Delta M}(x_0,\xi)| \le \liminf_{Z(F_{N+1})\ni\xi\to x_0+} |\boldsymbol{\Delta M}(x_0,\xi)| \\
&= \liminf_{Z(F_{N+1})\ni\xi\to x_0+} |\boldsymbol{\Delta} S_{N-1}(x_0,\xi) + \boldsymbol{\Delta} F_N(x_0,\xi)| \\
&\le \limsup_{Z(F_{N+1})\ni\xi\to x_0+} |\boldsymbol{\Delta} S_{N-1}(x_0,\xi)| + \liminf_{Z(F_{N+1})\ni\xi\to x_0+} |\boldsymbol{\Delta} F_N(x_0,\xi)| < +\infty
\end{aligned}$$

and

$$\begin{aligned}
&\limsup_{\xi\to x_0+} |\boldsymbol{\Delta M}(x_0,\xi)| \ge \liminf_{Z(F_{N+1})\ni\xi\to x_0+} |\boldsymbol{\Delta} S_{N-1}(x_0,\xi) + \boldsymbol{\Delta} F_N(x_0,\xi)| \\
&\ge \limsup_{Z(F_{N+1})\ni\xi\to x_0+} |\boldsymbol{\Delta} F_N(x_0,\xi) + \boldsymbol{\Delta} F_N(x_0,\xi)| - \liminf_{Z(F_{N+1})\ni\xi\to x_0+} |\boldsymbol{\Delta} S_{N-1}(x_0,\xi)| \\
&= +\infty.
\end{aligned}$$

Summarizing, we have proved that if $x_0 \in \bigcup_{n=0}^{\infty} Z(F_n)$, then

$$\liminf_{\xi\to x_0+} |\boldsymbol{\Delta M}(x_0,\xi)| < \limsup_{\xi\to x_0+} |\boldsymbol{\Delta M}(x_0,\xi)| = +\infty.$$

Step 2°. Let $x_0 \in \bigcap_{n=0}^{\infty} P(F_n)$. Fix an $n \in \mathbb{N}$ and choose a maximal interval $J_n = (\alpha_n, \beta_n) \subset P(F_n)$ with $x_0 \in J_n$. Then $\beta_n \in Z(F_n)$, and therefore $F_k(\beta_n) = 0$ for all $k \in \mathbb{N}_n$. As before, let m_n denote the midpoint of J_n.

Then by virtue of Corollary 11.1.10(a), we have $J_k \subset P(F_k) \subset P(F_{k+1}) = K(F_k)$, $0 \le k \le n-1$. Thus F_k is constant on J_n, and therefore, $F_k(m_n) = F_k(x_0) = F(\beta_n)$, $k = 0, \dots, n-1$. Hence, $S_{n-1}(\beta_n) = S_{n-1}(x_0)$.

If n is an even number, then

$$\begin{aligned}
&\boldsymbol{M}(\beta_n) - \boldsymbol{M}(x_0) \\
&\quad = S_{n-1}(\beta_n) - S_{n-1}(x_0) + F_n(\beta_n) - F_n(x_0) + R_n(\beta_n) - R_n(x_0) \\
&\quad = -F_n(x_0) - R_n(x_0) \le -F_n(x_0) - \sum_{k>n} (-1)^k F_k(x_0) \\
&\quad \le \Big(-1 + \frac{1}{\sqrt{2}} \Big) \sum_{k=0}^{\infty} F_{n+2k}(x_0) \le 0.
\end{aligned}$$

When n is odd, we get

$$\begin{aligned}
&\boldsymbol{M}(\beta_n) - \boldsymbol{M}(x_0) \\
&\quad = S_{n-1}(\beta_n) - S_{n-1}(x_0) - F_n(\beta_n) + F_n(x_0) + R_n(\beta_n) - R_n(x_0)
\end{aligned}$$

$$= F_n(x_0) - R_n(x_0) \geq F_n(x_0) - \sum_{k>n} (-1)^k F_k(x_0)$$

$$\geq \Big(1 - \frac{1}{\sqrt{2}}\Big) \sum_{k=0}^{\infty} F_{n+2k}(x_0) \geq 0.$$

Since the J_n's are chosen to be maximal, we have $J_{n+1} \subset J_n$. We saw just above that $\boldsymbol{M} - \boldsymbol{M}(x_0)$ has values of different signs at the points β_{n+1} and β_n. Hence, there are zeros $\xi_n \subset [\beta_{n+1}, \beta_n]$ of this function, i.e., $\boldsymbol{M}(\xi_n) = 0$. Moreover,

$$0 < \xi_n - x_0 \leq \beta_n - \xi_n \leq \eta_n - \alpha_n \leq 2(F_n(m_n))^2 \leq 2/2^n \underset{n\to\infty}{\longrightarrow} 0; \tag{11.1.1}$$

in particular, the ξ_n converge from the right to x_0. Hence we obtain the following inequality:

$$\liminf_{\xi \to x_0+} |\boldsymbol{\Delta M}(x_0, \xi)| \leq \liminf_{n\to\infty} |\boldsymbol{\Delta M}(x_0, \xi_n)| = 0.$$

Summarizing, we have proved that if $x_0 \in \bigcap_{n=0}^{\infty} P(F_n)$, then

$$\liminf_{\xi \to x_0+} |\boldsymbol{\Delta M}(x_0, \xi)| = 0.$$

Step 3°. Let $x_0 \in \bigcap_{\nu=1}^{\infty} P(F_\nu)$. Moreover, for $n \in \mathbb{N}$ we choose the interval $J_n = (\alpha_n, \beta_n) \subset P(F_n)$ as in Step 2°. Recall that with the same reasoning as in Step 2°, we have $F_k(m_n) - F_k(x_0) = 0$, $k = 0, \dots, n-1$. Thus, $S_{n-1}(m_n) - S_{n-1}(x_0) = 0$. Moreover, by virtue of Corollary 11.1.10(b) and Theorem 11.1.11(a), we know that $m_n \in H(F_k) = Z(F_{k+1})$. Hence, $F_k(m_n) = 0$, $k > n$, or $R_n(m_n) = 0$.

From now on, we assume n to be even.

First assume that $\alpha_n < x_0 \leq m_n$. Then by the remarks above, we conclude that

$$\begin{aligned}|\boldsymbol{M}(m_n) - \boldsymbol{M}(x_0)| &= |F_n(m_n) - F_n(x_0) + R_n(m_n) - R_n(x_0)| \\ &\geq F_n(m_n) - F_n(x_0) - R_n(x_0) =: A.\end{aligned}$$

Note that R_n is an alternating series, implying that $0 \geq -F_{n+1}(x_0) \geq R_n(x_0)$. Hence, $A \geq F_n(m_n) - \boldsymbol{M}(x_0)$. Then applying property (c) of Definition 11.1.5 leads to $F_n(x_0) = F_n(m_n)\theta(\frac{2x_0 - 2\alpha_n}{\beta_n - \alpha_n})$. Therefore,

$$A \geq F_n(m_n) - F_n(m_n)\theta\Big(\frac{2x_0 - 2\alpha_n}{\beta_n - \alpha_n}\Big) \geq F(m_n)\Big(1 - \frac{1}{4}\Big(\frac{2x_0 - 2\alpha_n}{\beta_n - \alpha_n} + 3\Big)\Big) =: B,$$

where the last inequality is a consequence of the inequality $\theta(x) \leq (x+3)/4$, $0 < x < 1$ (see Remark 11.1.3(d)).

A little calculation, Lemma 11.1.9(a), and the estimate (11.1.1) from the beginning of Step 2° lead to

$$\begin{aligned}B &\geq F_n(m_n)\frac{\beta_n + \alpha_n - 2x_0}{4(\beta_n - \alpha_n)} = F_n(m_n)\frac{m_n - x_0}{2\beta_n - 2\alpha_n} \\ &\geq \sqrt{\frac{\beta_n - \alpha_n}{2}}\,\frac{m_n - x_0}{2\beta_n - 2\alpha_n} = \frac{m_n - x_0}{(\sqrt{2})^3\sqrt{\beta_n - \alpha_n}} \geq 2^{(n-4)/2}(m_n - x_0).\end{aligned}$$

Hence, if there is a strongly increasing subsequence $(n_j)_{j=1}^{\infty} \subset 2\mathbb{N}$ such that $x_0 \leq m_{n_j}$, then m_{n_j} tends from the right to x_0, and therefore,

$$\limsup_{\xi \to x_0} |\boldsymbol{\Delta M}(x_0, \xi)| \geq \limsup_{j \to \infty} |\boldsymbol{\Delta M}(x_0, m_{n_j})| = +\infty.$$

Now we assume that $m_n \leq x_0 < \beta_n$. We will proceed in a way similar to that in the previous case. Recall that (α_n, β_n) is a maximal subinterval of $P(F_n)$, i.e., $F_n(\beta_n) = 0$. Thus, $F_k(\beta_n) = 0$ for all $k \geq n$. Moreover, according to Step 2° we have $S_{n-1}(\beta_n) = S_{n-1}(x_0)$. Therefore, using also (11.1.1) leads to

$$\begin{aligned} |\boldsymbol{M}(\beta_n) - \boldsymbol{M}(x_0)| &= |R_n(x_0) - F_n(x_0)| \geq F_n(x_0) - |R_n(x_0)| \\ &\geq F_n(x_0)(1 - 1/\sqrt{2}) \geq \frac{F_n(x_0)}{4} \\ &= \frac{1}{4} F_n(m_n) \theta\Big(\frac{2x_0 - 2\alpha_n}{\beta_n - \alpha_n}\Big) = \frac{1}{4} F_n(m_n) \theta\Big(\frac{2\beta_n - 2x_0}{\beta_n - \alpha_n}\Big) =: A, \end{aligned}$$

where the second equality is a consequence of Definition 11.1.5(d), and the last one follows from the fact that $\theta(x) = \theta(2 - x)$ for $x \in (1, 2)$.

To continue, recall Remark 11.1.3(d) and (11.1.1). Therefore,

$$A \geq \frac{1}{4} F_n(m_n) \frac{\beta_n - x_0}{\beta_n - \alpha_n} \geq \frac{1}{4\sqrt{2}} \frac{\beta_n - x_0}{\sqrt{\beta_n - \alpha_n}} \geq 2^{(n-6)/2}(\beta_n - x_0),$$

where the last inequality follows again from (11.1.1).

In the case that there is only a finite number of even n's with $x_0 \leq m_n$, there is a sequence $(n_j)_{j=1}^{\infty} \subset 2\mathbb{N}$ with $n_j \longrightarrow +\infty$ such that for all j, we have $m_{n_j} \leq x_0 \leq \beta_{n_j}$. Then β_{n_j} converges from the right to x_0. Therefore, as before, we end up with

$$\limsup_{\xi \to x_0+} \boldsymbol{\Delta M}(x_0, \xi) \geq \limsup_{j \to +\infty} \boldsymbol{\Delta M}(x_0, \beta_{n_j}) = +\infty.$$

Summarizing, we have proved that if $x_0 \in \bigcap_{n=0}^{\infty} P(F_n)$, then

$$\limsup_{\xi \to x_0+} |\boldsymbol{\Delta M}(x_0, \xi)| = +\infty. \qquad \square$$

11.2 Singh's Besicovitch Function

Besides the function presented in [Bes24] and the one discussed in the previous section (see [Mor38]), there are two other examples, found by A.N. Singh (see [Sin41, Sin43]). The main aim of this section is to discuss one of the latter examples.

11.2.1 A Representation of Numbers

Fix a sequence $\boldsymbol{k} := (k_j)_{j \in \mathbb{N}} \subset 2\mathbb{N} + 1$ such that $\sum_{j=1}^{\infty} \frac{1}{k_j} < +\infty$. Observe that $k_j \geq 5$ for sufficiently large values of j. Moreover, let $\ell_n := \frac{k_n - 1}{2}$ and $\mathfrak{p}_n := k_1 \cdots k_n$, $n \in \mathbb{N}$. Moreover, put $I_s := \{0, 1, \dots, k_s - 1\} \setminus \{\ell_s\}$, $s \in \mathbb{N}$. We define the *generalized Cantor set with respect to* $\boldsymbol{k}$ as

$$\mathfrak{C} := \Big\{ \sum_{j=1}^{\infty} \frac{c_j}{\mathfrak{p}_j} : (c_j)_{j=1}^{\infty} \in M_{\infty} \Big\},$$

where $M_\infty := \{(c_j)_{j=1}^\infty : c_j \in I_j,\ j \in \mathbb{N}\}$. Note that

$$\mathbb{I} \setminus \mathfrak{C} = \Big(\frac{\ell_1}{\mathfrak{p}_1}, \frac{\ell_1+1}{\mathfrak{p}_1}\Big) \cup \bigcup_{j=2}^{\infty} \bigcup_{c \in M_{j-1}} \Big(\sum_{s=1}^{j-1} \frac{c_s}{\mathfrak{p}_s} + \frac{\ell_j}{\mathfrak{p}_j}, \sum_{s=1}^{j-1} \frac{c_s}{\mathfrak{p}_s} + \frac{\ell_j+1}{\mathfrak{p}_j}\Big),$$

where $M_j := I_1 \times \cdots \times I_j$, $j \in \mathbb{N}$, and $M_0 := \varnothing$. Note that $\mathfrak{C}$ covers the whole interval $\mathbb{I}$ except the "holes" given in the previous formula. In particular, its Lebesgue measure is given by

$$\mathcal{L}(\mathfrak{C}) = 1 - \frac{1}{\mathfrak{p}_1} - \sum_{j=2}^{\infty} \frac{1}{\mathfrak{p}_j} \prod_{s=1}^{j-1} (k_s - 1) = \prod_{j=1}^{\infty} \Big(1 - \frac{1}{k_j}\Big) \in (0,1).$$

Remark 11.2.1. Recall that $\sum_{j=1}^\infty \frac{k_j - 1}{\mathfrak{p}_j} = 1$ (see Proposition A.1.1(c)). Therefore, $0, 1 \in \mathfrak{C}$.

Now take an arbitrary $y \in \mathbb{I} \setminus \mathfrak{C}$. There is a uniquely determined hole

$$\Big(\sum_{s=1}^{j-1} \frac{c_s}{\mathfrak{p}_s} + \frac{\ell_j}{\mathfrak{p}_j}, \sum_{s=1}^{j-1} \frac{c_s}{\mathfrak{p}_s} + \frac{\ell_j+1}{\mathfrak{p}_j}\Big),$$

which contains y. Recall that the empty sum is equal to zero. Therefore, $y = \sum_{s=1}^{j-1} \frac{c_s}{\mathfrak{p}_s} + \frac{\ell_j}{\mathfrak{p}_j} + y_1$, where $0 < y_1 < \frac{1}{\mathfrak{p}_j}$. Note that j and the c_s's are uniquely determined by y.

After this remark, we begin to construct a representation for an arbitrary $x \in (0,1)$: x may be written as $x = \frac{1}{2}(c_{1,0} + y_1)$ with uniquely determined $c_{1,0} \in \{0,1\}$ and $y_1 \in [0,1)$, where $y_1 > 0$ if $c_{1,0} = 0$. Then either y_1 belongs to $\mathfrak{C}$ or $y_1 \notin \mathfrak{C}$. In the first case, we have $x = \frac{1}{2}(c_{1,0} + \xi_1)$ with $\xi_1 \in \mathfrak{C} \cap [0,1)$ and ($\xi_1 > 0$ if $c_{1,0} = 0$). In the second case, we write

$$y_1 = \sum_{j=1}^{n_1 - 1} \frac{c_{1,j}}{\mathfrak{p}_j} + \frac{\ell_{n_1}}{\mathfrak{p}_{n_1}} + x_2$$

with $n_1 \in \mathbb{N}$, $c_{1,j} \in I_j$, and $0 < x_2 < \frac{1}{\mathfrak{p}_{n_1}}$ (see Remark 11.2.1).

In the second case, we continue as follows. First, we rewrite $\mathfrak{p}_{n_1} x_2$ as above as $\mathfrak{p}_{n_1} x_2 = \frac{1}{2}(c_{2,0} + y_2)$ with $c_{2,0} \in \{0,1\}$ and $y_2 \in [0,1)$ and ($y_2 > 0$ if $c_{2,0} = 0$). Again there are two cases: either $y_2 \in \mathfrak{C}$ (and then put $\xi_2 := y_2$) or $y_2 \notin \mathfrak{C}$, i.e.,

$$y_2 = \sum_{j=1}^{n_2 - 1} \frac{c_{2,j}}{\mathfrak{p}_j} + \frac{\ell_{n_2}}{\mathfrak{p}_{n_2}} + x_3$$

with $n_2 \in \mathbb{N}$, $c_{2,j} \in I_j$, and $0 < x_3 < \frac{1}{\mathfrak{p}_{n_2}}$ (see Remark 11.2.1). Hence we obtain the following two possible situations:

$$x = \frac{1}{2}\Big(c_{1,0} + \sum_{j=1}^{n_1 - 1} \frac{c_{1,j}}{\mathfrak{p}_j} + \frac{\ell_{n_1}}{\mathfrak{p}_{n_1}}\Big) + \frac{1}{2\mathfrak{p}_{n_1}} \frac{1}{2}\Big(c_{2,0} + \xi_2\Big)$$

$$\text{with } \xi_2 \in \mathfrak{C} \cap [0,1),\ (\xi_2 > 0 \text{ if } c_{2,0} = 0),$$

$$x = \frac{1}{2}\Big(c_{1,0} + \sum_{j=1}^{n_1-1} \frac{c_{1,j}}{\mathfrak{p}_j} + \frac{\ell_{n_1}}{\mathfrak{p}_{n_1}}\Big) + \frac{1}{2\mathfrak{p}_{n_1}} \frac{1}{2}\Big(c_{2,0} + \sum_{j=1}^{n_2-1} \frac{c_{2,j}}{\mathfrak{p}_j} + \frac{\ell_{n_2}}{\mathfrak{p}_{n_2}} + x_3\Big).$$

And then the process has to be repeated.

To be able to formulate the final representation in a simple way, let us first introduce some notation:

- for $s \in \mathbb{N}$, an $(s-1)$-tuple $\boldsymbol{c} = (c_1, \ldots, c_{s-1}) \in M_{s-1}$, and $c_0 \in \{0,1\}$ put

$$\boldsymbol{X}_s(c_0, \boldsymbol{c}) := \frac{1}{2}\Big(c_0 + \sum_{j=1}^{s-1} \frac{c_j}{\mathfrak{p}_j} + \frac{\ell_s}{\mathfrak{p}_s}\Big);$$

  note that the case $s = 1$ gives only the following two numbers:

$$\boldsymbol{X}_1(c_0) = \frac{1}{2}\Big(c_0 + \frac{\ell_1}{\mathfrak{p}_1}\Big), \quad c_0 \in \{0,1\};$$

- for $\boldsymbol{c} = (c_j)_{j\in\mathbb{N}} \in M_\infty$ and $c_0 \in \{0,1\}$, put

$$\boldsymbol{X}_\infty(c_0, \boldsymbol{c}) := \frac{1}{2}(c_0 + \xi), \text{ where } \xi := \sum_{j=1}^{\infty} \frac{c_j}{\mathfrak{p}_j} \in [0,1).$$

The expression $\boldsymbol{X}_s(c_0, \boldsymbol{c})$ (resp. $\boldsymbol{X}_\infty(c_0, \boldsymbol{c})$) is called a *term of finite* (resp. *of infinite*) *type*. Moreover, $\boldsymbol{X}_\infty(c_0, \boldsymbol{c})$ is a *special term of infinite type* if it satisfies the following additional conditions: if $c_0 = 0$, then $\boldsymbol{c} \neq (0)_{j\in\mathbb{N}}$, and if $c_0 \in \{0,1\}$, then $\boldsymbol{c} \neq (k_j - 1)_{j\in\mathbb{N}}$.

Then x can be written (EXERCISE) either as

$$x = \sum_{s=1}^{k} \frac{\boldsymbol{X}_{n_s}(c_{s,0}, \boldsymbol{c}_s)}{2^{s-1}\mathfrak{p}_{n_1} \cdots \mathfrak{p}_{n_{s-1}}} + \frac{\boldsymbol{X}_\infty(c_{k+1,0}, \boldsymbol{c})}{2^k \mathfrak{p}_{n_1} \cdots \mathfrak{p}_{n_k}}, \tag{11.2.1}$$

where $k \in \mathbb{N}_0$, $c_{s,0} \in \{0,1\}$, $n_s \in \mathbb{N}$, $\boldsymbol{c}_s \in M_{n_s-1}$, $s = 1, \ldots, k$, and $c_{k+1,0} \in \{0,1\}$, $\boldsymbol{c} \in M_\infty$ such that:

- if $c_{k+1,0} = 0$, then $\boldsymbol{c} \neq \boldsymbol{0} := (0)_{j\in\mathbb{N}}$,
- if $c_{k+1,0}$ is arbitrary, then $\boldsymbol{c} \neq (k_j - 1)_{j\in\mathbb{N}}$ (i.e., the term of infinite type in this representation is a special one),

or

$$x = \sum_{s=1}^{\infty} \frac{\boldsymbol{X}_{n_s}(c_{s,0}, \boldsymbol{c}_s)}{2^{s-1}\mathfrak{p}_{n_1} \cdots \mathfrak{p}_{n_{s-1}}}, \tag{11.2.2}$$

where $c_{s,0} \in \{0,1\}$, $n_s \in \mathbb{N}$, and $\boldsymbol{c}_s \in M_{n_s-1}$, $s \in \mathbb{N}$.

Thus x can be represented either by a finite sum of terms of finite type and one special term of infinite type (see (11.2.1)) or by an infinite sum of terms of finite type (see (11.2.2)). The representation (11.2.1) is said to be a *special representation* of x. When we do not make the special assumption in (11.2.1), we will speak of a *general representation* of x. Note that the representation (11.2.2) is uniquely determined.

Now let x be given by a general representation of the first type (i.e., the term of infinite type is not assumed to be special), where the last term $\boldsymbol{X}_\infty(c_{k+1,0}, \boldsymbol{c})$ contains infinitely many successive zeros, i.e., $c_j = 0$ for all $j \geq j_0$. Then there are four cases to be discussed:

- If $j_0 = 1$ (i.e., $\boldsymbol{c} = (0)_{j=1}^{\infty}$), $c_{k+1,0} = 0$, and $\boldsymbol{c}_k = (c_{k,1}, \ldots, c_{k,n_k-1})$, then

$$
\begin{aligned}
x &= \sum_{s=1}^{k} \frac{\boldsymbol{X}_{n_s}(c_{s,0}, \boldsymbol{c}_s)}{2^{s-1}\mathfrak{p}_{n_1}\cdots\mathfrak{p}_{n_{s-1}}} + \frac{\frac{1}{2}(0 + \sum_{j=1}^{\infty}\frac{0}{\mathfrak{p}_j})}{2^k\mathfrak{p}_{n_1}\cdots\mathfrak{p}_{n_k}} \\
&= \sum_{s=1}^{k-1} \frac{\boldsymbol{X}_{n_s}(c_{s,0}, \boldsymbol{c}_s)}{2^{s-1}\mathfrak{p}_{n_1}\cdots\mathfrak{p}_{n_{s-1}}} + \frac{\boldsymbol{X}_{n_k}(c_{k,0}, \boldsymbol{c}_k)}{2^{k-1}\mathfrak{p}_{n_1}\cdots\mathfrak{p}_{n_{k-1}}} + \frac{\frac{1}{2}(0 + \sum_{j=1}^{\infty}\frac{0}{\mathfrak{p}_j})}{2^k\mathfrak{p}_{n_1}\cdots\mathfrak{p}_{n_k}} \\
&= \sum_{s=1}^{k-1} \frac{\boldsymbol{X}_{n_s}(c_{s,0}, \boldsymbol{c}_s)}{2^{s-1}\mathfrak{p}_{n_1}\cdots\mathfrak{p}_{n_{s-1}}} + \frac{\frac{1}{2}(c_{k,0} + \sum_{j=1}^{n_k-1}\frac{c_{k,j}}{\mathfrak{p}_j} + \frac{\ell_{n_k}-1}{\mathfrak{p}_{n_k}} + \sum_{j=n_k+1}^{\infty}\frac{k_j-1}{\mathfrak{p}_j})}{2^{k-1}\mathfrak{p}_{n_1}\cdots\mathfrak{p}_{n_{k-1}}},
\end{aligned}
$$

i.e., the general representation with $k+1$ terms, where the last term is trivial, is changed into a representation with k terms, where now the last term is not trivial. Note that the first term of infinite type is not special, while the second term of infinite type is.

- If $j_0 = 1$ and $c_{k+1,0} = 1$, then

$$
\begin{aligned}
x &= \sum_{s=1}^{k} \frac{\boldsymbol{X}_{n_s}(c_{s,0}, \boldsymbol{c}_s)}{2^{s-1}\mathfrak{p}_{n_1}\cdots\mathfrak{p}_{n_{s-1}}} + \frac{\frac{1}{2}(1 + \sum_{j=1}^{\infty}\frac{0}{\mathfrak{p}_j})}{2^k\mathfrak{p}_{n_1}\cdots\mathfrak{p}_{n_k}} \\
&= \sum_{s=1}^{k} \frac{\boldsymbol{X}_{n_s}(c_{s,0}, \boldsymbol{c}_s)}{2^{s-1}\mathfrak{p}_{n_1}\cdots\mathfrak{p}_{n_{s-1}}} + \frac{\frac{1}{2}(0 + \sum_{j=1}^{\infty}\frac{k_j-1}{\mathfrak{p}_j})}{2^k\mathfrak{p}_{n_1}\cdots\mathfrak{p}_{n_k}},
\end{aligned}
$$

i.e., the representation with $k+1$ terms is changed into one with $k+1$ terms, where only the special term of infinite type has been changed into a nonspecial term of infinite type. See also the next case, which is similar.

- If $j_0 > 1$, i.e., $c_j = 0$ for all $j \geq j_0$, and $\ell_{j_0-1} + 1 \neq c_{j_0-1} > 0$, then

$$
\begin{aligned}
x &= \sum_{s=1}^{k} \frac{\boldsymbol{X}_{n_s}(c_{s,0}, \boldsymbol{c}_s)}{2^{s-1}\mathfrak{p}_{n_1}\cdots\mathfrak{p}_{n_{s-1}}} + \frac{\frac{1}{2}(c_{k+1,0} + \sum_{j=1}^{j_0-1}\frac{c_j}{\mathfrak{p}_j} + \sum_{j=j_0}^{\infty}\frac{0}{\mathfrak{p}_j})}{2^k\mathfrak{p}_{n_1}\cdots\mathfrak{p}_{n_k}} \\
&= \sum_{s=1}^{k} \frac{\boldsymbol{X}_{n_s}(c_{s,0}, \boldsymbol{c}_s)}{2^{s-1}\mathfrak{p}_{n_1}\cdots\mathfrak{p}_{n_{s-1}}} + \frac{\frac{1}{2}(c_{k+1,0} + \sum_{j=1}^{j_0-2}\frac{c_j}{\mathfrak{p}_j} + \frac{c_{j_0-1}-1}{\mathfrak{p}_{j_0-1}} + \sum_{j=j_0}^{\infty}\frac{k_j-1}{\mathfrak{p}_j})}{2^k\mathfrak{p}_{n_1}\cdots\mathfrak{p}_{n_k}},
\end{aligned}
$$

i.e., the last special term $\boldsymbol{X}_\infty(c_{k+1,0}, \boldsymbol{c})$ can be substituted by a new special term of infinite type containing infinitely many positive summands, which are defined via the new sequence $c_1, \ldots, c_{j_0-2}, c_{j_0-1} - 1, k_{j_0} - 1, \ldots, k_j - 1, \ldots$ (see the remark above).

- If $j_0 > 1$ and $0 < c_{j_0-1} = \ell_{j_0-1} + 1$, then

$$
\begin{aligned}
x &= \sum_{s=1}^{k} \frac{\boldsymbol{X}_{n_s}(c_{s,0}, \boldsymbol{c}_s)}{2^{s-1}\mathfrak{p}_{n_1}\cdots\mathfrak{p}_{n_{s-1}}} + \frac{\frac{1}{2}(c_{k+1,0} + \sum_{j=1}^{j_0-1}\frac{c_j}{\mathfrak{p}_j} + \sum_{j=j_0}^{\infty}\frac{0}{\mathfrak{p}_j})}{2^k\mathfrak{p}_{n_1}\cdots\mathfrak{p}_{n_k}} \\
&= \sum_{s=1}^{k} \frac{\boldsymbol{X}_{n_s}(c_{s,0}, \boldsymbol{c}_s)}{2^{s-1}\mathfrak{p}_{n_1}\cdots\mathfrak{p}_{n_{s-1}}} + \frac{\frac{1}{2}(c_{k+1,0} + \sum_{j=1}^{j_0-2}\frac{c_j}{\mathfrak{p}_j} + \frac{\ell_{j_0-1}+1}{\mathfrak{p}_{j_0-1}} + \sum_{j=j_0}^{\infty}\frac{0}{\mathfrak{p}_j})}{2^k\mathfrak{p}_{n_1}\cdots\mathfrak{p}_{n_k}}
\end{aligned}
$$

$$= \sum_{s=1}^{k} \frac{\boldsymbol{X}_{n_s}(c_{s,0}, \boldsymbol{c}_s)}{2^{s-1}\mathfrak{p}_{n_1} \cdots \mathfrak{p}_{n_{s-1}}} + \frac{\frac{1}{2}(c_{k+1,0} + \sum_{j=1}^{j_0-2} \frac{c_j}{\mathfrak{p}_j} + \frac{\ell_{j_0-1}}{\mathfrak{p}_{j_0-1}})}{2^k \mathfrak{p}_{n_1} \cdots \mathfrak{p}_{n_k}}$$
$$+ \frac{\frac{1}{2}(1 + \sum_{j=1}^{\infty} \frac{k_j - 1}{\mathfrak{p}_j})}{2^{k+1}\mathfrak{p}_{n_1} \cdots \mathfrak{p}_{n_k}\mathfrak{p}_{j_0-1}},$$

i.e., the former last special term $\boldsymbol{X}_\infty(c_{k+1,0}, \boldsymbol{c})$ of infinite type splits into a new term of finite type with $n_{k+1} = j_0 - 1$ and a new term of infinite type that is not special.

- To summarize: only in the third case does x have a *double* representation with special terms of infinite type. In all other cases, the representation (11.2.1) is uniquely determined (EXERCISE). Therefore, if a function $f : (0,1) \longrightarrow \mathbb{R}$ is introduced via the above representation, then to have a well-defined function on $(0,1)$ one has only to check whether the definition of f gives the same value for the double representations from above.

Remark 11.2.2. Let $x \in (0,1)$ be given by one of the following (general) representations:

$$x = \sum_{s=1}^{k} \frac{\boldsymbol{X}_{n_s}(c_{s,0}, \boldsymbol{c}_s)}{2^{s-1}\mathfrak{p}_{n_1} \cdots \mathfrak{p}_{n_{s-1}}} + \frac{\boldsymbol{X}_\infty(c_{k+1,0}, \boldsymbol{c})}{2^k \mathfrak{p}_{n_1} \cdots \mathfrak{p}_{n_k}} \quad \text{or} \quad x = \sum_{s=1}^{\infty} \frac{\boldsymbol{X}_{n_s}(c_{s,0}, \boldsymbol{c}_s)}{2^{s-1}\mathfrak{p}_{n_1} \cdots \mathfrak{p}_{n_{s-1}}}.$$

Then the above construction leads to (EXERCISE)

$$1 - x = \sum_{s=1}^{k} \frac{\boldsymbol{X}_{n_s}(\tilde{c}_{s,0}, \tilde{\boldsymbol{c}}_s)}{2^{s-1}\mathfrak{p}_{n_1} \cdots \mathfrak{p}_{n_{s-1}}} + \frac{\boldsymbol{X}_\infty(\tilde{c}_{k+1,0}, \tilde{\boldsymbol{c}})}{2^k \mathfrak{p}_{n_1} \cdots \mathfrak{p}_{n_k}}$$

with:

- $\{c_{s,0}, \tilde{c}_{s,0}\} = \{0, 1\}$, $s = 1, \ldots, k+1$,
- if $\boldsymbol{c}_s = (c_{s,1}, \ldots, c_{s,n_s-1})$, then $\widetilde{\boldsymbol{c}}_s = (k_1 - 1 - c_{s,1}, \ldots, k_{n_s-1} - 1 - c_{s,n_s-1})$, $s = 1, \ldots, k$,
- if $\boldsymbol{c} = (c_j)_{j=1}^{\infty}$, then $\widetilde{\boldsymbol{c}} = (k_j - 1 - c_j)_{j \in \mathbb{N}}$,

or

$$1 - x = \sum_{s=1}^{\infty} \frac{\boldsymbol{X}_{n_s}(\tilde{c}_{s,0}, \widetilde{\boldsymbol{c}}_s)}{2^{s-1}\mathfrak{p}_{n_1} \cdots \mathfrak{p}_{n_{s-1}}}$$

with:

- $\{c_{s,0}, \tilde{c}_{s,0}\} = \{0, 1\}$,
- if $\boldsymbol{c}_s = (c_{s,1}, \ldots, c_{s,n_s-1})$, then $\widetilde{\boldsymbol{c}}_s = (k_1 - 1 - c_{s,1}, \ldots, k_{n_s-1} - 1 - c_{s,n_s-1})$, $s \in \mathbb{N}$.

Note that the term of infinite type in the first representation of x is special if and only if the corresponding term in the representation of $1 - x$ is special. Thus the representations for x and $1 - x$ are very similar. This information will be needed when we discuss properties of the Besicovitch–Singh function that we intend to introduce.

11.2.2 Definition of Singh's Besicovitch Function

Recall that a given $x \in (0,1)$ has a special representation as in (11.2.1) or (11.2.2) given by the following data:

(a) $\Big(k \in \mathbb{N}_0,\ (c_{j,0})_{j=1}^{k+1} \subset \{0,1\},\ (n_s)_{s=1}^{k} \subset \mathbb{N},\ \boldsymbol{c}_1 = (c_{1,1},\dots,c_{1,n_1-1}) \in M_{n_1-1},\dots,\boldsymbol{c}_k = (c_{k,1},\dots,c_{k,n_k-1}) \in M_{n_k-1},\ \boldsymbol{c} = (c_j)_{j=1}^{\infty} \in M_\infty\Big)$ or

(b) $\Big(c_{s,0} \in \{0,1\},\ n_s \in \mathbb{N},\ \boldsymbol{c}_s = (c_{s,1},\dots,c_{s,n_s-1}) \in M_{n_s-1},\ s \in \mathbb{N}\Big)$.

If x be given via (a), then put

$$\boldsymbol{S}_4(x) := \boldsymbol{Y}_{n_1}(c_{1,0},\boldsymbol{c}_1) + \sum_{s=2}^{k}(-1)^{s-1}\frac{\boldsymbol{Y}_{n_s}(c_{s,0},\boldsymbol{c}_s)}{\mathfrak{q}_{n_1}\cdots\mathfrak{q}_{n_{s-1}}} + (-1)^k\frac{\boldsymbol{Y}_\infty(c_{k+1,0},\boldsymbol{c})}{\mathfrak{q}_{n_1}\cdots\mathfrak{q}_{n_k}},$$

where

- $\mathfrak{q}_s := 2^s\widetilde{\mathfrak{p}}_{s-1}$, $s \in \mathbb{N}_2$, $\mathfrak{q}_1 := 2$, with $\widetilde{\mathfrak{p}}_s := \ell_1\cdots\ell_s$, $s \in \mathbb{N}$;
- $\boldsymbol{Y}_{n_s}(c_{s,0},\boldsymbol{c}_s) := \sum_{j=1}^{n_s-1}\frac{b_{s,j}}{2^j\widetilde{\mathfrak{p}}_j} + \frac{1}{2^{n_s}\widetilde{\mathfrak{p}}_{n_s-1}}$, $s = 1,\dots,k$, with

$$b_{s,j} := \begin{cases} c_{s,j}, & \text{if } c_{s,0} = 0,\ c_{s,j} < \ell_j \\ c_{s,j}-1, & \text{if } c_{s,0} = 0,\ c_{s,j} > \ell_j \\ k_j - 2 - c_{s,j}, & \text{if } c_{s,0} = 1,\ c_{s,j} < \ell_j \\ k_j - 1 - c_{s,j}, & \text{if } c_{s,0} = 1,\ c_{s,j} > \ell_j \end{cases},\quad s = 1,\dots,k,\ j = 1,\dots,n_s-1;$$

- $\boldsymbol{Y}_\infty(c_{k+1,0},\boldsymbol{c}) := \sum_{j=1}^{\infty}\frac{b_j}{2^j\widetilde{\mathfrak{p}}_j}$ with

$$b_j := \begin{cases} c_j, & \text{if } c_{k+1,0} = 0,\ c_j < \ell_j \\ c_j-1, & \text{if } c_{k+1,0} = 0,\ c_j > \ell_j \\ k_j - 2 - c_j, & \text{if } c_{k+1,0} = 1,\ c_j < \ell_j \\ k_j - 1 - c_j, & \text{if } c_{k+1,0} = 1,\ c_j > \ell_j \end{cases},\quad j \in \mathbb{N}.$$

On the other hand, if x is given via (b), then put

$$\boldsymbol{S}_4(x) := \boldsymbol{Y}_{n_1}(c_{1,0},\boldsymbol{c}_1) + \sum_{s=2}^{\infty}(-1)^{s-1}\frac{\boldsymbol{Y}_{n_s}(c_{s,0},\boldsymbol{c}_s)}{\mathfrak{q}_{n_1}\cdots\mathfrak{q}_{n_{s-1}}},$$

where the $\boldsymbol{Y}_{n_s}(c_{s,0},\boldsymbol{c}_s)$, $s \in \mathbb{N}$, are defined as above. Note that the series inside of the above definition of $\boldsymbol{S}_4$ converges, i.e., the right-hand side is always defined.

Remark 11.2.3. Observe that the b_j's in the definition above are at most $k_j - 2$. Moreover, $2^j\widetilde{\mathfrak{p}}_j = (k_1-1)\cdots(k_j-1)$. Thus in fact, the series $\sum_{j=1}^{\infty}\frac{b_j}{2^j\widetilde{\mathfrak{p}}_j}$ is a Cantor series with respect to the sequence $(k_j-1)_{j=1}^{\infty}$. Therefore, using Proposition A.1.1(c), one gets

- $\frac{1}{\mathfrak{q}_{n_s}} \le \boldsymbol{Y}_{n_s}(c_{s,0},\boldsymbol{c}_s) \le \sum_{j=1}^{n_s-1}\frac{k_j-2}{2^j\widetilde{\mathfrak{p}}_j} + \frac{1}{\mathfrak{q}_{n_s}} = 1 - \frac{1}{2^{n_s-1}\widetilde{\mathfrak{p}}_{n_s-1}} + \frac{1}{\mathfrak{q}_{n_s}} = 1 - \frac{1}{\mathfrak{q}_{n_s}} < 1$;
- $\boldsymbol{Y}_\infty(c_{k+1,0},\boldsymbol{c}) \le 1$.

It remains to show that $\boldsymbol{S}_4$ is well defined, i.e., $\boldsymbol{S}_4(x)$ does not depend on the chosen representation for any x.

Lemma 11.2.4. *The function $\boldsymbol{S}_4$ is well defined on the interval* $(0,1)$.

Proof. Recall that there is only one double representation (11.2.1) of a point $x \in (0,1)$, namely

$$x = \sum_{s=1}^{k} \frac{\boldsymbol{X}_{n_s}(c_{s,0}, \boldsymbol{c}_s)}{2^{s-1}\mathfrak{p}_{n_1}\cdots\mathfrak{p}_{n_{s-1}}} + \frac{\boldsymbol{X}_\infty(c_{k+1,0}, \boldsymbol{c})}{2^k\mathfrak{p}_{n_1}\cdots\mathfrak{p}_{n_k}}$$

$$= \sum_{s=1}^{k} \frac{\boldsymbol{X}_{n_s}(c_{s,0}, \boldsymbol{c}_s)}{2^{s-1}\mathfrak{p}_{n_1}\cdots\mathfrak{p}_{n_{s-1}}} + \frac{\boldsymbol{X}_\infty(c_{k+1,0}, \widetilde{\boldsymbol{c}})}{2^k\mathfrak{p}_{n_1}\cdots\mathfrak{p}_{n_k}},$$

where $\boldsymbol{c} = (c_1, \ldots, c_{j_0-2}, c_{j_0-1}, 0, \ldots, 0 \ldots)$, $j_0 > 1$, $0 < c_{j_0-1} \neq \ell_{j_0-1} + 1$, and $\widetilde{\boldsymbol{c}} = (c_1, \ldots, c_{j_0-2}, c_{j_0-1} - 1, k_{j_0} - 1, k_{j_0+1} - 1, \ldots)$. Exploiting the definition of the b_j's in the associated terms $\boldsymbol{Y}_\infty(c_{k+1,0}, \boldsymbol{c})$ and $\boldsymbol{Y}_\infty(c_{k+1,0}, \widetilde{\boldsymbol{c}})$ leads to the equality of these terms. Hence the value $\boldsymbol{S}_4$ at x is independent of the above representations of x. □

Remark 11.2.5. If we also allow double general representations of x, then a simple calculation also gives that the value of $\boldsymbol{S}_4$ is independent of the representation used (EXERCISE).

Later, in discussing the nonexistence of one-sided derivatives of $\boldsymbol{S}_4$, the following observation will become important. It shows that it is enough to consider only right-sided derivatives.

Lemma 11.2.6. *Let $x \in (0,1)$. Then $\boldsymbol{S}_4(x) = \boldsymbol{S}_4(1-x)$.*

Proof. The proof is left as an EXERCISE for the reader to become more familiar with the definition of $\boldsymbol{S}_4$. □

11.2.3 Continuity of S_4

Theorem 11.2.7. *The function $\boldsymbol{S}_4$ is continuous on $(0,1)$.*

Proof. First we discuss continuity at a point $x \in (0,1)$ given via the representation (11.2.2), i.e.,

$$x = \sum_{s=1}^{\infty} \frac{\boldsymbol{X}_{n_s}(c_{s,0}, \boldsymbol{c}_s)}{2^{s-1}\mathfrak{p}_{n_1}\cdots\mathfrak{p}_{n_{s-1}}}.$$

Fix an $\varepsilon > 0$. Choose a $k \in \mathbb{N}$ sufficiently large. The precise value of k will be given later. Then

$$x = \sum_{s=1}^{k} \frac{\boldsymbol{X}_{n_s}(c_{s,0}, \boldsymbol{c}_s)}{2^{s-1}\mathfrak{p}_{n_1}\cdots\mathfrak{p}_{n_{s-1}}} + \sum_{s=k+1}^{\infty} \frac{\boldsymbol{X}_{n_s}(c_{s,0}, \boldsymbol{c}_s)}{2^{s-1}\mathfrak{p}_{n_1}\cdots\mathfrak{p}_{n_{s-1}}}$$

$$=: \xi + \frac{1}{2^k\mathfrak{p}_{n_1}\cdots\mathfrak{p}_{n_k}} \Big(\sum_{s=k+1}^{\infty} \frac{\boldsymbol{X}_{n_s}(c_{s,0}, \boldsymbol{c}_s)}{2^{s-k-1}\mathfrak{p}_{n_{k+1}}\cdots\mathfrak{p}_{n_{s-1}}} \Big).$$

Note that x is an interior point of the open interval $\xi + \frac{1}{2^k\mathfrak{p}_{n_1}\cdots\mathfrak{p}_{n_k}}(0,1) =: J$. We may assume that k is so large that $x \in J \subset (0,1)$.

Take now an arbitrary $\eta \in J$, i.e., $\eta = \xi + \frac{\widetilde{x}}{2^k\mathfrak{p}_{n_1}\cdots\mathfrak{p}_{n_k}}$, where $\widetilde{x} \in (0,1)$. Then there are two possibilities:

(a) $\widetilde{x} = \sum_{s=1}^{\infty} \frac{\boldsymbol{X}_{\widetilde{n}_s}(\widetilde{c}_{s,0}, \widetilde{\boldsymbol{c}}_s)}{2^{s-1}\mathfrak{p}_{\widetilde{n}_1}\cdots\mathfrak{p}_{\widetilde{n}_{s-1}}}$;

(b) $\widetilde{x} = \sum_{s=1}^{m} \frac{\boldsymbol{X}_{\widetilde{n}_s}(\widetilde{c}_{s,0}, \widetilde{\boldsymbol{c}}_s)}{2^{s-1}\mathfrak{p}_{\widetilde{n}_1}\cdots\mathfrak{p}_{\widetilde{n}_{s-1}}} + \frac{\boldsymbol{X}_\infty(\widetilde{c}_{m+1,0}, \widetilde{\boldsymbol{c}})}{2^m\mathfrak{p}_{\widetilde{n}_1}\cdots\mathfrak{p}_{\widetilde{n}_m}}$.

In case (a), we rewrite η as

$$\begin{aligned}\eta &= \sum_{s=1}^{k} \frac{\boldsymbol{X}_{n_s}(c_{s,0},\boldsymbol{c}_s)}{2^{s-1}\mathfrak{p}_{n_1}\cdots\mathfrak{p}_{n_{s-1}}} + \sum_{s=1}^{\infty} \frac{\boldsymbol{X}_{\widetilde{n}_s}(\widetilde{c}_{s,0},\widetilde{\boldsymbol{c}}_s)}{2^{k+s-1}\mathfrak{p}_{n_1}\cdots\mathfrak{p}_{n_k}\mathfrak{p}_{\widetilde{n}_1}\cdots\mathfrak{p}_{\widetilde{n}_{s-1}}} \\ &= \sum_{s=1}^{k} \frac{\boldsymbol{X}_{n_s}(c_{s,0},\boldsymbol{c}_s)}{2^{s-1}\mathfrak{p}_{n_1}\cdots\mathfrak{p}_{n_{s-1}}} + \sum_{\sigma=k+1}^{\infty} \frac{\boldsymbol{X}_{\widetilde{n}_{\sigma-k}}(\widetilde{c}_{\sigma-k,0},\widetilde{\boldsymbol{c}}_{\sigma-k})}{2^{\sigma-1}\mathfrak{p}_{n_1}\cdots\mathfrak{p}_{n_k}\mathfrak{p}_{\widetilde{n}_1}\cdots\mathfrak{p}_{\widetilde{n}_{\sigma-k-1}}}.\end{aligned}$$

Thus η is given by the representation (11.2.2). Therefore,

$$\begin{aligned}&|\boldsymbol{S}_4(x) - \boldsymbol{S}_4(\eta)| \\ &= \Big| \sum_{s=k+1}^{\infty} (-1)^s \frac{\boldsymbol{Y}_{n_s}(c_{s,0},\boldsymbol{c}_s)}{\mathfrak{q}_{n_1}\cdots\mathfrak{q}_{n_{s-1}}} - \sum_{\sigma=k+1}^{\infty} (-1)^\sigma \frac{\boldsymbol{Y}_{\widetilde{n}_{\sigma-k}}(\widetilde{c}_{\sigma-k,0},\widetilde{\boldsymbol{c}}_{\sigma-k})}{\mathfrak{q}_{n_1}\cdots\mathfrak{q}_{n_k}\mathfrak{q}_{\widetilde{n}_1}\cdots\mathfrak{q}_{\widetilde{n}_{\sigma-k-1}}} \Big| \\ &\le \frac{1}{\mathfrak{q}_{n_1}\cdots\mathfrak{q}_{n_k}} \sum_{s=k+1}^{\infty} \Big(\frac{1}{\mathfrak{q}_{n_{k+1}}\cdots\mathfrak{q}_{n_{s-1}}} + \frac{1}{\mathfrak{q}_{\widetilde{n}_1}\cdots\mathfrak{q}_{\widetilde{n}_{s-k-1}}} \Big) \le \frac{4}{2^k} < \varepsilon,\end{aligned}$$

if k is sufficiently large.

The remaining very similar case is left as an Exercise.

Finally, assume that x is given via the representation (11.2.1), i.e.,

$$x = \sum_{s=1}^{k} \frac{\boldsymbol{X}_{n_s}(c_{s,0},\boldsymbol{c}_s)}{2^{s-1}\mathfrak{p}_{n_1}\cdots\mathfrak{p}_{n_{s-1}}} + \frac{\boldsymbol{X}_\infty(c_{k+1,0},\boldsymbol{c})}{2^k\mathfrak{p}_{n_1}\cdots\mathfrak{p}_{n_k}} =: \xi + \frac{\frac{1}{2}(c_{k+1,0} + \sum_{j=1}^{\infty}\frac{c_j}{\mathfrak{p}_j})}{2^k\mathfrak{p}_{n_1}\cdots\mathfrak{p}_{n_k}}.$$

According to the above discussion on the double representation, we may assume that $c_j \neq 0$ for infinitely many j's. Fix a $j_0 > 1$ with $c_{j_0} > 0$. Then

$$\begin{aligned}x &= \xi + \frac{\frac{1}{2}(c_{k+1,0} + \frac{c_1}{\mathfrak{p}_1} + \cdots + \frac{c_{j_0-1}}{\mathfrak{p}_{j_0-1}} + \frac{0}{\mathfrak{p}_{j_0}} + \frac{\ell_{j_0+1}}{\mathfrak{p}_{j_0+1}})}{2^k\mathfrak{p}_{n_1}\cdots\mathfrak{p}_{n_k}} \\ &\qquad + \frac{\frac{c_{j_0}}{\mathfrak{p}_{j_0}} + \frac{c_{j_0+1}}{\mathfrak{p}_{j_0+1}} - \frac{\ell_{j_0+1}}{\mathfrak{p}_{j_0+1}} + \sum_{j=j_0+2}^{\infty}\frac{c_j}{\mathfrak{p}_j}}{2^{k+1}\mathfrak{p}_{n_1}\cdots\mathfrak{p}_{n_k}} \\ &= \xi + \frac{\boldsymbol{X}_{n_{k+1}}(c_{k+1,0},(c_1,\dots,c_{j_0-1},0))}{2^k\mathfrak{p}_{n_1}\cdots\mathfrak{p}_{n_k}} + \frac{\frac{c_{j_0}}{\mathfrak{p}_{j_0}} + \frac{c_{j_0+1}}{\mathfrak{p}_{j_0+1}} - \frac{\ell_{j_0+1}}{\mathfrak{p}_{j_0+1}} + \sum_{j=j_0+2}^{\infty}\frac{c_j}{\mathfrak{p}_j}}{2^{k+1}\mathfrak{p}_{n_1}\cdots\mathfrak{p}_{n_k}} \\ &=: \xi + \xi_1 + \frac{\Big(\frac{c_{j_0}}{k_{j_0}} + \frac{c_{j_0+1}}{k_{j_0}k_{j_0+1}} - \frac{\ell_{j_0+1}}{k_{j_0}k_{j_0+1}} + \frac{1}{k_{j_0}k_{j_0+1}}\sum_{j=j_0+2}^{\infty}\frac{c_j}{k_{j_0+1}\cdots k_j}\Big)}{2^{k+1}\mathfrak{p}_{n_1}\cdots\mathfrak{p}_{n_k}\mathfrak{p}_{j_0-1}},\end{aligned}$$

where $n_{k+1} := j_0 + 1$ and $\boldsymbol{c}_{k+1} := (c_1,\dots,c_{j_0-1},0)$. Note that the term inside the parantheses in the line before lies in $(0,1)$. Therefore,

$$x \in \xi + \xi_1 + \frac{1}{2^{k+1}\mathfrak{p}_{n_1}\cdots\mathfrak{p}_{n_k}\mathfrak{p}_{j_0-1}}(0,1) =: J,$$

i.e., x is an interior point of $J \cap (0,1)$. Then every $\zeta \in J \cap (0,1)$ can be written as $\zeta = \xi + \xi_1 + \frac{\widetilde{x}}{2^{k+1}\mathfrak{p}_{n_1}\cdots\mathfrak{p}_{n_k}\mathfrak{p}_{j_0-1}}$ with $\widetilde{x} \in (0,1)$. From now on, we may repeat the argument from above to verify continuity of $\boldsymbol{S}_4$ at the point x. Details are left as an Exercise. □

11.2.4 Nowhere Differentiability of S_4

Theorem 11.2.8. *The Singh function $\boldsymbol{S}_4$ allows nowhere on $(0,1)$ a finite or infinite one-sided derivative, i.e., it is a Besicovitch function.*

Proof. By virtue of Lemma 11.2.6, it suffices to prove that $\boldsymbol{S}_4$ possesses nowhere on $(0,1)$ an $\overline{\mathbb{R}}$-valued right derivative.

There are different cases of $x \in (0,1)$ to be discussed.

Case 1°. Let x be given via the representation (11.2.1), i.e.,

$$x = \sum_{s=1}^{k} \frac{\boldsymbol{X}_{n_s}(c_{s,0}, \boldsymbol{c}_s)}{2^{s-1}\mathfrak{p}_{n_1}\cdots\mathfrak{p}_{n_{s-1}}} + \frac{\boldsymbol{X}_\infty(c_{k+1,0}, \boldsymbol{c})}{2^k \mathfrak{p}_{n_1}\ldots\mathfrak{p}_{n_k}}$$

with the corresponding data.

(a) Assume in addition that there is an infinite subset $\mathfrak{M} \subset \mathbb{N}$ such that $c_m < k_m - 1$, $m \in \mathfrak{M}$. For an $m \in \mathfrak{M}$, put

$$n_{k+1}^{(m)} := m+1, \quad \widehat{\boldsymbol{c}}_{k+1}^{(m)} := (c_1, \ldots, c_{m-1}, c_m + s_m) \in M_{n_{k+1}^{(m)}-1} = M_m,$$

where $s_m := \begin{cases} 1, & \text{if } c_m \neq \ell_m - 1 \\ 2, & \text{if } c_m = \ell_m - 1 \end{cases}.$

To continue, define for each $m \in \mathfrak{M}$ two new points to the right of x:

$$x'_m := \sum_{s=1}^{k} \frac{\boldsymbol{X}_{n_s}(c_{s,0}, \boldsymbol{c}_s)}{2^{s-1}\mathfrak{p}_{n_1}\cdots\mathfrak{p}_{n_{s-1}}} + \frac{\boldsymbol{X}_{n_{k+1}^{(m)}}(c_{k+1,0}, \widehat{\boldsymbol{c}}_{k+1}^{(m)})}{2^k\mathfrak{p}_{n_1}\cdots\mathfrak{p}_{n_k}} + \frac{\boldsymbol{X}_\infty(1, \boldsymbol{0})}{2^{k+1}\mathfrak{p}_1\cdots\mathfrak{p}_{n_k}\mathfrak{p}_{n_{k+1}^{(m)}}};$$

$$x''_m := \sum_{s=1}^{k} \frac{\boldsymbol{X}_{n_s}(c_{s,0}, \boldsymbol{c}_s)}{2^{s-1}\mathfrak{p}_{n_1}\cdots\mathfrak{p}_{n_{s-1}}} + \frac{\boldsymbol{X}_{n_{k+1}^{(m)}}(c_{k+1,0}, \widehat{\boldsymbol{c}}_{k+1}^{(m)})}{2^k\mathfrak{p}_{n_1}\cdots\mathfrak{p}_{n_k}} + \frac{\boldsymbol{X}_\infty(1, \widetilde{\boldsymbol{c}})}{2^{k+1}\mathfrak{p}_{n_1}\cdots\mathfrak{p}_{n_k}\mathfrak{p}_{n_{k+1}^{(m)}}},$$

where $\widetilde{\boldsymbol{c}} := (k_j - 1)_{j\in\mathbb{N}}$. Note that

$$\begin{aligned}
&x'_m - x \\
&= \frac{1}{2^k\mathfrak{p}_{n_1}\cdots\mathfrak{p}_{n_k}}\Big(\boldsymbol{X}_{n_{k+1}^{(m)}}(c_{k+1,0}, \widehat{\boldsymbol{c}}_{k+1}^{(m)}) + \frac{1}{2\mathfrak{p}_{n_{k+1}^{(m)}}}\boldsymbol{X}_\infty(1,\boldsymbol{0}) - \boldsymbol{X}_\infty(c_{k+1,0}, \boldsymbol{c})\Big) \\
&= \frac{1}{2^k\mathfrak{p}_{n_1}\cdots\mathfrak{p}_{n_k}}\frac{1}{2\mathfrak{p}_{m+1}}\Big(s_m k_{m+1} + \frac{k_{m+1}-1}{2} + \frac{1}{2} - k_{m+1}\sum_{j=1}^{\infty}\frac{c_{m+j}}{k_{m+1}\cdots k_{m+j}}\Big) \\
&> \frac{1}{2^k\mathfrak{p}_{n_1}\cdots\mathfrak{p}_{n_k}}\frac{1}{2\mathfrak{p}_{m+1}}\frac{k_{m+1}}{2}(2s_m - 1) \geq 0.
\end{aligned}$$

Moreover,

$$x'_m - x \leq \frac{1}{2^k\mathfrak{p}_{n_1}\cdots\mathfrak{p}_{n_k}}\frac{1}{2\mathfrak{p}_{m+1}}\frac{5k_{m+1}}{2} \leq \frac{5}{\mathfrak{p}_m} \underset{\mathfrak{M}\ni m\to\infty}{\longrightarrow} 0,$$

i.e., the points x'_m converge from the right to x.

Let us first recall the value of $\boldsymbol{S}_4(x)$ and then calculate $\boldsymbol{S}_4(x'_m)$:

$$\boldsymbol{S}_4(x) = \boldsymbol{Y}_{n_1}(c_{1,0}, \boldsymbol{c}_1) + \sum_{s=2}^{k}(-1)^{s-1}\frac{\boldsymbol{Y}_{n_s}(c_{s,0}, \boldsymbol{c}_s)}{\mathfrak{q}_{n_1}\cdots\mathfrak{q}_{n_{s-1}}} + (-1)^k\frac{\boldsymbol{Y}_\infty(c_{k+1,0}, \boldsymbol{c})}{\mathfrak{q}_{n_1}\cdots\mathfrak{q}_{n_k}}$$

and

$$\boldsymbol{S}_4(x'_m) = \boldsymbol{Y}_{n_1}(c_{1,0}, \boldsymbol{c}_1) + \sum_{s=2}^{k}(-1)^{s-1}\frac{\boldsymbol{Y}_{n_s}(c_{s,0}, \boldsymbol{c}_s)}{\mathfrak{q}_{n_1}\cdots\mathfrak{q}_{n_{s-1}}}$$
$$+ (-1)^k\frac{\boldsymbol{Y}_{n_{k+1}^{(m)}}(c_{k+1,0}, (c_1, \dots, c_m + s_m))}{\mathfrak{q}_{n_1}\cdots\mathfrak{q}_{n_k}} + (-1)^{k+1}\frac{\boldsymbol{Y}_\infty(1, \boldsymbol{0})}{\mathfrak{q}_{n_1}\cdots\mathfrak{q}_{n_k}\mathfrak{q}_{m+1}}.$$

Therefore,

$$\boldsymbol{S}_4(x'_m) - \boldsymbol{S}_4(x) = (-1)^k\frac{\boldsymbol{Y}_{n_{k+1}^{(m)}}(c_{k+1,0}, (c_1, \dots, c_m + s_m))}{\mathfrak{q}_{n_1}\cdots\mathfrak{q}_{n_k}}$$
$$+ (-1)^{k+1}\frac{\boldsymbol{Y}_\infty(1, \boldsymbol{0})}{\mathfrak{q}_{n_1}\cdots\mathfrak{q}_{n_k}\mathfrak{q}_{m+1}} - (-1)^k\frac{\boldsymbol{Y}_\infty(c_{k+1,0}, \boldsymbol{c})}{\mathfrak{q}_{n_1}\cdots\mathfrak{q}_{n_k}}$$
$$= (-1)^k\frac{1}{\mathfrak{q}_{n_1}\cdots\mathfrak{q}_{n_k}}\Big(\sum_{j=1}^{m}\frac{\widetilde{b}_j}{2^j\widetilde{\mathfrak{p}}_j} + \frac{1}{2^{m+1}\widetilde{\mathfrak{p}}_m} - \frac{1}{\mathfrak{q}_{m+1}}\sum_{j=1}^{\infty}\frac{k_j - 2}{2^j\widetilde{\mathfrak{p}}_j} - \sum_{j=1}^{\infty}\frac{b_j}{2^j\widetilde{\mathfrak{p}}_j}\Big),$$

where $\widetilde{b}_j = b_j$, $j = 1, \dots, m-1$, and $\widetilde{b}_m = b_m + s'$ with

$$s' := \begin{cases} 1, & \text{if } c_{k+1,0} = 0 \\ -1, & \text{if } c_{k+1,0} = 1 \end{cases}$$

(use the definition of the b_j's). Observe that $\frac{1}{2^{m+1}\widetilde{\mathfrak{p}}_{m+1}}\Big(1 - \sum_{j=1}^{\infty}\frac{(k_j-1)-1}{(k_1-1)\cdots(k_j-1)}\Big) = 0$ (use Proposition A.1.1(c)). Therefore, we end up with

$$\boldsymbol{S}_4(x'_m) - \boldsymbol{S}_4(x) = (-1)^k\frac{1}{\mathfrak{q}_{n_1}\cdots\mathfrak{q}_{n_k}}\Big(\frac{s'}{2^m\widetilde{\mathfrak{p}}_m} - \sum_{j=m+1}^{\infty}\frac{b_j}{2^j\widetilde{\mathfrak{p}}_j}\Big).$$

Now we discuss the associated differential quotient

$$\boldsymbol{\Delta S}_4(x, x'_m) = (-1)^k\frac{2^k\mathfrak{p}_{n_1}\cdots\mathfrak{p}_{n_k}2}{\mathfrak{q}_{n_1}\cdots\mathfrak{q}_{n_k}}\frac{\mathfrak{p}_m}{2^m\widetilde{\mathfrak{p}}_m}\frac{s' - \sum_{j=1}^{\infty}\frac{b_{m+j}}{2^j\ell_{m+1}\cdots\ell_{m+j}}}{\frac{2s_m+1}{2} - \sum_{j=1}^{\infty}\frac{c_{m+j}}{k_{m+1}\cdots k_{m+j}}}.$$

Put

- $\mathfrak{m}_k := (-1)^k\frac{2^k\mathfrak{p}_{n_1}\cdots\mathfrak{p}_{n_k}2}{\mathfrak{q}_{n_1}\cdots\mathfrak{q}_{n_k}}$,
- $a_m := \frac{\mathfrak{p}_m}{2^m\widetilde{\mathfrak{p}}_m}$.

Thus

$$\boldsymbol{\Delta S}_4(x,x'_m) = \mathfrak{m}_k a_m \frac{s' - \sum_{j=1}^{\infty} \frac{b_{m+j}}{2^j \ell_{m+1}\cdots\ell_{m+j}}}{\frac{2s_m+1}{2} - \sum_{j=1}^{\infty} \frac{c_{m+j}}{k_{m+1}\cdots k_{m+j}}}$$

$$= \mathfrak{m}_k a_m \frac{s' - \frac{b_{m+1}}{2\ell_{m+1}} - \frac{1}{\ell_{m+1}} \sum_{j=2}^{\infty} \frac{b_{m+j}}{2^j \ell_{m+2}\cdots\ell_{m+j}}}{\frac{2s_m+1}{2} - \frac{c_{m+1}}{k_{m+1}} - \frac{1}{k_{m+1}} \sum_{j=2}^{\infty} \frac{c_{m+j}}{k_{m+2}\cdots k_{m+j}}}.$$

Note that $\mathfrak{m}_k$ is independent of m and that $a_m \underset{m\to\infty}{\longrightarrow} \frac{1}{\prod_{j=1}^{\infty}(1-\frac{1}{k_j})} =: P$ with $P := \frac{1}{\mathfrak{L}(\mathfrak{C})}$. Since the sequences $(\frac{b_{m+1}}{2\ell_{m+1}})_{m\in\mathfrak{M}}$ and $(\frac{c_{m+1}}{k_{m+1}})_{m\in\mathfrak{M}}$ are bounded, we may assume that there is a common subsequence, given by $\mathfrak{M}' \subset \mathfrak{M}$, such that $(\frac{b_{m+1}}{2\ell_{m+1}})_{m\in\mathfrak{M}'}$, resp. $(\frac{c_{m+1}}{k_{m+1}})_{m\in\mathfrak{M}'}$, converges to $-A$, resp. $-B$.

Put $\mathfrak{M}'' := \{m \in \mathfrak{M}' : c_m \neq \ell_m - 1\}$. Assume that $\mathfrak{M}''$ is infinite. Then, since $\frac{3}{2} + B \neq 0$, we get

$$\boldsymbol{\Delta S}_4(x,x'_m) \underset{\mathfrak{M}''\ni m\to\infty}{\longrightarrow} \mathfrak{m}_k P \frac{s'+A}{\frac{3}{2}+B}.$$

If $\mathfrak{M}''$ is finite, then since $\frac{5}{2} + B \neq 0$, we get

$$\boldsymbol{\Delta S}_4(x,x'_m) \underset{\mathfrak{M}'\setminus\mathfrak{M}''\ni m\to\infty}{\longrightarrow} \mathfrak{m}_k P \frac{s'+A}{\frac{5}{2}+B}.$$

To study $\boldsymbol{\Delta S}_4(x,x''_m)$, we proceed similarly to the above. Let us first calculate $x''_m - x$:

$$x''_m - x = \frac{\boldsymbol{X}_{n_{k+1}^{(m)}}(c_{k+1,0}, \widehat{\boldsymbol{c}}_{k+1}^{(m)})}{2^k \mathfrak{p}_{n_1}\cdots\mathfrak{p}_{n_k}} + \frac{\boldsymbol{X}_\infty(1,\widetilde{\boldsymbol{c}})}{2^{k+1}\mathfrak{p}_{n_1}\cdots\mathfrak{p}_{n_k}\mathfrak{p}_{n_{k+1}^{(m)}}} - \frac{\boldsymbol{X}_\infty(c_{k+1,0},\boldsymbol{c})}{2^k \mathfrak{p}_{n_1}\cdots\mathfrak{p}_{n_k}}$$

$$= \frac{1}{2^k \mathfrak{p}_{n_1}\cdots\mathfrak{p}_{n_k}\mathfrak{p}_m}\Big(s_m + \frac{\ell_{m+1}}{k_{m+1}} + \frac{1}{4k_{m+1}}\Big(1 + \sum_{j=1}^{\infty}\frac{k_j-1}{\mathfrak{p}_j}\Big) - \sum_{j=1}^{\infty}\frac{c_{m+j}}{k_{m+1}\cdots k_{m+j}}\Big).$$

Since the last summand inside of the outer parantheses is at most equal to 1, it follows that $x''_m - x \geq 0$. Moreover,

$$x''_m - x \leq \frac{5}{2^k \mathfrak{p}_{n_1}\cdots\mathfrak{p}_{n_k}\mathfrak{p}_m} \underset{m\to\infty}{\longrightarrow} 0.$$

Thus the points x''_m converge from the right to x.

Moreover, using the definition of $\boldsymbol{S}_4$, we get

$$\boldsymbol{S}_4(x''_m) = \boldsymbol{Y}_{n_1}(c_{1,0},\boldsymbol{c}_{n_1-1}) + \sum_{s=1}^{k}(-1)^{s-1}\frac{\boldsymbol{Y}_{n_s}(c_{s,0},\boldsymbol{c}_s)}{\mathfrak{q}_{n_1}\cdots\mathfrak{q}_{n_{s-1}}}$$
$$+ (-1)^k \frac{\boldsymbol{Y}_{n_{k+1}^{(m)}}(c_{k+1,0},(c_1,\ldots,c_{m-1},c_m+s_m))}{\mathfrak{q}_{n_1}\cdots\mathfrak{q}_{n_k}} + (-1)^{k+1}\frac{\boldsymbol{Y}_\infty(1,\widetilde{\boldsymbol{c}})}{\mathfrak{q}_{n_1}\cdots\mathfrak{q}_{n_k}\mathfrak{q}_{m+1}}.$$

Note that the last summand equals 0. Therefore,

$$\begin{aligned}
\boldsymbol{S}_4(x''_m) - \boldsymbol{S}_4(x) &= \boldsymbol{Y}_{n_1}(c_{1,0}, \boldsymbol{c}_{n_1-1}) + \sum_{s=1}^{k}(-1)^{s-1}\frac{\boldsymbol{Y}_{n_s}(c_{s,0}, \boldsymbol{c}_s)}{\mathfrak{q}_{n_1}\cdots\mathfrak{q}_{n_{s-1}}} \\
&+ (-1)^k \frac{\boldsymbol{Y}_{n_{k+1}^{(m)}}(c_{k+1,0}, (c_1, \ldots, c_{m-1}, c_m + s_m))}{\mathfrak{q}_{n_1}\cdots\mathfrak{q}_{n_k}} \\
&- \boldsymbol{Y}_{n_1}(c_{1,0}, \boldsymbol{c}_{n_1-1}) - \sum_{s=2}^{k}(-1)^{s-1}\frac{\boldsymbol{Y}_{n_s}(c_{s,0}, \boldsymbol{c}_s)}{\mathfrak{q}_{n_1}\cdots\mathfrak{q}_{n_{s-1}}} - (-1)^k\frac{\boldsymbol{Y}_\infty(c_{k+1,0}, \boldsymbol{c})}{\mathfrak{q}_{n_1}\cdots\mathfrak{q}_{n_k}} \\
&= (-1)^k \frac{\boldsymbol{Y}_{n_{k+1}^{(m)}}(c_{k+1,0}, (c_1, \ldots, c_{m-1}, c_m + s_m))}{\mathfrak{q}_{n_1}\cdots\mathfrak{q}_{n_k}} - (-1)^k\frac{\boldsymbol{Y}_\infty(c_{k+1,0}, \boldsymbol{c})}{\mathfrak{q}_{n_1}\cdots\mathfrak{q}_{n_k}} \\
&= (-1)^k \frac{1}{\mathfrak{q}_{n_1}\cdots\mathfrak{q}_{n_k}2^m\widetilde{\mathfrak{p}}_m}\left(s' + \frac{\ell_{m+1}}{2\ell_{m+1}} - \sum_{j=m+1}^{\infty}\frac{b_{m+1}}{2^j\ell_{m+1}\cdots\ell_{m+j}}\right).
\end{aligned}$$

Hence the differential quotient is given by

$$\boldsymbol{\Delta S}_4(x, x''_m) = \mathfrak{m}_k a_m \frac{s' + \frac{1}{2} - \frac{b_{m+1}}{2\ell_{m+1}} - \frac{1}{\ell_{m+1}}\sum_{j=m+2}^{\infty}\frac{b_{m+j}}{2^j\ell_{m+2}\cdots\ell_{m+j}}}{\frac{2s_m+1}{2} - \frac{c_{m+1}}{k_{m+1}} - \frac{1}{k_{m+1}}\sum_{j=2}^{\infty}\frac{c_{m+j}}{k_{m+2}\cdots k_{m+j}}}.$$

If $\mathfrak{M}''$ is infinite, then, taking a subsequence via $\mathfrak{M}''$, it follows that

$$\boldsymbol{\Delta S}_4(x, x''_m) \underset{\mathfrak{M}''\ni m\to\infty}{\longrightarrow} \mathfrak{m}_k P\frac{1 + \frac{1}{2} + A}{\frac{3}{2} + B}.$$

For the remaining case, we have

$$\boldsymbol{\Delta S}_4(x, x''_m) \underset{\mathfrak{M}'\setminus\mathfrak{M}''\ni m\to\infty}{\longrightarrow} \mathfrak{m}_k P\frac{s' + \frac{1}{2} + A}{\frac{5}{2} + B}.$$

What we have seen is that the two sequences of differential quotients (via a subsequence) have finite limits from the right, but these are different. Thus $\boldsymbol{S}_4$ has no right-sided finite or infinite derivative at the point x.

(b) In the remaining case, we have $c_m = k_m - 1$ for all $m \geq m_0 > 1$ and $c_{m_0-1} < k_{m_0-1} - 1$. Recall what was said above with respect to the double representation. Thus we may rewrite x as

$$\begin{aligned}
x &= \sum_{s=1}^{k}\frac{\boldsymbol{X}_{n_s}(c_{s,0}, \boldsymbol{c}_s)}{2^{s-1}\mathfrak{p}_{n_1}\cdots\mathfrak{p}_{n_{s-1}}} + \frac{\boldsymbol{X}_\infty(c_{k+1,0}, \boldsymbol{c})}{2^k\mathfrak{p}_{n_1}\cdots\mathfrak{p}_{n_k}} \\
&= \sum_{s=1}^{k}\frac{\boldsymbol{X}_{n_s}(c_{s,0}, \boldsymbol{c}_s)}{2^{s-1}\mathfrak{p}_{n_1}\cdots\mathfrak{p}_{n_{s-1}}} + \frac{\frac{1}{2}\left(c_{k+1,0} + \sum_{j=1}^{m_0-1}\frac{c_j}{\mathfrak{p}_j} + \frac{1}{\mathfrak{p}_{m_0-1}}\sum_{j=m_0}^{\infty}\frac{k_j-1}{k_{m_0}\cdots k_j}\right)}{2^k\mathfrak{p}_{n_1}\cdots\mathfrak{p}_{n_k}} \\
&= \sum_{s=1}^{k}\frac{\boldsymbol{X}_{n_s}(c_{s,0}, \boldsymbol{c}_s)}{2^{s-1}\mathfrak{p}_{n_1}\cdots\mathfrak{p}_{n_{s-1}}} + \frac{\frac{1}{2}\left(c_{k+1,0} + \sum_{j=1}^{m_0-2}\frac{c_j}{\mathfrak{p}_j} + \frac{c_{m_0-1}+1}{\mathfrak{p}_{m_0-1}} + \sum_{j=m_0}^{\infty}\frac{0}{\mathfrak{p}_j}\right)}{2^k\mathfrak{p}_{n_1}\cdots\mathfrak{p}_{n_k}}.
\end{aligned}$$

So we are back in the situation of (a) and can proceed as we did there.

Thus $\boldsymbol{S}_4$ has no right-sided finite or infinite derivative at any point x with the first representation.

Case 2°. Now we discuss a point $x \in (0,1)$ having the second representation (11.2.2), i.e.,

$$x = \sum_{s=1}^{\infty} \frac{\boldsymbol{X}_{n_s}(c_{s,0}, \boldsymbol{c}_s)}{2^{s-1}\mathfrak{p}_{n_1} \cdots \mathfrak{p}_{n_{s-1}}},$$

where $c_{s,0} \in \{0,1\}$, $n_s \in \mathbb{N}$, and $\boldsymbol{c}_s \in M_{n_s-1}$, $s \in \mathbb{N}$.
There are two cases to discuss.

(a) Assume that there is an infinite subset $\mathfrak{M} \subset \mathbb{N}$ such that $c_{m,0} = 0$, $m \in \mathfrak{M}$. As above, two sequences $(x'_m)_{m\in\mathfrak{M}}$ and $(x''_m)_{m\in\mathfrak{M}}$ will be discussed converging from the right to x. Namely, for $m \in \mathfrak{M}$, we put

$$x'_m := \sum_{s=1}^{m-1} \frac{\boldsymbol{X}_{n_s}(c_{s,0}, \boldsymbol{c}_s)}{2^{s-1}\mathfrak{p}_{n_1} \cdots \mathfrak{p}_{n_{s-1}}} + \frac{\frac{1}{2}\big(0 + \sum_{j=1}^{n_m-1} \frac{c_{m,j}}{\mathfrak{p}_j} + \frac{\ell_{n_m}+s_m}{\mathfrak{p}_{n_m}} + \sum_{j=n_m+1}^{\infty} \frac{0}{\mathfrak{p}_j}\big)}{2^{m-1}\mathfrak{p}_{n_1} \cdots \mathfrak{p}_{n_{m-1}}},$$

where $s_m \in \mathbb{N}$, $\ell_{n_m} + s_m \le k_{n_m} - 1$. The precise values of the s_m's will be chosen later. Observe that x'_m is given via the first representation. Moreover, let

$$x''_m := \sum_{s\in\mathbb{N}, s\ne m} \frac{\boldsymbol{X}_{n_s}(c_{s,0}, \boldsymbol{c}_s)}{2^{s-1}\mathfrak{p}_{n_1} \cdots \mathfrak{p}_{n_{s-1}}} + \frac{\frac{1}{2}\big(1 + \sum_{j=1}^{n_m-1} \frac{k_j-1-c_{m,j}}{\mathfrak{p}_{n_j}} + \frac{\ell_{n_m}}{\mathfrak{p}_{n_m}}\big)}{2^{m-1}\mathfrak{p}_{n_1} \cdots \mathfrak{p}_{n_{m-1}}}.$$

Then $x < x'_m$. Indeed, put

$$\lambda_m := \sum_{s=m+1}^{\infty} \frac{\frac{1}{2}\big(c_{s,0} + \sum_{j=1}^{n_s-1} \frac{c_{s,j}}{\mathfrak{p}_j} + \frac{\ell_{n_s}}{\mathfrak{p}_{n_s}}\big)}{2^{s-m-1}\mathfrak{p}_{n_{m+1}} \cdots \mathfrak{p}_{n_{s-1}}}.$$

Note that

$$\lambda_m \le \sum_{s=m+1}^{\infty} \frac{1 - \frac{k_{n_s}-1}{4\mathfrak{p}_{n_s}}}{2^{s-m-1}\mathfrak{p}_{n_{m+1}} \cdots \mathfrak{p}_{n_{s-1}}} \le 1 + \sum_{s=m+1}^{\infty} \frac{3 - k_{n_s}}{2^{s-m+1}\mathfrak{p}_{n_{m+1}} \cdots \mathfrak{p}_{n_s}} < 1;$$

recall that $k_j > 3$ for large $j \in \mathbb{N}$. Hence, we end up with

$$x'_m - x = \frac{s_m - \lambda_m}{2^m \mathfrak{p}_{n_1} \cdots \mathfrak{p}_{n_{m-1}}\mathfrak{p}_{n_m}} > 0.$$

Moreover,

$$\begin{aligned} x''_m - x'_m &\ge \frac{\frac{1}{2}\big(1 + \sum_{j=1}^{n_m-1} \frac{k_j-1-2c_{m,j}}{\mathfrak{p}_j} - \frac{s_m}{\mathfrak{p}_{n_m}}\big)}{2^{m-1}\mathfrak{p}_{n_1} \cdots \mathfrak{p}_{n_{m-1}}} \\ &\ge \frac{\sum_{j=1}^{n_m-1} \frac{k_j-1-c_{m,j}}{\mathfrak{p}_j} + \frac{k_{n_m}-1-s_m}{\mathfrak{p}_{n_m}}}{2^{m-1}\mathfrak{p}_{n_1} \cdots \mathfrak{p}_{n_{m-1}}} \ge 0. \end{aligned}$$

Thus $x \le x'_m \le x''_m$.

A short calculation leads to the additional fact that $x''_m \underset{\mathfrak{M}\ni m\to\infty}{\longrightarrow} x$ (Exercise), i.e., the two sequences $(x'_m)_{m\in\mathfrak{M}}$ and $(x''_m)_{m\in\mathfrak{M}}$ approach the point x from the right.

Using the definition of $\boldsymbol{S}_4$, we obtain for $m \in \mathfrak{M}$ that

$$\boldsymbol{S}_4(x''_m) - \boldsymbol{S}_4(x) = (-1)^{m-1}\frac{\boldsymbol{Y}_{n_m}(1, \widetilde{\boldsymbol{c}}_{n_m-1}) - \boldsymbol{Y}_{n_m}(0, \boldsymbol{c}_{n_m-1})}{\mathfrak{q}_{n_1}\cdots\mathfrak{q}_{n_{m-1}}},$$

where $\widetilde{\boldsymbol{c}}_{n_m-1} = (k_1 - 1 - c_{m,1}, \dots, k_{n_m-1} - 1 - c_{m,n_m-1})$. Thus,

$$\boldsymbol{S}_4(x''_m) - \boldsymbol{S}_4(x) = (-1)^{m-1}\frac{\sum_{j=1}^{n_m-1}\frac{\widetilde{b}_j - b_j}{2^j\widetilde{\mathfrak{p}}_j}}{\mathfrak{q}_{n_1}\cdots\mathfrak{q}_{n_{m-1}}} = 0;$$

here we have used the definition of the new factors $\widetilde{b}_j$ and b_j. Hence, $\boldsymbol{\Delta S}_4(x, x''_m) = 0$. Since $m \in \mathfrak{M}$ was arbitrarily chosen, it follows that $\lim_{\mathfrak{M}\ni m\to\infty}\boldsymbol{\Delta S}_4(x, x''_m) = 0$.

It remains to investigate $\boldsymbol{\Delta S}_4(x, x'_m)$, $m \in \mathfrak{M}$:

$$\begin{aligned}
&\boldsymbol{S}_4(x'_m) - \boldsymbol{S}_4(x)\\
&= (-1)^{m-1}\frac{\frac{\ell_{n_m}+s_m-1}{2^{n_m}\widetilde{\mathfrak{p}}_{n_m}} - \frac{1}{2^{n_m}\widetilde{\mathfrak{p}}_{n_m-1}}}{\mathfrak{q}_{n_1}\cdots\mathfrak{q}_{n_{m-1}}} - \sum_{s=m+1}^{\infty}(-1)^{s-1}\frac{\boldsymbol{Y}_{n_s}(c_{s,0}, \boldsymbol{c}_s)}{\mathfrak{q}_{n_1}\cdots\mathfrak{q}_{n_{s-1}}}\\
&= (-1)^{m-1}\frac{1}{\mathfrak{q}_{n_1}\cdots\mathfrak{q}_{n_{m-1}}\mathfrak{q}_{n_m}}\Big(\frac{s_m - 1}{\ell_{n_m}} - \sum_{s=m+1}^{\infty}(-1)^{s-m}\frac{\boldsymbol{Y}_{n_s}(c_{s,0}, \boldsymbol{c}_s)}{\mathfrak{q}_{n_{m+1}}\cdots\mathfrak{q}_{n_{s-1}}}\Big)\\
&= (-1)^{m-1}\frac{1}{\mathfrak{q}_{n_1}\cdots\mathfrak{q}_{n_{m-1}}}\frac{1}{2^{n_m-1}\widetilde{\mathfrak{p}}_{n_m-1}}\Big(\frac{s_m - 1}{k_{n_m} - 1} + \frac{1}{2}\theta_m\Big),
\end{aligned}$$

where

$$\theta_m := \sum_{s=m+1}^{\infty}(-1)^{s-m+1}\frac{\boldsymbol{Y}_{n_s}(c_{s,0}, \boldsymbol{c}_s)}{\mathfrak{q}_{n_{m+1}}\cdots\mathfrak{q}_{n_{s-1}}}.$$

Hence, the differential quotient is given by

$$\boldsymbol{\Delta S}_4(x, x'_m) = (-1)^{m-1}\frac{2^{m-1}\mathfrak{p}_{n_1}\cdots\mathfrak{p}_{n_{m-1}}}{\mathfrak{q}_{n_1}\cdots\mathfrak{q}_{n_{m-1}}}\frac{\mathfrak{p}_{n_m}}{2^{n_m-1}\widetilde{\mathfrak{p}}_{n_m-1}}\cdot\frac{\frac{s_m-1}{2\ell_{n_m}} + \frac{1}{2}\theta_m}{s_m - \lambda_m}.$$

The following estimate will show that $\theta_m \in (0, 1)$. Indeed, applying Remark 11.2.3, we obtain

$$\begin{aligned}
\theta_m &\le \sum_{\substack{s=m+1\\ s-m \text{ odd}}}^{\infty}\frac{1 - \frac{1}{\mathfrak{q}_{n_s}}}{\mathfrak{q}_{n_{m+1}}\cdots\mathfrak{q}_{n_{s-1}}} - \sum_{\substack{s=m+1\\ s-m \text{ even}}}^{\infty}\frac{\frac{1}{\mathfrak{q}_{n_s}}}{\mathfrak{q}_{n_{m+1}}\cdots\mathfrak{q}_{n_{s-1}}}\\
&= \sum_{t=0}^{\infty}\frac{1}{\mathfrak{q}_{n_{m+1}}\cdots\mathfrak{q}_{n_{m+2t}}} - \sum_{t=0}^{\infty}\frac{1}{\mathfrak{q}_{n_{m+1}}\cdots\mathfrak{q}_{n_{m+2t+1}}} - \sum_{t=1}^{\infty}\frac{1}{\mathfrak{q}_{n_{m+1}}\cdots\mathfrak{q}_{n_{m+2t}}}\\
&= 1 - \sum_{t=0}^{\infty}\frac{1}{\mathfrak{q}_{n_{m+1}}\cdots\mathfrak{q}_{n_{m+2t+1}}} < 1
\end{aligned}$$

and

$$\begin{aligned}\theta_m &\geq \sum_{\substack{s=m+1\\ s-m\text{ odd}}}^{\infty} \frac{\frac{1}{\mathfrak{q}_{n_s}}}{\mathfrak{q}_{n_{m+1}}\cdots\mathfrak{q}_{n_{s-1}}} - \sum_{\substack{s=m+1\\ s-m\text{ even}}}^{\infty} \frac{1-\frac{1}{\mathfrak{q}_{n_s}}}{\mathfrak{q}_{n_{m+1}}\cdots\mathfrak{q}_{n_{s-1}}}\\ &= \sum_{t=0}^{\infty}\frac{1}{\mathfrak{q}_{n_{m+1}}\cdots\mathfrak{q}_{n_{m+2t+1}}} - \sum_{t=1}^{\infty}\frac{1}{\mathfrak{q}_{n_{m+1}}\cdots\mathfrak{q}_{n_{m+2t-1}}} + \sum_{t=1}^{\infty}\frac{1}{\mathfrak{q}_{n_{m+1}}\cdots\mathfrak{q}_{n_{m+2t}}}\\ &= \sum_{t=1}^{\infty}\frac{1}{\mathfrak{q}_{n_{m+1}}\cdots\mathfrak{q}_{n_{m+2t}}} > 0.\end{aligned}$$

Note that

$$\frac{2^m\mathfrak{p}_{n_1}\cdots\mathfrak{p}_{n_{m-1}}}{\mathfrak{q}_{n_1}\cdots\mathfrak{q}_{n_{m-1}}} \underset{\mathfrak{M}\ni m\to\infty}{\longrightarrow} +\infty \quad\text{and}\quad \frac{\mathfrak{p}_{n_m}}{2^{n_m-1}\widetilde{\mathfrak{p}}_{n_m-1}} \geq 1.$$

We may take a subsequence $\mathfrak{M}'$ such that $\lambda_m \underset{\mathfrak{M}'\ni m\to\infty}{\longrightarrow} \lambda\in[0,1]$ and $\theta_m \underset{\mathfrak{M}'\ni m\to\infty}{\longrightarrow} \theta\in[0,1]$. When $\theta>0$, then for the s_m we may choose $s_m=1$, $m\in\mathfrak{M}'$. If $\theta=0$, then we choose $s_m:=\ell_{n_m}$, $m\in\mathfrak{M}'$. Thus in both cases, one obtains that $\lim_{\mathfrak{M}'\ni m\to\infty}|\boldsymbol{\Delta S}_4(x,x'_m)|=+\infty$. Hence we have shown that $\boldsymbol{S}_4$ has no finite or infinite right-sided derivative at the point x.

(b) Assume that there exists an $m_0\in\mathbb{N}$ such that $c_{m,0}=1$, $m\geq m_0$. For $m\geq m_0$, put

$$x'_m := \sum_{s=1}^{m-1}\frac{\boldsymbol{X}_{n_s}(c_{s,0},\boldsymbol{c}_s)}{2^{s-1}\mathfrak{p}_{n_1}\cdots\mathfrak{p}_{n_{s-1}}} + \frac{\frac{1}{2}(1+\sum_{j=1}^{n_m-1}\frac{c_{m,j}}{\mathfrak{p}_j}+\frac{\ell_{n_m}+s_m}{\mathfrak{p}_{n_m}}+\sum_{j=n_m+1}^{\infty}\frac{0}{\mathfrak{p}_j})}{2^{m-1}\mathfrak{p}_{n_1}\cdots\mathfrak{p}_{n_{m-1}}}.$$

The exact values of the s_m, $\ell_{n_m}+s_m\leq k_{n_m}-1$ will be chosen later. As above, the sequence $(x'_m)_{m\in\mathbb{N}_{m_0}}$ tends from the right to x. Moreover,

$$x'_m - x = \frac{s_m-\lambda_m}{2^{m-1}\mathfrak{p}_{n_1}\cdots\mathfrak{p}_{n_{m-1}}\mathfrak{p}_{n_m}},$$

where $\lambda_m\in(0,1)$ as above. Moreover, the difference of the $\boldsymbol{S}_4$-values is given by

$$\boldsymbol{S}_4(x'_m)-\boldsymbol{S}_4(x) = (-1)^{m-1}\frac{1}{\mathfrak{q}_{n_1}\cdots\mathfrak{q}_{n_{m-1}}}\frac{1}{2^{n_m-1}\widetilde{\mathfrak{p}}_{n_m-1}}\Big(\frac{-s_m+1}{k_{n_m}-1}+\frac{1}{2}\theta_m\Big).$$

Here we used the definition of the new coefficients b_j and again Proposition A.1.1(c) (Exercise). So for $m\geq m_0$, we end up with

$$\boldsymbol{\Delta S}_4(x,x'_m) = (-1)^{m-1}\frac{2^m\mathfrak{p}_{n_1}\cdots\mathfrak{p}_{n_{m-1}}}{\mathfrak{q}_{n_1}\cdots\mathfrak{q}_{n_{m-1}}}\frac{\mathfrak{p}_{n_m}}{2^{n_m-1}\widetilde{\mathfrak{p}}_{n_m-1}}\cdot\frac{\frac{-s_m+1}{2\ell_{n_m}}+\frac{1}{2}\theta_m}{s_m-\lambda_m}.$$

Again, two cases have to be discussed.

• Assume that $\sup_{m\in\mathbb{N}_{m_0}}\theta_m =: \theta < 1$. In this situation, put $s_m:=\ell_{n_m}$. Then the last factor can be estimated by

$$\frac{-s_m+1}{k_{n_m}-1}+\frac{1}{2}\theta_m \leq \frac{3-k_{n_m}}{2(k_{n_m}-1)}+\frac{1}{2}\theta = -\frac{1-\theta}{2}+\frac{1}{k_{n_m}} < -\frac{1-\theta}{3} =: \alpha < 0,$$

whenever m is sufficiently large. Recall that $k_m \longrightarrow +\infty$ as $m \longrightarrow +\infty$. Because of the sign $(-1)^{m-1}$, one obtains $\limsup_{m\to\infty} \boldsymbol{\Delta S}_4(x, x'_m) = +\infty$ and $\liminf_{m\to\infty} \boldsymbol{\Delta S}_4(x, x'_m) = -\infty$, implying that $\boldsymbol{S}_4$ has no finite or infinite right-sided derivative at the point x under discussion.

- Assume now that $\sup_{m\in\mathbb{N}_{m_0}} \theta_m = 1$. Then there exists a subsequence $(\theta_{m_k})_{k\in\mathbb{N}}$, $m_k \geq m_0$, such that $\theta_{m_k} \underset{k\to\infty}{\longrightarrow} 1$. We proceed, discussing only the sequence $(x_{m_k})_{k\in\mathbb{N}}$. Put $s_{m_k} = 1$. Then $\frac{-s_{m_k}+1}{k_{m_k}-1} + \frac{1}{2}\theta_{m_k} \longrightarrow \frac{1}{2}$. Therefore, $|\boldsymbol{\Delta S}_4(x, x'_{m_k})| \underset{k\to\infty}{\longrightarrow} +\infty$. If, in addition, the two sets

$$\mathfrak{M}_1 := \{k \in \mathbb{N} : m_k \text{ odd}\}, \quad \mathfrak{M}_2 := \{k \in \mathbb{N} : m_k \text{ even}\}$$

are infinite, then because of the signs $(-1)^{m-1}$, one gets

$$\limsup_{\mathfrak{M}_1 \ni k\to\infty} \boldsymbol{\Delta S}_4(x, x'_m) = \infty, \qquad \liminf_{\mathfrak{M}_2 \ni k\to\infty} \boldsymbol{\Delta S}_4(x, x'_m) = -\infty.$$

Therefore, in this case, the function $\boldsymbol{S}_4$ has also no finite or infinite right-sided derivative at x.

It remains to discuss the situation in which

(i) $m_k \in 2\mathbb{N}$, $k \geq k_1$, or
(ii) $m_k \in 2\mathbb{N}+1$, $k \geq k_1$.

In case (i) (resp. (ii)), it follows that $\lim_{k\to\infty} \boldsymbol{\Delta S}_4(x, x_{m_k}) = -\infty$ (resp. $\lim_{k\to\infty} \boldsymbol{\Delta S}_4(x, x_{m_k}) = +\infty$). For the remaining m's, $m \geq m_{k_1}$, take $s_m = 1$. In case (i) (resp. (ii)), these m's are odd (resp. even), and the corresponding terms $\frac{-s_m+1}{k_{n_m}-1} + \frac{1}{2}\theta_m$ in $\boldsymbol{\Delta S}_4(x, x'_m)$ are in both cases nonnegative. Therefore, if (i) (resp. (ii)), then $D_+\boldsymbol{S}_4(x) = -\infty$ and $D^+\boldsymbol{S}_4(x) \geq 0$ (resp. $D^+\boldsymbol{S}_4(x) = +\infty$ and $D_+\boldsymbol{S}_4(x) \geq 0$). Hence $\boldsymbol{S}_4$ has no finite or infinite derivative at the point x. □

11.3 $\mathfrak{BM}(\mathbb{I})$ Is Residual in a Certain Subspace of $\mathcal{C}(\mathbb{I})$

Recall that we already know that $\mathfrak{B}(\mathbb{I})$ is of first category in $\mathcal{C}(\mathbb{I})$, which does not give automatically the existence of a Besicovitch-type function. On the other hand, we have already discussed concrete functions belonging to $\mathfrak{BM}(\mathbb{I})$. In this section, a clever use of the categorial approach leads to the fact that $\mathfrak{BM}(\mathbb{I})$ is even a residual subset of a certain subspace of $\mathcal{C}(\mathbb{I})$ (see [Mal84]).

Put

$$K := \{f \in \mathcal{C}(\mathbb{I}) : f(0) = f(1) = 0,\ \mathrm{Lip}(f) \leq 1\},$$

where

$$\mathrm{Lip}(f) := \sup_{x,y\in\mathbb{I},\ x\neq y} |\boldsymbol{\Delta} f(x,y)|.$$

Note that K is a compact subset of the metric space $\mathcal{C}(\mathbb{I})$ (Exercise, use the theorem of Arzelà–Ascoli), where the metric is given by

$$d(f,g) := \|f - g\| = \sup\{|f(x) - f(y)| : x, y \in \mathbb{I}\}.$$

Finally, let E be the set of all $u = (u_n)_{n=1}^{\infty} \in K^{\mathbb{N}}$ such that:

$$\text{(a)} \quad u_n \geq u_{n+1} \geq 0 \ (n \in \mathbb{N}), \tag{11.3.1}$$

(b) if $u_{n+1} > 0$ on a subinterval $J \subset \mathbb{I}$, then $u_n|_J$ is constant $(n \in \mathbb{N})$. (11.3.2)

Note that E is a closed subset of the compact metric space $K^{\mathbb{N}}$ endowed with the product topology or the metric $d(u, v) = \sum_{n=1}^{\infty} \frac{1}{2^n} \frac{\|u_n - v_n\|}{1+\|u_n - v_n\|}$, where $u = (u_n)_{n=1}^{\infty}$, $v = (v_n)_{n=1}^{\infty} \in K^{\mathbb{N}}$ (Exercise).

In order to be able to continue, let us state the following lemma.

Lemma 11.3.1. *Let $u = (u_n)_{n=1}^{\infty} \in E$. Then $\|u_k\| < \frac{1}{k}$ for all $k \in \mathbb{N}$.*

Proof. Assume that there is a $k \in \mathbb{N}$ with $\|u_k\| \geq \frac{1}{k}$. Then there exists an $x_0 \in \mathbb{I}$ with $u_k(x_0) \geq \frac{1}{k}$. Taking into account that $u_k(0) = u_k(1) = 0$, we have $x_0 \in (0, 1)$. Moreover, $\frac{1}{k} \leq u_k(x_0) = u_k(x_0) - u_k(0) \leq x_0 - 0 = x_0$; in particular, $k \geq 2$.

Now choose a point $x_1 \in [0, x_0)$ such that $u_k(x_1) = 0$ and u_k is strictly positive on the interval $J_1 := (x_1, x_0]$. By assumption (11.3.2), it follows that

$$x_1 = x_1 - 0 \geq u_{k-1}(x_1) - u_{k-1}(0) = u_{k-1}(x_1) = u_{k-1}(x_0) \geq u_k(x_0) \geq \frac{1}{k},$$

and moreover,

$$\frac{1}{k} \leq u_k(x_0) = |u_k(x_1) - u_k(x_0)| \leq x_0 - x_1, \text{ i.e. } \frac{1}{k} \leq x_1 \leq x_0 - \frac{1}{k}.$$

In particular, $k \geq 3$. Repeating this argument leads to points $x_2 > x_3 > x_4 \geq x_m$ with $\frac{1}{k} \leq u_{k-m}(x_m) \leq x_m - \frac{1}{k} \leq x_0 - \frac{m}{k}$ and $m + 1 < k$. Since this process does not stop, we end up with a contradiction. □

Using the above lemma, we will introduce a continuous map from the compact space E into $\mathcal{C}(\mathbb{I})$. Let $\varphi \in \mathcal{C}(\mathbb{I})$ be an increasing function with $\varphi(0) = 0$ and $\varphi(1) \geq 1$. Then we define a map $A_\varphi : E \longrightarrow \mathcal{C}(\mathbb{I})$ via the following formula

$$A_\varphi(u)(x) := \sum_{k=1}^{\infty} (-1)^{k+1} \varphi(u_k(x)), \quad x \in \mathbb{I},\ u = (u_k)_{k=1}^{\infty} \in E.$$

Note that $\varphi(u_k(x)) \geq \varphi(u_{k+1}(x)) \searrow_{k\to\infty} 0$; thus, by the Leibniz criterion, this series is convergent for every $x \in \mathbb{I}$. So $A_\varphi(u)$ is well defined on $\mathbb{I}$.

Why is $A_\varphi(u)$ continuous? Fix a point $a \in \mathbb{I}$ and a positive ε. Now choose an $m \in \mathbb{N}$ such that $\varphi(\frac{1}{m}) < \varepsilon$ (use the fact that $\varphi(0) = 0$). Moreover, by the uniform continuity of φ, we may find a positive $\delta = \delta_\varepsilon$ such that $\varphi(t) - \varphi(\tau) < \varepsilon$ whenever $0 \leq \tau \leq t \leq \max\{\tau + \delta, 1\}$. Take an $\eta > 0$ such that for all $x \in \mathbb{I}$ with $|x - a| < \eta$, we have $|u_k(x) - u_k(a)| < \delta$, $k = 1, \ldots, m$ (use the fact that the u_k are continuous functions).

Now let $x \in \mathbb{I}$ with $|x - a| < \eta$. Then:

$$\begin{aligned}
&|A_\varphi(u)(x) - A_\varphi(u)(a)| \\
&\leq \Big|\sum_{k=1}^{m} (-1)^{k+1} (\varphi(u_k(x)) - \varphi(u_k(a))\Big| + \Big|\sum_{k=m+1}^{\infty} (-1)^{k+1} \varphi(u_k(x))\Big| \\
&+ \Big|\sum_{k=m+1}^{\infty} (-1)^{k+1} \varphi(u_k(a))\Big| \leq m\varepsilon + \varphi(u_{m+1}(x)) + \varphi(u_m(a)) < (m + 2)\varepsilon.
\end{aligned}$$

Hence $A_\varphi(u) \in \mathcal{C}(\mathbb{I})$.

Moreover, we have the following property of A_φ.

Lemma 11.3.2. *Let φ be as above. Then the mapping $A_\varphi : E \longrightarrow \mathcal{C}(\mathbb{I})$ is continuous. In particular, $A_\varphi(E)$ is a compact subset of the metric space $\mathcal{C}(\mathbb{I})$ and therefore a complete metric space.*

Proof. Fix a $u = (u_k)_{k=1}^\infty \in E$ and an $\varepsilon > 0$. The proof now is similar to the previous one. Choose an $m \in \mathbb{N}$ such that $\varphi(\frac{1}{m}) < \varepsilon$ and take a positive δ such that $\varphi(t) - \varphi(\tau) < \varepsilon$ whenever $0 \le \tau \le t \le \max\{\tau + \delta, 1\}$.

Now let $v \in E$ be such that $\|v_k - u_k\| < \delta$, $k = 1, \dots, m$. For $x \in \mathbb{I}$, by virtue of Lemma 11.3.1, we get

$$\begin{aligned}
&|A_\varphi(u)(x) - A_\varphi(v)(x)| \\
&\le \Big|\sum_{k=1}^{m}(-1)^{k+1}(\varphi(u_k(x)) - \varphi(v_k(x))\Big| + \Big|\sum_{k=m+1}^{\infty}(-1)^{k+1}\varphi(u_k(x))\Big| \\
&+ \Big|\sum_{k=m+1}^{\infty}(-1)^{k+1}\varphi(v_k(x))\Big| \le m\varepsilon + \varphi(u_{m+1}(x)) + \varphi(v_{m+1}(x)) < (m+2)\varepsilon.
\end{aligned}$$

Hence A_φ is continuous. □

Proposition 11.3.3. *Let φ be as above. Suppose, in addition, that φ is differentiable (with finite derivatives) on $(0,1)$ and that $D_+\varphi(0) = \liminf_{x\to 0+} \frac{\varphi(x)}{x} < \infty$. If $f \in A_\varphi(E)$, then f does not have an infinite right-sided derivative at any point in $[0,1)$.*

Proof. Let $f = A_\varphi(u)$, where $u = (u_k)_{k=1}^\infty \in E$, and fix a point $a \in [0,1)$. Three cases have to be discussed.

Case 1°. $u_k(a) > 0$ for all $k \in \mathbb{N}$.

Put

$$b_k := \inf\{x \in (a,1] : u_k(x) = 0\}.$$

Here b_k is the first zero of u_k to the right of a. Because of (11.3.1), it is clear that $b_{k+1} \le b_k$, $k \in \mathbb{N}$. Using (11.3.2), we know that u_{k-1} is identically constant on $[a, b_k]$, $k \ge 2$. Therefore,

$$\begin{aligned}
f(a) &\ge \sum_{k=1}^{2n}(-1)^{k+1}\varphi(u_k(a)) = \sum_{k=1}^{2n}(-1)^{k+1}\varphi(u_k(b_{2n+1})) = f(b_{2n+1}), \\
f(a) &\le \sum_{k=1}^{2n-1}(-1)^{k+1}\varphi(u_k(a)) = \sum_{k=1}^{2n-1}(-1)^{k+1}\varphi(u_k(b_{2n})) = f(b_{2n}), \quad n \in \mathbb{N}.
\end{aligned}$$

If $b_n \longrightarrow a$, then

$$\begin{aligned}
D_+f(a) &= \liminf_{x\to a+} \boldsymbol{\Delta} f(a,x) \le \liminf_{n\to\infty} \boldsymbol{\Delta} f(a, b_{2n+1}) \le 0, \\
D^+f(a) &= \limsup_{x\to a+} \boldsymbol{\Delta} f(a,x) \ge \limsup_{n\to\infty} \boldsymbol{\Delta} f(a, b_{2n}) \ge 0.
\end{aligned}$$

Thus in this case, no infinite right-sided derivative of f at a exists.

If $a < b := \lim_{n\to\infty} b_n$, then all the u_n are constant on $[a,b]$ (use condition (11.3.2)). Hence $D^+f(a) = D_+f(a) = 0$, i.e., f has a finite right-sided derivative $f'_+(a) = 0$ at the point a.

Case 2°. $u_1(a) = \cdots = u_{m-1}(a) > 0$, $u_k(a) = 0$ for $k \geq m$, and $u_1, \dots, u_{m-1}$ are identically constant on some interval $J = [a, b]$ with $a < b \leq 1$. In fact, because of (11.3.2), we need to assume only that u_{m-1} is identically constant on such an interval J.

Let $x \in J$. Then, using the Lipschitz property of u_k and the monotonicity of φ, we get

$$|f(x) - f(a)| = \Big|\sum_{k=1}^{\infty}(-1)^{k+1}(\varphi(u_k(x)) - \varphi(u_k(a))\Big|$$
$$\leq \varphi(u_m(x)) = \varphi(u_m(x) - u_m(a)) \leq \varphi(x - a).$$

Therefore, we have

$$\liminf_{x\to a+}|\boldsymbol{\Delta} f(a,x)| \leq \liminf_{x\to a+}\frac{\varphi(x-a)}{x-a} = D_+\varphi(0) < \infty.$$

Hence f has no infinite right-sided derivative at the point a.

Case 3°. $u_1(a) = \cdots = u_{m-1}(a) > 0$, $u_k(a) = 0$ for $k \geq m$, and u_{m-1} is not identically constant on any interval $(a, b]$ with $a < b \leq 1$.

By (11.3.2), we conclude that there is a sequence of points $x_j < 1$ with $x_j \underset{j\to\infty}{\searrow} a$ such that $u_m(x_j) = 0$ and $u_n(a) = u_n(x_j)$, $n = 1, \dots, m-2$ (use that $u_{m-1}(a) > 0$). Then we get

$$\limsup_{j\to\infty}\frac{|f(x_j) - f(a)|}{x_j - a} \leq \limsup_{j\to\infty}\frac{|\varphi(u_{m-1}(x_j)) - \varphi(u_{m-1}(a))|}{x_j - a}$$
$$\leq \varphi'(u_{m-1}(a))\limsup_{j\to\infty}\frac{|u_{m-1}(x_j) - u_{m-1}(a)|}{x_j - a}$$
$$\leq \varphi'(u_{m-1}(a)) < \infty.$$

Thus also in this case, f has no infinite right-sided derivative at the point a. □

The main statement is formulated by the following theorem.

Theorem 11.3.4. *Let φ, ψ be continuous strictly increasing functions on $\mathbb{I}$ with $\varphi(0) = \psi(0) = 0$, $\min\{\varphi(1), \psi(1)\} \geq 1$, and such that ψ is a concave function. If $\limsup_{x\to 0+}\frac{\varphi(x)}{\psi(x)} = \infty$, then the set*

$$M := \Big\{f \in A_\varphi(E) : \exists_{a\in[0,1)} : \limsup_{x\to a+}\frac{|f(x) - f(a)|}{\psi(x-a)} < \infty\Big\}$$

is of first category in the complete metric space $A_\varphi(E)$.

Proof. For $k \in \mathbb{N}$, we put

$$M_k := \{f \in A_\varphi(E) : \exists_{a\in[0,1-1/k]} : \ |f(x) - f(a)| \leq k\psi(x-a)$$
$$\text{for all } x \in [a, a + \frac{1}{k}]\}.$$

Obviously, the sets M_k are closed in $A_\varphi(E)$ and $M \subset \bigcup_{k=1}^{\infty} M_k$. It remains to verify that each of the sets M_k is nowhere dense in $A_\varphi(E)$.

Let $k \in \mathbb{N}$. Then take an $f = A_\varphi(u) \in M_k$, $u = (u_n)_{n=1}^\infty \in E$, and choose an arbitrary $\varepsilon \in (0, 1/2)$. We will show that the ε-neighborhood of u contains a sequence of functions $u^* = (u_n^*)_{n=1}^\infty \in E$ such that $A_\varphi(u^*)$ does not belong to M_k. Hence, f cannot be an interior point of M_k (use that A_φ is continuous), which will prove that M_k is nowhere dense.

For this reason, choose an $m \in \mathbb{N}$ with $\frac{1}{m} < \varepsilon' := \frac{\varepsilon}{5}$ and put

$$v_n := (1 - 2\varepsilon') \max\{0, u_n - nm^{-2}\}, \quad n \in \mathbb{N}.$$

Obviously, $v := (v_n)_{n=1}^\infty \in E$. Moreover, $v_n \equiv 0$ for $n > m$, since $\|u_n\| < \frac{1}{n}$, and $\mathrm{Lip}(v_n) \le 1 - 2\varepsilon'$, $n \in \mathbb{N}$. For later use, we mention also that if $n \in \mathbb{N}$ and $x \in \mathbb{I}$, then

$$\text{either } v_{n+1}(x) = 0 \text{ or } v_n(x) \ge v_{n+1}(x) + \frac{1 - 2\varepsilon'}{m^2}. \tag{11.3.3}$$

Indeed, if $v_{n+1}(x) > 0$, then $u_n(x) \ge u_{n+1}(x) > \frac{n+1}{m^2} > \frac{n}{m^2}$. Therefore,

$$\begin{aligned} v_n(x) &= (1 - 2\varepsilon')(u_n(x) - \frac{n}{m^2}) \\ &\ge (1 - 2\varepsilon')(u_{n+1}(x) - \frac{n+1}{m^2}) + (1 - 2\varepsilon')\frac{1}{m^2} \\ &= v_{n+1}(x) + \frac{1 - 2\varepsilon'}{m^2}. \end{aligned}$$

Let $n > m$. Then $\|u_n - v_n\| = \|u_n\| \le \frac{1}{n} < \frac{1}{m} < \varepsilon'$; if $n \le m$, then

$$\begin{aligned} &|u_n(x) - v_n(x)| = |u_n(x) - (1 - 2\varepsilon') \max\{0, u_n(x) - nm^{-2}\}| \\ &\le \begin{cases} |u_n(x)| \le nm^{-2} \le m^{-1}, & \text{if } u_n(x) \le nm^{-2} \\ |2\varepsilon' u_n(x) + (1 - 2\varepsilon')nm^{-2}| \le 2\varepsilon' + 1/m, & \text{if } u_n(x) > nm^{-2} \end{cases}; \end{aligned}$$

hence $\|u_n - v_n\| \le 2\varepsilon' + \frac{1}{m}$.

Using the assumption that ψ is concave, we get

$$\limsup_{x \to 0+} \frac{\varphi(\frac{\varepsilon' x}{4})}{\psi(x)} \ge \frac{1}{4}\varepsilon' \cdot \limsup_{x \to 0+} \frac{\varphi(\frac{\varepsilon' x}{4})}{\psi(\frac{\varepsilon' x}{4})} = \infty.$$

In particular, there exists a positive δ_0 such that if $\delta < \delta_0$, then

$$\sup\left\{\frac{\varphi(\frac{\varepsilon' x}{4})}{\psi(x)} : 0 < x < \delta\right\} > 8k.$$

Choose $q_0 \in \mathbb{N}$ with $q_0 \ge \max\{k, 2\}$, $\frac{\varepsilon'}{q_0} < \frac{1}{m^2}$, $\frac{1}{q_0} < \delta_0$, $q_0 \ge \frac{\varepsilon' m^2}{1-2\varepsilon'}$, and $q_0^2 > \frac{2m^2\varepsilon'^2}{1-2\varepsilon'}$. Then $\bigcup_{q=q_0}^\infty (\frac{1}{2q}, \frac{1}{q}) = (0, \frac{1}{q_0})$. Taking $\delta = \frac{1}{q_0}$, we fix a point $x \in (0, \delta)$ with $\frac{\varphi(\frac{\varepsilon' x}{4})}{\psi(x)} > 8k$. Then there exists a $q \ge q_0$ such that $\frac{1}{2q} < \frac{x}{4} < \frac{1}{q}$. Thus $\frac{\varepsilon' x}{4} \le \frac{\varepsilon'}{q}$ and $x \ge \frac{2}{q}$. By virtue of the monotonicity of φ and ψ, we end up with $8k\psi(\frac{2}{q}) < \varphi(\frac{\varepsilon'}{q})$, where $q \ge \max\{2, k\}$ and $\frac{\varepsilon'}{q} < \frac{1}{m^2}$.

Put $b_j := \frac{j}{q}$, $c_j := \frac{j-\varepsilon'}{q}$, and $a_j := \frac{j-2\varepsilon'}{q}$, $j = 0, \ldots, q$. Thus we have the following partition:

$$0 = b_0 < a_1 < c_1 < b_1 < a_2 < c_2 < b_2 < \cdots < a_q < c_q < b_q = 1.$$

Now we introduce the function $h : \mathbb{I} \longrightarrow \mathbb{I}$ via the following rule:

$$h(x) := \begin{cases} 0, & \text{if } x = 0 \\ b_j, & \text{if } x \in [a_j, b_j], 1 \le j \le q \\ b_{j-1} + \frac{x - b_{j-1}}{a_j - b_{j-1}}(b_j - b_{j-1}), & \text{if } x \in [b_{j-1}, a_j], 1 \le j \le q \end{cases}.$$

Note that

$$\mathrm{Lip}(h) = \max\left\{\frac{b_j - b_{j-1}}{a_j - b_{j-1}} : j = 1, \ldots, q\right\} = \frac{1}{1 - 2\varepsilon'}.$$

Put $w = (w_n)_{n=1}^{\infty}$ via $w_n := v_n \circ h$, $n \in \mathbb{N}$. Then $\mathrm{Lip}(w_n) \le \mathrm{Lip}(v_n)\,\mathrm{Lip}(h) \le 1$. Hence, $w \in E$ (recall that $w_n(0) = w_n(1) = 0$). Moreover,

$$|w_n(x) - v_n(x)| \le \mathrm{Lip}(v_n)|x - h(x)| \le \frac{1}{q}, \quad x \in \mathbb{I},\ n \in \mathbb{N} \quad (\text{EXERCISE}).$$

Finally, we define the desired family of functions in E near to u, namely $u^* = (u_n^*)_{n=1}^{\infty}$, via the following procedure. Let $n_j \in \mathbb{N}$ denote the smallest n with $v_n(b_j) < \frac{\varepsilon'}{q}$, i.e., $v_{n_j}(b_j) < \frac{\varepsilon'}{q} \le v_{n_j - 1}(b_j)$, $j = 1, \ldots, q$. Note that such an n_j exists, since $v_n(b_j) \underset{n \to \infty}{\longrightarrow} 0$. Moreover, we have that either $v_{n_j+1}(b_{j+1}) = 0$ or $v_{n_j}(b_{j+1}) \ge v_{n_j+1}(b_{j+1}) + \frac{1-2\varepsilon'}{m^2} > \frac{\varepsilon'}{q}$, i.e., either $v_{n_j+1}(b_{j+1}) = 0$ or $n_{j+1} > n_j$.

Then we define $u_n^*(x)$, $n \in \mathbb{N}$, $x \in \mathbb{I}$, by the following formulas:

- if $n \notin \{n_1, \ldots, n_q\}$, then $u_n^*(x) := w_n(x)$;
- if there exists a j such that $n = n_j$, $x \in [a_j, b_j]$, and $\varphi(v_{n_j}(b_j)) \le 4k\psi(\frac{2}{q})$, then $u_n^*(x) := v_n(b_j) + \frac{\varepsilon'}{q} - |x - c_j|$;
- if there exists a j such that $n = n_j$, $x \in [a_j, b_j]$, and $\varphi(v_{n_j}(b_j)) > 4k\psi(\frac{2}{q})$, then $u_n^*(x) := v_n(b_j) - \max\{v_{n_j}(b_j), \frac{\varepsilon'}{q} - |x - c_j|\}$;
- if $n \in \{n_1, \ldots, n_q\}$, $x \notin \bigcup_{j:\, n = n_j} [a_j, b_j]$, then $u_n^*(x) := w_n(x)$.

Note that u_n^* is well defined, since the intervals $[a_j, b_j]$ are pairwise disjoint. Obviously, $u_n^*(0) = w_n(0) = 0$ and $u_n^* \ge 0$. Moreover, for $n = n_j$ we have either

$$u_n^*(b_j) = v_n(b_j) + \frac{\varepsilon'}{q} - |b_j - c_j| = v_n(b_j) = v_n(h(b_j)) = w_n(b_j) \text{ or}$$

$$u_n^*(b_j) = v_n(b_j) - \min\{v_n(b_j), \frac{\varepsilon'}{q} - |b_j - c_j|\} = v_n(b_j) = w_n(b_j)$$

(in particular, $u_n^*(1) = 0$) and either

$$u_n^*(a_j) = v_n(b_j) + \frac{\varepsilon'}{q} - |a_j - c_j| = v_n(b_j) = v_n(h(a_j)) = w_n(a_j) \text{ or}$$

$$u_n^*(a_j) = v_n(b_j) - \min\{v_n(b_j), \frac{\varepsilon'}{q} - |a_j - c_j|\} = v_n(b_j) = w_n(a_j).$$

Hence, the function u_n^* is continuous in $\mathbb{I}$. It is clear that $\mathrm{Lip}(u_n^*) \le 1$.

Let $n = n_j$ and $x \in [a_j, b_j]$. Assume that $v_{n+1}(x) > 0$, then by virtue of (11.3.3), we have

$$\frac{1 - 2\varepsilon'}{m^2} < v_{n+1}(x) + \frac{1 - 2\varepsilon'}{m^2} \le v_n(x) \le v_n(b_j)(b_j - x) < \frac{\varepsilon'}{q}\frac{2\varepsilon'}{q};$$

a contradiction (because $q^2 > \frac{2m^2\varepsilon'^2}{1-2\varepsilon'}$). Thus, $v_{n+1}|_{[a_j,b_j]} \equiv 0$. Since $n+1 \neq n_j$, we have $u^*_{n+1}(x) = w_{n+1}(x) = v_{n+1}(h(x)) = v_{n+1}(b_j) = 0$, $x \in [a_j, b_j]$. Therefore, $u^*_{n+1}|_{[a_j,b_j]} \equiv 0 \leq u^*_n|_{[a_j,b_j]}$. Now let $n > 1$, $n = n_j$, and $x \in [a_j, b_j]$ as above. Then we get either

$$\begin{aligned} u^*_{n-1}(x) &= w_{n-1}(x) = v_{n-1}(b_j) \geq v_n(b_j) \\ &\geq v_n(x) - \min\{v_n(b_j), \varepsilon'/q - |x - c_j|\} = u^*_n(x) \text{ or} \\ v_{n_j-1}(b_j) &> v_{n_j}(b_j) + \frac{1-2\varepsilon'}{m^2} \geq v_{n_j}(b_j) + \frac{\varepsilon'}{q} - |x - c_j| = u_{n_j}(x) \end{aligned}$$

(recall that $q > \frac{\varepsilon' m^2}{1-2\varepsilon'}$). Hence $u^*_1 \geq u^*_2 \geq \cdots \geq u^*_n \geq \cdots \geq 0$ (recall that $u^*_n(x)$ differs from $w_n(x)$ only if $n = n_j$ and $x \in [a_j, b_j]$).

Since property (11.3.2) is also satisfied (EXERCISE), we have $u^* \in E$.

Finally, we discuss $|u^*_n(x) - w_n(x)|$, $x \in \mathbb{I}$. Let $n = n_j$ and $x \in [a_j, b_j]$. Then either

$$|u^*_n(x) - w_n(x)| = |v_n(b_j) + \frac{\varepsilon'}{q} - |x - c_j| - w_n(x)| \leq \frac{\varepsilon'}{q}$$

(use $w_n(x) = v_n(b_j)$) or

$$|u^*_n(x) - w_n(x)| = |v_n(b_j) - \min\{v_n(b_j), \frac{\varepsilon'}{q} - |x - c_j|\} - v(b_j)| \leq \frac{\varepsilon'}{q}.$$

All the other cases are trivial. Therefore, $\|u^*_n - w_n\| \leq \frac{\varepsilon'}{q} < \varepsilon'$. Finally, we have

$$\|u^*_n - u_n\| \leq \|u^*_n - w_n\| + \|w_n - v_n\| + \|v_n - u_n\| \leq 5\varepsilon' < \varepsilon,$$

i.e., the sequence u^* belongs to the ε-neighborhood of u.

It remains to verify that $f^* := A_\varphi(u^*) \notin M_k$. To prove this, fix an arbitrary point $a \in [0, 1-\frac{1}{k}]$. Since $1-\frac{1}{k} < 1-\frac{1}{q} = b_{q-1}$, there exists a $j \in \{2, \ldots, q\}$ with $b_{j-2} \leq a < b_{j-1}$. Recall that n_j is the smallest index l such that $v_l(b_j) < \frac{\varepsilon'}{q}$. Then $\varphi(u^*_{n_j}(b_j)) = \varphi(v_{n_j}(b_j)) < \varphi(\frac{\varepsilon'}{q})$.

If $\varphi(v_{n_j}(b_j)) \leq 4k\psi(\frac{2}{q})$, then

$$\begin{aligned} \varphi(u^*_{n_j}(c_j)) - \varphi(u^*_{n_j}(b_j)) &= \varphi\Big(v_{n_j}(b_j) + \frac{\varepsilon'}{q}\Big) - \varphi(v_{n_j}(b_j)) \\ &\geq \varphi\Big(\frac{\varepsilon'}{q}\Big) - 4k\psi\Big(\frac{2}{q}\Big) \geq 8k\psi\Big(\frac{2}{q}\Big) - 4k\psi\Big(\frac{2}{q}\Big) \\ &= 4k\psi\Big(\frac{2}{q}\Big). \end{aligned}$$

If $\varphi(v_{n_j}(b_j)) > 4k\psi(\frac{2}{q})$, then

$$\begin{aligned} |\varphi(u^*_{n_j}(c_j)) - \varphi(u^*_{n_j}(b_j))| &= |\varphi(0) - \varphi(u^*_{n_j}(b_j))| \\ &= \varphi(u^*_{n_j}(b_j)) = \varphi(v_{n_j}(b_j)) > 4k\psi\Big(\frac{2}{q}\Big). \end{aligned}$$

Recall that for $s \neq n_j$, we know that $u^*_s|_{[a_j,b_j]}$ is identically constant. Therefore,

$$|f^*(c_j) - f^*(b_j)| \geq 4\psi\Big(\frac{2}{q}\Big).$$

Hence we get

$$4k\psi\Big(\frac{2}{q}\Big) \leq |f^*(b_j) - f^*(c_j)| \leq |f^*(a) - f^*(c_j)| + |f^*(a) - f^*(b_j)|,$$

implying that there is an $x \in \{c_j, b_j\}$ such that $|f^*(x) - f^*(a)| \geq 2k\psi(\frac{2}{q})$. Moreover, we have $\psi(x-a) \leq \psi(\frac{2}{q})$ (recall that ψ is an increasing function). Hence

$$\frac{|f^*(a) - f^*(x)|}{\psi(x-a)} \geq 2k.$$

Since $b_j \in [a, a+\frac{1}{k}]$, it follows that the condition of the definition of M_k is not fulfilled. Since a was arbitrarily chosen, we see that $f^* \notin M_k$. □

Exercise 11.3.5. Verify the existence of a strictly increasing continuous function $\varphi : \mathbb{I} \longrightarrow \mathbb{I}$ with the following properties:

$$\varphi(0) = 0,\ \varphi(1) = 1,\ D_+\varphi(0) < \infty,\ D^+\varphi(0) = \infty,$$

such that $\varphi|_{(0,1)}$ is differentiable.

Now we are able to prove the existence of a Besicovitch–Morse function using Proposition 11.3.3 and Theorem 11.3.4.

Corollary 11.3.6. *There exists a Besicovitch–Morse function on* $\mathbb{I}$.

Proof. Take a continuous strictly increasing function $\varphi : \mathbb{I} \longrightarrow \mathbb{R}$ as in Exercise 11.3.5. Applying Proposition 11.3.3 yields that every function from $A_\varphi(E)$ has nowhere an infinite right-sided derivatives on $[0, 1)$.

Note that the condition $D^+\varphi(0) = \infty$ implies (take $\psi = \mathrm{id}_{\mathbb{I}}$) that the assumptions in Theorem 11.3.4 are satisfied. Therefore, there is a subset $S^+ \subset A_\varphi(E)$ of second category, where

$$S^+ := \Big\{g \in A_\varphi(E) : \forall_{a\in[0,1)} :\ \limsup_{x\to a+} |\boldsymbol{\Delta} f(a,x)| = \infty\Big\}.$$

By a similar argument, we see that also the set

$$S^- := \Big\{g \in A_\varphi(E) : \forall_{a\in(0,1]} :\ \limsup_{x\to a-} |\boldsymbol{\Delta} f(a,x)| = \infty\Big\}$$

is of second category in $A_\varphi(E)$. Hence, $\varnothing \neq S^+ \cap S^- \subset \mathfrak{BM}(\mathbb{I})$, which says that there exists a Besicovitch–Morse function in $A_\varphi(E)$. □

Remark 11.3.7. Note that the proof of the former corollary tells us even more than stated there, namely, that *the typical function in $A_\varphi(E)$ is a Besicovitch–Morse function.*

We conclude this section by adding some results on the existence of a Hölder continuous Besicovitch–Morse function.

Proposition 11.3.8. *For every $\alpha \in (0,1)$, there exists a Besicovitch–Morse function $f \in \mathfrak{H}^\alpha(\mathbb{I})$.*

Proof. [1] Step 1°. Fix an $\alpha \in (0,1)$ and put $\mu(t) := t^\alpha$. Then there is a decreasing function g such that $\mu(x) = \int_0^x g(t)dt$ (note that this representation is always true if μ is a concave

[1] We thank Professor Jan Maly for the idea of the proof.

function). In our case, $g(t) = \alpha t^{\alpha-1}$, $t \in (0,1]$. Now we choose a strictly decreasing sequence $(x_k)_{k=1}^{\infty} \in (0,1]$, which converges to 0 (the precise shape of the x_k's will be given later). Fix $x_1 := 1$, $x_2 := 1/2$. Put, for $t \in (0,1]$,

$$h(t) := \begin{cases} g(t), & \text{if } x_{2k+1} < t \le x_{2k},\ k \in \mathbb{N} \\ 0, & \text{if } x_{2k} < t \le x_{2k-1},\ k \in \mathbb{N} \end{cases}$$

and define $\varphi_1(t) := \int_0^t h(\tau)d\tau$, $t \in (0,1]$, $\varphi_1(0) := 0$. Then $\varphi_1 \le \mu$, and φ_1 is an increasing continuous function on $\mathbb{I}$, which is identically equal to $\varphi_1(x_{2k})$ on the intervals $[x_{2k}, x_{2k-1}]$, $k \in \mathbb{N}$.

Then

$$\frac{\varphi_1(x_{2k})}{x_{2k}} \ge \frac{x_{2k}^{\alpha} - x_{2k+1}^{\alpha}}{x_{2k}} \ge \frac{1}{x_{2k}^{1-\alpha}} - 1 \underset{k\to+\infty}{\longrightarrow} +\infty,$$

where $x_{2k+1} := x_{2k}^{1/\alpha} < x_{2k}$, $k \in \mathbb{N}$.

Moreover, for $k \in \mathbb{N}_3$, we have

$$\frac{\varphi_1(x_{2k+1})}{x_{2k+1}} = \frac{1}{x_{2k+1}} \sum_{\ell=k+1}^{\infty} (x_{2\ell}^{\alpha} - x_{2\ell+1}^{\alpha}) \le \frac{1}{x_{2k}^{1/\alpha}} \sum_{\ell=k+1}^{\infty} x_{2\ell}^{\alpha} =: A_k.$$

Now choose an $s \in \mathbb{N}$ with $s > 1/\alpha$ and put $x_{2(m+1)} := x_{2m}^{s^2}$, $m \in \mathbb{N}_2$. Note that $x_{2(m+1)} < x_{2m+1}$ for all m. On applying $\alpha s^{2\ell} - s \ge (\ell - 1)s$, $\ell \in \mathbb{N}$, it follows that

$$A_k = \frac{1}{x_{2k}^{1/\alpha}} \sum_{\ell=k+1}^{\infty} (x_{2k})^{\alpha s^{2(\ell-k)}} = \frac{1}{x_{2k}^{1/\alpha}} \sum_{\ell=1}^{\infty} (x_{2k})^{\alpha s^{2\ell}} = x_{2k}^{s-1/\alpha} \sum_{\ell=1}^{\infty} (x_{2k})^{\alpha s^{2\ell}-s}$$
$$\le x_{2k}^{s-1/\alpha} \sum_{\ell=0}^{\infty} (x_{2k}^{s})^{\ell} = x_{2k}^{s-1/\alpha} \frac{1}{1 - x_{2k}^{s}} \underset{k\to+\infty}{\longrightarrow} 0.$$

Hence we end up with $D^{+}\varphi_1(0) = +\infty$ and $D_{-}\varphi_1(0) = 0$.

Step 2°. Now we will smooth the function φ_1 to get a new function, called $\widetilde{\varphi}$, in such a way that $\varphi_1 \le \widetilde{\varphi} \le \mu$, $\lim_{k\to+\infty} \frac{\widetilde{\varphi}(x_{2k})}{x_{2k}} = +\infty$, and $\lim_{k\to+\infty} \frac{\widetilde{\varphi}(x_{2k+1})}{x_{2k+1}} = 0$. To do so, we choose $\varepsilon_{2k} > 0$ such that:

- $2\varepsilon_{2k} < x_{2k-1} - x_{2k}$,
- $\varepsilon_{2k} g(x_{2k}) < \frac{x_{2k+1}}{2^k}$,
- $g(x) \ge g(x_{2k}) \frac{x_{2k} + \varepsilon_{2k} - x}{\varepsilon_{2k}}$, $x_{2k} \le x \le x_{2k} + \varepsilon_{2k}$, $k \in \mathbb{N}$.

For $t \in \mathbb{I}$, put

$$\widetilde{h}(t) := \begin{cases} h(t), & \text{if } x_{2k+1} \le t \le x_{2k},\ k \in \mathbb{N}, \\ g(x_{2k}) \frac{x_{2k} + \varepsilon_{2k} - t}{\varepsilon_{2k}}, & \text{if } t \in [x_{2k}, x_{2k} + \varepsilon_{2k}],\ k \in \mathbb{N}, \\ g(x_{2k-1}) \frac{t - x_{2k-1} + \varepsilon_{2k}}{\varepsilon_{2k}}, & \text{if } t \in [x_{2k-1} - \varepsilon_{2k}, x_{2k-1}],\ k \in \mathbb{N}, \\ 0, & \text{otherwise} \end{cases}.$$

Then the function $\widetilde{\varphi}$ defined as $\widetilde{\varphi}(x) := \int_0^x \widetilde{h}(\tau)d\tau$, $x \in \mathbb{I}$, is an increasing continuous function on $\mathbb{I}$, differentiable on $(0,1)$, that satisfies $\varphi_1 \le \widetilde{\varphi} \le \mu$. Thus $\frac{\widetilde{\varphi}(x_{2k})}{x_{2k}} \ge \frac{\varphi_1(x_{2k})}{x_{2k}} \underset{k\to+\infty}{\longrightarrow} +\infty$.

Note that

$$\widetilde{\varphi}(x_{2k+1}) = \sum_{\ell=k+1}^{\infty} (x_{2\ell}^{\alpha} - x_{2\ell+1}^{\alpha}) + \sum_{\ell=k+1}^{\infty} g(x_{2\ell})\varepsilon_{2\ell}$$

$$\leq \sum_{\ell=k+1}^{\infty} (x_{2\ell}^{\alpha} - x_{2\ell+1}^{\alpha}) + \sum_{\ell=k+1}^{\infty} x_{2\ell+1}/2^{\ell}.$$

Therefore,

$$\frac{\widetilde{\varphi}(x_{2k+1})}{x_{2k+1}} \leq A_k + 1/2^k \underset{k\to+\infty}{\longrightarrow} 0.$$

Step 3°. Put $\varphi(t) := \widetilde{\varphi}(t) + t^2$, $t \in \mathbb{I}$. Then φ is strictly increasing with $\varphi(1) \geq 1$. Moreover, $D_+\varphi(0) = +\infty$ and $D_-\varphi(0) = 0$.

Step 4°. Finally, if $0 \leq x < y \leq 1$, then

$$\begin{aligned}\frac{\varphi(y) - \varphi(x)}{\mu(y-x)} &\leq \frac{\int_x^y \widetilde{h}(t)dt}{\mu(y-x)} + \frac{y^2 - x^2}{(y-x)^{\alpha}} \\ &\leq \frac{\int_x^y g(t)dt}{\mu(y-x)} + 2(y-x)^{1-\alpha} \leq \frac{\mu(y) - \mu(x)}{\mu(y-x)} + 2 \leq 3,\end{aligned}$$

because μ is a strictly increasing concave function with $\mu(0) = 0$ and therefore subadditive. Hence,

$$H_{\mu}(\varphi) := \sup\left\{\frac{|\varphi(y) - \varphi(x)|}{\mu(|y-x|)} : x, y \in \mathbb{I},\ x \neq y\right\} < \infty.$$

To finish the proof, we need the following lemma.

Lemma 11.3.9. *Let μ be as above and let $\varphi \in \mathcal{C}(\mathbb{I})$ be an increasing function with $\varphi(0) = 0$. If $H_{\mu}(\varphi) \leq M$, then $H_{\mu}(f) \leq 2M$ for every function $f \in A_{\varphi}(E)$.*

Proof. Let $f \in A_{\varphi}(E)$, i.e., $f = A_{\varphi}(u)$, where $u \in E$. Fix points $x < y$ in $\mathbb{I}$ and put $m := \min\{n \in \mathbb{N} : u_n(x) \neq u_n(y)\}$. By virtue of the definition of E, we find (since u_m is not identically constant on the interval $J := [x, y]$) a point $z \in J$ with $u_{m+1}(z) = 0$.

Assume that m is an odd number. Then

$$f(y) = \sum_{n=1}^{\infty} (-1)^{n+1} \varphi \circ u_n(y) \leq \sum_{n=1}^{m} (-1)^{n+1} \varphi \circ u_n(y),$$

$$f(x) \geq \sum_{n=1}^{m+1} (-1)^{n+1} \varphi \circ u_n(x).$$

Recalling that $u_n(x) = u_n(y)$ for $n < m$ and the Lipschitz property of the u_n's, we get

$$\begin{aligned}f(y) - f(x) &\leq \varphi(u_m(y)) - \varphi(u_m(x)) + \varphi(u_{m+1}(x)) - \varphi(u_{m+1}(z)) \\ &\leq H_{\mu}(\varphi)\mu(|u_m(x) - u_m(y)|) + H_{\mu}(\varphi)\mu(|u_{m+1}(x) - u_{m+1}(z)|) \\ &\leq 2M\mu(|x-y|).\end{aligned}$$

Similarly, if m is even, then $f(x) - f(y) \leq 2M\mu(|y-x|)$ (EXERCISE). □

Hence, by virtue of Proposition 11.3.3 and Theorem 11.3.4, we get α-Hölder continuous Besicovitch–Morse functions. □

Chapter 12
Linear Spaces of Nowhere Differentiable Functions

Summary. This chapter gives some ideas for studying linear structures within the nonlinear set $\mathcal{ND}(\mathbb{I})$.

12.1 Introduction

A theorem of S. Banach and S. Mazur (see [AK06a], Theorem 1.4.3) states that every separable Banach space[1] X is isometrically embedded as a closed subspace of $\mathcal{C}(\mathbb{I})$. The theorem tells us that $\mathcal{C}(\mathbb{I})$ is a "really big" space, big enough to contain every possible separable Banach space. So one can ask whether we can require more properties of the functions in the image of the embedding or when these properties place restrictions on the Banach space X to be embedded. In this direction, the following results are known:

- If E is a closed linear subspace of $\mathcal{C}(\mathbb{I})$ such that every function $f \in E$ has bounded variation, then E is necessarily finite-dimensional [LM40].
- If E is a closed linear subspace of $\mathcal{C}(\mathbb{I})$ such that every function $f \in E$ is differentiable at every point of $\mathbb{I}$, then E is finite-dimensional (cf. [Gur67]).
- If E is infinite-dimensional and every function in E has a derivative at every point of $(0,1]$, then E must contain an isomorphic copy of c_0.[2] (See [Gur67].) In fact, E must be isomorphic to a subspace of c_0.
- If ℓ^1 can be embedded in $\mathcal{C}(\mathbb{I})$ as a linear subspace,[3] then there exists a function in the image of the embedding that is nondifferentiable at every point of a perfect subset[4] of $\mathbb{I}$ (see [PT84]).
- In [Gur91], using trigonometric sums, an infinite-dimensional subspace E of $\mathcal{C}(\mathbb{I})$ is constructed such that every $f \in E \setminus \{0\}$ is nowhere differentiable on $\mathbb{I}$.
- In [FGK99], the authors used van der Waerden's functions to give a closed subspace of $\mathcal{C}(\mathbb{I})$ that is isomorphic to ℓ^1.

[1] That is, the space contains a countable dense subset.

[2] Recall that $c_0 := \{(a_n)_{n=1}^{\infty} \subset \mathbb{R} : a_n \longrightarrow 0\}$ with the supremum norm.

[3] Recall that $\ell^1 := \{a = (a_n)_{n=1}^{\infty} \subset \mathbb{R} : \|a\|_{\ell^1} := \sum_{n=1}^{\infty} |a_n| < +\infty\}$.

[4] A closed set $A \subset \mathbb{I}$ is called *perfect* if each of its points is an accumulation point of A.

M. Jarnicki, P. Pflug, *Continuous Nowhere Differentiable Functions*, Springer Monographs in Mathematics, DOI 10.1007/978-3-319-12670-8_12

- There is a stronger version of the Banach–Mazur result by Rodriguez-Piazza (see [RP95]), namely that every separable Banach space can be isometrically embedded as a subspace E of $\mathcal{C}(\mathbb{I})$ such that $f \in \mathcal{ND}(\mathbb{I})$ whenever $f \in E \setminus \{0\}$.

Our aim in this section is to study the $\mathfrak{m}$-lineability (resp. $\mathfrak{m}$-spaceability) of $\mathcal{ND}_{\pm}(\mathbb{I})$, where $\mathfrak{m} \in \{\aleph_0, \mathfrak{c}\}$ ($\mathfrak{c}$ stands for the continuum). A set $M \subset B$ of an infinite-dimensional Banach space B is called $\mathfrak{m}$-*lineable* (resp. $\mathfrak{m}$-*spaceable*) if there is an $\mathfrak{m}$-dimensional (resp. a closed $\mathfrak{m}$-dimensional) subspace $E \subset B$ such that $E \setminus \{0\} \subset M$. These notions were first introduced in an unpublished paper by Enflo and Gurariy (see [EG]) that circulated among specialists at the beginning of this century (see also the final and extended version of this preprint [EGSS14]).[5] In the sequel, the Banach space B will be given by $\mathcal{C}(\mathbb{I})$.

12.2 $\mathfrak{c}$-Lineability of $\mathcal{ND}^{\infty}(\mathbb{R})$

In [Gur91], Gurariy presented a nonconstructive proof of the fact that $\mathcal{ND}(\mathbb{R})$ is $\aleph_0$-lineable. The aim here is to give a constructive proof of the following stronger result due to Jiménez-Rodriguez et al. (see [JRMFSS13]).

Theorem 12.2.1. *The set $\mathcal{ND}^{\infty}(\mathbb{R}) \cap \mathcal{ND}_{\pm}(\mathbb{R})$ is $\mathfrak{c}$-lineable.*

Proof. For $a \in (0,1)$, we put $f_a(x) := \boldsymbol{C}_{a,9}(x) = \sum_{n=0}^{\infty} a^n \cos(2\pi 9^n x)$, $x \in \mathbb{R}$. Now the proof consists in proving the following two lemmas.

Lemma 12.2.2. *Let $0 < a_1 < a_2 < \cdots < a_k < 1$. Then the functions $f_j := f_{a_j}$, $j = 1, \ldots, k$, are linearly independent.*

Proof. Let $(\lambda_1, \ldots, \lambda_k) \in \mathbb{R}^k$ be such that $\sum_{j=1}^{k} \lambda_j f_j = 0$. To conclude that all the λ_j vanish, it suffices (together with the Vandermonde determinant) to prove the following claim:

$$\sum_{j=1}^{k} \lambda_j a_j^n = 0 = \sum_{j=1}^{k} \lambda_j \frac{a_j^{n+1}}{1-a_j}, \qquad n \in \mathbb{N}_0.$$

We will use induction. If $n = 0$, then

$$\begin{aligned} 0 &= \sum_{j=1}^{k} \lambda_j f_j\Big(\frac{1}{6}\Big) = \sum_{j=1}^{k} \lambda_j \Big(\cos\Big(\frac{\pi}{3}\Big) + \sum_{n=1}^{\infty} a_j^n \cos(3^n 3^{n-1} \pi) \Big) \\ &= \sum_{j=1}^{k} \lambda_j \Big(\cos\Big(\frac{\pi}{3}\Big) - \frac{a_j}{1-a_j} \Big) \end{aligned}$$

and

$$0 = \sum_{j=1}^{k} \lambda_j f_j\Big(\frac{1}{18}\Big) = \sum_{j=1}^{k} \lambda_j \Big(\cos\Big(\frac{\pi}{9}\Big) - \frac{a_j}{1-a_j} \Big).$$

Therefore, $0 = (\cos(\frac{\pi}{3}) - \cos(\frac{\pi}{9})) \sum_{j=1}^{k} \lambda_j$, which immediately implies the claim in the case $n = 0$.

[5] We thank Professor L. Bernal-González for informing us about the history of [EG].

Assume now that the claim is true for all $m \in \mathbb{N}_0$, $0 \le m \le n$. Similarly to the argument above, we calculate

$$
\begin{aligned}
0 &= \sum_{j=1}^k \lambda_j f_j\Big(\frac{1}{2\cdot 9^{n+2}}\Big) = \sum_{j=1}^k \lambda_j\Big(\cos\Big(\frac{\pi}{9^{n+2}}\Big) + a_j\cos\Big(\frac{\pi}{9^{n+1}}\Big)\\
&\quad + \cdots + a_j^n\cos\Big(\frac{\pi}{9^2}\Big) + a_j^{n+1}\cos\Big(\frac{\pi}{9}\Big) - \frac{a_j^{n+2}}{1-a_j}\Big)\\
&\underset{\text{induction assumption}}{\overset{\text{use twice}}{=}} \sum_{j=1}^k \lambda_j\Big(a_j^{n+1}\cos\Big(\frac{\pi}{9}\Big) - \frac{a_j^{n+2}}{1-a_j} + \frac{a_j^{n+1}}{1-a_j}\Big)\\
&= \sum_{j=1}^k \lambda_j a_j^{n+1}\Big(\cos\Big(\frac{\pi}{9}\Big)+1\Big),
\end{aligned}
$$

implying the first part of the claim and then the second part. □

Put $E := \operatorname{span}(\{f_a : 7/9 < a < 1\})$. Then $\dim E = \mathfrak{c}$.

Lemma 12.2.3. *Let $7/9 < a_k < \cdots < a_1 < 1$ and $(\lambda_1,\ldots,\lambda_k) \in \mathbb{R}^k \setminus \{0\}$ be given. Then $f := \sum_{j=1}^k \lambda_j f_j \in \mathcal{ND}_\pm(\mathbb{R}) \cap \mathcal{ND}^\infty(\mathbb{R})$, where $f_j := f_{a_j}$.*

Proof. We may assume that $\lambda_1 > 0$. Fix an $x \in \mathbb{R}$. The proof of the nonexistence of derivatives at x will be based on the estimates from the proof of Theorem 3.5.1. Recall the essential part:

$$
\boldsymbol{\Delta} f_j(x, x_m^\pm) = \mp(-1)^{\alpha_m} 2(9a_j)^m\Big(\frac{\pi}{9a_j-1}V_{j,m,\pm} + \frac{2}{3}U_{j,m,\pm}\Big),
$$

where $\alpha_m \in \mathbb{Z}$ is such that

$$
h_m := 2\cdot 9^m x - \alpha_m \in (-\tfrac12, \tfrac12], \quad x_m^\pm := \frac12(\alpha_m \pm 1)9^{-m},
$$
$$
U_{j,m,\pm} \ge 1, \quad |V_{j,m,\pm}| \le 1.
$$

Note that

$$
H_{j,m} := \Big(\frac{\pi}{9a_j-1}V_{j,m,\pm} + \frac23 U_{j,m,\pm}\Big) \ge \frac23 - \frac{\pi}{9a_1-1} > \frac{4-\pi}{6} =: \beta.
$$

Therefore, $\operatorname{sgn} \boldsymbol{\Delta} f_j(x, x_m^+) = -\operatorname{sgn} \boldsymbol{\Delta} f_j(x, x_m^-)$.

Putting all this information together, we obtain

$$
\begin{aligned}
\boldsymbol{\Delta} f(x, x_m^+) &= \sum_{j=1}^k \lambda_j \boldsymbol{\Delta} f_j(x, x_m^+) = \sum_{j=1}^k \lambda_j (-1)^{\alpha_m+1} 2(9a_j)^m H_{j,m}\\
&= (-1)^{\alpha_m+1}2(9a_1)^m\Big(\lambda_1 H_{1,m} + \sum_{j=2}^k \lambda_j\Big(\frac{a_j}{a_1}\Big)^m H_{j,m}\Big),
\end{aligned}
$$

where $\lambda_1 H_{1,m} + \sum_{j=2}^k \lambda_j(\frac{a_j}{a_1})^m H_{j,m} \ge \lambda_1\beta/2$ for sufficiently large m. Therefore, $\max\{|D^+f(x)|, |D_-f(x)|\} = \max\{|D^-f(x)|, |D_-f(x)|\} = +\infty$. Moreover, if, e.g., $\boldsymbol{\Delta} f(x, x_{m_k}^+) \longrightarrow \pm\infty$, then $\boldsymbol{\Delta} f(x, x_{m_k}^-) \longrightarrow \mp\infty$. Thus an infinite derivative $f'(x)$ does not exist. □ □

12.3 Spaceability of $\mathcal{ND}_\pm(\mathbb{I})$

The aim of this section is to construct a closed linear subspace E of $\mathcal{C}(\mathbb{I})$, norm-isomorphic to ℓ^1, such that $E \setminus \{0\} \subset \mathcal{ND}_\pm(\mathbb{I})$.

Theorem 12.3.1 (Theorem of Berezhnoi (see [Ber03])). *There exists a closed linear subspace E of $\mathcal{C}(\mathbb{I})$, $\dim E = \infty$, such that $E \setminus \{0\} \subset \mathcal{ND}_\pm(\mathbb{I})$. Moreover, E is norm-isomorphic to the sequence space ℓ^1.*

Remark 12.3.2. A weaker result is contained in [FGK99], namely that there exist a closed linear subspace $E \subset \mathcal{C}(\mathbb{I})$, $\dim E = \infty$, and a subset $A \subset \mathbb{I}$ with $\mathcal{L}^1(A) = 1$ such that if $f \in E \setminus \{0\}$, then $f \in \mathcal{ND}((0,1))$, and moreover, f has neither a finite right-sided nor a finite left-sided derivative at any point of A.

The proof of Theorem 12.3.1 will be done in several steps.

12.3.1 Two Matrices

Let $D = (d_{j,k})_{j,k\in\mathbb{N}}$ (resp. $H = (n_{j,k})_{j,k\in\mathbb{N}}$) be an upper-triangular matrix, i.e., $d_{j,k} = 0$ (resp. $n_{j,k} = 0$) for $k < j$, such that $d_{j,k} > 0$ (resp. $n_{j,k} \in \mathbb{N}$), $k \ge j \in \mathbb{N}$. Moreover, we assume that these matrices satisfy the following conditions:

$$\text{if } j \in \mathbb{N}, k \in \mathbb{N}_{j+1}, \quad \text{then} \quad d_{j,k} > 2^{3+k} d_{j+1,k}; \tag{12.3.1}$$
$$\text{if } j \in \mathbb{N}, k \in \mathbb{N}_j, \quad \text{then} \quad d_{j,k+1} > 2^3 d_{j,k}; \tag{12.3.2}$$
$$\text{if } k \in \mathbb{N}, \quad \text{then} \quad \frac{n_{k+1,k+1}}{8n_{1,k}} \in \mathbb{N}; \tag{12.3.3}$$
$$\text{if } j \in \mathbb{N}, k \in \mathbb{N}_{j+1}, \quad \text{then} \quad \frac{n_{j,k}}{8n_{j+1,k}} \in \mathbb{N}; \tag{12.3.4}$$
$$\text{if } j \in \mathbb{N}, \quad \text{then} \quad 1 < \frac{2d_{j,j}}{n_{j,j}} < 2; \tag{12.3.5}$$
$$\text{if } j \in \mathbb{N}, k \in \mathbb{N}_j, \quad \text{then} \quad \frac{d_{j,k+1}}{n_{j,k+1}} \le \frac{d_{j,k}}{8n_{j,k}}. \tag{12.3.6}$$

Proposition 12.3.3. *There exist matrices D and H as above with properties* (12.3.1)–(12.3.6).

Proof. Fix an arbitrary $n_{1,1} \in \mathbb{N}$. Then choose a positive number $d_{1,1}$ such that (12.3.5) is true. Next we find an $n_{2,2} \in \mathbb{N}$ such that (12.3.3) holds, and then we take a positive $d_{2,2}$ satisfying (12.3.5). Now choose a positive $d_{1,2}$ with (12.3.1) and (12.3.2). Finally, fix an $n_{1,2} \in \mathbb{N}$ such that (12.3.4) and (12.3.6) hold. Thus the 2×2 left upper parts of D and H are constructed. What remains is to use an induction argument to complete the proof (Exercise). □

Corollary 12.3.4. *Let H and D be as in Proposition 12.3.3 and put $h_{j,k} := \frac{1}{n_{j,k}}$, $j \in \mathbb{N}$, $k \in \mathbb{N}_j$. Then:*

$$8^{j-1} h_{1,k} \le h_{j,k}, \quad \text{if } j \in \mathbb{N},\ k \in \mathbb{N}_j; \tag{12.3.7}$$
$$h_{1,k} \ge 8h_{k+1,k+1}, \quad \text{if } k \in \mathbb{N}; \tag{12.3.8}$$
$$n_{j,k} \ge 8^{k-j} d_{j,k}, \quad \text{if } j \in \mathbb{N},\ k \in \mathbb{N}_j; \tag{12.3.9}$$
$$n_{j,\ell} \in 8n_{j,k+1}\mathbb{N}, \quad \text{if } j \in \mathbb{N},\ k \in \mathbb{N}_j,\ \ell \in \mathbb{N}_{k+1};$$

$$
\begin{aligned}
&n_{j,k+1} \in 2n_{1,k}\mathbb{N}, && \textit{if } j \in \mathbb{N},\ k \in \mathbb{N}_j; && (12.3.10)\\
&n_{j,m} \in n_{k+1,k+1}\mathbb{N}, && \textit{if } j \in \mathbb{N},\ k \in \mathbb{N}_j,\ m \in \mathbb{N}_{k+1} \textit{ or} \\
& && k \in \mathbb{N},\ j \in \mathbb{N}_{k+1},\ m \in \mathbb{N}_j; && (12.3.11)\\
&d_{j,k} \geq 8^{k-j-s} d_{j,j+s}, && \textit{if } j \in \mathbb{N},\ k \in \mathbb{N}_j,\ s \in \mathbb{N}_0 \setminus \mathbb{N}_{k-j+1};\\
&\sum_{s=0}^{k-1-j} \tfrac{d_{j,j+s}}{d_{j,k}} \leq \tfrac{1}{7}, && \textit{if } j \in \mathbb{N},\ k \in \mathbb{N}_{j+1}. && (12.3.12)
\end{aligned}
$$

Proof. The proof is left as an easy EXERCISE. □

12.3.2 Auxiliary Functions

From now on, we fix matrices D and H as in Proposition 12.3.3. Recall the function $\psi(x) = \operatorname{dist}(x, \mathbb{Z})$ (see Chap. 4). Put $\varphi_0(x) := -\frac{1}{4} + \psi(x)$, $x \in \mathbb{R}$. Obviously, φ_0 has period 1 and satisfies $|\varphi_0(x) - \varphi_0(y)| \leq |x - y|$, $x, y \in \mathbb{R}$. Define

$$\varphi_{j,k}(t) = \tfrac{1}{n_{j,k}} \varphi_0(n_{j,k} t), \quad t \in \mathbb{R},\ j \in \mathbb{N},\ k \in \mathbb{N}_j.$$

Then the functions $\varphi_{j,k}$ are continuous and satisfy $|\varphi_{j,k}(x) - \varphi_{j,k}(y)| \leq |x - y|$, $x, y \in \mathbb{R}$.

In the sequel, we will use the abbreviations $h_{j,k} := \frac{1}{n_{j,k}}$ (see Corollary 12.3.4) and $\tau_{j,k,\ell} := \frac{\ell h_{j,k}}{2}$, $\ell \in \mathbb{Z}$.

Lemma 12.3.5. *Let $s, t \in \mathbb{R}$, $j \in \mathbb{N}$, $k \in \mathbb{N}_j$, and $\ell \in \mathbb{Z}$.*

(a) *If $\tau_{j,k,\ell} \leq s, t \leq \tau_{j,k,\ell+1}$, then $|\varphi_{j,k}(t) - \varphi_{j,k}(s)| = |t - s|$.*
(b) *If $t < \tau_{j,k,\ell} < t + h_{k+1,k+1}$, then $|\varphi_{j,k}(t + h_{1,k}/2) - \varphi_{j,k}(t)| \geq h_{1,k}/4$.*

Proof. (a) Step 1°. If $\ell = 2m$, $m \in \mathbb{Z}$, then $m \leq n_{j,k}s, n_{j,k}t \leq m + 1/2$. Recall that φ_0 is a linear function on $[0, 1/2]$. Therefore,

$$|\varphi_{j,k}(s) - \varphi_{j,k}(t)| = h_{j,k}|\varphi_0(n_{j,k}s - m) - \varphi_0(n_{j,k}t - m)| = |s - t|.$$

Step 2°. If $\ell = 2m + 1$, then $m + 1/2 \leq n_{j,k}s, n_{j,k}t \leq m + 1$. Thus (a) follows with the same argument as in Step 1°, since φ_0 is also linear on $[1/2, 1]$.

(b) By the assumption and (12.3.7), we have

$$t + h_{1,k}/2 \leq \tau_{j,k,\ell} + h_{1,k}/2 \leq \ell h_{j,k}/2 + h_{1,k}/2 \leq (\ell + 1)h_{j,k}/2 = \tau_{j,k,\ell+1}.$$

On the other hand, using (12.3.8), we get

$$t + h_{1,k}/2 \geq h_{1,k}/2 - h_{k+1,k+1} + \tau_{j,k,\ell} \geq \tau_{j,k,\ell}.$$

Thus $t + h_{1,k}/2 \in [\tau_{j,k,\ell}, \tau_{j,k,\ell+1}]$. Applying (a), it follows that

$$
\begin{aligned}
&|\varphi_{j,k}(t + h_{1,k}/2) - \varphi_{j,k}(t)| \\
&\quad \geq |\varphi_{j,k}(t + h_{1,k}/2) - \varphi_{j,k}(\tau_{j,k,\ell})| - |\varphi_{j,k}(\tau_{j,k,\ell}) - \varphi_{j,k}(t)| \\
&\quad \geq |t + h_{1,k}/2 - \tau_{j,k\ell}| - |\tau_{j,k,\ell} - t| \geq h_{1,k}/2 - 2|\tau_{j,k,\ell} - t|
\end{aligned}
$$

$$\geq h_{1,k}/2 - 2h_{k+1,k+1} = h_{1,k}(1 - 4\tfrac{h_{k+1,k+1}}{h_{1,k}})/2 \geq h_{1,k}(1 - 4/8)/2$$
$$= h_{1,k}/4,$$

where we used (12.3.8).

□

12.3.3 The Closed Linear Subspace $E \subset \mathcal{ND}_{\pm}(\mathbb{I})$

For $j \in \mathbb{N}$, put

$$\psi_j(t) := \sum_{k=j}^{\infty} d_{j,k}\varphi_{j,k}(t), \quad t \in \mathbb{R}.$$

By virtue of (12.3.9), note that

$$\sum_{k=j}^{\infty} d_{j,k}|\varphi_{j,k}(t)| \leq \tfrac{1}{4}\sum_{k=j}^{\infty} d_{j,k}h_{j,k} \leq \tfrac{1}{4}\sum_{k=j}^{\infty} 8^{-k+j} = \tfrac{2}{7};$$

thus ψ_j is a continuous function on $\mathbb{R}$.

Let now $j, k, m \in \mathbb{N}$ with $j \leq k < m$. Then by virtue of (12.3.11), we get

$$\varphi_{j,m}(t + h_{k+1,k+1}) = h_{j,m}\varphi_0(n_{j,m}t + n_{j,m}h_{k+1,k+1}) = h_{j,m}\varphi_0(n_{j,m}t) = \varphi_{j,m}(t).$$

Thus all the functions $\varphi_{j,m}$ have period $h_{k+1,k+1}$ whenever $j \leq k < m$. In particular, using (12.3.10) and (12.3.11), we see that $\varphi_{j,m}$ has also period $h_{1,k}/2$.

Lemma 12.3.6. *Let $j \in \mathbb{N}$, $k \in \mathbb{N}_{j+1}$, $\ell \in \mathbb{Z}$, and $t \in \mathbb{R}$.*

(a) *If $\tau_{j,k,\ell} \leq t < t + h_{k+1,k+1} < \tau_{j,k,\ell+1}$, then*

$$|\psi_j(t + h_{k+1,k+1}) - \psi_j(t)| \geq \tfrac{6}{7}d_{j,k}h_{k+1,k+1}.$$

(b) *If $t < \tau_{j,k,\ell} < t + h_{k+1,k+1}$, then*

$$|\psi_j(t + h_{1,k}/2) - \psi_j(t)| \geq \tfrac{5}{28}d_{j,k}h_{1,k}.$$

Proof. (a) Recall that $\varphi_{j,m}$ has period $h_{k+1,k+1}$ as long as $m \geq k+1$, $k \in \mathbb{N}_j$. Thus, using Lemma 12.3.5 and (12.3.12), we obtain

$$|\psi_j(t + h_{k+1,k+1}) - \psi_j(t)| = \Big|\sum_{m=j}^{k} d_{j,m}(\varphi_{j,m}(t + h_{k+1,k+1}) - \varphi_{j,m}(t))\Big|$$
$$\geq d_{j,k}|\varphi_{j,k}(t + h_{k+1,k+1}) - \varphi_{j,k}(t)| - \sum_{m=j}^{k-1} d_{j,m}|\varphi_{j,m}(t + h_{k+1,k+1}) - \varphi_{j,m}(t)|$$
$$\geq h_{k+1,k+1}\Big(d_{j,k} - \sum_{m=j}^{k-1} d_{j,m}\Big) \geq \tfrac{6}{7}d_{j,k}h_{k+1,k+1}.$$

(b) Using that $\varphi_{j,\ell}$, $\ell \geq k+1$, $k \in \mathbb{N}_j$, has period $h_{1,k}/2$ (apply (12.3.3)), we get, using the same argument as before,

$$|\psi_j(t+h_{1,k}/2)-\psi_j(t)| = \Big|\sum_{m=j}^{k} d_{j,m}(\varphi_{j,m}(t+h_{1,k}/2)-\varphi_{j,m}(t))\Big|$$

$$\geq d_{j,k}|\varphi_{j,k}(t+h_{1,k}/2)-\varphi_{j,k}(t)| - \sum_{m=j}^{k-1} d_{j,m}|\varphi_{j,m}(t+h_{1,k}/2)-\varphi_{j,m}(t)|$$

$$\geq \tfrac{1}{4}d_{j,k}h_{1,k}\Big(1-2\sum_{m=j}^{k-1}\frac{d_{j,m}}{d_{j,k}}\Big) = \tfrac{5}{4\cdot 7}d_{j,k}h_{1,k}. \qquad \square$$

Put now $e_j := \frac{\psi_j}{\|\psi_j\|_{\mathbb{I}}}$, $j \in \mathbb{N}$. These will be the functions generating the linear subspace E we are looking for.

Lemma 12.3.7. *For $j \in \mathbb{N}$, we have $\frac{3}{28} \leq \frac{3}{14}d_{j,j}h_{j,j} \leq \|\psi_j\|_{\mathbb{I}} \leq \frac{2}{7}d_{j,j}h_{j,j} \leq \frac{2}{7}$.*

Proof. Fix a $t \in \mathbb{R}$ and a $j \in \mathbb{N}$. Then

$$|\psi_j(t)| \leq \sum_{m=j}^{\infty}\frac{1}{4}d_{j,m}h_{j,m} = \frac{1}{4}d_{j,j}h_{j,j}\sum_{m=j}^{\infty}\frac{d_{j,m}h_{j,m}}{d_{j,j}h_{j,j}} \leq \frac{2}{7}d_{j,j}h_{j,j},$$

where we have used (12.3.6).

Moreover,

$$\|\psi_j\|_{\mathbb{I}} \geq \tfrac{1}{4}\Big(d_{j,j}h_{j,j} - \sum_{m=j+1}^{\infty} d_{j,m}h_{j,m}\Big) \geq \tfrac{1}{4}\Big(2-\tfrac{8}{7}\Big) = \tfrac{3}{14}d_{j,j}h_{j,j},$$

which completes the proof. $\square$

Note that if $N \in \mathbb{N}$ and $a_1,\dots,a_N \in \mathbb{R}$, then $\|\sum_{j=1}^{N} a_j e_j\|_{\mathbb{I}} \leq \sum_{j=1}^{N}|a_j|$. So it remains to find an estimate of $\|\sum_{j=1}^{N} a_j e_j\|_{\mathbb{I}}$ from below.

Let $j \in \mathbb{N}$ and $\ell \in \mathbb{Z}$. Put

$$D(j,\ell)_{\pm} := \{t \in [\tau_{j,\ell}, \tau_{j,\ell+1}] : \pm\varphi_0(n_{j,j}) \geq \alpha/4\},$$

where $\alpha \in (1/7, 1/4)$. Then by a simple estimate (Exercise), we have

$$\begin{aligned}
&\text{if } t \in D(j,\ell)_+, \quad \text{then} \quad \psi_j(t) \geq \tfrac{d_{j,j}h_{j,j}}{4}(\alpha - \tfrac{1}{7}),\\
&\text{if } t \in D(j,\ell)_-, \quad \text{then} \quad \psi_j(t) \leq -\tfrac{d_{j,j}h_{j,j}}{4}(\alpha - \tfrac{1}{7}).
\end{aligned}$$

Note that if $\ell = 2m$, then

$$\begin{aligned}
&t \in D(j,\ell)_+ \quad \text{if and only if} \quad \tfrac{1}{4}(1+\alpha) \leq n_{j,j}t - m \leq \tfrac{1}{2},\\
&t \in D(j,\ell)_- \quad \text{if and only if} \quad 0 \leq t \leq n_{j,j}t - m \leq \tfrac{1}{4}(1-\alpha).
\end{aligned}$$

Moreover, if $\varepsilon := \pm$, then there exists an $\widetilde{m} = \widetilde{m}(m,\varepsilon) \in \mathbb{Z}$ such that

$$[\tau_{j+1,j+1,2\widetilde{m}}, \tau_{j+1,j+1,2\widetilde{m}+1}] \subset D(j,2m)_{\pm}.$$

Indeed, let us discuss in detail the case $\varepsilon = +$. Choose $\widetilde{m}$ such that

$$\frac{2\widetilde{m}-1}{2n_{j+1,j+1}} \leq \frac{1+\alpha+4m}{4n_{j,j}} \leq \frac{2\widetilde{m}}{2n_{j+1,j+1}} \leq \frac{1+\alpha+4m}{4n_{j,j}} + \frac{2}{2n_{j+1,j+1}}.$$

Then by virtue of $\alpha \le 1/4$ and $\frac{n_{j,j}}{n_{j+1,j+1}} \le 8^{-j}$, we get

$$\frac{2\widetilde{m}+1}{2n_{j+1,j+1}} \le \frac{1+\alpha+4m}{4n_{j,j}} + \frac{3}{2n_{j+1,j+1}} \le \frac{1}{4n_{j,j}}\Big(1+\alpha+4m+6\frac{n_{j,j}}{n_{j+1,j+1}}\Big)$$
$$\le \frac{1+\alpha+4m+\frac{6}{8}}{4n_{j,j}} \le \frac{2+4m}{4n_{j,j}}.$$

The second case is left as an EXERCISE.

Putting together all the information obtained thus far, we get the following result.

Proposition 12.3.8. *Let $N \in \mathbb{N}$, $a_j \in \mathbb{R}$ $(j = 1, \dots, N)$. Then*

$$\frac{7\alpha-1}{8}\sum_{j=1}^{N}|a_j| \le \Big\|\sum_{j=1}^{N} a_j e_j\Big\|_{\mathbb{I}} \le \sum_{j=1}^{N}|a_j|.$$

Proof. We may assume that $a_1 \ge 0$. Put $T_1 := D(1,0)_+$. If now $a_2 \ge 0$, then choose $m_2 \in \mathbb{Z}$ such that $T_2 := D(2,2m_2)_+ \subset T_1$. If $a_2 < 0$, take an $m_2 \in \mathbb{Z}$ such that $T_2 := D(2,2m_2)_- \subset T_1$, etc.

Fix a $t \in \bigcap_{j=1}^{N} T_j$, and recall that $a_j\psi_j(t) \ge 0$. Then

$$\Big|\sum_{j=1}^{n} a_j e_j(t)\Big| \ge \sum_{j=1}^{n} |a_j| \frac{\frac{1}{4}d_{j,j}h_{j,j}(\alpha-\frac{1}{7})}{\frac{2}{7}d_{j,j}h_{j,j}} = \frac{7\alpha-1}{8}\sum_{j=1}^{N}|a_j|. \qquad \square$$

Put $E := \overline{\operatorname{span}\{e_j : j \in \mathbb{N}\}}$. Then the previous proposition says that the mapping $\ell^1 \ni (a_j)_{j\in\mathbb{N}} \longmapsto \sum_{j=1}^{\infty} a_j e_j|_{\mathbb{I}}$ is a norm-isomorphism between ℓ^1 and the closed linear subspace E of $\mathcal{C}(\mathbb{I})$. Finally, it remains to verify that $E \setminus \{0\} \subset \mathcal{ND}_{\pm}(\mathbb{I})$.

Proposition 12.3.9. *If $f \in E \setminus \{0\}$, then $f \in \mathcal{ND}_{\pm}(\mathbb{I})$.*

Proof. Let $f = \sum_{j=j_0}^{\infty} a_j e_j \in E$ with $a_{j_0} \ne 0$, $j_0 \in \mathbb{N}$, and $t_0 \in [0,1)$. Recall that $\|(a_j)_{j=1}^{\infty}\|_{\ell^1} < +\infty$ and $\|\psi_j\|_{\mathbb{I}} \ge \frac{3}{28}$. Therefore, we may assume that $f = \sum_{j=j_0}^{\infty} a_j\psi_j$ with $\sum_{j=j_0}^{\infty}|a_j| \le 1$. We will show that f has no finite right-sided derivative at t_0.

Recall that $n_{k+1,k+1} \ge 8^k \underset{k\to+\infty}{\longrightarrow} \infty$. Thus there exists a $k_0 \in \mathbb{N}$ such that $h_{k,k} < 1 - t_0$.

Step 1°. Fix a $k \ge k_0$ and assume that there is an $\ell \in \mathbb{Z}$ such that $t_0, t_0 + h_{k+1,k+1} \in [\tau_{j_0,k,\ell}, \tau_{j_0,k,\ell+1}]$. We are going to estimate $A_k := |f(t_0 + h_{k+1,k+1}) - f(t_0)|$:

Using (12.3.11), if $j \le k < m$ or $k+1 \le j \le m$, then

$$\varphi_{j,m}(t + h_{k+1,k+1}) = h_{j,m}\varphi_0(n_{j,m}t + n_{j,m}h_{k+1,k+1}) = \varphi_{j,m}(t).$$

Therefore,

$$A_k = \Big|\sum_{j=j_0}^{k} a_j(\psi_j(t + h_{k+1,k+1}) - \psi_j(t_0))\Big|$$
$$\ge |a_{j_0}||\psi_{j_0}(t_0 + h_{k+1,k+1}) - \psi_{j_0}(t_0)|$$
$$- \sum_{j=j_0+1}^{k}|a_j|\sum_{m=j}^{k} d_{j,m}|\varphi_{j,m}(t_0 + h_{k+1,k+1}) - \varphi_{j,m}(t_0)|$$

$$\begin{aligned}
&\overset{\text{Lemma 12.3.6}}{\geq} \frac{6}{7}|a_{j_0}|d_{j_0,k}h_{k+1,k+1} - \sum_{j=j_0+1}^{k}|a_j|\sum_{m=j}^{k} d_{j,m}h_{k+1,k+1}\\
&\geq \frac{6}{7}h_{k+1,k+1}\Big(|a_{j_0}|d_{j_0,k} - \frac{7}{6}\sum_{j=j_0+1}^{k}|a_j|(d_{j,j}+d_{j,j+1}+\cdots+d_{j,k})\Big)\\
&\geq \frac{6}{7}h_{k+1,k+1}\Big(|a_{j_0}|d_{j_0,k} - \frac{4}{3}\sum_{j=j_0}^{k}|a_j|d_{j,k}\Big)\\
&\geq \frac{6}{7}d_{j_0,k}h_{k+1,k+1}\Big(|a_{j_0}| - \frac{4}{3}\sum_{j=j_0}^{k}\frac{d_{j,k}}{d_{j_0,k}}\Big)\\
&\geq \frac{6}{7}d_{j_0,k}h_{k+1,k+1}\Big(|a_{j_0}| - \frac{4}{3}\frac{8}{7}\frac{d_{j_0+1,k}}{d_{j_0,k}}\Big)\\
&\geq \frac{6}{7}d_{j_0,k}h_{k+1,k+1}\Big(|a_{j_0}| - \frac{32}{21}\frac{1}{2^{3+k}}\Big) \text{ and if } k \geq k_1 \geq k_0, \text{then}\\
&\geq \frac{6}{14}d_{j_0,k}h_{k+1,k+1}|a_{j_0}| \underset{k\to+\infty}{\longrightarrow} +\infty.
\end{aligned}$$

Thus, if there is a sequence $(s_k)_{k\in\mathbb{N}} \subset \mathbb{N}_{k_1}$ such that $t_0, t_0 + h_{s_k+1,s_k+1} \in [\tau_{j_0,s_k,\ell_k}, \tau_{j_0,s_k,\ell_k+1}]$, then

$$\frac{|f(t_0 + h_{s_k+1,s_k+1}) - f(t_0)|}{h_{s_k+1,s_k+1}} \underset{k\to+\infty}{\longrightarrow} +\infty,$$

meaning that f allows no finite right-sided derivative at t_0.

Step 2°. On the other hand, there is a $k_2 \geq k_0$ such that

$$t_0, t_0 + h_{k+1,k+1} \notin [\tau_{j_0,k,\ell}, \tau_{j_0,k,\ell+1}], \quad k \geq k_2,\ \ell \in \mathbb{Z}.$$

Thus there exists a sequence $(\ell_k)_{k\in\mathbb{N}_{k_2}}$ such that $t_0 \leq \tau_{j_0,k,\ell_k} \leq t_0 + h_{k+1,k+1}$ whenever $k \geq k_2$.

Similarly to how we proceeded above, we would like to estimate $B_k := |f(t_0 + \frac{h_{1,k}}{2}) - f(t_0)|$, $k \geq k_2$:

$$\begin{aligned}
B_k &= \Big|\sum_{j=j_0}^{\infty} a_j(\psi(t_0 + \tfrac{h_{1,k}}{2}) - \psi_j(t_0))\Big|\\
&= \Big|\sum_{j=j_0}^{\infty} a_j \sum_{m=j}^{\infty} d_{j,m}(\varphi_{j,m}(t_0 + \tfrac{h_{1,k}}{2}) - \varphi_{j,m}(t_0))\Big|.
\end{aligned}$$

Recall that if $k < j \leq m$, then $\frac{n_{j,m}}{2n_{1,k}} \in \mathbb{N}$. Therefore,

$$\begin{aligned}
B_k &= \Big|\sum_{j=j_0}^{k} a_j \sum_{m=j}^{k} d_{j,m}(\varphi_{j,m}(t_0 + \tfrac{h_{1,k}}{2}) - \varphi_{j,m}(t))\Big|\\
&\geq |a_{j_0}||\psi_{j_0}(t_0 + \tfrac{h_{1,k}}{2}) - \psi_{j_0}(t_0)|\\
&\quad - \sum_{j=j_0+1}^{k}|a_j|\sum_{m=j}^{k} d_{j,m}|\varphi_{j,m}(t_0 + \tfrac{h_{1,k}}{2}) - \varphi_{j,m}(t_0)| =: \widetilde{B}_k.
\end{aligned}$$

Using Lemma 12.3.6, it follows that

$$\begin{aligned}
B_k &\geq \frac{5}{28}|a_{j_0}|d_{j_0,k}h_{1,k} - \sum_{j=j_0+1}^{k} |a_j| \sum_{m=j}^{k} d_{j,m}\frac{h_{1,k}}{2} \\
&\geq \frac{5}{14}\frac{h_{1,k}}{2}\Big(|a_{j_0}|d_{j_0,k} - \frac{14}{5}\sum_{j=j_0+1}^{k}\sum_{m=j}^{k} d_{j,m}\Big) \\
&\geq \frac{5}{14}\frac{h_{1,k}}{2}\Big(|a_{j_0}|d_{j_0,k} - \frac{14}{5}\sum_{j=j_0+1}^{k} d_{j,k}\sum_{m=j}^{k}\Big(\frac{1}{8}\Big)^{k-m}\Big) \\
&\geq \frac{5}{14}\frac{h_{1,k}}{2}\Big(|a_{j_0}|d_{j_0,k} - \frac{16}{5}\sum_{j=j_0+1}^{k} d_{j,k}\Big) \\
&\geq \frac{5}{14}\frac{h_{1,k}}{2}d_{j_0,k}\Big(|a_{j_0}| - \frac{16}{5}\frac{8}{7}\frac{d_{j_0+1,k}}{d_{j_0,k}}\Big) \\
&\geq \frac{5}{14}\frac{h_{1,k}}{2}d_{j_0,k}\Big(|a_{j_0}| - \frac{16\cdot 8}{35\cdot 2^{3+k}}\Big) \geq \frac{5}{14}\frac{h_{1,k}}{2}d_{j_0,k}\frac{|a_{j_0}|}{2},
\end{aligned}$$

if $k \geq k^* \geq k_2$, k^* sufficiently large. Hence,

$$\frac{|f(t_0 + \frac{h_{1,k}}{2}) - f(t_0)|}{\frac{h_{1,k}}{2}} \geq \frac{5}{14}\frac{|a_{j_0}|d_{j_0,k}}{2} \geq \frac{5}{28}|a_{j_0}|2^{3(k-j_0)}d_{j_0,j_0} \underset{k\to+\infty}{\longrightarrow} +\infty,$$

implying again that f allows no finite right-sided derivative at t_0. □

Remark 12.3.10. In a recent paper (see [Bob14]), Bobok has shown that even $\mathfrak{ND}^{\infty}_{\pm}(\mathbb{I})$ is a spaceable set. The proof of this result is based on the construction of Besicovitch (see [Bes24]) and its description by Pepper (see [Pep28]). For earlier results, see also [Bob05, Bob07].

Part IV
Riemann Function

Chapter 13
Riemann Function

13.1 Introduction

The aim of this chapter is to discuss the problem of differentiability of the classical *Riemann function*

$$\boldsymbol{R}(x) := \sum_{n=1}^{\infty} \frac{\sin(\pi n^2 x)}{n^2}, \quad x \in \mathbb{R}.$$

To get some feeling of the behavior of $\boldsymbol{R}$ see Fig. 13.1.

At the outset, we should notice that $\boldsymbol{R} \notin \mathcal{ND}(\mathbb{R})$ (cf. Remark 13.1.1). Thus in fact, the function $\boldsymbol{R}$ is not in line with this book. Nevertheless, for many years, mathematicians believed that $\boldsymbol{R} \in \mathcal{ND}(\mathbb{R})$. The first to claim that $\boldsymbol{R} \in \mathcal{ND}(\mathbb{R})$ was B. Riemann (cf. [BR74], p. 28, and [Wei95], p. 71).

Remark 13.1.1. (a) Obviously, $\boldsymbol{R}(x+2) = \boldsymbol{R}(x)$, $\boldsymbol{R}(-x) = -\boldsymbol{R}(x)$, $x \in \mathbb{R}$. In particular, the differentiability of $\boldsymbol{R}$ may be checked only for $x \in (0, 2]$.

(b) Hardy proved in [Har16] that $\boldsymbol{R}$ has no finite derivative at irrational x nor at $x = \frac{2p+1}{2q}$ (see also [Ita81]) or $x = \frac{2p}{4q+1}$ ($p \in \mathbb{Z}$, $q \in \mathbb{N}$).

(c) Hardy's result was extended by Gerver, who proved in [Ger70] that $\boldsymbol{R}'(x) = -\frac{\pi}{2}$ for $x = \frac{2p+1}{2q+1}$ ($p \in \mathbb{Z}$, $q \in \mathbb{N}$). Moreover, he proved in [Ger71] that $\boldsymbol{R}$ has no finite derivative at points $x = \frac{2p}{2q+1}$, $x = \frac{2p+1}{2q}$ ($p \in \mathbb{Z}$, $n \in \mathbb{N}$); see also [Moh80].

(d) Our presentation will be based on [Smi72, Smi83]. Theorem 13.3.1 gives the full characterization of finite or infinite one-sided derivatives $\boldsymbol{R}'_{\pm}(x)$ for $x \in \mathbb{Q}$, and shows that a finite derivative $\boldsymbol{R}'(x)$ at $x \notin \mathbb{Q}$ does not exist.

It remains an open question

$\boxed{?}$ whether an infinite derivative $\boldsymbol{R}'(x)$ exists for $x \notin \mathbb{Q}$ $\boxed{?}$ and, more generally,

$\boxed{?}$ whether finite or infinite one-sided derivatives $\boldsymbol{R}'_{\pm}(x)$ exist for $x \notin \mathbb{Q}$ $\boxed{?}$

M. Jarnicki, P. Pflug, *Continuous Nowhere Differentiable Functions*, Springer Monographs in Mathematics, DOI 10.1007/978-3-319-12670-8_13

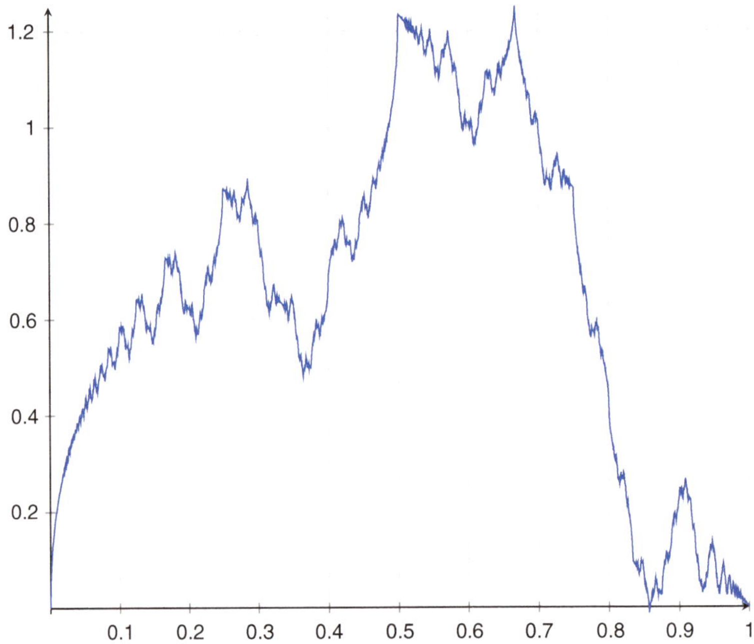

Fig. 13.1 Riemann function $\mathbb{I} \ni x \longmapsto \sum_{n=1}^{\infty} \frac{\sin(\pi n^2 x)}{n^2}$

13.2 Auxiliary Lemmas

Lemma 13.2.1. *Let $\varphi \in \mathcal{C}(\mathbb{R}, \mathbb{C}) \cap L^1(\mathbb{R})$ be such that:*

(a) *for every $h > 0$, the series $\sum_{k=-\infty}^{\infty} h\varphi(h(t+k))$ is locally uniformly convergent for $t \in \mathbb{R}$,*
(b) *there exist $\beta > 1$, $C > 0$ such that $|\tau|^\beta |\widehat{\varphi}(\tau)| \le C$, $\tau \in \mathbb{R}$, where $\widehat{\varphi}$ is the Fourier transform of φ (cf. § A.3).*

Then

$$\sum_{n=-\infty}^{\infty} h\varphi(hn + h\alpha) = \widehat{\varphi}(0) + A(h, \alpha)h^\beta, \quad h > 0,\ \alpha \in \mathbb{R},$$

where $|A(h, \alpha)| \le 2C \sum_{n=1}^{\infty} \frac{1}{n^\beta}$.

Proof. Fix $h > 0$ and $\alpha \in \mathbb{R}$. Define $f(t) := h\varphi(h(t + \alpha))$, $t \in \mathbb{R}$. Obviously, $f \in \mathcal{C}(\mathbb{R}, \mathbb{C}) \cap L^1(\mathbb{R})$. Moreover, the series $\sum_{n=-\infty}^{\infty} f(t+n) = \sum_{n=-\infty}^{\infty} h\varphi(h(t+n+\alpha))$ is uniformly convergent for $t \in \mathbb{I}$. Observe that

$$\widehat{f}(\tau) = \int_{\mathbb{R}} h\varphi(h(t + \alpha))e^{-2\pi i\tau t} dt = \int_{\mathbb{R}} \varphi(u)e^{-2\pi i\tau(\frac{u}{h} - \alpha)} du = e^{2\pi i\alpha\tau} \widehat{\varphi}\Big(\frac{\tau}{h}\Big).$$

Consequently, $|\tau|^\beta |\widehat{f}(\tau)| \le Ch^\beta$. Thus, we may apply Proposition A.5.1, and we get

$$\sum_{n=-\infty}^{\infty} h\varphi(hn + h\alpha) = \sum_{n=-\infty}^{\infty} f(n) = \sum_{n=-\infty}^{\infty} \widehat{f}(n) = \sum_{n=-\infty}^{\infty} e^{2\pi i\alpha n} \widehat{\varphi}\Big(\frac{n}{h}\Big).$$

It remains to observe that

$$\Big|\sum_{n\in\mathbb{Z}_*} e^{2\pi i\alpha n}\widehat{\varphi}\Big(\frac{n}{h}\Big)\Big| \le \sum_{n\in\mathbb{Z}_*} C\Big(\frac{h}{|n|}\Big)^{\beta} = 2C\Big(\sum_{n=1}^{\infty}\frac{1}{n^{\beta}}\Big)h^{\beta}. \qquad \square$$

Lemma 13.2.2. *Let*

$$\psi_1(x) := \begin{cases} \frac{\sin(\pi x^2)}{\pi x^2}, & \text{if } x \neq 0 \\ 1, & \text{if } x = 0 \end{cases}, \quad \psi_2(x) := \begin{cases} \frac{1-\cos(\pi x^2)}{\pi x^2}, & \text{if } x \neq 0 \\ 0, & \text{if } x = 0 \end{cases}.$$

Then

$$\sum_{k=-\infty}^{\infty} h\psi_j(hk + h\alpha) = \sqrt{2} + A_j(h,\alpha)h^2, \quad h > 0,\ \alpha \in \mathbb{R},\ j = 1, 2,$$

where the functions A_1, A_2 are bounded.

Proof. We are going to apply Lemma 13.2.1. First observe that for every $h, R > 0$, if $|t| < R$, then

$$\sum_{|k|>R} |h\psi_j(h(t+k))| \le \sum_{|k|>R} \frac{h}{\pi h^2(t+k)^2} \le \frac{1}{\pi h}\sum_{|k|>R}\frac{1}{(|k|-R)^2} < +\infty.$$

Thus for every $h > 0$, the series $\sum_{k=-\infty}^{\infty} h\psi_j(h(t+k))$ is locally uniformly convergent for $t \in \mathbb{R}$. Let $\varphi := \psi_1 + i\psi_2$. We have proved that for every $h > 0$, the series $\sum_{k=-\infty}^{\infty} h\varphi(h(t+k))$ is locally uniformly convergent for $t \in \mathbb{R}$. Now, in view of Lemma 13.2.1, we have only to show that

$$\exists_{C>0} :\ y^2|\widehat{\varphi}(y)| \le C, \quad y \in \mathbb{R}, \tag{13.2.1}$$

$$\widehat{\varphi}(0) = \sqrt{2}(1+i). \tag{13.2.2}$$

We have

$$\widehat{\varphi}(y) = \int_{\mathbb{R}} \frac{e^{i\pi x^2} - 1}{i\pi x^2} e^{-2\pi i x y}dx = 2\int_0^{\infty} \frac{e^{i\pi x^2} - 1}{i\pi x^2}\cos(2\pi x y)dx, \quad y \in \mathbb{R}.$$

Proof of (13.2.1). Observe that $\widehat{\varphi}(-y) = \widehat{\varphi}(y)$. So it suffices to prove that the function $(0, +\infty) \ni y \longmapsto y^2\widehat{\varphi}(y)$ is bounded. We have

$$\widehat{\varphi}(y) = \frac{2}{i\sqrt{\pi}}\int_0^{\infty}\frac{e^{it^2} - 1}{t^2}\cos(2\sqrt{\pi}ty)dt.$$

Consequently, it suffices to prove that the function $u \longmapsto u^2F(u)$, where

$$(0, +\infty) \ni u \overset{F}{\longmapsto} \int_0^{\infty}\frac{e^{it^2} - 1}{t^2}\cos(tu)dt,$$

is bounded. Observe that the function $Q(t) := \frac{e^{it^2}-1}{t^2}$, $t \in \mathbb{R}_*$, $Q(0) := i$, is real analytic and $Q'(0) = 0$. Obviously, $\lim_{t\to+\infty} Q(t) = 0$. Moreover,

$$Q'(t) = 2\frac{e^{it^2}(it^2 - 1) + 1}{t^3}.$$

In particular, $\lim_{t\to+\infty} Q'(t) = 0$. Integration by parts gives

$$\begin{aligned}F(u) &= Q(t)\frac{\sin(tu)}{u}\Big|_0^\infty - \frac{1}{u}\int_0^\infty Q'(t)\sin(tu)dt = -\frac{1}{u}\int_0^\infty Q'(t)\sin(tu)dt\\ &= \frac{1}{u}\Big(Q'(t)\frac{\cos(tu)}{u}\Big|_0^\infty - \frac{1}{u}\int_0^\infty Q''(t)\cos(tu)dt\Big)\\ &= -\frac{1}{u^2}\int_0^\infty Q''(t)\cos(tu)dt.\end{aligned}$$

Thus we need to show only that the function $u \longmapsto \int_0^\infty Q''(t)\cos(tu)dt$ is bounded. We have

$$Q''(t) = -4e^{it^2} - \frac{6}{t^4}\Big(e^{it^2}(it^2-1)+1\Big).$$

It is clear that the function $u \longmapsto \int_0^\infty \frac{1}{t^4}(e^{it^2}(it^2-1)+1)\cos(tu)dt$ is bounded. Thus, it remains to prove that the function $u \longmapsto \int_0^\infty e^{it^2}\cos(tu)dt$ is bounded. Fix a $u > 0$ and let $q := \frac{u}{2}$. Using Fresnel integrals (cf. § A.4), we have

$$\begin{aligned}&\int_0^\infty e^{it^2}\cos(tu)dt = \frac{1}{2}\Big(\int_0^\infty e^{it^2+itu}dt + \int_0^\infty e^{it^2-itu}dt\Big)\\ &= \frac{e^{-iq^2}}{2}\Big(\int_0^\infty e^{i(t+q)^2}dt + \int_0^\infty e^{i(t-q)^2}dt\Big) = \frac{e^{-iq^2}}{2}\Big(\int_q^\infty e^{it^2}dt + \int_{-q}^\infty e^{it^2}dt\Big)\\ &= \frac{e^{-iq^2}}{2}\sqrt{\frac{\pi}{2}}\Big(\int_{q\sqrt{\frac{2}{\pi}}}^\infty e^{i\frac{1}{2}\pi t^2}dt + \int_{-q\sqrt{\frac{2}{\pi}}}^\infty e^{i\frac{1}{2}\pi t^2}dt\Big)\\ &= \frac{e^{-iq^2}}{2}\sqrt{\frac{\pi}{2}}\Big(\frac{1}{2}(1+i) - \mathrm{Fr}\Big(q\sqrt{\frac{2}{\pi}}\Big) + \frac{1}{2}(1+i) - \mathrm{Fr}\Big(-q\sqrt{\frac{2}{\pi}}\Big)\Big)\\ &= \frac{e^{-iq^2}}{2}\sqrt{\frac{\pi}{2}}(1+i).\end{aligned}$$

Thus, $|\int_0^\infty e^{it^2}\cos(tu)dt| = \frac{\sqrt{\pi}}{2}$.

Proof of (13.2.2). Let $g(z) := \frac{e^{i\pi z^2}-1}{i\pi z^2}$, $z \in \mathbb{C}_*$, $g(0) = 1$. Note that g is holomorphic on $\mathbb{C}$. For $R > 0$, consider the contour $\Gamma_R := [0, R] \cup \gamma_R \cup [e^{i\frac{\pi}{4}}, 0] \subset \mathbb{C}$, where γ_R is the arc $[0, \frac{\pi}{4}] \ni \theta \longmapsto Re^{i\theta}$. By Cauchy's theorem, we get

$$\begin{aligned}0 &= \int_{\Gamma_R} f(z)dz\\ &= \int_0^R \frac{e^{i\pi x^2}-1}{i\pi x^2}dx - e^{i\frac{\pi}{4}}\int_0^R \frac{1-e^{-\pi x^2}}{\pi x^2}dx + \int_0^{\pi/4}\frac{e^{i\pi R^2 e^{2i\theta}}-1}{Re^{i\theta}}d\theta.\end{aligned}$$

Observe that

$$\Big|\int_0^{\pi/4}\frac{e^{i\pi R^2 e^{2i\theta}}-1}{Re^{i\theta}}d\theta\Big| \le \frac{1}{R}\int_0^{\pi/4}(e^{-R^2\sin 2\theta}+1)d\theta \le \frac{\pi}{2R} \underset{R\to+\infty}{\longrightarrow} 0.$$

Thus

$$\widehat{\varphi}(0) = 2e^{i\frac{\pi}{4}} \int_0^\infty \frac{1-e^{-\pi x^2}}{\pi x^2}dx = 2e^{i\frac{\pi}{4}} \int_0^\infty \int_0^1 e^{-\pi x^2 t} dt dx$$
$$= 2e^{i\frac{\pi}{4}} \int_0^1 \int_0^\infty e^{-\pi x^2 t} dx dt = 2e^{i\frac{\pi}{4}} \int_0^1 \frac{1}{2}\sqrt{\frac{1}{t}}dt = 2e^{i\frac{\pi}{4}} = \sqrt{2}(1+i). \qquad \square$$

13.3 Differentiability of the Riemann Function

To simplify notation, instead of $\boldsymbol{R}$ (cf. § 13.1), we will study the function

$$\boldsymbol{f}(x) := x + \frac{2}{\pi}\boldsymbol{R}(x) = x + 2\sum_{n=1}^{\infty} \frac{\sin(\pi n^2 x)}{\pi n^2}, \quad x \in \mathbb{R}.$$

The following theorem characterizes the differentiability of the Riemann function.

Theorem 13.3.1. (a) *If* $x = \frac{r}{s}$, $r \in \mathbb{Z}$, $s \in \mathbb{N}$, $(r,s) = 1$, *then:*
—*if* $rs = 0 \pmod 2$, *then:*

(i) *if* $r = 0$, *then* $\boldsymbol{f}'(0) = +\infty$;
(ii) *if* $r \neq 0$, $r \equiv 0 \pmod 2$, *and* $s \equiv 1 \pmod 4$, *then* $\boldsymbol{f}'(x) = \left\{\frac{\frac{1}{2}r}{s}\right\} \cdot (+\infty)$;
(iii) *if* $r \neq 0$, $r \equiv 0 \pmod 2$, *and* $s \equiv 3 \pmod 4$, *then* $\boldsymbol{f}'_{\pm}(x) = \mp\left\{\frac{\frac{1}{2}r}{s}\right\} \cdot (+\infty)$;
(iv) *if* $s \equiv 0 \pmod 2$, *and* $r \equiv 1 \pmod 4$, *then* $\boldsymbol{f}'_{+}(x) = 0$, $\boldsymbol{f}'_{-}(x) = \left\{\frac{\frac{1}{2}s}{r}\right\} \cdot (+\infty)$;
(v) *if* $s \equiv 0 \pmod 2$, *and* $r \equiv 3 \pmod 4$, *then* $\boldsymbol{f}'_{+}(x) = \left\{\frac{\frac{1}{2}s}{r}\right\} \cdot (+\infty)$, $\boldsymbol{f}'_{-}(x) = 0$.

—*if* $rs = 1 \pmod 2$, *then* $\boldsymbol{f}'(x) = 0$.

(b) *If* $x \in \mathbb{R} \setminus \mathbb{Q}$, *then a finite derivative* $\boldsymbol{f}'(x)$ *does not exist.*

Thus, for $x = \frac{r}{s}$ with $r \in \mathbb{Z}$, $s \in \mathbb{N}$, $(r,s) = 1$, the above result may be written in the following tabular form.

<table>
<tr><th></th><th>$s \in 2\mathbb{N}$</th><th>$s \in 4\mathbb{N}_0 + 1$</th><th>$s \in 4\mathbb{N}_0 + 3$</th></tr>
<tr><td>$r = 0$</td><td colspan="3">$\boldsymbol{f}'(x) = +\infty$</td></tr>
<tr><td>$r \in 2\mathbb{Z}_*$</td><td>×</td><td>$\boldsymbol{f}'(x) \in \{-\infty, +\infty\}$</td><td>$\boldsymbol{f}'_{\pm}(x) \in \{-\infty,+\infty\}$
$\boldsymbol{f}'_{+}(x) \neq \boldsymbol{f}'_{-}(x)$</td></tr>
<tr><td>$r \in 4\mathbb{Z} + 1$</td><td>$\boldsymbol{f}'_{+}(x) = 0$
$\boldsymbol{f}'_{-}(x) \in \{-\infty,+\infty\}$</td><td colspan="2" rowspan="2">$\boldsymbol{f}'(x) = 0$</td></tr>
<tr><td>$r \in 4\mathbb{Z} + 3$</td><td>$\boldsymbol{f}'_{+}(x) \in \{-\infty,+\infty\}$
$\boldsymbol{f}'_{-}(x) = 0$</td></tr>
</table>

Proof of Theorem 13.3.1. (a) Case 1°: $x = \frac{r}{s}$, $r \in \mathbb{Z}$, $s \in \mathbb{N}$, $(r,s) = 1$, $rs \equiv 0 \pmod 2$.
Then for $h > 0$, we get (cf. Lemma 13.2.2)

$$\frac{1}{2}\Big(\boldsymbol{f}(x+h^2) + \boldsymbol{f}(x-h^2)\Big) = x + 2\sum_{n=1}^{\infty} \frac{\sin(\pi n^2 x)}{\pi n^2} \cos(\pi n^2 h^2)$$
$$= \boldsymbol{f}(x) - h^2 \sum_{n=-\infty}^{\infty} \sin(\pi n^2 x) \psi_2(nh).$$

Let $n = ks + m$, $k \in \mathbb{Z}$, $m \in \{0, \ldots, s-1\}$. Then

$$\begin{aligned}
&\frac{1}{2}\Big(\boldsymbol{f}(x+h^2) + \boldsymbol{f}(x-h^2)\Big) \\
&\qquad = \boldsymbol{f}(x) - h^2 \sum_{m=0}^{s-1} \sum_{k=-\infty}^{\infty} \sin(\pi(ks+m)^2 x)\psi_2(khs+hm) \\
&\qquad \overset{x=\frac{r}{s},\ rs\equiv 0 \ (\mathrm{mod}\, 2)}{=} \boldsymbol{f}(x) - h^2 \sum_{m=0}^{s-1} \sum_{k=-\infty}^{\infty} \sin(\pi m^2 x)\psi_2(khs+hm) \\
&\qquad \overset{\text{Lemma 13.2.2}}{=} \boldsymbol{f}(x) - \frac{h}{s} \sum_{m=0}^{s-1} \sin(\pi m^2 x)\Big(\sqrt{2} + A_2\Big(hs, \frac{m}{s}\Big)(hs)^2\Big) \\
&\qquad = \boldsymbol{f}(x) - \sqrt{2}S(x)\frac{h}{s} + B_{x,+}(h)h^3,
\end{aligned}$$

where the function $h \longmapsto B_{x,+}(h) := -s\sum_{m=0}^{s-1} \sin(\pi m^2 x) A_2(hs, \frac{m}{s})$ is bounded. Note that $|B_{x,+}| \le cs^2$, where c is independent of x and $h > 0$. Analogously,

$$\begin{aligned}
&\frac{1}{2}\Big(\boldsymbol{f}(x+h^2) - \boldsymbol{f}(x-h^2)\Big) = 2\sum_{n=1}^{\infty} \frac{\cos(\pi n^2 x)}{\pi n^2} \sin(\pi n^2 h^2) \\
&= h^2 \sum_{n=-\infty}^{\infty} \cos(\pi n^2 x)\psi_1(nh) = h^2 \sum_{m=0}^{s-1} \sum_{k=-\infty}^{\infty} \cos(\pi m^2 x)\psi_1(khs+hm) \\
&= \frac{h}{s} \sum_{m=0}^{s-1} \cos(\pi m^2 x)\Big(\sqrt{2} + A_1\Big(hs, \frac{m}{s}\Big)(hs)^2\Big) = \sqrt{2}C(x)\frac{h}{s} + B_{x,-}(h)h^3,
\end{aligned}$$

where the function $h \longmapsto B_{x,-}(h) := s\sum_{m=0}^{s-1} \cos(\pi m^2 x) A_1(hs, \frac{m}{s})$ is bounded, $|B_{x,-}| \le cs^2$. Consequently,

$$\boldsymbol{f}(x \pm h^2) = \boldsymbol{f}(x) - \sqrt{2}\Big(S(x) \mp C(x)\Big)\frac{h}{s} + P_{x,\pm}(h)h^3,$$

where the functions $h \longmapsto P_{x,\pm}(h) := B_{x,+}(h) \pm B_{x,-}(h)$ are bounded and $|P_{x,\pm}| \le 2cs^2$. Thus

$$\boldsymbol{f}'_{\pm}(x) = \operatorname{sgn}(C(x) \mp S(x)) \cdot (+\infty) \in \{-\infty, 0, +\infty\}$$

with $0 \cdot \pm\infty := 0$. Using Lemma A.7.3, we get (i)–(v).

Case 2°: $x = \frac{r}{s}$, $r \in \mathbb{Z}$, $s \in \mathbb{N}$, $(r,s) = 1$, $rs \equiv 1 \pmod 2$.
First observe that

$$\begin{aligned}
\boldsymbol{f}(x) + \boldsymbol{f}(x+1) &= 2x + 1 + 2\sum_{n=1}^{\infty} \frac{\sin(\pi n^2 x) + \sin(\pi n^2 (x+1))}{\pi n^2} \\
&= 2x + 1 + 2\sum_{k=1}^{\infty} \frac{2\sin(\pi (2k)^2 x)}{\pi (2k)^2} = 1 + \frac{1}{2}\boldsymbol{f}(4x), \quad x \in \mathbb{R}.
\end{aligned}$$

Moreover, $4x = \frac{4r}{s}$ and $x+1 = \frac{r+s}{s}$ are as in Case 1°. Thus for $h > 0$, we have

$$\begin{aligned} \boldsymbol{f}(x \pm h^2) &= 1 + \frac{1}{2}\boldsymbol{f}(4x \pm (2h)^2) - \boldsymbol{f}(x+1 \pm h^2) \\ &= \boldsymbol{f}(x) - \sqrt{2}\Big(S(4x) - S(x+1) \mp (C(4x) - C(x+1))\Big)\frac{h}{s} + Q_{x,\pm}(h)h^3, \end{aligned}$$

where the functions $h \longmapsto Q_{x,\pm}(h) := 2P_{4x,\pm}(2h) \mp P_{x+1,\pm}(h)$ are bounded. Recall (cf. Remark A.6.2(f)) that

$$\left\{\frac{2r}{s}\right\} = \left\{\frac{\frac{r+s}{2}}{s}\right\}.$$

Hence by Lemma A.7.3, $G(4x) = G(x+1)$, which gives

$$\boldsymbol{f}(x \pm h^2) = \boldsymbol{f}(x) + Q_{x,\pm}(h)h^3,$$

and therefore $\boldsymbol{f}'(x) = 0$, and even more, namely

$$\lim_{h \to 0+} \frac{\boldsymbol{f}(x+h) - \boldsymbol{f}(x-h)}{h} = 0.$$

(b) Since $\boldsymbol{f}(-x) = -\boldsymbol{f}(x)$, we may assume that $x > 0$. By Proposition A.8.3, there exists a sequence of rational numbers $x_n = \frac{r_n}{s_n}$ such that $r_n, s_n \in \mathbb{N}$, $(r_n, s_n) = 1$, $r_n s_n \equiv 0 \pmod 2$, and $|x - x_n| < \frac{1}{s_n^2}$, $n \in \mathbb{N}$.

Using Case 1° with $h_n = s_n^{-3/2}$ and Lemma A.7.3, we get

$$\begin{aligned} \boldsymbol{f}(x_n \pm s_n^{-3}) &= \boldsymbol{f}(x_n) - \sqrt{2}\Big(S(x_n) \mp C(x_n)\Big)s_n^{-5/2} + P_{x_n,\pm}(s_n^{-3/2})s_n^{-9/2} \\ &= \boldsymbol{f}(x_n) + T_\pm(x_n)s_n^{-2} + U_\pm(x_n)s_n^{-5/2}, \end{aligned} \tag{13.3.1}$$

where $|T_\pm(x_n)| \le 2$ and $|U_\pm(x_n)| \le 2c$. Moreover, $|T_{\varepsilon_n}(x_n)| \ge 1$ in each of the following cases:

- $r_n \equiv 0 \pmod 2$ and $\varepsilon_n \in \{-,+\}$;
- $s_n \equiv 0 \pmod 2$, $r_n \equiv 1 \pmod 4$, and $\varepsilon_n = -$;
- $s_n \equiv 0 \pmod 2$, $r_n \equiv 3 \pmod 4$, and $\varepsilon_n = +$.

For each $n \in \mathbb{N}$, let us fix ε_n as above. We will identify $\varepsilon_n = -$ with $\varepsilon_n = -1$ and $\varepsilon_n = +$ with $\varepsilon_n = +1$.

Suppose that $\boldsymbol{f}'(x)$ exists and is finite. Let

$$\boldsymbol{f}(x+q) = \boldsymbol{f}(x) + \boldsymbol{f}'(x)q + \alpha(q)q,$$

where $\lim_{q \to 0} \alpha(q) = 0$. Put $q_n := x_n - x$. Then (13.3.1) gives

$$\begin{aligned} &\boldsymbol{f}(x) + \boldsymbol{f}'(x)(q_n + \varepsilon_n s_n^{-3}) + \alpha(q_n + \varepsilon_n s_n^{-3})(q_n + \varepsilon_n s_n^{-3}) \\ &\qquad = \boldsymbol{f}(x) + \boldsymbol{f}'(x)q_n + \alpha(q_n)q_n + T_{\varepsilon_n}(x_n)s_n^{-2} + U_{\varepsilon_n}(x_n)s_n^{-5/2}. \end{aligned}$$

Hence

$$\boldsymbol{f}'(x)\varepsilon_n s_n^{-1} + \alpha(q_n + \varepsilon_n s_n^{-3})(s_n^2 q_n + \varepsilon_n s_n^{-1})$$
$$= \alpha(q_n) s_n^2 q_n + T_{\varepsilon_n}(x_n) + U_{\varepsilon_n}(x_n) s_n^{-1/2}.$$

Recall that $s_n^2 |q_n| < 1$. Letting $n \longrightarrow +\infty$ gives $T_{\varepsilon_n}(x_n) \longrightarrow 0$; a contradiction. □

Appendix A

We collect here various auxiliary results that may help the reader.

A.1 Cantor Representation

Fix a sequence $(\mathfrak{q}_n)_{n=1}^{\infty} \subset \mathbb{N}_2$. A series of the form

$$\sum_{n=1}^{\infty} \frac{a_n}{\mathfrak{q}_1 \cdots \mathfrak{q}_n},$$

where $a_n \in \{0, \dots, \mathfrak{q}_n - 1\}$, is called a *Cantor series*.

Proposition A.1.1 (cf. [Can69]). (a) *Every number $x \in \mathbb{I}$ may be represented in the form of a Cantor series*

$$x = \sum_{n=1}^{\infty} \frac{a_n(x)}{\mathfrak{q}_1 \cdots \mathfrak{q}_n}.$$

It is called the Cantor representation *of x. In the case $\mathfrak{q}_n := \mathfrak{b}$, $n \in \mathbb{N}$, the Cantor representation reduces to the $\mathfrak{b}$*-adic representation. *Write*

$$S_k(x) := \sum_{n=1}^{k} \frac{a_n(x)}{\mathfrak{q}_1 \cdots \mathfrak{q}_n}, \qquad k \in \mathbb{N}.$$

(b) *If $x = S_k(x)$ with $a_k(x) \geq 1$, then x may be also represented in the form*

$$x = S_k(x) + \frac{a_k(x) - 1}{\mathfrak{q}_1 \cdots \mathfrak{q}_k} + \sum_{n=k+1}^{\infty} \frac{\mathfrak{q}_n - 1}{\mathfrak{q}_1 \cdots \mathfrak{q}_n}.$$

The above situation is the only one in which x has a double representation.

(c) $1 = \sum_{n=1}^{\infty} \frac{\mathfrak{q}_n - 1}{\mathfrak{q}_1 \cdots \mathfrak{q}_n}$.

(d) *For $x, x' \in \mathbb{I}$, if $S_k(x) = S_k(x')$, then $a_n(x) = a_n(x')$, $n = 1, \dots, k$.*

(e) $x \leq S_k(x) + \frac{1}{\mathfrak{q}_1 \cdots \mathfrak{q}_k}$, $k \in \mathbb{N}$.

(f) *If $S_k(x) < S_k(x')$, then $S_k(x) + \frac{1}{\mathfrak{q}_1 \cdots \mathfrak{q}_k} \leq S_k(x')$.*

M. Jarnicki, P. Pflug, *Continuous Nowhere Differentiable Functions*, Springer Monographs in Mathematics, DOI 10.1007/978-3-319-12670-8

A.2 Harmonic and Holomorphic Functions

Let $\mathbb{D}$ denote the open unit disk in the complex plane $\mathbb{C}$, i.e.,

$$\mathbb{D} = \{z = x + iy \in \mathbb{C} : |z| < 1\},$$

where $|z| := \sqrt{x^2 + y^2}$. Then $\mathcal{O}(\mathbb{D})$ is the set of all holomorphic functions on $\mathbb{D}$. Put $\mathcal{A}(\mathbb{D}) := \mathcal{C}(\overline{\mathbb{D}}, \mathbb{C}) \cap \mathcal{O}(\mathbb{D})$; $\mathcal{A}(\mathbb{D})$ is called to be the *disk algebra*.

Proposition A.2.1. *$\mathcal{A}(\mathbb{D})$ equipped with the supremum norm is a Banach space.*

Proof. Use that $\mathcal{C}(\overline{\mathbb{D}}, \mathbb{C})$ with the above norm is a Banach space and the fact (Weierstrass theorem) that the uniform limit of a sequence of holomorphic functions is again a holomorphic function. □

Recall that a function $u \in \mathcal{C}^2(\mathbb{D}, \mathbb{R})$ is called *harmonic* if $u_{x,x} + u_{y,y} = 0$ on $\mathbb{D}$, where $u_{x,x} := \frac{\partial^2 u}{\partial x \partial x}$ and $u_{y,y} = \frac{\partial^2 u}{\partial y \partial y}$. It is easy to see that if $f \in \mathcal{O}(\mathbb{D})$, then $\operatorname{Re} f$ (the real part of f) and $\operatorname{Im} f$ (the imaginary part of f) are harmonic functions on $\mathbb{D}$. Therefore, a function $f \in \mathcal{A}(\mathbb{D})$ leads to functions $\operatorname{Re} f, \operatorname{Im} f \in \mathcal{C}(\overline{\mathbb{D}}) \cap \mathcal{H}(\mathbb{D})$, where $\mathcal{H}(\mathbb{D})$ denotes the set of all real-valued harmonic functions on the open unit disk.

Recall the maximum principle for harmonic functions.

Proposition A.2.2. *If $u \in \mathcal{C}(\overline{\mathbb{D}}) \cap \mathcal{H}(\mathbb{D})$, then $u(z) \le \max_{\mathbb{T}} u$, $z \in \mathbb{D}$, where $\mathbb{T} := \partial\mathbb{D}$.*

Moreover, we have the following solution of a *Dirichlet problem* on $\overline{\mathbb{D}}$.

Proposition A.2.3. *If $u \in \mathcal{C}(\mathbb{T})$, then there exists a unique function $\widehat{u} \in \mathcal{C}(\overline{\mathbb{D}}) \cap \mathcal{H}(\mathbb{D})$ with $\widehat{u}|_{\mathbb{T}} = u$.*

Proof. Put

$$\widehat{u}(re^{it}) := \begin{cases} u(e^{it}), & \text{if } r = 1 \\ P(u)(re^{it}), & \text{if } r \in [0, 1) \end{cases},$$

where

$$P(u)(re^{it}) := \frac{1}{2\pi} \int_0^{2\pi} \frac{1 - r^2}{1 - 2r\cos(t - \theta) + r^2} d\theta.$$

Then $\widehat{u}$ satisfies the properties of the proposition. Details may be found in any standard book on complex analysis. □

Moreover, every harmonic function on $\mathbb{D}$ is the real part of a function $f \in \mathcal{O}(\mathbb{D})$. To be more precise, we state the following result.

Proposition A.2.4. *Let $u \in \mathcal{H}(\mathbb{D})$. Then there exists a unique (up to an additive constant) $v \in \mathcal{H}(\mathbb{D})$ such that $u + iv \in \mathcal{O}(\mathbb{D})$.*

Proof. Put

$$v(z) := -\int_0^x u_y(t, 0)dt + \int_0^y u_x(x, t)dt, \quad z = x + iy \in \mathbb{D}.$$

Further details are left to the reader. □

The function v is called the *harmonic conjugate* to u.

Example A.2.5. For later use, we will construct a function $u^* \in \mathcal{C}_{2\pi}(\mathbb{R})$, i.e., u^* is a continuous function on $\mathbb{R}$ having period 2π, and a function $h_{u^*} \in \mathcal{C}(\overline{\mathbb{D}}) \cap \mathcal{H}(\mathbb{D})$ with $h_{u^*}(e^{i\theta}) = u^*(\theta)$, $\theta \in \mathbb{R}$, such that its conjugate harmonic function $\widetilde{h}_{u^*}$ is unbounded on $\mathbb{D}$.

Indeed, put

$$\Omega := \{z = x + iy \in \mathbb{C} : 0 < |x| < 1,\ 0 < y < 1/x^2\} \cup \{z = 0 + iy : 0 < y\}.$$

Obviously, Ω is a simply connected domain in $\mathbb{C}$, and therefore, by virtue of the Riemann mapping theorem, there exists a biholomorphic mapping $f : \mathbb{D} \longrightarrow \Omega$. Applying a general theorem on the boundary behavior of biholomorphic mappings (see [Pom92], Theorem 2.1), one concludes that f extends to a topological mapping $\widetilde{f} : \overline{\mathbb{D}} \longrightarrow \overline{\Omega} \subset \overline{\mathbb{C}}$ ($\overline{\mathbb{C}}$ denotes the Riemann sphere endowed with the spherical distance). In particular, one finds a point $a = e^{i\theta_0} \in \mathbb{T}$ such the function $\widetilde{f}|_{\overline{\mathbb{D}}\setminus\{a\}}$ gives a homeomorphism from $\overline{\mathbb{D}} \setminus \{a\}$ onto $\overline{\Omega} \cap \mathbb{C}$ (now with respect to the Euclidean metric on both sides). Observe that $\operatorname{Re} \widetilde{f}|_{\mathbb{T}\setminus\{a\}}$ extends to a continuous function on $\mathbb{T}$ by putting $\operatorname{Re} \widetilde{f}(a) = 0$. Define $u^*(\theta) := \operatorname{Re} \widetilde{f}(e^{i\theta})$, $\theta \in \mathbb{R}$. Then u^* is continuous on $\mathbb{R}$ having period 2π. Solving the Dirichlet problem (see Proposition A.2.3), there exists $h_{u^*} \in \mathcal{C}(\overline{\mathbb{D}}) \cap \mathcal{H}(\mathbb{D})$ such that $h(e^{i\theta}) = u^*(\theta)$, $\theta \in \mathbb{R}$. Using the maximum principle gives that $h_{u^*} = \operatorname{Re} \widetilde{f}$. Then its conjugate harmonic function $\widetilde{h}_{u^*}$ is given as $\widetilde{h}_{u^*} = \operatorname{Im} f$, which is unbounded on $\mathbb{D}$.

Moreover, we recall Schwarz's lemma: if $f : \mathbb{D} \longrightarrow \mathbb{D}$, $f(0) = 0$, then $|f(z)| \le |z|$, $z \in \mathbb{D}$. Applying this result leads to the so-called Carathéodory inequality.

Proposition A.2.6. *Let f be a holomorphic function in a neighborhood of $\overline{\mathbb{D}}(R)$, $R > 0$, where $\mathbb{D}(R) := \{z \in \mathbb{C} : |z| < R\}$. Then*

$$|f(re^{i\theta})| \le |f(0)| + \frac{2r}{1-r}\Big(A(R) - \operatorname{Re} f(0)\Big), \quad 0 < r < R,\ \theta \in \mathbb{R},$$

where $A(R) := \sup\{\operatorname{Re} f(w) : |w| = R\}$.

Proof. There is nothing is to prove if f is a constant. If f is not a constant, then we may start with the case $f(0) = 0$. Put

$$g := \frac{f}{2A(R) - f}.$$

Then g is holomorphic in a neighborhood of $\overline{\mathbb{D}}(R)$, $g(0) = 0$. A simple estimate leads to $|g(z)| \le 1$ on $\mathbb{D}(R)$. Applying Schwarz's lemma gives $|g(z)| \le \frac{|z|}{R}$ for $z \in \mathbb{D}(R)$, which finally proves the proposition. For more details see [Boa10]. □

Finally, we present a result due to L. Fejér.

Proposition A.2.7. *If $f \in \mathcal{C}(\mathbb{T})$ and $\varepsilon > 0$, then there exists a complex polynomial $p \in \mathbb{C}[z]$ such that $\| \operatorname{Re} p - f\|_{\mathbb{T}} < \varepsilon$.*

Proof. The proof is based on the Stone–Weierstrass theorem. Put $A := \mathcal{C}(\mathbb{T}, \mathbb{C})$ and $A' = \mathbb{C}[z, \overline{z}] =:$ the algebra of all complex-valued polynomials in z and $\overline{z}$. Then A' is a subalgebra of A, A' is closed under complex conjugation, i.e., $\overline{A'} \subset A'$, and A' separates points in $\mathbb{T}$, i.e., the assumptions for the Stone–Weierstrass theorem are satisfied. Hence, A' is dense in A. Therefore, there exists a polynomial

$$p(z, \overline{z}) = \sum_{j=0}^{N} a_j z^j + \sum_{j=0}^{N} b_j \overline{z}^j$$

with $\|p - f\|_{\mathbb{T}} < \varepsilon$. Thus, $\|\operatorname{Re} p - f\|_{\mathbb{T}} < \varepsilon$. Using that $\overline{z} = \frac{1}{z}$, $z \in \mathbb{T}$, allows one to rewrite p and to get a complex polynomial $q(z) = \sum_{j=0}^{N} c_j z^j$ with $\operatorname{Re} q = \operatorname{Re} p$ on $\mathbb{T}$ (EXERCISE). Hence the proposition is proved. □

A.3 Fourier Transform

Let $L^p(\mathbb{R})$ denote the space of all p-integrable (with respect to the Lebesgue measure) functions $f : \mathbb{R} \longrightarrow \mathbb{C}$ with the norm $\| \ \|_{L^p}$ $(1 \le p \le +\infty)$.

Definition A.3.1. For $f \in L^1(\mathbb{R})$, we define its *Fourier transform* $\widehat{f} : \mathbb{R} \longrightarrow \mathbb{C}$,

$$\widehat{f}(\tau) := \int_{\mathbb{R}} f(t) e^{-2\pi i t \tau} dt, \quad \tau \in \mathbb{R}.$$

Put $\mathfrak{X}(\mathbb{R}) := \{f \in \mathcal{C}(\mathbb{R}, \mathbb{C}) \cap L^\infty(\mathbb{R}) \cap L^1(\mathbb{R}) : \widehat{f} \in L^1(\mathbb{R})\}$.

To simplify notation, set $\overset{\vee}{f}(t) := f(-t)$.

Remark A.3.2. (a) $\widehat{f}$ is uniformly continuous.
(b) $\widehat{f} \in L^\infty(\mathbb{R})$ and $\|\widehat{f}\|_{L^\infty} \le \|f\|_{L^1}$.
(c) The operator $L^1(\mathbb{R}) \ni f \longmapsto \widehat{f} \in \mathcal{C}(\mathbb{R}, \mathbb{C}) \cap L^\infty(\mathbb{R})$ is $\mathbb{C}$-linear.
(d) If $t^k f \in L^1(\mathbb{R})$ for some $k \in \mathbb{N}$, then the differentiation under the integral gives $\widehat{f} \in \mathcal{C}^k(\mathbb{R})$ and $\widehat{f}^{(k)} = (-2\pi i)^k \widehat{t^k f}$. Notice that here and in the sequel, $t^k f$ denotes the function $t \longmapsto t^k f(t)$. In particular, if $f \in \mathcal{C}_0^\infty(\mathbb{R})$, then $\widehat{f} \in \mathcal{C}^\infty(\mathbb{R})$.
(e) If $f \in \mathcal{C}_0^k(\mathbb{R})$ for some $k \in \mathbb{N}$, then integration by parts gives $\tau^k \widehat{f} = \frac{1}{(2\pi i)^k} \widehat{f^{(k)}}$. In particular, if $f \in \mathcal{C}_0^k(\mathbb{R})$ for some $k \in \mathbb{N}_2$, then $\tau^{k-2} \widehat{f} \in L^1(\mathbb{R})$.
(f) $\mathfrak{X}(\mathbb{R})$ is a complex vector space.
(g) $\mathcal{C}_0^2(\mathbb{R}) \subset \mathfrak{X}(\mathbb{R})$.
(h) $\widehat{\overset{\vee}{f}} = \overset{\vee}{\widehat{f}}, \quad f \in L^1(\mathbb{R})$.

Proposition A.3.3. $\widehat{(\widehat{f})} = \overset{\vee}{f}$, $f \in \mathfrak{X}(\mathbb{R})$. *In particular, the operator*

$$\mathfrak{X}(\mathbb{R}) \ni f \overset{\mathcal{F}}{\longmapsto} \widehat{f} \in \mathfrak{X}(\mathbb{R})$$

is bijective and

$$\mathcal{F}^{-1}(g) = \overset{\vee}{\widehat{g}}, \ \textit{i.e.,} \ (\mathcal{F}^{-1}(g))(t) = \int_{\mathbb{R}} \widehat{g}(\tau) e^{2\pi i t \tau} d\tau, \quad t \in \mathbb{R}, \quad g \in \mathfrak{X}(\mathbb{R}).$$

We need the following two lemmas to prove Proposition A.3.3.

Lemma A.3.4. *For $f, g \in L^1(\mathbb{R})$ we have*

$$\widehat{(g \cdot \widehat{f})}(\tau) = \int_{\mathbb{R}} \widehat{g}(u + \tau) f(u) du, \quad \tau \in \mathbb{R}.$$

Proof.

$$\widehat{(g\cdot \widehat{f})}(\tau) = \int_{\mathbb{R}} g(t)\widehat{f}(t)e^{-2\pi it\tau}dt = \int_{\mathbb{R}} g(t)\Big(\int_{\mathbb{R}} f(u)e^{-2\pi iut}du\Big)e^{-2\pi it\tau}dt$$

$$\overset{\text{Fubini}}{=} \int_{\mathbb{R}}\Big(\int_{\mathbb{R}} g(t)e^{-2\pi it(u+\tau)}dt\Big)f(u)du = \int_{\mathbb{R}} \widehat{g}(u+\tau)f(u)du. \qquad \square$$

Lemma A.3.5.

$$\widehat{e^{-at^2}}(\tau) = \sqrt{\frac{\pi}{a}}e^{-\frac{\pi^2}{a}t^2}, \quad \tau\in\mathbb{R},\ a>0.$$

Note that $\widehat{e^{-at^2}}$ is the Fourier transform of the function $t \longmapsto e^{-at^2}$.

Proof. Let

$$F(\tau) := \widehat{e^{-at^2}}(\tau) = \int_{\mathbb{R}} e^{-at^2}e^{-2\pi it\tau}dt, \quad \tau\in\mathbb{R}.$$

Then $F\in\mathcal{C}^\infty(\mathbb{R})$ and

$$F'(\tau) = \int_{-\infty}^{\infty} e^{-at^2}(-2\pi it)e^{-2\pi it\tau}dt$$

$$= e^{-at^2}\frac{\pi i}{a}e^{-2\pi it\tau}\Big|_{-\infty}^{\infty} - \int_{-\infty}^{\infty} e^{-at^2}\frac{\pi i}{a}(-2\pi i\tau)e^{-2\pi it\tau}dt = -\frac{2\pi^2}{a}\tau F(\tau).$$

Hence

$$F(\tau) = Ce^{-\frac{\pi^2}{a}\tau^2}, \quad \tau\in\mathbb{R},$$

where

$$C = F(0) = \int_{-\infty}^{\infty} e^{-at^2}dt = \sqrt{\frac{\pi}{a}}. \qquad \square$$

Proof of Proposition A.3.3. Let $g_\varepsilon(t) := e^{-\pi\varepsilon^2t^2}$, $t\in\mathbb{R}$, $\varepsilon>0$. Note that $g_\varepsilon\in L^1(\mathbb{R})$ and $g_\varepsilon \overset{\text{pointwise}}{\longrightarrow} 1$ when $\varepsilon\longrightarrow 0+$. In particular, by Lebesgue's theorem,

$$\widehat{(\widehat{f})}(\tau) = \int_{\mathbb{R}} \widehat{f}(t)e^{-2\pi it\tau}dt = \lim_{\varepsilon\to 0+}\int_{\mathbb{R}} g_\varepsilon(t)\widehat{f}(t)e^{-2\pi it\tau}dt, \quad \tau\in\mathbb{R}.$$

By Lemmas A.3.4 and A.3.5, we get

$$\int_{\mathbb{R}} g_\varepsilon(t)\widehat{f}(t)e^{-2\pi it\tau}dt = \widehat{(g_\varepsilon\widehat{f})}(\tau) = \int_{\mathbb{R}} \widehat{g_\varepsilon}(t+\tau)f(t)dt = \int_{\mathbb{R}} \widehat{g_\varepsilon}(t)f(t-\tau)dt$$

$$= \int_{\mathbb{R}} \frac{1}{\varepsilon}e^{-\frac{\pi}{\varepsilon^2}t^2}f(t-\tau)dt = \int_{\mathbb{R}} e^{-\pi t^2}f(\varepsilon t-\tau)dt, \quad \tau\in\mathbb{R}.$$

Using once again Lebesgue's theorem, we easily conclude that the last integral tends to $f(-\tau)$ when $\varepsilon\longrightarrow 0+$. $\square$

A.4 Fresnel Function

Definition A.4.1 (cf. [OMS09]). We define the *Fresnel* function $\mathrm{Fr}:\mathbb{R}\longrightarrow\mathbb{C}$,

$$\mathrm{Fr}(x) := \int_0^x e^{\frac{i}{2}\pi t^2}dt, \quad x\in\mathbb{R}.$$

Observe that $\mathrm{Fr}(-x) = -\mathrm{Fr}(x)$, $x\in\mathbb{R}$.

Proposition A.4.2. $\lim_{x\to+\infty}\mathrm{Fr}(x) = \frac{1}{2}(1+i)$.

Proof. Consider the holomorphic function $f(z) := e^{\frac{\pi}{2}z^2}$, $z\in\mathbb{C}$. Take an $R>0$ and consider the contour $\varGamma_R := [0, e^{i\pi/4}R]\cup\gamma_R\cup[iR,0]$, where γ_R stands for the arc $[\pi/4,\pi/2]\ni\theta\longmapsto e^{i\theta}R$. By Cauchy's theorem, we have

$$\begin{aligned}0 = \int_{\varGamma_R} f(z)dz &= \int_{[0,e^{i\pi/4}R]} f(z)dz + \int_{\gamma_R} f(z)dz + \int_{[iR,0]} f(z)dz \\ &= e^{i\pi/4}\,\mathrm{Fr}(R) + \int_0^{\pi/4} f(e^{i(\frac{\pi}{4}+\theta)}R)e^{i(\frac{\pi}{4}+\theta)}iRd\theta - i\int_0^R e^{-\frac{\pi}{2}t^2}dt.\end{aligned}$$

Then,

$$\begin{aligned}\left|\int_0^{\pi/4} f(e^{i(\frac{\pi}{4}+\theta)}R)e^{i(\frac{\pi}{4}+\theta)}iRd\theta\right| &\le R\int_0^{\pi/4}\left|e^{\frac{\pi}{2}e^{2i(\frac{\pi}{4}+\theta)}R^2}\right|d\theta \\ &= R\int_0^{\pi/4} e^{-\frac{\pi}{2}R^2\sin(2\theta)}d\theta \le R\int_0^{\pi/4} e^{-2R^2\theta}d\theta = \frac{1}{2R}(1-e^{-\frac{\pi}{4}R^2}) \underset{R\to+\infty}{\longrightarrow} 0\end{aligned}$$

and

$$\lim_{R\to+\infty}\int_0^R e^{-\frac{\pi}{2}t^2}dt = \int_0^\infty e^{-\frac{\pi}{2}t^2} = \frac{\sqrt{2}}{2}.$$

Thus,

$$0 = \frac{\sqrt{2}}{2}(1+i)\lim_{R\to+\infty}\mathrm{Fr}(R) - i\frac{\sqrt{2}}{2},$$

which implies that $\lim_{R\to+\infty}\mathrm{Fr}(R) = \frac{1}{2}(1+i)$. □

A.5 Poisson Summation Formula

Proposition A.5.1 (Poisson Summation Formula, cf. [Zyg02], p. 68). *Let $f\in\mathcal{C}(\mathbb{R},\mathbb{C})\cap L^1(\mathbb{R})$ be such that:*

(a) *the series $\sum_{k=-\infty}^{\infty} f(t+k)$ is uniformly convergent on $\mathbb{I}$,*
(b) *there exist $\beta>1$, $C>0$ such that $|\tau|^\beta|\widehat{f}(\tau)|\le C$, $\tau\in\mathbb{R}$,*

where $\widehat{f}$ stands for the Fourier transform of f (cf. § A.3). Then

$$\sum_{n=-\infty}^{\infty} f(n) = \sum_{n=-\infty}^{\infty}\widehat{f}(n).$$

Proof. Let $g(t) := \sum_{k=-\infty}^{\infty} f(t+k)$, $t \in \mathbb{I}$. Note that g is continuous. Let $c_n = c_n(g)$ be the nth Fourier coefficient of g, i.e.,

$$c_n := \int_0^1 g(t)e^{-2\pi int}dt, \quad n \in \mathbb{Z}.$$

Using (a), we get

$$\begin{aligned} c_n &= \int_0^1 \sum_{k=-\infty}^{\infty} f(t+k)e^{-2\pi int}dt = \sum_{k=-\infty}^{\infty} \int_0^1 f(t+k)e^{-2\pi int}dt \\ &= \sum_{k=-\infty}^{\infty} \int_k^{k+1} f(t)e^{-2\pi int}dt = \int_{\mathbb{R}} f(t)e^{-2\pi int}dt = \widehat{f}(n). \end{aligned}$$

In view of (b), the function $h(t) := \sum_{n=-\infty}^{\infty} \widehat{f}(n)e^{2\pi int}$, $t \in \mathbb{I}$, is well defined and continuous. Moreover, $c_n(h) = \widehat{f}(n) = c_n(g)$, $n \in \mathbb{Z}$. Hence $h \equiv g$. In particular, for $t = 0$ we get the required equality. □

A.6 Legendre, Jacobi, and Kronecker Symbols

Definition A.6.1 (Legendre, Jacobi, and Kronecker Symbols; cf. [HW79]). For $n, p \in \mathbb{Z}$, we define the symbol $\{\frac{n}{p}\}$ in the following four steps.

Step 1°: $n \in \mathbb{Z}$, $p \in \mathbb{N}_3$, p prime.

Then the *Legendre symbol* $\{\frac{n}{p}\}_L$ is defined by the formula

$$\left\{\frac{n}{p}\right\}_L := \begin{cases} 0, & \text{if } p|n \\ 1, & \text{if } p\not| n \text{ and there exists an } m \in \mathbb{N}\text{: } n \equiv m^2 \ (\mathrm{mod}\, p) \\ -1, & \text{if } p\not| n \text{ and for all } m \in \mathbb{N}\text{: } n \not\equiv m^2 \ (\mathrm{mod}\, p) \end{cases}.$$

Step 2°: $n \in \mathbb{Z}$, $p \in \mathbb{N}_3$, p odd.

If $p = p_1^{k_1} \cdots p_s^{k_s}$, where $p_1, \dots, p_s \in \mathbb{N}_3$, $p_1, \dots, p_s$ are distinct primes, $k_1, \dots, k_s \in \mathbb{N}$, then we define the *Jacobi symbol* $\{\frac{n}{p}\}_J$:

$$\left\{\frac{n}{p}\right\}_J := \left\{\frac{n}{p_1}\right\}_L^{k_1} \cdots \left\{\frac{n}{p_s}\right\}_L^{k_s}.$$

Moreover, we put $\{\frac{n}{1}\}_J := 1$. It is clear that $\{\frac{n}{p}\}_J = \{\frac{n}{p}\}_L$, provided that p is prime.

Step 3°: $n \in \mathbb{Z}$, $p \in \mathbb{Z}_*$.

If $p = up_1^{k_1} \cdots p_s^{k_s}$, where $u \in \{-1, +1\}$, $p_1, \dots, p_s \in \mathbb{N}_2$, $p_1, \dots, p_s$ are distinct primes, $k_1, \dots, k_s \in \mathbb{N}$, then we define the *Kronecker symbol* $\{\frac{n}{p}\}_K$:

$$\left\{\frac{n}{p}\right\}_K := \left\{\frac{n}{u}\right\}_K \left\{\frac{n}{p_1}\right\}_K^{k_1} \cdots \left\{\frac{n}{p_s}\right\}_K^{k_s},$$

where $\left\{\frac{n}{p_j}\right\}_K := \left\{\frac{n}{p_j}\right\}_L$ for $p_j \in \mathbb{N}$, p_j odd,

$$\left\{\frac{n}{-1}\right\}_K := \begin{cases} -1, & \text{if } n < 0 \\ 1, & \text{if } n \geq 0 \end{cases}, \text{ and } \left\{\frac{n}{2}\right\}_K := \begin{cases} 0, & \text{if } n \equiv 2 \pmod 2 \\ 1, & \text{if } n \equiv \pm 1 \pmod 8 \\ -1, & \text{if } n \equiv \pm 3 \pmod 8 \end{cases}.$$

It is clear that $\left\{\frac{n}{p}\right\}_K = \left\{\frac{n}{p}\right\}_J$, provided that $p \in \mathbb{N}_3$ is odd.

Step 4°. Finally, we put

$$\left\{\frac{n}{0}\right\}_K := \begin{cases} 1, & \text{if } n \in \{-1, +1\} \\ 0, & \text{otherwise} \end{cases}.$$

In the sequel, the subindices in $\left\{\frac{n}{p}\right\}_L, \left\{\frac{n}{p}\right\}_J, \left\{\frac{n}{p}\right\}_K$ will be skipped, and we will simply write $\left\{\frac{n}{p}\right\}$.

The following remark collects some properties of the Legendre, Jacobi, and Kronecker symbols.

Remark A.6.2. Assume that $n, m \in \mathbb{Z}$, $p \in \mathbb{N}_3$, p is odd.

(a) (Euler's criterion; cf. [HW79], Theorem 83) If p is prime, then $\left\{\frac{n}{p}\right\} \equiv n^{\frac{p-1}{2}} \pmod p$.

(b) $\left\{\frac{nm}{p}\right\} = \left\{\frac{n}{p}\right\} \cdot \left\{\frac{m}{p}\right\}$.
Indeed, if p is prime, then we use (a). The general case follows directly from the definition of $\left\{\frac{n}{p}\right\}_J$.

(c) $\left\{\frac{n}{p}\right\} = 0$ iff n and p are not relatively prime.
This follows directly from the definitions of $\left\{\frac{n}{p}\right\}_L$ and $\left\{\frac{n}{p}\right\}_J$.

(d) If $n \equiv m \pmod p$, then $\left\{\frac{n}{p}\right\} = \left\{\frac{m}{p}\right\}$.
Indeed, if p is prime, then we use (a). In the general case, if $n \equiv m \pmod p$, then $n \equiv m \pmod{p_i}$. Hence $\left\{\frac{n}{p_i}\right\} = \left\{\frac{m}{p_i}\right\}$, $i = 1, \ldots, s$, and we have only to use the definition of $\left\{\frac{n}{p}\right\}_J$.

(e) $\left\{\frac{-1}{p}\right\} = (-1)^{\frac{p-1}{2}} = \begin{cases} 1, & \text{if } p \equiv 1 \pmod 4 \\ -1, & \text{if } p \equiv 3 \pmod 4 \end{cases}$.
Indeed, if p is prime, then we use (a). To get the general case, we have only to observe that if $p, q \in \mathbb{N}_3$ are odd, then $\frac{pq-1}{2} \equiv \frac{p-1}{2} + \frac{q-1}{2} \pmod 2$. In fact, $a - 1 \equiv 0 \pmod 2$, $b - 1 \equiv 0 \pmod 2$. Hence $ab - a - b + 1 = (a-1)(b-1) \equiv 0 \pmod 4$. Thus $ab - 1 \equiv (a-1) + (b-1) \pmod 4$.
In particular, by (b), $\left\{\frac{-n}{p}\right\} = (-1)^{\frac{p-1}{2}}\left\{\frac{n}{p}\right\} = \begin{cases} \left\{\frac{n}{p}\right\}, & \text{if } p \equiv 1 \pmod 4 \\ -\left\{\frac{n}{p}\right\}, & \text{if } p \equiv 3 \pmod 4 \end{cases}$.

(f) If $r \in \mathbb{Z}$, $s \in \mathbb{N}$ are relatively prime and $rs \equiv 1 \pmod 2$, then

$$\left\{\frac{2r}{s}\right\} \overset{\text{(d)}}{=} \left\{\frac{2r+2s}{s}\right\} = \left\{\frac{4\frac{r+s}{2}}{s}\right\} \overset{\text{(b)}}{=} \left\{\frac{4}{s}\right\}\left\{\frac{\frac{r+s}{2}}{s}\right\} = \left\{\frac{\frac{r+s}{2}}{s}\right\}.$$

(g) If $s \in \mathbb{Z}$, $r \in \mathbb{N}$, then $\left\{\frac{s}{-r}\right\} = \begin{cases} -\left\{\frac{s}{r}\right\}, & \text{if } s < 0 \\ \left\{\frac{s}{r}\right\}, & \text{if } s \geq 0 \end{cases}$.
This follows directly from the definition of $\left\{\frac{s}{-r}\right\}_K$.

A.7 Gaussian Sums

Definition A.7.1 (cf. [BE81]). For $n \in \mathbb{Z}$, $p \in \mathbb{N}$ with $(n,p)=1$, we define the *Gaussian sum*

$$G(n,p) := \sum_{m=0}^{p-1} e^{\pi i m^2 n/p}.$$

Remark A.7.2 (cf. [BE81]). (a) $G(0,p)=p$.
(b) $G(2n,p)=\{\frac{n}{p}\}G(2,p)$.
(c) $G(2,p)=\begin{cases}\sqrt{p}, & \text{if } p\equiv 1 \pmod 4\\ i\sqrt{p}, & \text{if } p\equiv 3 \pmod 4\end{cases}$.
(d) (Schaar identity) If $n,p\in\mathbb{N}$, $(n,p)=1$, and n,p are of opposite parity, then

$$G(n,p)=e^{i\pi/4}\sqrt{\frac{p}{n}}G(-p,n).$$

Lemma A.7.3. *Let $x=\frac{r}{s}$, $r\in\mathbb{Z}_*$, $s\in\mathbb{N}$, $(r,s)=1$,*

$$G(x) := G(r,s) = \sum_{n=0}^{s-1} e^{i\pi n^2 x} = \sum_{n=0}^{s-1}\cos(\pi n^2 x) + i\sum_{n=0}^{s-1}\sin(\pi n^2 x)$$
$$=: C(x)+iS(x).$$

Then:
(a) *if $r\equiv 0 \pmod 2$, $s\equiv 1 \pmod 4$, then $G(x)=\left\{\frac{\frac12 r}{s}\right\}\sqrt{s}$;*
(b) *if $r\equiv 0 \pmod 2$, $s\equiv 3 \pmod 4$, then $G(x)=\left\{\frac{\frac12 r}{s}\right\}\sqrt{s}i$;*
(c) *if $s\equiv 0 \pmod 2$, $r\equiv 1 \pmod 4$, then $G(x)=\left\{\frac{\frac12 s}{r}\right\}\sqrt{\frac{s}{2}}(1+i)$;*
(d) *if $s\equiv 0 \pmod 2$, $r\equiv 3 \pmod 4$, then $G(x)=\left\{\frac{\frac12 s}{r}\right\}\sqrt{\frac{s}{2}}(1-i)$;*
(e) *if $rs\equiv 0 \pmod 2$, then $|G(x)|=\sqrt{s}$.*

Proof. (a, b) Using Remark A.7.2, we have

$$G(x)=G(r,s)=\left\{\frac{\frac12 r}{s}\right\}G(2,s)=\left\{\frac{\frac12 r}{s}\right\}\sqrt{s}\begin{cases}1, & \text{if } s\equiv 1 \pmod 4\\ i, & \text{if } s\equiv 3 \pmod 4\end{cases}.$$

(c, d) If $r>0$, then using (a, b) and Remarks A.6.2, A.7.2, we get

$$\begin{aligned} G(x)=G(r,s)&=e^{i\pi/4}\sqrt{\frac{s}{r}}G(-s,r)\\ &=\frac{1+i}{\sqrt2}\sqrt{\frac{s}{r}}\left\{\frac{-\frac12 s}{r}\right\}\sqrt{r}\begin{cases}1, & \text{if } r\equiv 1 \pmod 4\\ i, & \text{if } r\equiv 3 \pmod 4\end{cases}\end{aligned}$$

$$= (1+i)\sqrt{\frac{s}{2}}\left\{\frac{\frac{1}{2}s}{r}\right\}\left\{\frac{-1}{r}\right\}\begin{cases}1, & \text{if } r \equiv 1 \pmod 4\\ i, & \text{if } r \equiv 3 \pmod 4\end{cases}$$

$$= \sqrt{\frac{s}{2}}\left\{\frac{\frac{1}{2}s}{r}\right\}\begin{cases}1+i, & \text{if } r \equiv 1 \pmod 4\\ 1-i, & \text{if } r \equiv 3 \pmod 4\end{cases}.$$

If $r < 0$, then using (a, b) and Remarks A.6.2, A.7.2, we get

$$G(x) = G(r,s) = G(-(-r),s) = e^{-i\pi/4}\sqrt{\frac{s}{-r}}G(s,-r)$$

$$= \frac{1-i}{\sqrt{2}}\sqrt{\frac{s}{-r}}\left\{\frac{\frac{1}{2}s}{-r}\right\}\sqrt{-r}\begin{cases}1, & \text{if } -r \equiv 1 \pmod 4\\ i, & \text{if } -r \equiv 3 \pmod 4\end{cases}$$

$$= (1-i)\sqrt{\frac{s}{2}}\left\{\frac{\frac{1}{2}s}{r}\right\}\begin{cases}1, & \text{if } r \equiv 3 \pmod 4\\ i, & \text{if } r \equiv 1 \pmod 4\end{cases}$$

$$= \sqrt{\frac{s}{2}}\left\{\frac{\frac{1}{2}s}{r}\right\}\begin{cases}1+i, & \text{if } r \equiv 3 \pmod 4\\ 1-i, & \text{if } r \equiv 1 \pmod 4\end{cases}.$$

(e) is obvious. □

A.8 Farey Fractions

Definition A.8.1 (cf. [HW79]). We say that $\frac{a}{b}$ is a *Farey fraction of order* n if $0 \le a \le b \le n$, $b > 0$, and $(a,b) = 1$. The Farey fractions of order n form an increasing sequence, e.g., for $n = 5$, we have

$$\frac{0}{1}, \frac{1}{5}, \frac{1}{4}, \frac{1}{3}, \frac{2}{5}, \frac{1}{2}, \frac{3}{5}, \frac{2}{3}, \frac{3}{4}, \frac{4}{5}, \frac{1}{1}.$$

Remark A.8.2 (cf. [HW79], Theorems 28–31, 36). (a) If $\frac{a}{b}$, $\frac{a'}{b'}$ are two successive Farey fractions of order n, then $a'b - ab' = 1$.

(b) If $\frac{a}{b}$, $\frac{a''}{b''}$, $\frac{a'}{b'}$ are three successive Farey fractions of order n, then $\frac{a''}{b''} = \frac{a+a'}{b+b'}$.

(c) If $\frac{a}{b}$, $\frac{a'}{b'}$ are two successive Farey fractions of order n, then $b + b' > n$.

(d) If $\frac{a}{b}$, $\frac{a'}{b'}$ are two successive Farey fractions of order $n \ge 2$, then $b \ne b'$.

(e) For every $x \in \mathbb{R} \setminus \mathbb{Q}$ and $n \in \mathbb{N}_2$, there exists an irreducible fraction $\frac{r}{s}$ with $0 < s \le n$ such that $|x - \frac{r}{s}| \le \frac{1}{s(n+1)}$.

Proposition A.8.3. *For every $x \in \mathbb{R} \setminus \mathbb{Q}$, we have*

$$\forall_{n\in\mathbb{N}}\ \exists_{r_n\in\mathbb{Z},\ s_n\in\mathbb{N}} : (r_n, s_n) = 1,\ r_n s_n \equiv 0 \pmod 2,\ \left|x - \frac{r_n}{s_n}\right| < \frac{1}{s_n^2}. \tag{A.8.1}$$

Note that such an approximation *without* the condition $r_n s_n \equiv 0 \pmod 2$, $n \in \mathbb{N}$, follows directly from Remark A.8.2(e).

Proof. First observe that if x satisfies (A.8.1), then so does $-x$. Moreover, if x satisfies (A.8.1), then $x + 2k$ also satisfies (A.8.1) ($k \in \mathbb{Z}$). Consequently, we may assume that $0 < x < 1$. Suppose that for some $n \in \mathbb{N}_2$, we have $\frac{a}{b} < x < \frac{a'}{b'}$, where $\frac{a}{b}$, $\frac{a'}{b'}$ are successive Farey fractions of order n. We know (cf. the proof of Theorem 36 in [HW79]) that either $\frac{r}{s} := \frac{a}{b}$ or $\frac{r}{s} := \frac{a'}{b'}$ satisfies the condition $|x - \frac{r}{s}| < \frac{1}{s^2}$. We also know that $a'b - ab' = 1$ (Remark A.8.2(a)). Thus, there are the following three possibilities.

(1) $ab \equiv 0 \pmod 2$ and $a'b' \equiv 0 \pmod 2$: Then we continue with Farey fractions of order $n+1$.

(2) $ab \equiv 1 \pmod 2$ and $a'b' \equiv 0 \pmod 2$: Consider the intervals

$$I_k := \Big(\frac{ka + a'}{kb + b'}, \frac{(k-1)a + a'}{(k-1)b + b'}\Big) = \Big(\frac{a_k}{b_k}, \frac{a'_k}{b'_k}\Big), \quad k \in \mathbb{N}.$$

Observe that:

- $\frac{a'_1}{b'_1} = \frac{a'}{b'}$ and $\frac{a_k}{b_k} \searrow \frac{a}{b}$. Hence there exists a $k \in \mathbb{N}$ such that $\frac{a_k}{b_k} < x < \frac{a'_k}{b'_k}$.
- $a_k b_k \equiv 0 \pmod 2$ and $a'_k b'_k \equiv 0 \pmod 2$.
- $(a_k, b_k) = 1 = (a'_k, b'_k)$. Thus $\frac{a_k}{b_k}$, $\frac{a'_k}{b'_k}$ are Farey fractions of order $kb + b' > n$ (cf. Remark A.8.2(c)).
- $a'_k b_k - a_k b'_k = 1$, and hence $\frac{a_k}{b_k}$, $\frac{a'_k}{b'_k}$ are successive Farey fractions.

Consequently, we are in the situation of (1) (with new $n = kb + b'$), and we may continue.

(3) $ab \equiv 0 \pmod 2$ and $a'b' \equiv 1 \pmod 2$: Then we use the intervals $I_k := (\frac{a+(k-1)a'}{b+(k-1)b'}, \frac{a+ka'}{b+kb'})$, $k \in \mathbb{N}$—EXERCISE.

□

A.9 Normal Numbers

Definition A.9.1. (a) We say that a number $x \in \mathbb{I}$ is a *dyadic rational* ($x \in \mathfrak{D}$) if $x = \frac{k}{2^m}$ with $m \in \mathbb{N}$, $k \in \mathbb{N}_0$, $k \leq 2^m$ (note that $\mathfrak{D}$ is countable).

(b) For a number $x = \sum_{k=1}^{\infty} \frac{\varepsilon_k(x)}{2^k} \in \mathfrak{D}' := \mathbb{I} \setminus \mathfrak{D}$, where $\varepsilon_k(x) \in \{0, 1\}$, let

$$d_1(x) := \lim_{n \to +\infty} \frac{1}{n} \sum_{k=1}^{n} \varepsilon_k(x),$$

if the limit exists. We say that $x \in \mathfrak{D}'$ is *normal* if $d_1(x) = \frac{1}{2}$.

The following proposition shows that almost all numbers are normal.

Proposition A.9.2 (Borel Theorem). *The set of all normal numbers is a full-measure set.*

Proof. (Cf. [Kac59]) For $x = \sum_{k=1}^{\infty} \frac{\varepsilon_k(x)}{2^k} \in \mathfrak{D}'$, let

$$D_n(x) := X_1(x) + \cdots + X_n(x),$$

where $X_k(x) := 1 - 2\varepsilon_k(x) = (-1)^{\varepsilon_k(x)} \in \{-1, 1\}$. Obviously, $D_n(x) = n - 2I_n(x) = O_n(x) - I_n(x)$, where $I_n(x) = \varepsilon_1(x) + \cdots + \varepsilon_n(x)$ and $O_n(x) := n - I_n(x)$ stands for the number of 1's and 0's, respectively. Observe that x is normal iff $\lim_{n \to +\infty} \frac{1}{n} D_n(x) = 0$. Note that each of the functions $\varepsilon_k : \mathfrak{D}' \longrightarrow \{0, 1\}$ is piecewise constant, and therefore Borel measurable. First, we prove that

$$\int_0^1 X_{s_1}(t)\cdots X_{s_p}(t)dt = 0, \quad s_1 < \cdots < s_p,\ p \in \mathbb{N}_2. \tag{A.9.1}$$

Indeed, let $T := \frac{1}{2^{s_1}}$. Observe that for $t \in (0, 1-T) \cap \mathfrak{D}'$, we have:

- $X_{s_1}(t+T) = -X_{s_1}(t)$,
- $X_{s_j}(t+T) = X_{s_j}(t)$, $j = 2, \ldots, p$.

Thus

$$\int_0^1 X_{s_1}(t)\cdots X_{s_p}(t)dt = \sum_{j=0}^{\frac{1}{2T}-1} \int_{2jT}^{2(j+1)T} X_{s_1}(t)\cdots X_{s_p}(t)dt$$
$$= \sum_{j=0}^{\frac{1}{2T}-1} \Big(\int_{2jT}^{(2j+1)T} X_{s_1}(t)\cdots X_{s_p}(t)dt + \int_{(2j+1)T}^{2(j+1)T} X_{s_1}(t)\cdots X_{s_p}(t)dt \Big)$$
$$= \sum_{j=0}^{\frac{1}{2T}-1} \int_{2jT}^{(2j+1)T} \Big(X_{s_1}(t)\cdots X_{s_p}(t) + X_{s_1}(t+T)\cdots X_{s_p}(t+T) \Big) dt = 0.$$

Using (A.9.1), we easily get

$$\int_0^1 (D_n(t))^4 dt = n + \binom{n}{2}\frac{4!}{2!2!}.$$

Hence

$$\sum_{n=1}^{\infty} \int_0^1 \Big(\frac{1}{n} D_n(t)\Big)^4 dt < +\infty.$$

Consequently,

$$\sum_{n=1}^{\infty} \Big(\frac{1}{n} D_n(t)\Big)^4 < +\infty \text{ for almost all } t.$$

In particular, $\lim_{n\to+\infty} \frac{1}{n} D_n(x) = 0$ for almost all x. □

Proposition A.9.3. *The set*

$$\{x \in \mathfrak{D}' : \sup_{n\in\mathbb{N}} D_n(x) < +\infty \text{ or } \inf_{n\in\mathbb{N}} D_n(x) > -\infty\}$$

is of measure zero.

Proof. Observe that $D_n(1-x) = -D_n(x)$, $x \in \mathfrak{D}'$. Thus, it suffices to show that the set $Z := \{x \in \mathfrak{D}' : \sup_{n\in\mathbb{N}} D_n(x) < +\infty\}$ is of measure zero. Obviously, $Z = \bigcup_{k\in\mathbb{Z}} Z_k$, where $Z_k := \{x \in \mathfrak{D}' : \sup_{n\in\mathbb{N}} D_n(x) = k\}$. Hence, we have only to prove that for arbitrary $(k,\ell) \in \mathbb{Z}\times\mathbb{N}$, the set $Z_{k,\ell} := \{x \in \mathfrak{D}' : \sup_{n\in\mathbb{N}} D_n(x) = D_\ell(x) = k\}$ is of measure zero. We have $Z_{k,\ell} = \bigcup_{c\in C_{k,\ell}} Z_{k,\ell,c}$, where

$$Z_{k,\ell,c} := \{x \in \mathfrak{D}' : \sup_{n\in\mathbb{N}} D_n(x) = D_\ell(x) = k,\ X_j(x) = c_j,\ j = 1, \ldots, \ell\},$$
$$C_{k,\ell} := \{c = (c_1, \ldots, c_\ell) \in \{-1,1\}^\ell :$$
$$c_1 + \cdots + c_\ell = k,\ c_1 + \cdots + c_j \le k,\ j = 1, \ldots, \ell\}.$$

We have to prove that each set $Z_{k,\ell,c}$ is of measure zero. Fix k, ℓ, c. Let

$$\Phi(x) := 2^\ell\Big(x - \sum_{j=1}^{\ell} \frac{1-c_j}{2^{j+1}}\Big), \quad x \in \mathbb{R};$$

Φ is an affine isomorphism. We have

$$\Phi(Z_{k,\ell,c}) \subset Q := \{x \in \mathfrak{D}' : D_n(x) \le 0,\ n \in \mathbb{N}\}.$$

Thus it suffices to prove that Q is of measure zero. Observe that $Q = \bigcap_{p=1}^\infty Q_p$, where $Q_p := \{x \in \mathfrak{D}' : D_n(x) \le 0,\ n = 1, \dots, 2p\}$. We have

$$Q_p = \bigcup_{c \in C_p} \Big[x_c, x_c + \frac{1}{2^{2p}}\Big) \cap \mathfrak{D}',$$

where $x_c := \sum_{j=1}^{2p} \frac{1-c_j}{2^{j+1}}$ and

$$C_p := \{c = (c_1, \dots, c_{2p}) \in \{-1, 1\}^{2p} : c_1 + \dots + c_j \le 0,\ j = 1, \dots, 2p\}.$$

Our aim is to show that $\mathcal{L}(Q_p) \longrightarrow 0$. Obviously, $\mathcal{L}(Q_p) \le \frac{1}{2^{2p}} \# C_p$. First, we will show that $\# C_p = \binom{2p}{p}$. Fix a $p \in \mathbb{N}$ and let

$$B_d := \{c = (c_1, \dots, c_{2p}) \in \{-1, 1\}^{2p} : \#\{j \in \{1, \dots, 2p\} : c_j = 1\} = d\}.$$

Since $\# C_p + \sum_{d=0}^{2p} \#(B_d \setminus C_p) = 2^{2p}$, it suffices to prove that

$$\sum_{d=0}^{2p} \#(B_d \setminus C_p) = 2^{2p} - \binom{2p}{p}.$$

Obviously, $\#(B_d \setminus C_p) = \binom{2p}{d}$ for $d < p$, since $B_d \cap C_p = \varnothing$. Clearly, $\#(B_{2p} \setminus C_p) = 0$, since $B_{2p} \subseteq C_p$. It suffices to show that $\#(B_d \setminus C_p) = \binom{2p}{d+1}$ for $p \le d < 2p$ (then $\sum_{d=0}^{2p} \#(B_d \setminus C_p) = \sum_{d=0}^{p-1} \binom{2p}{d} + \sum_{d=p}^{2p-1} \binom{2p}{d+1} = 2^{2p} - \binom{2p}{p}$).

Fix a $d \in \{p, \dots, 2p-1\}$. To prove that $\#(B_d \setminus C_p) = \binom{2p}{d+1}$, we will apply *André's reflection method* [And87]:

For $c \in B_d \setminus C_p$, let

$$\mu(c) := \min\{j \in \{1, \dots, 2p\} : c_1 + \dots + c_j = -1\}.$$

Define $\iota(c) := (-c_1, \dots, -c_{\mu(c)}, c_{\mu(c)+1}, \dots, c_{2p})$. Then $\iota : B_d \setminus C_p \longrightarrow B_{d+1}$ is bijective (Exercise), which immediately gives the required result.

Finally, using Stirling's formula, we get

$$\mathcal{L}(Q_p) \approx \frac{1}{2^{2p}} \frac{(\frac{2p}{e})^{2p}\sqrt{2\pi 2p}}{((\frac{p}{e})^p \sqrt{2\pi p})^2} = \frac{1}{\sqrt{\pi p}}. \qquad \square$$

Appendix B
List of Symbols

B.1 General Symbols

$\mathbb{N} :=$ the set of natural numbers, $0 \notin \mathbb{N}$;
$\mathbb{N}_0 := \mathbb{N} \cup \{0\}$;
$\mathbb{N}_k := \{n \in \mathbb{N} : n \geq k\}$;
$\mathbb{Z} :=$ the ring of integers;
$\mathbb{Q} :=$ the field of rational numbers;
$\mathbb{R} :=$ the field of real numbers;

$$\operatorname{sgn} : \mathbb{R} \longrightarrow \{-1, 0, +1\},\ \operatorname{sgn}(x) := \begin{cases} -1, & \text{if } x < 0 \\ 0, & \text{if } x = 0; \\ +1, & \text{if } x > 0 \end{cases}$$

$\mathbb{I} := [0, 1] \subset \mathbb{R}$;
$A_+ := \{x \in A : x \geq 0\}$, $A_{>0} := \{x \in A : x > 0\}$ $(A \subset \mathbb{R})$, e.g. $\mathbb{R}_+$, $\mathbb{R}_{>0}$;
$\lfloor t \rfloor := \sup\{k \in \mathbb{Z} : k \leq t\} =$ the least-integer part of $t \in \mathbb{R}$;
$\overline{\mathbb{R}} := \mathbb{R} \cup \{-\infty, +\infty\}$;
$\mathbb{C} :=$ the field of complex numbers;
$\operatorname{Re} z := x =$ the real part of $z = x + iy \in \mathbb{C}$ $(x, y \in \mathbb{R})$;
$\operatorname{Im} z := y =$ the imaginary part of $z = x + iy \in \mathbb{C}$ $(x, y \in \mathbb{R})$;
$\overline{z} := x - iy =$ the conjugate of $z = x + iy \in \mathbb{C}$ $(x, y \in \mathbb{R})$;
$|z| := \sqrt{x^2 + y^2} =$ the modulus of $z = x + iy \in \mathbb{C}$ $(x, y \in \mathbb{R})$;
$\mathbb{D} := \{z \in \mathbb{C} : |z| < 1\} =$ the unit disk;
$\mathbb{T} := \partial\mathbb{D}$;
$\mathbb{D}(z_0, r) := \{z \in \mathbb{C} : |z - z_0| < r\} = z_0 + r\mathbb{D} =$ the disk centered at $z_0 \in \mathbb{C}$ with radius $r > 0$;
$A_* := A \setminus \{0\}$, e.g., $\mathbb{C}_*$;
$\overline{\mathbb{C}} := \mathbb{C} \cup \{\infty\} =$ the Riemann sphere;
$\arg z := \{\varphi \in \mathbb{R} : z = |z|e^{i\varphi}\} =$ the argument of $z \in \mathbb{C}$ $(\arg 0 = \mathbb{R})$;
$\#A :=$ the number of elements of A;
$\operatorname{diam} A :=$ the diameter of the set $A \subset \mathbb{C}^n$ with respect to the Euclidean distance;
$A \subset\subset X$ $:\Longleftrightarrow$ A is relatively compact in X;
$\|f\|_A := \sup\{|f(a)| : a \in A\}$, $f\colon A \longrightarrow \mathbb{C}$;
$\operatorname{id} : X \longrightarrow X$, $\operatorname{id}(x) := x$, $x \in X$;
$f_k \overset{K}{\underset{k\to\infty}{\Longrightarrow}} f :\Longleftrightarrow f_k \longrightarrow f$ locally uniformly;
$f_k \underset{k\to\infty}{\Longrightarrow} f :\Longleftrightarrow f_k \longrightarrow f$ uniformly;

M. Jarnicki, P. Pflug, *Continuous Nowhere Differentiable Functions*, Springer Monographs in Mathematics, DOI 10.1007/978-3-319-12670-8

$\operatorname{supp} f := \overline{\{x : f(x) \neq 0\}}$ = the support of f;
$\mathcal{C}^k(X,Y)$:= the space of all $\mathcal{C}^k$-mappings $f : X \longrightarrow Y$;
$\mathcal{C}^k(X) := \mathcal{C}^k(X,\mathbb{R})$;
$\mathcal{C}_0^k(X) := \{f \in \mathcal{C}^k(X) : \operatorname{supp} f \subset\subset X\}$;
$\mathcal{L}$:= the Lebesgue measure in $\mathbb{R}$.

B.2 Symbols in Individual Chapters

Chapter 2

Chapter 3

Chapter 4

Chapter 5

Appendix C
List of Problems

M. Jarnicki, P. Pflug, *Continuous Nowhere Differentiable Functions*, Springer Monographs in Mathematics, DOI 10.1007/978-3-319-12670-8

List of Figures

M. Jarnicki, P. Pflug, *Continuous Nowhere Differentiable Functions*, Springer Monographs in Mathematics, DOI 10.1007/978-3-319-12670-8

References

[AFS09] E. de Amo, J. Fernández-Sánchez, Takagi's function revisited from an arithmetical point of view. Int. J. Pure Appl. Math. **54**, 407–427 (2009) 52 and 54

[AK12] P.C. Allaart, K. Kawamura, The Takagi function: a survey. Real Anal. Exch. **37**, 1–54 (2011/2012) 52 and 53

[AK06a] F.A. Albiac, N.J. Kalton, in *Topics in Banach Space Theory*. Graduate Texts in Mathematics (Springer, New York, 2006) 245

[AK06b] P.C. Allaart, K. Kawamura, Extreme values of some continuous nowhere differentiable functions. Math. Proc. Camb. Philos. Soc. **140**, 269–295 (2006) 52, 52, 52, and 52

[AK10] P.C. Allaart, K. Kawamura, The improper infinite derivatives of Takagi's nowhere-differentiable function. J. Math. Anal. Appl. **372**, 656–665 (2010) 52, 192, and 193

[All13] P.C. Allaart, Level sets of signed Takagi functions. arXiv:1209.6120v2 (2013) 52

[Als81] J. Alsina, The Peano curve of Schoenberg is nowhere differentiable. J. Approx. Theory **33**, 28–42 (1981) 99

[Amp06] A.M. Ampère, Recherches sur quelques points de la théorie des fonctions dérivées qui condiusent à une nouvelle démonstration de la série de Taylor, et à l'expression finie des termes qu'on néglige lorsqu'on arrête cette série à un terme quelconque. J. École Polytech. **6**, 148–181 (1806) 2

[And87] D. André, Solution directe du problème résolu par M. Bertrand. C. R. Acad. Sci. Paris **105**, 436–437 (1887) 277

[Ani93] V. Anisiu, Porosity and continuous, nowhere differentiable functions. Ann. Fac. Sci. Toulouse **2**, 5–14 (1993) 125

[BA36] E.G. Begle, W.L. Ayres, On Hildebrandt's example of a function without a finite derivative. Am. Math. Mon. **43**, 294–296 (1936) 191

[Bab84] Y. Baba, On maxima of Takagi–van der Waerden functions. Proc. Am. Math. Soc. **91**, 373–376 (1984) 53

[Ban22] S. Banach, Sur les fonctions dérivées des fonctions mesurables. Fundam. Math. **3**, 128–132 (1922) 12

[Ban31] S. Banach, Über die bairesche Kategorie gewisser Funktionenmengen. Stud. Math. **3**, 174–179 (1931) 108

[BB05] B. Bailland, H. Bourget, *Correspondence d'Hermite et de Stieltjes*, vol. 2 (Gauthier–Villas, Paris, 1905) 3

M. Jarnicki, P. Pflug, *Continuous Nowhere Differentiable Functions*, Springer Monographs in Mathematics, DOI 10.1007/978-3-319-12670-8

[BD92] A. Baouche, S. Dubuc, La non-dérivabilité de la fonction de Weierstrass. Enseign. Math. **38**, 89–94 (1992) 48, 143, and 143

[BD94] A. Baouche, S. Dubuc, A unified approach for nondifferentiable functions. J. Math. Anal. Appl. **182**, 134–142 (1994) 53 and 187

[BE81] B.C. Berndt, R.J. Evans, The determination of Gauss sums. Bull. Am. Math. Soc. **5**, 107–129 (1981) 273 and 273

[Beh49] F.A. Behrend, Some remarks on the construction of continuous non-differentiable functions. Proc. Lond. Math. Soc. **50**, 463–481 (1949) 22, 25, and 39

[Bel71] A.S. Belov, Everywhere divergent trigonometric series (Russian). Mat. Sb. (N.S.) **85**(127), 224–237 (1971) 161

[Bel73] A.S. Belov, A study of certain trigonometric series (Russian). Mat. Zametki **13**, 481–492 (1973) 161

[Bel75] A.S. Belov, The sum of a lacunary series (Russian). Trudy Moskov. Mat. Obšč **33**, 107–153 (1975) 161

[Ber03] E.I. Berezhnoi, The subspace of $\mathcal{C}[0,1]$ consisting of functions having finite one-sided derivatives nowhere. Math. Notes **73**, 321–327 (2003) 248

[Bes24] A.S. Besicovitch, An investigation of continuous functions in connection with the question of their differentiability (Russian). Mat. Sb. **31**, 529–556 (1924) 121, 220, and 254

[BKY06] R. Balasubramanian, S. Kanemitsu, M. Yoshimoto, Euler products, Farey series and the Riemann hypothesis II. Publ. Math. Debrecen **69**, 1–16 (2006) 52

[Boa69] P. van Emde Boas, Nowhere differentiable continuous functions. Technical Report, Mathematisch Centrum Amsterdam (1969) 5 and 46

[Boa10] R.P. Boas, in *Invitation to Complex Analysis (Second Edition Revised by Harold P. Boas)*. MAA Textbooks (Mathematical Association of America, Washington, DC, 2010) 161 and 267

[Bob05] J. Bobok, On a space of Besicovitch functions. Real Anal. Exch. **30**, 173–182 (2005) 254

[Bob07] J. Bobok, Infinite dimensional Banach space of Besicovitch functions. Real Anal. Exch. **32**, 319–333 (2007) 254

[Bob14] J. Bobok, On spaceability of Besicovitch functions. Preprint (2014) 254

[Bou04] N. Bourbaki, *Functions of a Real Variable* (Springer, Berlin, 2004) 80

[BR74] P. Bois-Reymond, Versuch einer Classification der willkürlichen Functionen reeller Argumente nach ihren Aenderungen in den kleinsten Intervallen. J. Reine Angew. Math. **79**, 21–37 (1874) 2, 21, 27, 27, and 257

[Bro08] T.J.I'a. Bromwich, *An Introduction to the Theory of Infinite Series* (Macmillan, London/New York, 1908) 38

[Bru84] A. Bruckner, Some new simple proofs of old difficult theorems. Real Anal. Exch. **9**, 74–75 (1984) 121

[Brž49] V.F. Bržečka, On Bolzano's function (Russian). Uspehi Matem. Nauk (N.S.) **2**(30), 15–21 (1949) 203 and 203

[Bus52] K.A. Bush, Continuous functions without derivatives. Am. Math. Mon. **59**, 222–225 (1952) 90

[Can69] G. Cantor, Über die einfachen Zahlensysteme. Schlömilch Z. **XIV**, 121–128 (1869) 265

[Cat03] F.S. Cater, Constructing nowhere differentiable functions from convex functions. Real Anal. Exch. **28**, 617–621 (2002/03) 53, 59, and 59

[Cat83] F.S. Cater, A typical nowhere differentiable function. Canad. Math. Bull. **26**, 149–151 (1983) 164

[Cat84] F.S. Cater, On van der Waerden's nowhere differentiable function. Am. Math. Mon. **91**, 307–308 (1984) 59

[Cat86] F.S. Cater, An elementary proof of a theorem on unilateral derivatives. Can. Math. Bull. **29**, 341–343 (1986) 121

[Cat94] F.S. Cater, Remark on a function without unilateral derivatives. J. Math. Anal. Appl. **182**, 718–721 (1994) 53 and 58

[Cel90] M.Ch. Cellérier, Note sur les principes fondamentaux de l'analyse. Darboux Bull. **14**, 142–160 (1890) 2, 35, and 35

[Dar75] G. Darboux, Mémoir sur les fonctions discontinues. Ann. Sci. École Norm. Sup. Sér. 2 **4**, 57–112 (1875) 3

[Dar79] G. Darboux, Addition au mémoire sur les fonctions discontinues. Ann. Sci. École Norm. Sup. Sér. 2 **8**, 195–202 (1879) 3 and 163

[Den15] A. Denjoy, Mémoire sur les nombres dérivés des fonctions continues. J. Math. **1**, 105–240 (1915) 13

[DFK89] W.F. Darsow, M.J. Frank, H.-H. Kairies, Errata: "Functional equations for a function of van der Waerden type". Rad. Mat. **5**, 179–180 (1989) 51 and 56

[Din77] U. Dini, Su alcune funzioni che in tutto un intervallo non hanno mai derivata. Annali di Matematica **8**, 122–137 (1877) 3

[Din92] U. Dini, *Grundlagen für eine Theorie der Functionen einer veränderlichen reelen Grösse* (B.G. Teubner, Leipzig, 1892) 5, 11, 36, 37, and 37

[Dol67] E.P. Dolženko, Boundary properties of arbitrary functions. Izv. Akad. Nauk SSSR Ser. Mat. **31**, 3–14 (1967) 125

[DS00] L. Dirichlet, P.L. Seidel, *Die Darstellung ganz willkürlicher Functionen durch Sinus- und Cosinusreihen* (Wilhelm Engelmann, Leipzig, 1900) 1 and 1

[EG] P. Enflo, V.I. Gurariy, On lineability and spaceability of sets in function spaces. Preprint 246 and 246

[EGSS14] P. Enflo, V.I. Gurariy, J.B. Seoane-Sepúlveda, Some results and open questions on spaceability in function spaces. Trans. Am. Math. Soc. **366**, 611–625 (2014) 246

[EM14] A. Eskenazis, K. Makridis, Topological genericity of nowhere differentiable functions in the disc and polydisc algebras. J. Math. Anal. Appl. **420**, 435–446 (2014) 113 and 113

[Esk14] A. Eskenazis, Topological genericity of nowhere differentiable functions in the disc algebra. Arch. Math. **103**, 85–92 (2014) 113, 113, and 114

[Eul48] L. Euler, *Introductio in Analysin in Infinitorum* (Appud Marcum-Michaelem Bousquet & Socios, Lausanne, 1748) 1

[Fab07] G. Faber, Einfaches Beispiel einer stetigen nirgends differentiierbaren Funktion. Jahresber. Deutschen Math. Verein. **16**, 538–540 (1907) 60

[Fab08] G. Faber, Über stetige Funktionen. Math. Ann. **66**, 81–94 (1908) 60, 147, and 149

[Fab10] G. Faber, Über stetige Funktionen. Math. Ann. **69**, 372–443 (1910) 60, 147, and 149

[FGK99] V.P. Fonf, V.I. Gurariy, M.I. Kadets, An infinite-dimensional subspace of $\mathcal{C}[0,1]$ consisting of nowhere differentiable functions. C. R. Acad. Bulg. **52**, 13–16 (1999) 6, 245, and 248

[Fil69] L. Filipczak, Exemple d'une fonction continue privée symétrique partout. Colloq. Math. **20**, 249–253 (1969) 112

[Gar70] K.M. Garg, On a residual set of continuous functions. Czechoslov. Math. J. **20**, 537–543 (1970) 121

[Ger70] J. Gerver, The differentiability of the Riemann function at certain rational multiples of π. Am. J. Math. **92**, 33–55 (1970) 257

[Ger71] J. Gerver, More on the differentiability of the Riemann function. Am. J. Math. **93**, 33–41 (1971) 257

[Gir94] R. Girgensohn, Nowhere differentiable solutions of a system of functional equations. Aequationes Math. **47**, 89–99 (1994) 145

[Gur67] V.I. Gurariy, Subspaces of differentiable functions in the space of continuous functions (Russian). Funktsii Funktsional. Anal. Prilozhen **4**, 161–121 (1967) 6, 245, and 245

[Gur91] V.I. Gurariy, Linear spaces composed of everywhere nondifferentiable functions (Russian). C. R. Acad. Bulg. Sci. **44**, 13–16 (1991) 6, 245, and 246

[Hah17] H. Hahn, Über stetige Funktionen ohne Ableitung. Deutsche Math. Ver. **26**, 281–284 (1917) 73

[Hai76] B. Hailpern, Continuous non-differentiable functions. Pi Mu Epsilon J. **6**, 249–260 (1976) 53

[Han34] E.H. Hanson, A new proof of a theorem of Denjoy, Young, and Saks. Bull. Am. Math. Soc. **40**, 691–694 (1934) 13

[Har16] G.H. Hardy, Weierstrass's non-differentiable function. Trans. Am. Math. Soc. **17**, 301–325 (1916) 22, 24, 25, 134, 134, 140, 143, 167, and 257

[Hat88a] M. Hata, On Weierstrass's non-differentiable function. C. R. Acad. Sci. Paris **307**, 119–123 (1988) 171

[Hat88b] M. Hata, Singularities of the Weierstrass type functions. J. Anal. Math. **51**, 62–90 (1988) 25, 171, and 171

[Hat91] M. Hata, Topological aspects of self-similar sets and singular functions, in *Fractal Geometry and Analysis*, ed. by J. Bélais, S. Dubuc (Kluwer, Dordrecht, 1991), pp. 255–276 51

[Hat94] M. Hata, Correction to: "Singularities of the Weierstrass type functions" [J. Anal. Math. 51 (1988)]. J. Anal. Math. **64**, 347 (1994) 171

[Her79] K. Hertz, On functions having no differential quotient (Polish). Diary of the Society of Sciences in Paris **11**, 1–24 (1879) 27 and 27

[Hil33] T.H. Hildebrandt, A simple continuous function with a finite derivative at no point. Am. Math. Mon. **40**, 547–548 (1933) 51

[Hob26] E.W. Hobson, *The Theory of Functions of a Real Variable and the Theory of Fourier Series*, vol. II (Cambridge University Press, Cambridge, 1926) 5 and 25

[Hov09] R.I. Hovsepyan, On trigonometric series and primitive functions. J. Contemp. Math. Anal. Armen. Acad. Sci. **44**, 205–211 (2009) 117

[HSY92] B.R. Hunt, T. Sauer, J. Yorke, Prevalence: a translation-invariant "almost every" on infinite-dimensional spaces. Bull. Am. Math. Soc. **27**, 217–238 (1992) 127

[Hun94] B.R. Hunt, The prevalence of continuous nowhere differentiable functions. Proc. Am. Math. Soc. **122**, 711–717 (1994) 127

[HW41] P. Hartman, A. Wintner, On the law of the iterated logarithm. Am. J. Math. **63**, 169–176 (1941) 82

[HW79] G.H. Hardy, E.M. Wright, *An Introduction to the Theory of Numbers* (Clarendon, Oxford, 1979) 271, 272, 274, 274, and 275

[HY84] M. Hata, M. Yamaguti, The Takagi function and its generalization. Japan J. Appl. Math. **1**, 183–199 (1984) 58

[Ita81] S. Itatsu, Differentiability of Riemann's function. Proc. Jpn. Acad. Ser. A Math. Sci. **57**, 492–495 (1981) 257

[Jar22] V. Jarník, O funkci Bolzanově. Časopis pro pěstování matematiky a fysiky **51**, 248–264 (1922) 5 and 203

[Jar33] V. Jarník, Über die Differenzierbarkeit stetiger Funktionen. Fundam. Math. **21**, 48–58 (1933) 108 and 112

[Jar34] V. Jarník, Sur les nombres dérivés approximatifs. Fundam. Math. **22**, 4–16 (1934) 118

[Jar81] V. Jarník, On Bolzano's function, in *Bolzano and the Foundations of Mathematical Analysis (On the Occasion of the Bicentennial of Bernard Bolzano)* (Society of Czechoslovak Mathematicians and Physicists, Prague, 1981), pp. 67–81 203, 203, and 204

[Joh10] J. Johnsen, Simple proofs of nowhere-differentiability for Weierstrass's function and cases of slow growth. J. Fourier Anal. Appl. **16**, 17–33 (2010) 161, 161, and 162

[JRMFSS13] P. Jiménez-Rodriguez, G.A. Muñoz-Fernández, J.B. Seoane-Sepúlveda, On Weierstrass monsters and lineability. Bull. Belg. Math. Soc. Simon Stevin **20**, 577–586 (2013) 246

[Kac59] M. Kac, in *Statistical Independence in Probability, Analysis and Number Theory*. The Carus Mathematical Monographs, vol. 12 (Wiley, New York, 1959) 275

[Kal73] P. Kalášek, Note to a paper by V. Petrůva (Czech). Časopis Pěst. Mat. **98**, 156–158 (1973) 112

[Kat91] H. Katsuura, Nowhere differentiable functions—an application of contraction mappings. Am. Math. Mon. **98**, 411–416 (1991) 80

[KDF88] H.-H. Kairies, W.F. Darsow, M.J. Frank, Functional equations for a function of van der Waerden type. Rad. Mat. **4**, 361–374 (1988) 51 and 56

[Kie66] K. Kiesswetter, Ein einfaches Beispiel einer Funktion, welche überall stetig und nicht differenzierbar ist. Math. Phys. Semesterber. **13**, 216–221 (1966) 73

[KK96] R. Kannan, C.K. Krueger, *Advanced Analysis on the Real Line* (Springer, Berlin, 1996) 13, 13, 16, and 117

[Kle02] F. Klein, *Anwendung der Differential- und Integralrechnung auf Geometrie, eine Revision der Prinzipien* (B.G. Teubner, Leipzig, 1902) 5

[Kno18] K. Knopp, Ein einfaches Verfahren zur Bildüng stetiger nirgends differenzierbarer Funktionen. Math. Z. **2**, 1–26 (1918) 53

[Kob09] K. Kobayashi, On the critical case of Okamoto's continuous non-differentiable functions. Proc. Jpn. Acad. **85**, 101–104 (2009) 79 and 80

[Kôn87] N. Kôno, On generalized Takagi functions. Acta Math. Hung. **49**, 315–324 (1987) 52

[Kos72] P. Kostyrko, On the symmetric derivative. Colloq. Math. **25**, 265–267 (1972) 112

[Kow23] G. Kowalewski, Über Bolzanos nichtdifferenzierbare stetige Funktion. Acta Math. **44**, 315–319 (1923) 203 and 203

[Krü07] M. Krüppel, On the extrema and the improper derivatives of Takagi's continuous nowhere differentiable function. Rostock. Math. Kolloq. **62**, 41–59 (2007) 52, 192, and 193

[Krü08] M. Krüppel, Takagi's continuous nowhere differentiable function and binary digital sums. Rostock. Math. Kolloq. **63**, 37–54 (2008) 52

[Krü10] M. Krüppel, On the improper derivatives of Takagi's continuous nowhere differentiable function. Rostock. Math. Kolloq. **65**, 3–13 (2010) 52

[KSZ48] M. Kac, R. Salem, A. Zygmund, A gap theorem. Trans. Am. Math. Soc. **63**, 235–243 (1948) 177

[Kuc14] A. Kucharski, Math's beautiful monsters; how a destructive idea paved the way to modern math. Nautilus Q. **11** (2014) 3

[KV02] H.-H. Kairies, P. Volkmann, Ein Vektorraumisomorphismus vom Summentyp. Preprint, University of Karlsruhe (2002) 51

[KY00] S. Kanemitsu, M. Yoshimoto, Euler products, Farey series and the Riemann hypothesis. Publ. Math. Debr. **56**, 431–449 (2000) 52

[Lag12] J.C. Lagarias, The Takagi function and its properties, in *Functions in Number Theory and Their Probabilistic Aspects* (Research Institute of Mathematical Sciences, Kyoto, 2012), pp. 153–189 52 and 53

[Lan08] G. Landsberg, On the differentiability of continuous functions (Über die Differentiierbarkeit stetiger Funktionen). Jahresber. Deutschen Math. Verein. **17**, 46–51 (1908) 61

[Leb40] H. Lebesgue, Une fonction continue sans dérivée. Enseign. Math. **38**, 212–213 (1939–1940) 60

[Ler88] M. Lerch, Über die Nichtdifferentiirbarkeit gewisser Functionen. J. Math. **103**, 126–138 (1888) 31, 31, and 32

[LM40] B. Levine, D. Milman, On linear sets in space C consisting of functions of bounded variation (Russian). Comm. Inst. Sci. Math. Mecan. Univ. Kharkoff Soc. Math. Kharkoff IV. Ser. **16**, 102–105 (1940) 245

[LM14] D. Li, J. Miao, Generalized Kiesswetter's functions. Preprint (2014) 78

[Lut86] W. Luther, The differentiability of Fourier gap series and "Riemann's example" of a continuous, nondifferentiable function. J. Approx. Theory **48**, 303–321 (1986) 167

[Lyc40] R.T. Lyche, Une fonction continue sans dérivée. Enseign. Math. **38**, 208–211 (1939/1940) 54

[Mal84] J. Malý, Where the continuous functions without unilateral derivatives are typical. Trans. Am. Math. Soc. **283**, 169–175 (1984) 235

[Mal09] L. Maligranda, Karol Hertz (1843–1904) — absolwent Szkoły Głównej Warszawskiej (Polish). Antiquit. Math. **3**, 65–87 (2009) 27

[Mar35] J. Marcinkiewicz, Sur les nombres dérivés. Fundam. Math. **24**, 305–308 (1935) 115, 115, and 117

[Mau79] R.D. Mauldin, The set of continuous nowhere differentiable functions. Pac. J. Math. **83**, 199–205 (1979) 127 and 127

[Maz31] S. Mazurkiewicz, Sur les fonctions non dérivables. Stud. Math. **3**, 92–94 (1931) 108 and 127

[McC53] J. McCarthy, An everywhere continuous nowhere differentiable function. Am. Math. Mon. **60**, 709 (1953) 60

[MG84] D.P. Minassian, J.W. Gaisser, A simple Weierstrass function. Am. Math. Mon. **91**, 254–256 (1984) 53

[Mik56] M. Mikolás, Construction des familles de fonctions partout continues non dérivables. Acta Sci. Math. **17**, 49–62 (1956) 59

[Moh80] E. Mohr, Wo ist die Riemannsche Funktion $\sum_{n=1}^{\infty} \frac{\sin n^2 x}{n^2}$ nichtdifferenzierbar? Ann. Mat. Pura Appl. **123**, 93–104 (1980) 257

[Mor38] A.P. Morse, A continuous function with no unilateral derivatives. Trans. Am. Math. Soc. **44**, 496–507 (1938) 209 and 220

[Mor87] M. Morayne, On differentiability of Peano type functions. Colloq. Math. **53**, 129–132 (1987) 102 and 102

[Muk34] B.N. Mukhopadhyay, On some generalisations of Weierstraß nondifferentiable functions. Bull. Calcutta M.S. **25**, 179–184 (1934) 27 and 164

[Oka05] H. Okamoto, A remark on continuous, nowhere differentiable functions. Proc. Jpn. Acad. **81**, 47–50 (2005) 79, 80, and 83

[OMS09] K. Oldham, J. Myland, J. Spanier, *An Atlas of Functions* (Springer, Berlin, 2009) 270

[OW07] H. Okamoto, M. Wunsch, A geometric construction of continuous, strictly increasing singular functions. Proc. Jpn. Acad. **83**, 114–118 (2007) 79 and 83

[Pas14] M. Pasch, *Veränderliche und Funktion* (B.G. Teubner, Leipzig/Berlin, 1914) 5 and 73

[Pep28] E.D. Pepper, On continuous functions without a derivative. Fundam. Math. **12**, 244–253 (1928) 209 and 254

[Per16] J.B. Perrin, *Atoms* (D. van Nostrand Company, New York, 1916) 4

[Per27] F.W. Perkins, An elementary example of a continuous non-differentiable function. Am. Math. Mon. **34**, 476–478 (1927) 80

[Per29] F.W. Perkins, A note on certain continuous non-differentiable functions. Bull. Am. Math. Soc. **35**, 239–247 (1929) 84

[Pet20] K. Petr, Example of a continuous function without derivative at any point (Czech). Časopis Pěst. Mat. **49**, 25–31 (1920) 87 and 88

[Pet58] V. Petrův, On the symmetric derivative of continuous functions (Czech). Časopis Pěst. Mat. **83**, 336–342 (1958) 112

[Poi99] H. Poincaré, La logique et l'intuition dans la sciences mathématique et dans l'enseigment. Enseign. Math. **1**, 157–162 (1899) 4

[Pom92] C. Pommerenke, in *Boundary Behaviour of Conformal Maps.* Grundlehren Der Mathematischen Wissenschaften, vol. 299 (Springer, Berlin, 1992) 267

[Por19] M.B. Porter, Derivativeless continuous functions. Bull. Am. Math. Soc. **25**, 176–180 (1919) 33, 34, and 34

[PT84] P.P. Petrushev, S.L. Troyanski, On the Banach–Mazur theorem on the universality of $w[0,1]$ (Russian). C. R. Acad. Bulg. Sci. **37**, 283–285 (1984) 245

[PV13] M. Pratsiovytyi, N. Vasylenko, Fractal properties of functions defined in terms of Q-representation. Int. J. Math. Anal. **7**, 3155–3167 (2013) 69, 86, and 87

[Rha57a] G. de Rham, Sur quelques courbes définies par des équations fonctionelles. Rend. Sem Mat. Univ. Torino **16**, 101–113 (1957) 55

[Rha57b] G. de Rham, Sur un exemple de fonction continue sans dérivée. Enseign. Math. **3**, 71–72 (1957) 53

[RP95] L. Rodriguez-Piazza, Every separable Banach space is isometric to a space of continuous nowhere differentiable functions. Proc. Am. Math. Soc. **123**, 3650–3654 (1995) 246

[RS02] R. Remmert, G. Schumacher, *Funktionentheorie 2.* Springer-Lehrbuch (Springer, Berlin/Heidelberg, 2002) 161

[Rud74] W. Rudin, *Real and Complex Analysis* (McGraw-Hill, New York, 1974) 116

[Ryc23] K. Rychlik, Eine stetige nicht differenzierbare Funktion im Gebiete der Henselschen Zahlen. J. Math. **152**, 178–179 (1923) 5 and 88

[Sag92] H. Sagan, An elementary proof that Schoenberg's space-filling curve is nowhere differentiable. Math. Mag. **65**, 125–128 (1992) 99

[Sag94] H. Sagan, *Space-Filling Curves.* Universitext (Springer, New York, 1994) 101

[Sak24] S. Saks, Sur les nombres derivées des fonctions. Fundam. Math. **5**, 98–104 (1924) 13

[Sak32] S. Saks, On the functions of Besicovitch in the space of continuous functions. Fundam. Math. **19**, 211–219 (1932) 121

[Sch38] I.J. Schoenberg, On the Peano curve of Lebesgue. Bull. Am. Math. Soc. **44**, 8 (1938) 101

[Sie14a] W. Sierpiński, An arithmetic example of a continuous non-differentiable function (Polish). Wektor **3**, 337–343 (1914) 85

[Sie14b] W. Sierpiński, Sur deux problémes de la théorie des fonctions non dérivables. Bull. Int Acad. Sci. Cracovie 162–182 (1914) 73

[Sie35] W. Sierpiński, Sur une propriété des fonctions quelconques d'une variable reéle. Fundam. Math. **25**, 1–4 (1935) 115

[Sin27] A.N. Singh, On some new types of non-differentiable functions. Ann. Math. **28**, 472–476 (1927) 91

[Sin28] A.N. Singh, On Bolzano's non-differentiable function. Bull. Acad. Pol. A **1928**, 191–200 (1928) 73 and 203

[Sin30] A.N. Singh, Arithmetic non-differentiable functions. Ann. Math. **31**, 657–659 (1930) 93 and 97

[Sin35] A.N. Singh, The theory and construction of non-differentiable functions, in *Squaring the Circle and Other Monographs*. (Lucknow University Studies, Chelsea Publishing Company) (1935) v, 5, 88, 93, 94, 95, and 97

[Sin41] A.N. Singh, On functions without one-sided derivatives. Proc. Benares Math. Soc. **3**, 55–69 (1941) 220

[Sin43] A.N. Singh, On functions without one-sided derivatives ii. Proc. Benares Math. Soc. **4**, 95–108 (1943) 220

[Smi72] A. Smith, The differentiability of Riemann's functions. Proc. Am. Math. Soc. **34**, 463–468 (1972) 257

[Smi83] A. Smith, Correction to: "The differentiability of Riemann's function" [Proc. Am. Math. Soc. 34 (1972)]. Proc. Am. Math. Soc. **89**, 567–568 (1983) 257

[Spu04] K.G. Spurrier, Continuous nowhere differentiable functions. Master's thesis, South Carolina Honors College (2004) 53

[SS86] A. Shidfar, K. Sabetfakhri, On the continuity of van der Waerden's function in the Hölder sense. Am. Math. Mon. **93**, 375–376 (1986) 51

[Ste99] E. Steinitz, Stetigkeit und Differentialquotient. Math. Ann. **52**, 58–69 (1899) 73

[Sti14] T.J. Stieltjes, Quelques remarques à propos des dérivées d'une fonction d'une seule variable, in *Œuvres Complètes de Thomas Jan Stieltjes*, vol. I (Groningen, P. Noordhoff, 1914), pp. 67–72 10

[Swi61] W.C. Swift, Simple constructions of nondifferentiable functions and space-filling curves. Am. Math. Mon. **68**, 653–655 (1961) 90

[Tak03] T. Takagi, A simple example of the continuous function without derivative. Proc. Phys. Math. Soc. Jpn. **1**, 176–177 (1903). See also: The Collected Papers of Teiji Takagi (Springer, New York, 1990), pp. 4–5 51 and 54

[TH] A.L. Thomson, J.N. Hagler, Between a parabola and a nowhere differentiable place. Preprint 61

[Thi03] J. Thim, Continuous nowhere differentiable functions. Master's thesis, Luleå University of Technology (2003) v

[Vas13] N. Vasylenko, Nowhere differentiable Takagi function in problems and applications, in *Fifth International Conference on Analytic Number Theory and Spatial Tessellations* (2013), pp. 54–55 52

[Vol1987] K. Volkert, Die Geschichte der pathologischen Funktionen. Ein Beitrag zur Entstehung der mathematischen Methodologie. Arch. Hist. Exact Sci. **37**, 193–232 (1987) 3

[Vol1989] K. Volkert, Zur Differenzierbarkeit stetiger Funktionen — Ampère's Beweis und seine Folgen. Arch. Hist. Exact Sci. **40**, 37–112 (1989) 3

[Wae30] B.L. van der Waerden, Ein einfaches Beispiel einer nicht-differenzierbaren stetigen Funktion. Math. Z. **32**, 474–475 (1930) 53

[Wei86] K. Weierstrass, *Abhandlungen aus der Funktionenlehre* (Julius Springer, Berlin, 1886) 19, 21, 27, 30, and 30

[Wei95] K. Weierstrass, *Mathematische Werke* (Mayer & Müller, Berlin, 1895) 257

[Wei23] K. Weierstrass, Briefe von K. Weierstrass an Paul du Bois-Reymond. Acta Math. **39**, 199–225 (1923) 3

[Wen00] L. Wen, A nowhere differentiable continuous function. Am. Math. Mon. **107**, 450–453 (2000) 90

[Wen01] L. Wen, A nowhere differentiable continuous function constructed using Cantor series. Math. Mag. **74**, 400–402 (2001) 91 and 92

[Wen02] L. Wen, A nowhere differentiable continuous function constructed by infinite products. Am. Math. Mon. **109**, 378–380 (2002) 102

[Wie81] C. Wiener, Geometrische und analytische Untersuchung der Weierstrassschen Function. J. Reine Angew. Math. **90**, 221–252 (1881) 28 and 29

[Wun52] W. Wunderlich, Eine überall stetige und nirgends differenzierbare Funktion. Elem. der Math. **7**, 73–79 (1952) 90

[YG04] T. Yong, Y. Guangjun, Construction of some Kiesswetter-like functions—the continuous but non-differentiable function define by quinary decimal. Anal. Theory Appl. **20**, 58–68 (2004) 78 and 78

[Yon10] T. Yong, Construction and generalization of a class of Kiesswetter-like functions. J. Kunming Univ. Sci. Technol. **35**, 104–110 (2010) 78 and 79

[You16a] G.C. Young, On infinite derivates. Q. J. Pure Appl. Math. **47**, 127–175 (1916) 4 and 29

[You16b] G.C. Young, On the derivates of a function. Lond. M. S. Proc. **15**, 360–384 (1916) 13

[Zaj87] L. Zajiček, Porosity and σ-porosity. Real Anal. Exch. **13**, 314–350 (1987) 125, 125, and 125

[Zaj05] L. Zajiček, On σ-porous sets in abstract spaces. Abstr. Appl. Anal. **2005**(5), 309–334 (2005) 125

[Zyg02] A. Zygmund, *Trigonometric Series* (Cambridge University Press, Cambridge, 2002) 270

Index

M. Jarnicki, P. Pflug, *Continuous Nowhere Differentiable Functions*, Springer Monographs in Mathematics, DOI 10.1007/978-3-319-12670-8

MIX
Papier aus verantwortungsvollen Quellen
Paper from responsible sources
FSC® C105338

FSC
www.fsc.org

If you have any concerns about our products,
you can contact us on
ProductSafety@springernature.com

In case Publisher is established outside the EU,
the EU authorized representative is:
Springer Nature Customer Service Center GmbH
Europaplatz 3, 69115 Heidelberg, Germany

Printed by Libri Plureos GmbH
in Hamburg, Germany